Linear Functional Analysis for Scientists and Engineers

Balmohan V. Limaye

Linear Functional Analysis for Scientists and Engineers

Springer

Balmohan V. Limaye
Department of Mathematics
Indian Institute of Technology Bombay
Powai, Mumbai
India

ISBN 978-981-10-9298-5 ISBN 978-981-10-0972-3 (eBook)
DOI 10.1007/978-981-10-0972-3

Library of Congress Control Number: 2016939047

© Springer Science+Business Media Singapore 2016, corrected publication 2023
This work is subject to copyright. All rights are reserved by the Publisher, whether the whole or part of the material is concerned, specifically the rights of translation, reprinting, reuse of illustrations, recitation, broadcasting, reproduction on microfilms or in any other physical way, and transmission or information storage and retrieval, electronic adaptation, computer software, or by similar or dissimilar methodology now known or hereafter developed.
The use of general descriptive names, registered names, trademarks, service marks, etc. in this publication does not imply, even in the absence of a specific statement, that such names are exempt from the relevant protective laws and regulations and therefore free for general use.
The publisher, the authors and the editors are safe to assume that the advice and information in this book are believed to be true and accurate at the date of publication. Neither the publisher nor the authors or the editors give a warranty, express or implied, with respect to the material contained herein or for any errors or omissions that may have been made.

Printed on acid-free paper

This Springer imprint is published by Springer Nature
The registered company is Springer Science+Business Media Singapore Pte Ltd.

Preface

The aim of this book is to provide a short and simple introduction to the charming subject of linear functional analysis. The adjective ‘linear’ is used to indicate our focus on linear maps between linear spaces. The applicability of the topics covered to problems in science and engineering is kept in mind.

In principle, this book is accessible to anyone who has completed a course in linear algebra and a course in real analysis. Relevant topics from these subjects are collated in Chap. 1 for a ready reference. A familiarity with measure theory is not required for the development of the main results in this book, but it would help appreciate them better. For this reason, a sketch of rudimentary results about the Lebesgue measure on the real line is included in Chap. 1. To keep the prerequisites minimal, we have restricted to metric topology, and to the Lebesgue measure, that is, neither general topological spaces nor arbitrary measure spaces would be considered.

Year after year, several students from the engineering branches, who took my course on functional analysis at the Indian Institute of Technology Bombay, have exclaimed ‘Why were we not introduced to this subject in the early stages of our work? Many things with which we had to struggle hard would have fallen in place right at the start.’ Such comments have prompted me to write this book. It is also suitable for an introductory course in functional analysis as a part of a masters’ program in mathematics.

The treatment of each topic is brief, and to the point. This book is not meant to be a compendium of all the relevant results in a given topic. On the other hand, we give a lot of examples of results proved in the book. The entire book can be followed as a text for a course in linear functional analysis without having to make any specific selection of topics treated here. Teaching courses in analysis for decades has convinced me that if a student correctly grasps the notion of convergence of a sequence in a metric space, then (s)he can sail through the course easily. We define most concepts in metric topology in terms of sequences; these include total boundedness, compactness and uniform continuity. Also, among the ℓ^p and L^p spaces, $1 \le p \le \infty$, we consider only the cases $p = 1, 2$ and ∞, since they represent all essential features, and since they can be treated easily.

A novelty of this book is the inclusion of a result of Zabreiko which states that every countably subadditive seminorm on a Banach space is continuous. Once this result is proved using the Baire theorem, several important theorems can be deduced quickly. They include the uniform boundedness principle, the closed graph theorem, the bounded inverse theorem and the open mapping theorem. Surprisingly, not many textbooks on functional analysis have followed this efficient path. Another noteworthy feature of this book is that the spectral theory is treated at a single location. It deals with the eigenspectrum, the approximate eigenspectrum and the spectrum of a bounded operator, of a compact operator and of their transposes and adjoints. The spectral theorem gives a characterization of a compact self-adjoint operator.

The main body of this book consists of Chaps. 2–5, each of which is divided into four sections. Forty exercises, roughly ten on each section, are given at the end of each of these chapters. They are of varying levels of difficulty; some of them are meant to reassure the reader that (s)he has indeed grasped the core ideas developed in the chapter, while others demand some ingenuity on part of the reader. The exercises follow the same order as the text on which they are based. All exercises are in the form of statements to be justified. Their solutions are given at the end.

When a new concept is introduced, it appears in boldface. The symbol $:=$ is used for defining various objects. Definitions are not numbered, but can be located easily by using the index given at the end. All other things (lemmas, propositions, theorems, remarks, examples) are numbered serially in each chapter, and so are the exercises on it. Before the chapters begin, a list of symbols and abbreviations (along with their descriptions and the page numbers where they appear for the first time) is given in the order in which they appear in the text.

The Department of Mathematics of the Indian Institute of Technology Bombay deserves credit for providing excellent infrastructure. I thank Peeter Oja for bringing the Zabreiko theorem to my attention. I am indebted to Ameer Athavale, Anjan Chakrabarty and Venkitesh Iyer for critically reading the book and making useful suggestions. I am grateful to my wife Nirmala Limaye for her wholehearted support, both material and moral. She has drawn all the figures in this book using PSTricks, and has also figured out, without using personal tricks, how to keep me in good shape.

I would appreciate receiving comments, suggestions and corrections. A dynamic errata together with all relevant information about this book will be posted at http://www.math.iitb.ac.in/~bvl/. I encourage readers to visit this webpage for updates concerning this book.

Mumbai, India
January 2016

Balmohan V. Limaye

The original version of this book was revised: For detailed information, please see Correction. The correction to this book is available at https://doi.org/10.1007/978-981-10-0972-3_6

Contents

About the Author

Balmohan V. Limaye is Adjunct Professor in the Department of Mathematics of the Indian Institute of Technology Bombay, where he has worked for more than 40 years. Earlier, he worked at the University of California, Irvine, for 1 year, and at the Tata Institute of Fundamental Research, Mumbai, for 6 years. Professor Limaye has made research visits to the University of Grenoble (France), the Australian National University at Canberra, the University of Saint-Étienne (France), the University of Tübingen (Germany), Oberwolfach Research Institute for Mathematics (Germany), the University of California at Irvine (USA), the University of Porto (Portugal), and the Technical University of Berlin (Germany).

Professor Limaye earned his Ph.D. in mathematics from the University of Rochester, New York, in 1969. His research interests include algebraic analysis, numerical functional analysis and linear algebra. He has published more than 50 articles in refereed journals. In 1995, he was invited by the Indian Mathematical Society to deliver the Sixth Srinivasa Ramanujan Memorial Award Lecture. In 1999 and in 2014, he received an 'Award for Excellence in Teaching' from the Indian Institute of Technology Bombay. An International Conference on 'Topics in Functional Analysis and Numerical Analysis' was held in his honour in 2005, and its proceedings were published in a special issue of The Journal of Analysis in 2006. He is an emeritus member of the American Mathematical Society and a life member of the Indian Mathematical Society.

He has authored/coauthored the following books: (i) *Textbook of Mathematical Analysis* (Tata McGraw-Hill, 1980), (ii) *Functional Analysis* (Wiley Eastern, 1981, and New Age International, 1996), (iii) *Spectral Perturbation and Approximation with Numerical Experiments* (Australian National University, 1987), (iv) *Real Function Algebras* (Marcel Dekker, 1992), (v) *Spectral Computations for Bounded Operators* (CRC Press, 2001), (vi) *A Course in Calculus and Real Analysis* (Springer, 2006), and (vii) *A Course in Multivariable Calculus and Analysis* (Springer, 2010).

Symbols and Abbreviations

$d_1(x, y)$	$\sum_{j=1}^{\infty} \lvert x(j) - y(j) \rvert$, where $x, y \in \ell^1$	9
$d_2(x, y)$	$\left(\sum_{j=1}^{\infty} \lvert x(j) - y(j) \rvert^2\right)^{1/2}$, where $x, y \in \ell^2$	9
$d_\infty(x, y)$	$\sup\{\lvert x(j) - y(j) \rvert : j \in \mathbb{N}\}$, where $x, y \in \ell^\infty$	9
$B(T)$	Set of all $\mathbb{K}$-valued bounded functions on T	10
(x_n)	Sequence with x_n as its nth term	10
$x_n \to x$	Sequence (x_n) converges to x	10
e_n	$(0, \ldots, 0, 1, 0, 0, \ldots)$ with 1 only in the nth entry	10
$\overline{E}$	Closure of a subset E	10
$U(x, r)$	Open ball about x of radius r	11
$d(x, E)$	Distance of a point x from a subset E	16
$C(T)$	Set of all $\mathbb{K}$-valued bounded continuous functions on T	18
$m(E)$	Lebesgue measure of a subset E of $\mathbb{R}$	20
a.e.	Almost everywhere	20
x^+	$\max\{x, 0\}$, where $x : \mathbb{R} \to [-\infty, \infty]$	21
x^-	$-\min\{x, 0\}$, where $x : \mathbb{R} \to [-\infty, \infty]$	21
$\int_E x\, dm$	Lebesgue integral of x over E	22
$\int_a^b x(t)\, dm(t)$	Lebesgue integral of x over $[a, b]$	22
$m \times m$	Lebesgue measure on $\mathbb{R}^2$	23
$k(\cdot, \cdot)$	k is a function of two variables	23
$\int_a^b x(t)dt$	Riemann integral of x over $[a, b]$	24
$V(x)$	Total variation of a function x of bounded variation on $[a, b]$	25
$L^1(E)$	Set of all equivalence classes of integrable functions on a subset E of $\mathbb{R}$	25
$d_1(x, y)$	$\int_E \lvert x - y \rvert dm$, where $x, y \in L^1(E)$	26
$L^2(E)$	Set of all equivalence classes of square-integrable functions on a subset E of $\mathbb{R}$	26
$d_2(x, y)$	$\left(\int_E \lvert x - y \rvert^2 dm\right)^{1/2}$, where $x, y \in L^2(E)$	27
$\text{ess sup}_E \lvert x \rvert$	Essential supremum of $\lvert x \rvert$ on a subset E of $\mathbb{R}$	27
$L^\infty(E)$	Set of all equivalence classes of essentially bounded functions on a subset E of $\mathbb{R}$	28
$d_\infty(x, y)$	$\text{ess sup}_E \lvert x - y \rvert$, where $x, y \in L^\infty(E)$	28
$\hat{x}(k)$	k-th Fourier coefficient of x, where $k \in \mathbb{Z}$	29
$\lVert x \rVert$	Norm of x	33
$(X, \lVert\cdot\rVert)$	Linear space X with norm $\lVert\cdot\rVert$ on it	34
$\overline{U}(x, r)$	Closed ball about x of radius r	34
$\lVert x \rVert_1$	$\sum_j \lvert x(j) \rvert$, 1-norm of x in $\mathbb{K}^n$ or in ℓ^1	34
$\lVert x \rVert_2$	$\left(\sum_j \lvert x(j) \rvert^2\right)^{1/2}$, 2-norm of x in $\mathbb{K}^n$ or in ℓ^2	34
$\lVert x \rVert_\infty$	$\sup_j\{\lvert x(j) \rvert\}$, ∞-norm of x in $\mathbb{K}^n$ or in ℓ^∞	34
c	Set of all convergent sequences in $\mathbb{K}$	35
c_0	Set of all sequences in $\mathbb{K}$ converging to 0	35

c_{00}	Set of all sequences in $\mathbb{K}$ having only a finite number of nonzero terms	35
$\\|x\\|_\infty$	Sup norm of $x \in B(T)$	36
$C_0(T)$	Set of all $\mathbb{K}$-valued continuous functions on T vanishing at infinity	36
$C_{00}(T)$	Set of all $\mathbb{K}$-valued continuous functions on T having compact supports	36
$\\|x\\|_1$	$\int_E \|x\|dm$, 1-norm of $x \in L^1(E)$	36
$\\|x\\|_2$	$\left(\int_E \|x\|^2\, dm\right)^{1/2}$, 2-norm of $x \in L^2(E)$	36
$\\|x\\|_\infty$	ess $\sup_E \|x\|$, ∞-norm of $x \in L^\infty(E)$	36
	Quotient norm of $x+Y$	38
$\langle x, y\rangle$	Inner product of x with y	44
$(X, \langle\cdot,\cdot\rangle)$	Linear space X with inner product $\langle\cdot,\cdot\rangle$ on it	44
$x \perp y$	x is orthogonal to y	48
$E^\perp$	Set of all elements orthogonal to a set E	48
$\sum_n$	Sum over $n \in \mathbb{N}$, or over $n \in \{1, \ldots, n_0\}$	50
$\{u_\alpha\}$	$\{u_\alpha : \alpha \in A\}$, where A is an index set	52
$C^1([a, b])$	Set of all $\mathbb{K}$-valued continuously differentiable functions on $[a, b]$	55
$\\|x\\|_{1,\infty}$	$\max\{\\|x\\|_\infty, \\|x'\\|_\infty\}$	55
$\\|x\\|_{1,1}$	$\\|x\\|_1 + \\|x'\\|_1$	57
$W^{1,1}([a, b])$	Sobolev space of order $(1, 1)$ on $[a, b]$	58
$\langle x, y\rangle_{1,2}$	$\int_a^b x(t)\bar{y}(t)dt + \int_a^b x'(t)\bar{y}'(t)dm(t)$	62
$\\|x\\|_{1,2}$	$\left(\\|x\\|_2^2 + \\|x'\\|_2^2\right)^{1/2}$	62
$W^{1,2}([a, b])$	Sobolev space of order (1, 2) on $[a, b]$	62
$Y \oplus Z$	Direct sum of the subspaces Y and Z	68
$C^k([a, b])$	Set of all $\mathbb{K}$-valued k times continuously differentiable functions on $[a, b]$	73
$W^{k,1}([a, b])$	Sobolev space of order $(k, 1)$ on $[a, b]$	73
$W^{k,2}([a, b])$	Sobolev space of order $(k, 2)$ on $[a, b]$	73
$BL(X, Y)$	Set of all bounded linear maps from X to Y	76
$G \circ F, GF$	Composition of a map G with a map F	76
$BL(X)$	Set of all bounded operators on X	76
$\\|F\\|$	Operator norm of $F \in BL(X, Y)$	79
I	Identity operator	81
sgn	Signum function	81
$\|M\|$	Matrix whose entries are the absolute values of the entries of matrix M	83
$\alpha_1(j)$	1-norm of the jth column of an infinite matrix	83
α_1	$\sup\{\alpha_1(j) : j \in \mathbb{N}\}$	83
$\beta_1(i)$	1-norm of the ith row of an infinite matrix	84
β_1	$\sup\{\beta_1(i) : i \in \mathbb{N}\}$	84

$\gamma_{2,2}$	$\left(\sum_{i=1}^{\infty}\sum_{j=1}^{\infty}\|k_{i,j}\|^2\right)^{1/2}$	85
α_1	$\text{ess sup}_{[a,b]}\int_a^b \|k(s,\cdot)\|dm(s)$	87
β_1	$\text{ess sup}_{[a,b]}\int_a^b \|k(\cdot,t)\|dm(t)$	87
$\gamma_{2,2}$	$\left(\int_a^b\int_a^b \|k(s,t)\|^2\,dm(t)dm(s)\right)^{1/2}$	88
$\mathrm{Gr}(F)$	Graph of F	96
$CL(X,Y)$	Set of all compact linear maps from X to Y	107
$CL(X)$	Set of all compact operators on X	107
$\alpha_r(j)$	r-norm of the jth column of an infinite matrix, where $r\in\{1,2,\infty\}$	117
$\beta_q(i)$	q-norm of the ith row of an infinite matrix, where $q\in\{1,2,\infty\}$	117
α_r	$\sup\{\alpha_r(j): j\in\mathbb{N}\}$, where $r\in\{1,2,\infty\}$	117
β_q	$\sup\{\beta_q(i): i\in\mathbb{N}\}$, where $q\in\{1,2,\infty\}$	117
$\mathrm{diag}(k_1,k_2,\ldots)$	Diagonal matrix with entries $k_1,k_2,\ldots$	117
$\gamma_{1,1}$	$\sum_{i=1}^{\infty}\beta_1(i)$	118
$\gamma_{1,2}$	$\left(\sum_{i=1}^{\infty}\beta_1(i)^2\right)^{1/2}$	118
$\gamma_{2,1}$	$\sum_{i=1}^{\infty}\beta_2(i)$	118
X'	$BL(X,\mathbb{K})$, dual of a normed space X	128
X''	$(X')'$, second dual of X	129
J	Canonical embedding of a normed space X into X''	129
X_c	Completion of a normed space X	129
y_f	Representer of $f\in H'$ in a Hilbert space H	133
$\langle f,g\rangle'$	Inner product of f with g in H'	133
$BV([a,b])$	Set of all $\mathbb{K}$-valued functions of bounded variation on $[a, b]$	135
$\|y\|_{BV}$	$y(a)+V(y)$	136
$NBV([a,b])$	Set of all $\mathbb{K}$-valued normalized functions of bounded variation on $[a, b]$	136
X''	Transpose of $F\in BL(X,Y)$	137
M^t	Transpose of a matrix M	137
A^*	Adjoint of $A\in BL(H,G)$	141
$\overline{M^t}$	Conjugate-transpose of matrix M	143
$A\geq 0$	A is a positive operator	146
$\omega(A)$	Numerical range of $A\in BL(X)$	149
m_A	$\inf\omega(A)$	150
M_A	$\sup\omega(A)$	150
$x_n\xrightarrow{w}x$	(x_n) converges weakly to x	155
F''	$(F')'$, transpose of the transpose of F	156
$\sigma(A)$	Spectrum of $A\in BL(X)$	160
$\sigma_e(A)$	Eigenspectrum of $A\in BL(X)$	160
$\sigma_a(A)$	Approximate eigenspectrum of $A\in BL(X)$	160

Chapter 1
Prerequisites

In this chapter, we gather definitions and results that will be used in the sequel. Most of these are covered in courses on linear algebra and real analysis. We shall prove some of the nontrivial statements to give a flavour of the kind of arguments that are involved. The second section of this chapter on linear spaces and the third section on metric spaces constitute the main prerequisites for this book. The last section on Lebesgue measure and integration is not required for developing the main results given in this book, but it is very much useful for illustrating them. Readers familiar with the contents of this chapter can directly go to the next chapter and look up the relevant material in this chapter as and when a reference is made. There are no exercises at the end of this chapter.

We denote the set $\{1, 2, \ldots\}$ of all **natural numbers** by $\mathbb{N}$, the set of all **integers** by $\mathbb{Z}$, the set of all **rational numbers** by $\mathbb{Q}$, the set of all **real numbers** by $\mathbb{R}$ and the set of all **complex numbers** by $\mathbb{C}$. For z in $\mathbb{C}$, $\operatorname{Re} z$ and $\operatorname{Im} z$ will denote the **real part** and the **imaginary part** of z, respectively, so that $z = \operatorname{Re} z + i \operatorname{Im} z$, and we let $\overline{z} := \operatorname{Re} z - i \operatorname{Im} z$. Also, let $\mathbb{K} := \mathbb{R}$ or $\mathbb{C}$.

1.1 Relations on a Set

We begin with some concepts in the set theory. Let X be a set and E be a subset of X. The **complement** $X \setminus E$ of E is the set of all elements of X that do not belong to E. The empty set is the set having no elements in it. We shall denote it by $\emptyset$. Let X and Y be nonempty sets. We shall assume that the reader is familiar with the notion of a **function** or a **map** f from X to Y, denoted by $f : X \to Y$. If $f : X \to \mathbb{K}$, then $\overline{f} : X \to \mathbb{K}$ is the function defined by $\overline{f}(x) := \overline{f(x)},\ x \in X$. For $x \in X$, the element $f(x) \in Y$ is called the **value** of f at x. The subset $R(f) := \{f(x) : x \in X\}$ of Y is called the **range** of f. We say that a function f is from X **onto** Y if $R(f) = Y$. A function $f : X \to Y$ is called **one-one** if $f(x_1) \neq f(x_2)$ whenever $x_1, x_2 \in X$ and $x_1 \neq x_2$. In this case, the function $f^{-1} : R(f) \to X$ defined by $f^{-1}(y) := x$

© Springer Science+Business Media Singapore 2016, corrected publication 2023
B.V. Limaye, *Linear Functional Analysis for Scientists and Engineers*,
https://doi.org/10.1007/978-981-10-0972-3_1

when $y := f(x)$, is called the **inverse** of f. A function $f : X \to Y$ is said to give a **one-to-one correspondence** if f is one-one as well as onto.

A set is called **finite** if it is either empty or if it is in a one-to-one correspondence with $\{1, \ldots, n\}$ for some $n \in \mathbb{N}$, **denumerable** if it has a one-to-one correspondence with $\mathbb{N}$, **countable** if it is either finite or denumerable and **uncountable** if it is not countable. It is well known that $\mathbb{Q}$ is denumerable and $\mathbb{R}$ is uncountable.

A **relation** on a set X is a subset of the Cartesian product $X \times X := \{(x, y) : x, y \in X\}$. We consider two important relations.

(i) A relation R on a set X is called an **equivalence relation** if it is **reflexive** (that is, $(x, x) \in R$ for every $x \in X$), **symmetric** (that is, $(y, x) \in R$ whenever $(x, y) \in R$) and **transitive** (that is, $(x, z) \in R$ whenever $(x, y), (y, z) \in R$). In this case, '$(x, y) \in R$' is denoted by $x \sim y$. The **equivalence class** of $x \in X$ is the set of all $y \in X$ such that $x \sim y$. The equivalence classes are mutually disjoint and their union is X. For example, let $X := \mathbb{Z}$ and $R := \{(m, n) \in \mathbb{Z} \times \mathbb{Z} : m - n \text{ is divisible by } 2\}$. Then R is an equivalence relation on $\mathbb{Z}$, and it partitions $\mathbb{Z}$ into two equivalence classes, one consisting of all odd integers and the other consisting of all even integers.

(ii) A relation R on a set X is called a **partial order** if R is reflexive, **antisymmetric** (that is, $x = y$ whenever $(x, y), (y, x) \in R$) and transitive. In this case, we write $x \le y$ if $(x, y) \in R$. A **partially ordered set** is a set X together with a partial order on it. For $Y \subset X$ and $x \in X$, if $y \le x$ for every $y \in Y$, then x is said to be an **upper bound** for Y in X. If $x \in X$ and $x \le y$ implies that $x = y$ for every $y \in X$, then x is called a **maximal element** of X. A maximal element may not exist, and if it exists, it may not be unique. For example, the set $\{t \in \mathbb{R} : 0 \le t < 1\}$ with the usual partial order does not have a maximal element. On the other hand, if $X := \{(s, |s|) : -1 \le s \le 1\}$, and $R := \big\{\big((s, |s|), (t, |t|)\big) \in X \times X : s\,t \ge 0 \text{ and } |s| \le |t|\big\}$, then R is a partial order on X, and both $(-1, 1)$ and $(1, 1)$ are maximal elements of X. A subset Y of X is called **totally ordered** if $x, y \in Y$ implies that $x \le y$ or $y \le x$, that is, if any two elements of Y are comparable. For example, let $X := \mathbb{N}$ and $R := \{(m, n) \in \mathbb{N} \times \mathbb{N} : m \text{ divides } n\}$. Then R is a partial order on $\mathbb{N}$. The subset of all even natural numbers is not totally ordered, but the subset of all powers of 2 is totally ordered.

The **Zorn lemma** states that if X is a nonempty partially ordered set such that every totally ordered subset of X has an upper bound in X, then X contains a maximal element. This maximal element may not be unique, and it may not be possible to construct a maximal element. The Zorn lemma is only an existential statement. Although it is called a 'lemma', it is in fact an axiom. We shall assume it to be valid. It will be used several times in this book: in Sect. 1.2 to prove the existence of a basis for a linear space, in Sect. 2.2 to prove the existence of an orthonormal basis for an inner product space, and in Sect. 4.1 to prove the existence of a Hahn–Banach extension.

1.2 Linear Spaces and Linear Maps

A **linear space** X over $\mathbb{K}$ is a nonempty set X along with an operation called **addition**, denoted by $(x, y) \in X \times X \longmapsto x + y \in X$, and an operation called **scalar multiplication**, denoted by $(k, x) \in \mathbb{K} \times X \longmapsto kx \in X$, satisfying the standard properties. For subsets E_1 and E_2 of X, define $E_1 + E_2 := \{x_1 + x_2 : x_1 \in E_1 \text{ and } x_2 \in E_2\}$.

A nonempty subset Y of X is called **subspace** of X if $x + y, kx \in Y$ whenever $x, y \in Y$ and $k \in \mathbb{K}$. If E is a nonempty subset of X, then the **span** of E consists of all **linear combinations** $k_1x_1 + \cdots + k_nx_n$, where $x_1, \ldots, x_n \in E$, and $k_1, \ldots, k_n \in \mathbb{K}$. It is denoted by span E. If span $E = X$, then we say that E is a **spanning** subset of X, or that E **spans** X.

A subset E of X is called **linearly independent** if $n \in \mathbb{N}$, $x_1, \ldots, x_n \in E$, $k_1, \ldots, k_n \in \mathbb{K}$ and $k_1x_1 + \cdots + k_nx_n = 0$ imply that $k_1 = \cdots = k_n = 0$. A subset E of X is called **linearly dependent** if it is not linearly independent, that is, if there exist $n \in \mathbb{N}$, $x_1, \ldots, x_n \in E$ and $k_1, \ldots, k_n \in \mathbb{K}$ such that $k_1x_1 + \cdots + k_nx_n = 0$, where at least one of $k_1, \ldots, k_n$ is nonzero.

A subset E of X is called a **Hamel basis** (or simply a **basis**) for X if it is spanning as well as linearly independent.

Proposition 1.1 *Let X be a linear space over $\mathbb{K}$. Suppose L is a linearly independent subset of X, and S is a spanning subset of X such that $L \subset S$. Then there is a basis B for X such that $L \subset B \subset S$.*

Proof Let $\mathcal{E} := \{E : L \subset E \subset S \text{ and } E \text{ is linearly independent}\}$. Then $\mathcal{E}$ is nonempty since $L \in \mathcal{E}$. The inclusion relation $\subset$ is a partial order on $\mathcal{E}$. Also, for a totally ordered subfamily $\mathcal{F}$ of $\mathcal{E}$, let E denote the union of all members of $\mathcal{F}$. Clearly, $L \subset E \subset S$. To show that E is linearly independent, let $x_1, \ldots, x_n \in E$. There are $F_1, \ldots, F_n$ in $\mathcal{F}$ such that $x_1 \in F_1, \ldots, x_n \in F_n$. Since $\mathcal{F}$ is totally ordered, there is $j \in \{1, \ldots, n\}$ such that all $x_1, \ldots, x_n$ are in F_j. Since F_j is linearly independent, we are through. Thus E is an upper bound for $\mathcal{F}$ in $\mathcal{E}$. By the Zorn lemma stated in Sect. 1.1, $\mathcal{E}$ contains a maximal element B.

Since $B \in \mathcal{E}$, we see that $L \subset B \subset S$ and B is linearly independent. To show that B is spanning, we prove that $S \subset$ span B. Assume for a moment that there is $x \in S \setminus \text{span } B$. Then $B \cup \{x\}$ is a linearly independent subset of X. To see this, let $kx + k_1b_1 + \cdots + k_nb_n = 0$, where $k, k_1, \ldots, k_n \in \mathbb{K}$ and $b_1, \ldots, b_n \in B$. Then $kx = -k_1b_1 - \cdots - k_nb_n \in \text{span } B$. Since $x \notin \text{span } B$, we obtain $k = 0$. But then $k_1b_1 + \cdots + k_nb_n = 0$, and since B is linearly independent, we obtain $k_1 = \cdots = k_n = 0$. This shows that $B \cup \{x\}$ is an element of $\mathcal{E}$ such that $B \subset B \cup \{x\}$, but $B \neq B \cup \{x\}$. This is a contradiction to the maximality of B in $\mathcal{E}$. Hence $S \subset \text{span } B$. Since span $S = X$, we see that span $B = X$. Thus B is a basis for X such that $L \subset B \subset S$. □

The empty set $\emptyset$ is linearly independent, and by convention, we let span $\emptyset := \{0\}$. Thus $\emptyset$ is a basis for the linear space $\{0\}$.

Corollary 1.2 *Let X be a linear space over $\mathbb{K}$.*

(i) *There exists a basis for X.*
(ii) *If Y is a subspace of X, and C is a basis for Y, then there is a basis B for X such that $C \subset B$.*

Proof (i) Let $L := \emptyset$ and $S := X$ in Proposition 1.1.
(ii) Let $L := C$ and $S := X$ in Proposition 1.1. □

It follows that a basis for a linear space X is a maximal linearly independent subset as well as a minimal spanning subset of X. Corollary 1.2(ii) shows that a basis for a subspace of X can be extended to a basis for X. Using this result, one can show that if a linear space X has a basis consisting of a finite number of elements, then every basis for X has the same number of elements. This number is called the **dimension** of X, and X is called **finite dimensional**. In particular, the linear space $\{0\}$ has dimension 0. If a linear space X has an infinite linearly independent subset, then X is called **infinite dimensional**, and it has dimension ∞. The dimension of a linear space X will be denoted by $\dim X$.

For example, let $n \in \mathbb{N}$, and

$$\mathbb{K}^n := \{(x(1), \ldots, x(n)) : x(j) \in \mathbb{K} \text{ for } j = 1, \ldots, n\}.$$

Then $\mathbb{K}^n$ is a linear space over $\mathbb{K}$ if we define addition and scalar multiplication componentwise. For $j = 1, \ldots, n$, define $e_j := (0, \ldots, 0, 1, 0, \ldots, 0)$, where 1 occurs only in the jth entry. Then $\{e_1, \ldots, e_n\}$ is a basis for $\mathbb{K}^n$. It will be called the **standard basis** for $\mathbb{K}^n$. Thus $\mathbb{K}^n$ is of dimension n. Next, let $\mathcal{P}$ denote the set of all polynomial functions defined on $\mathbb{R}$ with coefficients in $\mathbb{K}$. Then $\mathcal{P}$ is a linear space over $\mathbb{K}$ if we define addition and scalar multiplication pointwise. Let $x_0(t) := 1$, and for $j \in \mathbb{N}$, let $x_j(t) := t^j$, $t \in \mathbb{R}$. Then $\{x_0, x_1, x_2, \ldots\}$ is an infinite linearly independent subset of $\mathcal{P}$. Thus $\mathcal{P}$ is an infinite dimensional linear space.

Let Y be a subspace of a linear space X, and for $x_1, x_2 \in X$, let $x_1 \sim x_2$ if $x_1 - x_2 \in Y$. Then $\sim$ is an **equivalence relation** on X. For $x \in X$, the equivalence class of x is the set $x + Y := \{x + y : y \in Y\}$. Let X/Y denote the set of all these equivalence classes, that is, $X/Y := \{x + Y : x \in X\}$. Thus, in considering X/Y, we ignore differences between elements of Y.

For $x_1 + Y$, $x_2 + Y$ in X/Y and $k \in \mathbb{K}$, define

$$(x_1 + Y) + (x_2 + Y) := (x_1 + x_2) + Y \quad \text{and} \quad k(x_1 + Y) := k\,x_1 + Y.$$

It is easy to see that these operations are well-defined, and X/Y is a linear space over $\mathbb{K}$. It is called the **quotient space** of X by Y. An element $x + Y$ of X/Y is called the **coset** of x in X/Y. If $Y := X$, then X/Y consists only of the zero coset $0 + X$, and if $Y := \{0\}$, then a coset $x + \{0\}$ of X/Y has x as its only element.

Next, let $X_1, \ldots, X_n$ be linear spaces over $\mathbb{K}$, and let

$$X_1 \times \cdots \times X_n := \{(x_1, \ldots, x_n) : x_j \in X_j \text{ for } j = 1, \ldots, n\}.$$

For $(x_1, \ldots, x_n), (y_1, \ldots, y_n) \in X_1 \times \cdots \times X_n$, and $k \in \mathbb{K}$, let

$$(x_1, \ldots, x_n) + (y_1, \ldots, y_n) := (x_1 + y_1, \ldots, x_n + y_n),$$

and

$$k(x_1, \ldots, x_n) := (kx_1, \ldots, kx_n).$$

Then $X_1 \times \cdots \times X_n$, along with these operations, is a linear space over $\mathbb{K}$. It is called the **product space** of $X_1, \ldots, X_n$. The most common example of a product space is $\mathbb{K}^n = \mathbb{K} \times \overset{n \text{ times}}{\cdots} \times \mathbb{K}$, where $n \in \mathbb{N}$.

Linear Maps

Let X and Y be linear spaces over $\mathbb{K}$. A **linear map** from X to Y is a map $F : X \to Y$ such that $F(k_1 x_1 + k_2 x_2) = k_1 F(x_1) + k_2 F(x_2)$ for all $x_1, x_2 \in X$ and $k_1, k_2 \in \mathbb{K}$.

Let X_0 be a subspace of X, and F_0 be a linear map from X_0 to Y. Let C be a basis for X_0. By Corollary 1.2(ii), there is a basis B of X such that $C \subset B$. For $x \in X$ with $x = x_0 + x_1$, where $x_0 \in X_0$ and $x_1 \in \text{span}\,(B \setminus C)$, define $F(x) := F_0(x_0)$. Then $F : X \to Y$ is linear and $F(x_0) = F_0(x_0)$ for all $x_0 \in X_0$. Thus a linear map from X_0 to Y can be extended to a linear map from X to Y.

Two important subspaces are associated with a linear map $F : X \to Y$. One is the subspace $R(F) = \{y \in Y : \text{there is } x \in X \text{ such that } y = F(x)\}$ of Y, which we shall call the **range space** of F. The other is the subspace $Z(F) := \{x \in X : F(x) = 0\}$ of X, which we shall call the **zero space** of F. It is also called the **null space** or the **kernel** of F. It is easy to see that F is one-one if and only if $Z(F) = \{0\}$. Also, if F is one-one, then the inverse function $F^{-1} : R(F) \to X$ is a linear map. To see this, consider $y_1, y_2 \in R(F)$, and let $x_1 := F^{-1}(y_1)$, $x_2 := F^{-1}(y_2)$. Then $F(x_1) = y_1$, $F(x_2) = y_2$, and so $F(x_1 + x_2) = F(x_1) + F(x_2) = y_1 + y_2$, that is, $F^{-1}(y_1 + y_2) = x_1 + x_2 = F^{-1}(y_1) + F^{-1}(y_2)$.

The dimension of the subspace $R(F)$ is called the **rank** of F. If $R(F)$ is finite dimensional, then we say that the linear map F is of **finite rank**. The dimension of the subspace $Z(F)$ is called the **nullity** of F. The **rank-nullity theorem** states that rank (F) + nullity $(F) = \dim X$.[1] Hence if X is finite dimensional, then a linear map $F : X \to X$ is one-one if and only if it is onto.

A linear map from X to the one-dimensional linear space $\mathbb{K}$ is called a **linear functional** on X. Let f be a nonzero linear functional on X, and let $x_1 \in X$ be such that $f(x_1) \neq 0$. Then for every $x \in X$, there are unique $x_0 \in Z(f)$ and $k \in \mathbb{K}$ such that $x = x_0 + k\,x_1$. In fact

$$x_0 = x - \frac{f(x)}{f(x_1)} x_1 \quad \text{and} \quad k = \frac{f(x)}{f(x_1)}.$$

Hence if g is a nonzero linear functional on X such that $Z(g) = Z(f)$ and $g(x_1) = f(x_1)$, then $g = f$. Also, the subset $Z(f) \cup \{x_1\}$ of X is spanning. In other words,

[1] Here we have adopted the convention $\infty + n = n + \infty = \infty + \infty = \infty$ for $n \in \mathbb{N}$.

$Z(f)$ is proper subspace of the linear space X, and it is maximal in the following sense: If Y is a subspace of X containing $Z(f)$, then either $Y = Z(f)$ or $Y = X$. A maximal proper subspace of X is called a **hyperspace** in X. Thus the zero space of a nonzero linear functional on X is a hyperspace in X. If Z is a hyperspace in X and $x \in X$, then $x + Z := \{x + z : z \in Z\}$ is called a **hyperplane** in X.

Examples 1.3 (i) Let X and Y be finite dimensional linear spaces of dimensions n and m, respectively. Let $x_1, \ldots, x_n$ constitute a basis for X, and let $y_1, \ldots, y_m$ constitute a basis for Y. Fix $j \in \{1, \ldots, n\}$. If $x \in X$ and $x = a_1 x_1 + \cdots + a_n x_n$, where $a_1, \ldots, a_n \in \mathbb{K}$, then define $f_j(x) := a_j$. Then f_j is a linear functional on X, and $f_j(x_i) = \delta_{i,j}$, $i = 1, \ldots, n$, where the **Kronecker symbol** $\delta_{i,j}$ is defined by $\delta_{i,j} := 1$ if $i = j$ and $\delta_{i,j} := 0$ if $i \neq j$. Thus $x = f_1(x)x_1 + \cdots + f_n(x)x_n$ for all $x \in X$. Similarly, for every fixed $i \in \{1, \ldots, m\}$, there is a linear functional g_i on Y such that $g_i(y_j) = \delta_{i,j}$, $j = 1, \ldots, m$. Thus $y = g_1(y)y_1 + \cdots + g_m(y)y_m$ for all $y \in Y$.

Consider an $m \times n$ matrix $M := [k_{i,j}]$ having $k_{i,j} \in \mathbb{K}$ as the element in the ith row and the jth column for $i = 1, \ldots, m$, $j = 1, \ldots, n$. For $x := f_1(x)x_1 + \cdots + f_n(x)x_n \in X$, let

$$y := \sum_{i=1}^{m} \Big(\sum_{j=1}^{n} k_{i,j} f_j(x) \Big) y_i,$$

so that, $g_i(y) = \sum_{j=1}^{n} k_{i,j} f_j(x)$ for $i = 1, \ldots, m$. This definition is inspired by the matrix multiplication

$$\begin{bmatrix} k_{1,1} & \ldots & k_{1,j} & \ldots & k_{1,n} \\ \vdots & & \vdots & & \vdots \\ k_{i,1} & \ldots & k_{i,j} & \ldots & k_{i,n} \\ \vdots & & \vdots & & \vdots \\ k_{m,1} & \ldots & k_{m,j} & \ldots & k_{m,n} \end{bmatrix} \begin{bmatrix} f_1(x) \\ \vdots \\ f_j(x) \\ \vdots \\ f_n(x) \end{bmatrix} = \begin{bmatrix} \sum_{j=1}^{n} k_{1,j} f_j(x) \\ \vdots \\ \sum_{j=1}^{n} k_{i,j} f_j(x) \\ \vdots \\ \sum_{j=1}^{n} k_{m,j} f_j(x) \end{bmatrix} = \begin{bmatrix} g_1(y) \\ \vdots \\ g_i(y) \\ \vdots \\ g_m(y) \end{bmatrix}.$$

If we define $F(x) := y$ for $x \in X$, then $F : X \to Y$ is a linear map, and $g_i(F(x_j)) = k_{i,j}$ for all $i = 1, \ldots, m$, $j = 1, \ldots, n$. We say that the matrix M defines the linear map F with respect to the basis $\{x_1, \ldots, x_n\}$ for X and the basis $\{y_1, \ldots, y_m\}$ for Y.

Conversely, let $F : X \to Y$ be a linear map. Then

$$F(x_j) = g_1(F(x_j))y_1 + \cdots + g_m(F(x_j))y_m \quad \text{for all } j = 1, \ldots, n.$$

Let $k_{i,j} := g_i(F(x_j))$ for $i = 1, \ldots, m$, $j = 1, \ldots, n$, and $M := [k_{i,j}]$. For $x = f_1(x)x_1 + \cdots + f_n(x)x_n \in X$, we obtain

$$F(x) = \sum_{j=1}^{n} f_j(x) F(x_j) = \sum_{j=1}^{n} f_j(x) \Big(\sum_{i=1}^{m} k_{i,j} y_i \Big) = \sum_{i=1}^{m} \Big(\sum_{j=1}^{n} k_{i,j} f_j(x) \Big) y_i.$$

Thus the $m \times n$ matrix $M := [g_i(F(x_j))]$ defines the given linear map F with respect to the basis $\{x_1, \ldots, x_n\}$ for X and the basis $\{y_1, \ldots y_m\}$ for Y.

In particular, let $X := \mathbb{K}^n$ and $Y := \mathbb{K}$. For $j = 1, \ldots, n$, let $k_j \in \mathbb{K}$. For $x := (x(1), \ldots, x(n)) \in \mathbb{K}^n$, define

$$f(x) := k_1 x(1) + \cdots + k_n x(n).$$

Clearly, f is a linear functional on $\mathbb{K}^n$. In fact, every linear functional on $\mathbb{K}^n$ is of this form. If $f \neq 0$, then the subspace $Z(f) = \{(x(1), \ldots, x(n)) \in \mathbb{K}^n : k_1 x(1) + \cdots + k_n x(n) = 0\}$ is of dimension $n - 1$, and it is a hyperspace in $\mathbb{K}^n$.

Let $\mathbb{K} := \mathbb{R}$. If $n = 2$, then the hyperspaces in X are the straight lines passing through the origin $(0, 0)$, and if $n = 3$, then the hyperspaces in X are the planes passing through the origin $(0, 0, 0)$.

(ii) Let us give examples of linear maps on some infinite dimensional linear spaces. Let X denote the set of all functions from the interval $[0, 1]$ to $\mathbb{K}$. Let $X_1 := \{x \in X : x$ is Riemann integrable on $[0, 1]\}$, $Y_1 := \{x \in X : x$ is continuous on $[0, 1]\}$ and $X_2 := \{x \in X : x$ is differentiable on $[0, 1]\}$. Consider $F_1 : X_1 \to Y_1$ and $F_2 : X_2 \to X$ defined by

$$F_1(x)(s) := \int_0^s x(t)\,dt \quad \text{and} \quad F_2(x)(s) := x'(s) \quad \text{whenever } s \in [0, 1].$$

Then F_1 and F_2 are linear maps. Now fix $s_0 \in [0, 1]$, and define

$$f_1(x) := \int_0^{s_0} x(t)dt \text{ for } x \in X_1 \quad \text{and} \quad f_2(x) := x'(s_0) \text{ for } x \in X_2.$$

Then f_1 and f_2 are linear functionals on X_1 and X_2, respectively.

These examples show that many important concepts in analysis such as integration and differentiation can be treated in a unified manner by considering appropriate linear spaces and linear maps. ◊

1.3 Metric Spaces and Continuous Functions

Let X be a nonempty set. A **metric** on X is a function $d : X \times X \to \mathbb{R}$ such that for all $x, y, z \in X$,

(i) **(triangle inequality)** $d(x, y) \leq d(x, z) + d(z, y)$,
(ii) **(symmetry)** $d(y, x) = d(x, y)$,
(iii) $d(x, y) \geq 0$ and $d(x, y) = 0$ if and only if $x = y$.

A **metric space** is a nonempty set along with a metric on it. A nonempty subset of a metric space is itself a metric space with the induced metric.

Before giving examples of metric spaces, we mention a crucial property of real numbers. If a subset E of real numbers is bounded above, and $\alpha \in \mathbb{R}$ is an upper bound of E, that is, if $x \le \alpha$ for all $x \in E$, then the set E has a (unique) upper bound in $\mathbb{R}$ which is less than or equal to every upper bound of E; it is called the **supremum** of E, and we shall denote it by $\sup E$. It follows that if a subset E of real numbers is bounded below, and $\beta \in \mathbb{R}$ is a lower bound of E, that is, if $x \ge \beta$ for all $x \in E$, then the set E has a (unique) lower bound in $\mathbb{R}$ which is greater than or equal to every lower bound of E; it is called the **infimum** of E, and we shall denote it by $\inf E$.

We prove two important inequalities involving nonnegative real numbers.

Lemma 1.4 *Let $a_1, \ldots, a_n, b_1, \ldots, b_n$ be nonnegative real numbers. Then*

(i) **(Schwarz inequality for numbers)**

$$\sum_{j=1}^{n} a_j b_j \le \Big(\sum_{j=1}^{n} a_j^2\Big)^{1/2} \Big(\sum_{j=1}^{n} b_j^2\Big)^{1/2},$$

(ii) **(Minkowski inequality for numbers)**

$$\Big(\sum_{j=1}^{n} (a_j + b_j)^2\Big)^{1/2} \le \Big(\sum_{j=1}^{n} a_j^2\Big)^{1/2} + \Big(\sum_{j=1}^{n} b_j^2\Big)^{1/2}.$$

Proof Let $\alpha := \Big(\sum_{j=1}^{n} a_j^2\Big)^{1/2}$ and $\beta := \Big(\sum_{j=1}^{n} b_j^2\Big)^{1/2}$.

(i) If either $\alpha = 0$ or $\beta = 0$, then both sides of the inequality in (i) are equal to zero. Now assume that $\alpha \ne 0$ and $\beta \ne 0$. For $j = 1, \ldots, n$, the geometric mean of a_j^2/α^2 and b_j^2/β^2 is less than or equal to their arithmetic mean:

$$\frac{a_j}{\alpha}\frac{b_j}{\beta} \le \frac{1}{2}\Big(\frac{a_j^2}{\alpha^2} + \frac{b_j^2}{\beta^2}\Big), \quad \text{and so} \quad \sum_{j=1}^{n} a_j b_j \le \frac{\alpha\beta}{2}\Big(\frac{\sum_{j=1}^{n} a_j^2}{\alpha^2} + \frac{\sum_{j=1}^{n} b_j^2}{\beta^2}\Big) = \alpha\beta,$$

as desired.

(ii) By the Schwarz inequality in (i) above,

$$\sum_{j=1}^{n} (a_j + b_j)^2 = \sum_{j=1}^{n} a_j^2 + 2\sum_{j=1}^{n} a_j b_j + \sum_{j=1}^{n} b_j^2 \le \alpha^2 + 2\alpha\beta + \beta^2 = (\alpha + \beta)^2,$$

as desired. □

Examples 1.5 (i) Let $n \in \mathbb{N}$ and $X := \mathbb{K}^n$. For $x := (x(1), \ldots, x(n))$ and $y := (y(1), \ldots, y(n))$ in $\mathbb{K}^n$, define

$$d_1(x, y) := |x(1) - y(1)| + \cdots + |x(n) - y(n)|,$$

$$d_2(x, y) := \big(|x(1) - y(1)|^2 + \cdots + |x(n) - y(n)|^2\big)^{1/2},$$

$$d_\infty(x, y) := \max\{|x(1) - y(1)|, \ldots, |x(n) - y(n)|\}.$$

Let $x, y, z \in \mathbb{K}^n$. Since $|x(j) - y(j)| \le |x(j) - z(j)| + |z(j) - y(j)|$ for all $j = 1, \ldots, n$, it is easy to see that d_1 and d_∞ are metrics on $\mathbb{K}^n$. To see that d_2 is also a metric on $\mathbb{K}^n$, let $a_j := |x(j) - z(j)|$, $b_j := |z(j) - y(j)|$ for $j = 1, \ldots, n$. By the Minkowski inequality for numbers (Lemma 1.4(ii)),

$$d_2(x, y) \le \Big(\sum_{j=1}^n (a_j + b_j)^2\Big)^{1/2} \le \Big(\sum_{j=1}^n a_j^2\Big)^{1/2} + \Big(\sum_{j=1}^n b_j^2\Big)^{1/2} = d_2(x, z) + d_2(z, y).$$

If $n = 1$, that is, if $X := \mathbb{K}$, then the three metrics d_1, d_2 and d_∞ reduce to the **usual metric** on $\mathbb{K}$ given by $d(x, y) := |x - y|$ for $x, y \in \mathbb{K}$.

(ii) Let $x := (x(1), x(2), \ldots)$ and

$$\ell^1 := \Big\{x : x(j) \in \mathbb{K} \text{ for each } j \in \mathbb{N} \text{ and } \textstyle\sum_{j=1}^\infty |x(j)| < \infty\Big\},$$
$$\ell^2 := \Big\{x : x(j) \in \mathbb{K} \text{ for each } j \in \mathbb{N} \text{ and } \textstyle\sum_{j=1}^\infty |x(j)|^2 < \infty\Big\},$$
$$\ell^\infty := \Big\{x : x(j) \in \mathbb{K} \text{ for each } j \in \mathbb{N} \text{ and } \sup_{j\in\mathbb{N}} |x(j)| < \infty\Big\}.$$

Clearly, ℓ^1 and ℓ^∞ are linear spaces over $\mathbb{K}$. Also, the Minkowski inequality for numbers (Lemma 1.4(ii)) shows that ℓ^2 is a linear space over $\mathbb{K}$. Define

$$d_1(x, y) := \sum_{j=1}^\infty |x(j) - y(j)| \text{ for } x, y \in \ell^1,$$
$$d_2(x, y) := \Big(\sum_{j=1}^\infty |x(j) - y(j)|^2\Big)^{1/2} \quad \text{for } x, y \in \ell^2,$$
$$d_\infty(x, y) := \sup\{|x(j) - y(j)| : j \in \mathbb{N}\} \text{ for } x, y \in \ell^\infty.$$

Since $|x(j) - y(j)| \le |x(j) - z(j)| + |z(j) - y(j)|$ for all $j \in \mathbb{N}$, it is easy to see that d_1 is a metric on ℓ^1, and d_∞ is a metric on ℓ^∞. To see that d_2 is a metric on ℓ^2, consider $x := (x(1), x(2), \ldots)$, $y := (y(1), y(2), \ldots)$ and $z := (z(1), z(2), \ldots)$ in ℓ^2, and let $a_j := |x(j) - z(j)|$, $b_j := |z(j) - y(j)|$ for $j \in \mathbb{N}$. Letting $n \to \infty$ in the Minkowski inequality proved in Lemma 1.4(ii), we obtain

$$d_2(x, y) \le \Big(\sum_{j=1}^\infty (a_j + b_j)^2\Big)^{1/2} \le \Big(\sum_{j=1}^\infty a_j^2\Big)^{1/2} + \Big(\sum_{j=1}^\infty b_j^2\Big)^{1/2} = d_2(x, z) + d_2(z, y).$$

(iii) Let T be a nonempty set. A function $x : T \to \mathbb{K}$ is called **bounded** on T if there is $\alpha > 0$ such that $|x(t)| \le \alpha$ for all $t \in T$. Let $B(T)$ denote the set of all $\mathbb{K}$-valued bounded functions on T. For $x, y \in B(T)$, let

$$d_\infty(x, y) := \sup\{|x(t) - y(t)| : t \in T\}.$$

It is easy to see that d_∞ is a metric on $B(T)$. It is known as the **sup metric** on $B(T)$. If $T := \mathbb{N}$, then $B(T) = \ell^\infty$, introduced in (ii) above. ◊

Sequences

Let X be a nonempty set. A **sequence** in X is a function from $\mathbb{N}$ to X. We shall denote a typical sequence in X by (x_n), where x_n is its **value** at $n \in \mathbb{N}$; x_n is also called the nth **term** of the sequence. We emphasize that $E := \{x_n : n \in \mathbb{N}\}$ is a *subset* of X, whereas (x_n) is a *function* with values in X. For example, if $X := \mathbb{R}$ and $x_n := (-1)^n$ for $n \in \mathbb{N}$, then the sequence (x_n) can be written out as $-1, 1, -1, 1, \ldots$, but $E = \{-1, 1\}$.

Let d be a metric on X. We say that a sequence (x_n) in X is **convergent** in X if there is $x \in X$ satisfying the following condition: For every $\epsilon > 0$, there is $n_0 \in \mathbb{N}$ such that $d(x_n, x) < \epsilon$ for all $n \geq n_0$. In this case, we say that (x_n) **converges** to x, or that x is a **limit** of (x_n), and we write $x_n \to x$. Thus $x_n \to x$ in the metric d on X if and only if $d(x_n, x) \to 0$ in the usual metric on $\mathbb{R}$. The convergence of a sequence is the most basic notion for us.

If (x_n) is a sequence in X, and if $n_1, n_2, \ldots$ in $\mathbb{N}$ satisfy $n_1 < n_2 < \cdots$, then (x_{n_k}) is called a **subsequence** of (x_n). It is clear that $x_n \to x$ in X if and only if $x_{n_k} \to x$ for every subsequence (x_{n_k}) of (x_n).

Examples 1.6 (i) Let $m \in \mathbb{N}$ and $X := \mathbb{K}^m$ with one of the metrics d_1, d_2, d_∞ given in Example 1.5(i). It is easy to see that $x_n \to x$ in $\mathbb{K}^m$ if and only if $x_n(j) \to x(j)$ in $\mathbb{K}$ for each $j = 1, \ldots, m$. This convergence is known as **componentwise convergence**.

(ii) Let $p \in \{1, 2, \infty\}$, and $X := \ell^p$ with the metric d_p (Example 1.5(ii)). Then

$$x_n \to x \text{ in } \ell^1 \text{ means } \textstyle\sum_{j=1}^{\infty} |x_n(j) - x(j)| \to 0,$$
$$x_n \to x \text{ in } \ell^2 \text{ means } \textstyle\sum_{j=1}^{\infty} |x_n(j) - x(j)|^2 \to 0,$$
$$x_n \to x \text{ in } \ell^\infty \text{ means } \sup\{|x_n(j) - x(j)| : j \in \mathbb{N}\} \to 0.$$

Let $x_n \to x$ in ℓ^p. It follows that $x_n(j) \to x(j)$ in $\mathbb{K}$ for each $j \in \mathbb{N}$, that is, componentwise convergence holds. However, componentwise convergence does not imply convergence in ℓ^p. For example, let $e_n := (0, \ldots, 0, 1, 0, 0, \ldots)$, where 1 occurs only in the nth entry, and let $x := 0$. Then $e_n(j) \to 0$ for each $j \in \mathbb{N}$, but $e_n \not\to 0$ in ℓ^p, since $d_p(e_n, 0) = 1$ for each $n \in \mathbb{N}$.

(iii) Let T be a nonempty set, and $X := B(T)$ with the sup metric. Then $x_n \to x$ in $B(T)$ means $\sup\{|x_n(t) - x(t)| : t \in T\} \to 0$, that is, for every $\epsilon > 0$, there is $n_0 \in \mathbb{N}$ such that for every $n \geq n_0$, $|x_n(t) - x(t)| < \epsilon$ for all $t \in T$. This convergence is known as **uniform convergence** on T. ◊

Let X be a metric space. A subset E of X is called a **closed set** if every sequence in E that is convergent in X converges to an element in E. The set of all limits of sequences in E that are convergent in X is called the **closure** of E in X; it will be denoted by $\overline{E}$. Clearly, $E \subset \overline{E}$, and the subset E is closed if and only if $\overline{E} = E$.

A subset E of X is called an **open set** if its complement $X \setminus E$ is a closed set. For $x \in X$ and $r > 0$, consider the ball

$$U(x, r) := \{y \in X : d(x, y) < r\}$$

about x of radius r. It is easy to see that E is open if and only if for every $x \in E$, there is $r > 0$ such that $U(x, r) \subset E$. Since $U(x, r)$ is itself an open subset of X, it is called the **open ball** about x of radius r. Clearly, $\emptyset$ and X are open subsets of X. Also, an arbitrary union of open subsets of X is open, and a finite intersection of open subsets of X is open.

A subset E of X is called **dense** in X if $\overline{E} = X$, that is, for every $x \in X$, there is a sequence (x_n) in E such that $x_n \to x$; equivalently, $E \cap U(x, r) \neq \emptyset$ for every $x \in X$ and $r > 0$. Further, X is called **separable** if it has a countable dense subset. It can be seen that a nonempty subset of a separable metric space is separable, and the closure of a separable subset is separable.

Examples 1.7 (i) The set $\mathbb{R}$ of all real numbers with the usual metric is separable, since the subset $\mathbb{Q}$ of all rational numbers is denumerable and it is dense in $\mathbb{R}$. Also, the set $\mathbb{C}$ of all complex numbers with the usual metric is separable, since the subset $\{p + iq : p, q \in \mathbb{Q}\}$ is denumerable and is dense in $\mathbb{C}$. Thus $\mathbb{K}$ with the usual metric is separable. Let $m \in \mathbb{N}$. It follows that $\mathbb{K}^m$ with any of the metrics d_1, d_2, d_∞ is separable.

(ii) Let $p \in \{1, 2\}$. To see that the metric space ℓ^p is separable, consider $e_j := (0, \ldots, 0, 1, 0, 0, \ldots)$, where 1 occurs only in the jth entry, $j \in \mathbb{N}$. Let

$$E := \{k_1 e_1 + \cdots + k_n e_n : n \in \mathbb{N}, \text{ and } \operatorname{Re} k_j, \operatorname{Im} k_j \in \mathbb{Q} \text{ for } j = 1, \ldots, n\}.$$

E is a countable set, since $\mathbb{Q}$ is countable. We show that E is dense in ℓ^p. Let $x \in \ell^p$ and $r > 0$. Since $\sum_{j=1}^{\infty} |x(j)|^p$ is finite, there is $n \in \mathbb{N}$ such that

$$\sum_{j=n+1}^{\infty} |x(j)|^p < \frac{r^p}{2}.$$

Since $\mathbb{Q}$ is dense in $\mathbb{R}$, there are $k_1, \ldots, k_n$ in $\mathbb{K}$ with $\operatorname{Re} k_j, \operatorname{Im} k_j \in \mathbb{Q}$ such that

$$|x(j) - k_j|^p < \frac{r^p}{2n} \quad \text{for } j = 1, \ldots, n.$$

Define $y := k_1 e_1 + \cdots + k_n e_n \in E$. Then

$$\big(d_p(x, y)\big)^p = \sum_{j=1}^{n} |x(j) - k_j|^p + \sum_{j=n+1}^{\infty} |x(j)|^p < \frac{r^p}{2} + \frac{r^p}{2} = r^p.$$

Hence $y \in U(x, r)$. Thus $E \cap U(x, r) \neq \emptyset$ for every $x \in \ell^p$ and $r > 0$, that is, E is dense in ℓ^p.

On the other hand, the metric space ℓ^∞ is not separable. To see this, let $S := \{x \in \ell^\infty : x(j) = 0 \text{ or } 1 \text{ for all } j \in \mathbb{N}\}$. Then $d_\infty(x, y) = 1$ for all $x \neq y$ in S. Let

$\{x_1, x_2, \ldots\}$ be any countable subset of ℓ^∞. Then $U(x_n, 1/2)$ contains at most one element of S for each $n \in \mathbb{N}$. Since S is uncountable, there is $x \in S$ such that x does not belong to $U(x_n, 1/2)$ for all $n \in \mathbb{N}$. In other words, $\{x_1, x_2, \ldots\} \cap U(x, 1/2) = \emptyset$. Hence the set $\{x_1, x_2, \ldots\}$ is not dense in ℓ^∞.

(iii) The metric space $B(T)$ is separable if and only if the set T is finite. Suppose T is finite, that is, there is $n \in \mathbb{N}$ such that $T := \{t_1, \ldots, t_n\}$. Then $B(T) = \mathbb{K}^n$, and the sup metric on $B(T)$ is the d_∞ metric on $\mathbb{K}^n$. By (i) above, $B(T)$ is separable. Next, suppose T is infinite. Then there is a denumerable subset $\{t_1, t_2, \ldots\}$ of T. We can then show that $B(T)$ is not separable just as we showed that ℓ^∞ is not separable in (ii) above. ◊

A subset E of X is called a **bounded set** if there are $x \in X$ and $r > 0$ such that $E \subset U(x, r)$, and it is called a **totally bounded set** if for every $\epsilon > 0$, there are $x_1, \ldots, x_n$ in X such that $E \subset U(x_1, \epsilon) \cup \cdots \cup U(x_n, \epsilon)$.[2] If E is totally bounded, then it is bounded, since $E \subset U(x_1, 1) \cup \cdots \cup U(x_n, 1)$ implies that $E \subset U(x_1, r)$, where $r := 1 + \max\{d(x_1, x_j) : j = 2, \ldots, n\}$.

But a bounded set need not be totally bounded. For example, let $X := \mathbb{R}$ with the metric d given by $d(x, y) = \min\{1, |x - y|\}$ for $x, y \in \mathbb{R}$. Then $\mathbb{R}$ is clearly bounded. But it is not totally bounded, since for any $x_1, \ldots, x_n \in \mathbb{R}$, the real number $x := \max\{|x_1|, \ldots, |x_n|\} + 1/2$ does not belong to the union of open balls of radius $1/2$ about $x_1, \ldots, x_n$. However, if $X := \mathbb{K}^m$ with any of the metrics d_1, d_2, d_∞, then it can be seen that every bounded subset of $\mathbb{K}^m$ is in fact totally bounded. We note that if E is totally bounded, then for every $\epsilon > 0$, we can find $x_1, \ldots, x_n$ in E itself such that $E \subset U(x_1, \epsilon) \cup \cdots \cup U(x_n, \epsilon)$. It can then be seen that a subset of a totally bounded set is totally bounded, and the closure of a totally bounded set is totally bounded.

Cauchy Sequences

Let d be a metric on a nonempty set X. A sequence (x_n) in X is called a **bounded sequence** if there are $x \in X$ and $\alpha > 0$ such that $d(x_n, x) \le \alpha$ for all $n \in \mathbb{N}$, that is, if $\{x_n : n \in \mathbb{N}\}$ is a bounded subset of X. A sequence (x_n) in X is called a **Cauchy sequence** if for every $\epsilon > 0$, there is $n_0 \in \mathbb{N}$ such that $d(x_n, x_m) < \epsilon$ for all $n, m \ge n_0$.

Every Cauchy sequence is bounded, but a bounded sequence in X need not be a Cauchy sequence. For example, let $X := \mathbb{R}$ with the usual metric, and let $x_n := (-1)^n$ for $n \in \mathbb{N}$. Further, every convergent sequence is a Cauchy sequence, but a Cauchy sequence need not be convergent. For example, let $X := (0, 1]$ with the usual metric, and let $x_n := 1/n$ for $n \in \mathbb{N}$. A Cauchy sequence is convergent if it has a convergent subsequence.

We characterize total boundedness in terms of Cauchy sequences as follows.

Proposition 1.8 *Let E be a subset of a metric space X. Then E is totally bounded if and only if every sequence in E has a Cauchy subsequence.*

---

[2] If a set E represents a town, and a point x represents a watchman, then we may say that a town is 'totally bounded' if it can be guarded by a finite number of watchmen having an arbitrarily short sight.

Proof Suppose E is totally bounded. Let (x_n) be a sequence in E. Since E can be covered by a finite number of open balls of radius 1, there is an open ball U_1 of radius 1 such that $E \cap U_1$ contains infinitely many terms of the sequence (x_n). Let $E_0 := E$, $n_0 := 1$ and $E_1 := E_0 \cap U_1$. Then there is $n_1 \in \mathbb{N}$ such that $n_1 > n_0$ and $x_{n_1} \in E_1$. Since E_1 is a subset of E, it is totally bounded. Since E_1 can be covered by a finite number of open balls of radius $1/2$, there is an open ball U_2 of radius $1/2$ such that $E_1 \cap U_2$ contains infinitely many terms of the sequence (x_n). Let $E_2 := E_1 \cap U_2$. Then there is $n_2 \in \mathbb{N}$ such that $n_2 > n_1$ and $x_{n_2} \in E_2$. Continuing in this manner, for each $k \in \mathbb{N}$, there is an open ball U_k of radius $1/k$ and there is $n_k \in \mathbb{N}$ such that $n_k > n_{k-1}$ and $x_{n_k} \in E_k$, where $E_k := E_{k-1} \cap U_k$. Then (x_{n_k}) is a Cauchy subsequence of the sequence (x_n) since $d(x_{n_i}, x_{n_j}) < 2/k$ for all $i, j \geq k$.

Conversely, suppose E is not totally bounded. Then there is $\epsilon > 0$ such that E cannot be covered by finitely many open balls of radius ϵ about elements of E. Let $x_1 \in E$. Then there is $x_2 \in E$ such that $x_2 \notin U(x_1, \epsilon)$. Having chosen $x_1, \ldots, x_n \in E$, we can find $x_{n+1} \in E$ such that $x_{n+1} \notin U(x_1, \epsilon) \cup \cdots \cup U(x_n, \epsilon)$ for $n \in \mathbb{N}$. Then $d(x_n, x_m) \geq \epsilon$ for all $n \neq m$ in $\mathbb{N}$. Hence the sequence (x_n) in E cannot have a Cauchy subsequence. □

Completeness

A metric space X is called **complete** if every Cauchy sequence in X converges in X. Loosely speaking, the completeness of a metric space X means the following: If the elements of a sequence in X come arbitrarily close to each other, then they find an element of X to come arbitrarily close to!

Suppose E is a subset of a metric space X. If E is complete, then E is closed. Conversely, if X is complete and E is closed, then E is complete.

Examples 1.9 (i) $\mathbb{K}$ with the usual metric is complete. To see this, first consider a bounded sequence (x_n) in $\mathbb{R}$. Interestingly, every sequence in $\mathbb{R}$ has a monotonic subsequence. (See, for instance, [12, Proposition 2.14].) Hence (x_n) has a bounded monotonic subsequence, which must converge in $\mathbb{R}$. If (x_n) is bounded sequence in $\mathbb{C}$, then by considering the bounded sequences $(\text{Re}\, x_n)$ and $(\text{Im}\, x_n)$ in $\mathbb{R}$, we see that (x_n) has a convergent subsequence. This is known as the **Bolzano–Weierstrass theorem** for $\mathbb{K}$. Now let (x_n) be a Cauchy sequence in $\mathbb{K}$. Then it is bounded, and so it has a convergent subsequence. Consequently, (x_n) itself converges in $\mathbb{K}$.

Let $m \in \mathbb{N}$, and $X := \mathbb{K}^m$ with the metric d_1, d_2 or d_∞. Then a sequence (x_n) is Cauchy in $\mathbb{K}^m$ if and only if the sequence $(x_n(j))$ is Cauchy in $\mathbb{K}$ for each j in $\{1, \ldots, m\}$. In view of the componentwise convergence in $\mathbb{K}^m$ (Example 1.6(i)), we see that $\mathbb{K}^m$ is complete.

(ii) The metric spaces $\ell^1, \ell^2, \ell^\infty$ introduced in Example 1.5(ii) are complete. This will be proved in Example 2.24(ii).

(iii) The metric space $B(T)$ introduced in Example 1.5(iii) is complete. This will be proved in Example 2.24(iii). ◊

We now prove an important result regarding complete metric spaces.

Theorem 1.10 (Baire, 1899) *Let X be a complete metric space. Then the intersection of a countable number of dense open subsets of X is dense in X.*

Proof Let $D_1, D_2, \ldots$ be dense open subsets of X. Let $x_0 \in X$ and $r_0 > 0$. Since D_1 is dense in X, there is $x_1 \in D_1 \cap U(x_0, r_0)$. Also, since $D_1 \cap U(x_0, r_0)$ is open in X, there is $r_1 > 0$ such that $U(x_1, r_1) \subset D_1 \cap U(x_0, r_0)$. Proceeding inductively, given $m \in \mathbb{N}$, there are $x_1, \ldots, x_m$ in X and positive numbers $r_1, \ldots, r_m$ such that $U(x_n, r_n) \subset D_n \cap U(x_{n-1}, r_{n-1})$ for $n = 1, \ldots, m$. Clearly, $x_m \in \left(\bigcap_{n=1}^m D_n\right) \cap U(x_0, r_0)$. Thus $\left(\bigcap_{n=1}^m D_n\right) \cap U(x_0, r_0) \neq \emptyset$. Now suppose the sets $D_1, D_2, \ldots$ are infinitely many. By decreasing r_n, if needed, we assume that $r_n \leq 1/n$ and $\overline{U}(x_n, r_n) \subset D_n \cap U(x_{n-1}, r_{n-1})$ for all $n \in \mathbb{N}$.

Let $\epsilon > 0$, and find $n_0 \in \mathbb{N}$ such that $(2/n_0) \leq \epsilon$. For $n, m \geq n_0$, we see that $x_n, x_m \in U(x_{n_0}, r_{n_0})$, and so

$$d(x_n, x_m) \leq d(x_n, x_{n_0}) + d(x_{n_0}, x_m) < 2\,r_{n_0} \leq \frac{2}{n_0} \leq \epsilon.$$

Hence (x_n) is a Cauchy sequence in X. Since X is complete, there is $x \in X$ such that $x_n \to x$ in X. Now for each fixed $n \in \mathbb{N}$, $x_m \in U(x_n, r_n)$ for all $m \geq n$, and so $x \in \overline{U}(x_n, r_n)$. Since $\overline{U}(x_n, r_n) \subset D_n \cap U(x_0, r_0)$ for all $n \in \mathbb{N}$, $x \in \left(\bigcap_{n=1}^\infty D_n\right) \cap U(x_0, r_0)$. Thus $\left(\bigcap_{n=1}^\infty D_n\right) \cap U(x_0, r_0) \neq \emptyset$. Since $x_0 \in X$ and $r_0 > 0$ are arbitrary, $\bigcap_{n=1}^\infty D_n$ is dense in X. □

We remark that if a metric space X is not complete, then the intersection of a denumerable number of dense open subsets of X need not be dense in X; in fact, it may be empty. For example, let $X := \mathbb{Q}$ along with the metric induced by the usual metric on $\mathbb{R}$. Let $\mathbb{Q} = \{q_1, q_2, \ldots\}$, and let $D_n := \mathbb{Q} \setminus \{q_n\}$ for $n \in \mathbb{N}$. Then D_n is dense and open in X for each $n \in \mathbb{N}$, but $\bigcap_{n=1}^\infty D_n = \emptyset$.

Compactness

Closed and bounded subsets of $\mathbb{K}^n$ are very important in many branches of analysis. We seek an analogue in arbitrary metric spaces. Let E be a subset of a metric space X. Then E is called **compact** if every sequence in E has a subsequence which converges in E. If E is compact, then clearly E is closed in X. Conversely, if X is compact and E is closed in X, then E is compact.

We characterize compactness in terms of total boundedness and completeness as follows.

Theorem 1.11 *Let E be a subset of a metric space X. Then E is compact if and only if it is totally bounded and complete.*

Proof Suppose E is compact. Let (x_n) be a sequence in E. Then (x_n) has a subsequence which converges in E. Since a convergent subsequence is a Cauchy subsequence, Proposition 1.8 shows that E is totally bounded. Also, if (x_n) is a Cauchy sequence, then it is converges in E since it has a subsequence that converges in E. Hence E is complete.

Conversely, let E be totally bounded and complete. Let (x_n) be a sequence in E. Since E is totally bounded, (x_n) has a Cauchy subsequence by Proposition 1.8. But since E is complete, this Cauchy subsequence converges in E. Thus E is compact. □

Corollary 1.12 *Let X be a complete metric space and $E \subset X$.*

(i) *E is compact if and only if E is closed and totally bounded.*
(ii) *The closure $\overline{E}$ of E is compact if and only if E is totally bounded.*

Proof (i) A subset of a complete metric space is closed if and only if it is complete. Hence the desired result follows from Theorem 1.11.

(ii) A subset of a totally bounded set is totally bounded, and its closure is also totally bounded. Hence the desired result follows from (i) above. □

Corollary 1.13 (Heine–Borel theorem) *Let $p \in \{1, 2, \infty\}$, and consider the metric d_p on $\mathbb{K}^m$. A subset of $\mathbb{K}^m$ is compact if and only if it is closed and bounded.*

Proof Note that $\mathbb{K}^m$ is complete, and a subset of $\mathbb{K}^m$ is totally bounded if and only if it is bounded. The result follows from Corollary 1.12(i). □

Continuous Functions

Let X and Y be nonempty sets, and let d_X and d_Y be metrics on X and Y, respectively. Consider a function $F : X \to Y$. Let $x_0 \in X$. We say that F is **continuous** at x_0, if $F(x_n) \to F(x_0)$ in Y whenever $x_n \to x_0$ in X. It is easy to see that this happens if and only if the following ϵ-δ condition holds: For every $\epsilon > 0$, there is $\delta > 0$ such that $d_Y(F(x), F(x_0)) < \epsilon$ whenever $x \in X$ and $d_X(x, x_0) < \delta$. Further, we say that F is **continuous** on X if it is continuous at every $x \in X$, that is, $d_Y(F(x_n), F(x)) \to 0$ whenever $d_X(x_n, x) \to 0$ for $x \in X$ and a sequence (x_n) in X. We can see that F is continuous on X if and only if the set $F^{-1}(E)$ is open in X for every open subset E of Y.

Let X be a compact metric space, Y be a metric space, and let $F : X \to Y$ be continuous. Then the range $R(F)$ of F is a compact subset of Y. To see this, let (y_n) be a sequence in $R(F)$, and let $x_n \in X$ be such that $y_n = F(x_n)$ for each $n \in \mathbb{N}$. Since X is compact, there is a subsequence (x_{n_k}) of the sequence (x_n) and there is $x \in X$ such that $x_{n_k} \to x$. Then $y_{n_k} = F(x_{n_k}) \to F(x)$ by the continuity of F. Thus (y_n) has a subsequence which converges in $R(F)$.

If $F : X \to Y$ is one-one and continuous, and if $F^{-1} : R(F) \to X$ is also continuous, then F is called a **homeomorphism**. If there is a homeomorphism from X onto Y, then the metric spaces X and Y are called **homeomorphic**. If $F : X \to Y$ satisfies $d_Y(F(x_1), F(x_2)) = d_X(x_1, x_2)$ for all $x_1, x_2 \in X$, then F is called an **isometry**. If there is an isometry from X onto Y, then the metric spaces X and Y are called **isometric**. An isometry is clearly a homeomorphism, but a homeomorphism may not be an isometry. For example, let $X := \mathbb{R}$ and $Y := (-1, 1)$ with the usual metrics. Let us define $F(t) := t/(1 + |t|)$ for $t \in \mathbb{R}$. Then F is a homeomorphism from X onto Y, but it is not an isometry. Isometric spaces share the same 'metric' properties. This may not hold for homeomorphic spaces. For example, $\mathbb{R}$ with the usual metric is complete but not totally bounded, while $(-1, 1)$ is totally bounded but not complete in the induced metric.

We shall also use the concept of the distance of a point from a subset of a metric space. Let d be a metric on a nonempty set X, and consider a nonempty subset E of X. For $x \in X$, define

$$d(x, E) := \inf\{d(x, y) : y \in E\}.$$

Clearly, $d(x, E) = 0$ if and only if $x \in \overline{E}$. Also, for all $x_1, x_2 \in X$, $d(x_1, x_2) \geq |d(x_1, E) - d(x_2, E)|$. Hence the function given by $x \longmapsto d(x, E)$ from X to $\mathbb{R}$ is continuous. Suppose E_1 and E_2 are disjoint nonempty closed subsets of a metric space X. Let $E := E_1 \cup E_2$, and define $G : E \to \mathbb{R}$ by $G := 0$ on E_1 and $G := 1$ on E_2. Then G is continuous on the closed set E. Let

$$F(x) := \frac{d(x, E_1)}{d(x, E_1) + d(x, E_2)} \quad \text{for } x \in X.$$

Then F is a continuous function from X to $[0, 1]$ such that $F = G$ on E, that is, $F = 0$ on E_1 and $F = 1$ on E_2. This is known as the **Urysohn lemma** for a metric space. A more general result says that if E is a nonempty closed subset of a metric space X, and if a function $G : E \to \mathbb{K}$ is continuous, then there is a continuous function $F : X \to \mathbb{K}$ such that $F = G$ on E. It is known as the **Tietze extension theorem**. For a proof which uses the function $x \longmapsto d(x, E)$ from X to $\mathbb{R}$, we refer the reader to [8, 4.5.1]. Further, if $|G(y)| \leq \alpha$ for all $y \in E$, we can require that $|F(x)| \leq \alpha$ for all $x \in X$.

A function $F : X \to Y$ is called **uniformly continuous** on X if $d_Y(F(u_n), F(v_n)) \to 0$ whenever (u_n) and (v_n) are sequences in X and $d_X(u_n, v_n) \to 0$. It is easy to see that this happens if and only if the following ϵ-δ condition holds: For every $\epsilon > 0$, there is $\delta > 0$ such that $d_Y(F(u), F(v)) < \epsilon$ whenever $u, v \in X$ and $d_X(u, v) < \delta$. A uniformly continuous function on X is continuous on X. To see this, consider $u_n \to u$ in X, and let $v_n := u$ for all $n \in \mathbb{N}$. But the converse does not hold. For example, let $X = Y := \mathbb{R}$ with the usual metrics, and let $F(x) := x^2$ for $x \in \mathbb{R}$, or let $X := (0, 1]$ and $Y := \mathbb{R}$ with the usual metrics, and let $F(x) := 1/x$ for $x \in (0, 1]$.

Proposition 1.14 *Let X be a compact metric space, Y be a metric space, and $F : X \to Y$ be continuous. Then F is uniformly continuous on X.*

Proof Assume for a moment that F is not uniformly continuous on X. Then there are sequences (u_n) and (v_n) in X such that $d_X(u_n, v_n) \to 0$, but $d_Y(F(u_n), F(v_n)) \not\to 0$. Consequently, there exist $\epsilon > 0$ and positive integers $n_1 < n_2 < \cdots$ such that $d_Y(F(u_{n_k}), F(v_{n_k})) \geq \epsilon$ for all $k \in \mathbb{N}$. Since X is compact, there is a convergent subsequence, say, $(u_{n_{k_j}})$ of (u_{n_k}). Let us denote the sequences $(u_{n_{k_j}})$ and $(v_{n_{k_j}})$ by $(\tilde{u}_j)$ and $(\tilde{v}_j)$ for simplicity. Let $\tilde{u}_j \to u$ in X. Then $\tilde{v}_j \to u$ as well, since $d_X(u_n, v_n) \to 0$. Because F is continuous at u, we obtain $F(\tilde{u}_j) \to F(u)$ and

$F(\tilde{v}_j) \to F(u)$. Thus $d_Y(F(\tilde{u}_j), F(\tilde{v}_j)) \to 0$. But this is a contradiction since $d_Y(F(\tilde{u}_j), F(\tilde{v}_j)) \geq \epsilon$ for all $j \in \mathbb{N}$. □

Let us now consider a sequence (F_n) of continuous functions from X to Y. We say that (F_n) **converges pointwise** on X if there is a function $F : X \to Y$ such that $F_n(x) \to F(x)$ for every $x \in X$. In this case, F is called the **pointwise limit** of (F_n). It may not be continuous on X. For example, let $X := [0, 1]$ and $Y := \mathbb{R}$ with the usual metrics, and for $n \in \mathbb{N}$, let

$$F_n(x) := \begin{cases} nx & \text{if } x \in [0, 1/n], \\ 1 & \text{if } x \in (1/n, 1] \end{cases} \quad \text{and} \quad F(x) := \begin{cases} 0 & \text{if } x = 0, \\ 1 & \text{if } x \in (0, 1]. \end{cases}$$

Then each F_n is continuous on $[0, 1]$, $F_n(x) \to F(x)$ for every $x \in [0, 1]$, but F is not continuous on $[0, 1]$.

Further, we say that (F_n) **converges uniformly** on X if there is a function $F : X \to Y$ such that for every $\epsilon > 0$, there is $n_0 \in \mathbb{N}$ satisfying

$$n \geq n_0,\ x \in X \Longrightarrow d_Y(F_n(x), F(x)) < \epsilon.$$

We have already come across this concept in Example 1.6(iii).

Proposition 1.15 *Let X and Y be metric spaces, and let (F_n) be a sequence of continuous functions from X to Y. If (F_n) converges uniformly to a function $F : X \to Y$, then F is continuous on X.*

Proof Let $\epsilon > 0$. There is $n_0 \in \mathbb{N}$ such that $d_Y(F(x), F_{n_0}(x)) < \epsilon/3$ for all $x \in X$. Let $x_0 \in X$. Since F_{n_0} is continuous at x_0, there is $\delta > 0$ such that $d_Y(F_{n_0}(x), F_{n_0}(x_0)) < \epsilon/3$ for all $x \in X$ with $d_X(x, x_0) < \delta$. Hence for $x \in X$ with $d_X(x, x_0) < \delta$, the distance $d_Y(F(x), F(x_0))$ is less than or equal to

$$d_Y(F(x), F_{n_0}(x)) + d_Y(F_{n_0}(x), F_{n_0}(x_0)) + d_Y(F_{n_0}(x_0), F(x_0)),$$

which is less than $\epsilon/3 + \epsilon/3 + \epsilon/3 = \epsilon$. Hence F is continuous at x_0. Since $x_0 \in X$ is arbitrary, we see that F is continuous on X. □

The above result says that a **uniform limit** of a sequence of continuous functions is continuous. Conversely, we may ask whether every continuous function is the uniform limit of a sequence of functions of a special kind. We give below a very important and well-known result in this regard. Another such result will be given in Sect. 1.4 (Theorem 1.27).

Theorem 1.16 (Weierstrass, 1885) *Every continuous function from a closed and bounded interval $[a, b]$ to $\mathbb{K}$ is the uniform limit of a sequence of polynomials with coefficients in $\mathbb{K}$.*

Proof First let $\mathbb{K} := \mathbb{R}$. Suppose $a := 0$, $b := 1$, and let x be a real-valued continuous function on $[0, 1]$. For $n \in \mathbb{N}$, define the nth **Bernstein polynomial** corresponding to x by

$$B_n(x)(t) := \sum_{k=0}^{n} x\left(\frac{k}{n}\right)\binom{n}{k} t^k(1-t)^{n-k}, \quad t \in [0, 1].$$

Then each B_n is a linear map from the linear space of all real-valued continuous functions on $[0, 1]$ to itself. Further, if x is a nonnegative continuous function on $[0, 1]$, then $B_n(x)$ is also a nonnegative function. Let $x_0(t) := 1$, $x_1(t) := t$ and $x_2(t) := t^2$ for $t \in [0, 1]$. It can be seen that

$$B_n(x_0) = x_0, \quad B_n(x_1) = x_1 \text{ and } B_n(x_2) = \left(1 - \frac{1}{n}\right)x_2 + \frac{1}{n}x_1 \quad \text{for } n \in \mathbb{N}.$$

Hence for $j = 0, 1, 2$, the sequence $(B_n(x_j))$ converges uniformly to x_j on $[0, 1]$. These remarkable properties of the Bernstein polynomials enable us to show that for every real-valued continuous function x on $[0, 1]$, the sequence $(B_n(x))$ converges uniformly to x on $[0, 1]$. For details, see [19, p. 28] or [2, p. 118]. For another proof, see [25, Theorem 7.26].

By considering the change of variable given by $\phi(t) := (b - a)t + a$, $t \in [0, 1]$, and $\phi^{-1}(s) := (s - a)/(b - a)$, $s \in [a, b]$, we can prove the above result for functions on the interval $[a, b]$. Also, the case $\mathbb{K} := \mathbb{C}$ can be easily deduced from the case $\mathbb{K} := \mathbb{R}$. □

Let T be a metric space with metric d_T. Consider $C(T)$, the set of all $\mathbb{K}$-valued bounded continuous functions on T. Thus $C(T)$ is contained in the metric space $B(T)$ introduced in Example 1.5(iii). Let E be a subset of $C(T)$. We say that E is **pointwise bounded** on T if for each $t \in T$, there is $\alpha_t > 0$ such that $|x(t)| \le \alpha_t$ for all $x \in E$. We say that E is **uniformly bounded** on T if there is $\alpha > 0$ such that $|x(t)| \le \alpha$ for all $x \in E$ and all $t \in T$, that is, if E is bounded in the sup metric on $B(T)$. Clearly, if E is uniformly bounded, then it is pointwise bounded. But the converse does not hold. For example, let $T := [0, 1]$ with the usual metric, and let

$$x_n(t) := \begin{cases} n^2 t & \text{if } t \in [0, 1/n], \\ 1/t & \text{if } t \in (1/n, 1] \end{cases} \quad \text{for } n \in \mathbb{N}.$$

Then $x_n \in C([0, 1])$, $x_n(0) = 0$, $|x_n(t)| \le 1/t$ for each fixed $t \in (0, 1]$, and for all $n \in \mathbb{N}$. But $x(1/n) = n$ for all $n \in \mathbb{N}$. Thus the subset $E := \{x_n : n \in \mathbb{N}\}$ of $C([0, 1])$ is pointwise bounded, but it is not uniformly bounded. However, every pointwise bounded subset of $C(T)$ is in fact uniformly bounded under a special condition introduced below.

Let E be a subset of $C(T)$. We say that E is **equicontinuous** on T if for every $\epsilon > 0$, there is $\delta > 0$ such that $|x(s) - x(t)| < \epsilon$ for all $x \in E$ and all $s, t \in T$ with $d_T(s, t) < \delta$. It is easy to see that this happens if and only if the following condition holds: For every $\epsilon > 0$, and sequences (s_n), (t_n) in T with $d_T(s_n, t_n) \to 0$, there is $n_0 \in \mathbb{N}$ such that $|x(s_n) - x(t_n)| < \epsilon$ for all $x \in E$ and $n \ge n_0$. If E is equicontinuous on T, then each element of E is uniformly continuous on E. The converse does not hold. For example, let $T := [0, 1]$, and let $x_n(t) := t^n$ for $n \in \mathbb{N}$ and $t \in [0, 1]$. Then each x_n is uniformly continuous on T, but the set $\{x_n : n \in \mathbb{N}\}$ is not equicontinuous.

Theorem 1.17 *Let T be a totally bounded metric space, and E be a subset of $C(T)$. Suppose that E is pointwise bounded and equicontinuous on T. Then*

(i) **(Ascoli, 1883)** *E is uniformly bounded on T. In fact, E is totally bounded in the sup metric on $C(T)$.*
(ii) **(Arzelà, 1889)** *Every sequence in E contains a subsequence which converges uniformly to a bounded continuous function on T.*

Proof (i) Let $\epsilon > 0$. Since E is equicontinuous on T, and since T is totally bounded, there are $t_1, \ldots, t_n$ in T, and $\delta > 0$ such that $T = \bigcup_{i=1}^{n} U(t_i, \delta)$, and for $i = 1, \ldots, n$, $|x(t) - x(t_i)| < \epsilon$ for all $x \in E$ and $t \in U(t_i, \delta)$. Since E is pointwise bounded on T, there are $\alpha_1, \ldots, \alpha_n > 0$ such that $|x(t_i)| \leq \alpha_i$ for all $x \in E$ and $i \in \{1, \ldots, n\}$. Let $\alpha := \max\{\alpha_1, \ldots, \alpha_n\} + \epsilon$. Then $|x(t)| < \alpha$ for all $x \in E$ and $t \in T$. Hence E is uniformly bounded on T.

Next, let $\mathbb{K}_\alpha := \{k \in \mathbb{K} : |k| < \alpha\}$. For $x \in E$, define

$$e(x) := (x(t_1), \ldots, x(t_n)) \in \mathbb{K}_\alpha \times \overset{n \text{ times}}{\cdots} \times \mathbb{K}_\alpha.$$

Since $\mathbb{K}_\alpha \times \overset{n \text{ times}}{\cdots} \times \mathbb{K}_\alpha$ is the open ball about $(0, \ldots, 0)$ of radius α in the d_∞ metric on $\mathbb{K}^n$, it is totally bounded. Hence there is $m \in \mathbb{N}$ and there are $x_1, \ldots, x_m$ in E such that the set $\{e(x) : x \in E\}$ is contained in the union of $V_1, \ldots, V_m$, where V_j is the open ball of radius ϵ about $e(x_j)$ for $j = 1, \ldots, m$.

Let $x \in E$. There is $j \in \{1, \ldots, m\}$ such that $e(x) \in V_j$. Since V_j is of radius ϵ, we see that $|x(t_i) - x_j(t_i)| < \epsilon$ for all $i = 1, \ldots, n$. Let $t \in T$. There is $i \in \{1, \ldots, n\}$ such that $t \in U(t_i, \delta)$, and so

$$|x(t) - x_j(t)| \leq |x(t) - x(t_i)| + |x(t_i) - x_j(t_i)| + |x_j(t_i) - x_j(t)| < \epsilon + \epsilon + \epsilon.$$

Hence $\sup\{|x(t) - x_j(t)| : t \in T\} \leq 3\epsilon < 4\epsilon$. Thus E is contained in the union of m open balls of radius 4ϵ in $C(T)$. Since $\epsilon > 0$ is arbitrary, E is totally bounded in the sup metric on $C(T)$.

(ii) We shall show in Example 2.24(iii) that the metric space $C(T)$ with the sup metric is complete. Since E is a totally bounded subset of $C(T)$, Corollary 1.12 shows that $\overline{E}$ is compact. Hence every sequence in E has a subsequence which converges in $\overline{E}$, and so in $C(T)$, that is, the subsequence converges uniformly to a bounded continuous function on T. □

1.4 Lebesgue Measure and Integration

Let I be an interval in $\mathbb{R}$, and let $\ell(I)$ denote its length. In an effort to generalize the concept of 'length', we proceed as follows.

The **Lebesgue outer measure** of a subset E of $\mathbb{R}$ is defined by

$$m^*(E) := \inf\left\{\sum_{n=1}^{\infty} \ell(I_n) : E \subset \bigcup_{n=1}^{\infty} I_n\right\},$$

where I_n is an open interval in $\mathbb{R}$ for each $n \in \mathbb{N}$.

It is easy to see that $m^*(I) = \ell(I)$ for every interval I in $\mathbb{R}$. Also, $m^*(\emptyset) = 0$, $m^*(E) \geq 0$ for all $E \subset \mathbb{R}$, $m^*(E_1) \leq m^*(E_2)$ for $E_1 \subset E_2 \subset \mathbb{R}$, and

$$m^*\left(\bigcup_{n=1}^{\infty} E_n\right) \leq \sum_{n=1}^{\infty} m^*(E_n)$$

for subsets $E_1, E_2, \ldots$ of $\mathbb{R}$. The last property is known as the **countable subadditivity** of the Lebesgue outer measure. In case the subsets $E_1, E_2, \ldots$ of $\mathbb{R}$ are pairwise disjoint, we would like an equality to hold here. Keeping this in mind, we introduce the following **Carathéodory condition**.

A subset E of $\mathbb{R}$ is called **(Lebesgue) measurable** if

$$m^*(A) = m^*(A \cap E) + m^*(A \cap (\mathbb{R} \setminus E)) \quad \text{for every } A \subset \mathbb{R}.$$

If E is measurable, then $m^*(E)$ is called the **Lebesgue measure** of E, and it is denoted simply by $m(E)$.

It is easy to see that $\mathbb{R}$ and $\emptyset$ are measurable subsets, and that complements and countable unions of measurable subsets are measurable. Also, every open set in $\mathbb{R}$ is measurable. It can be checked that the Lebesgue measure m is **countably additive** on measurable sets in $\mathbb{R}$, that is, if $E_1, E_2, \ldots$ are pairwise disjoint measurable sets, then $m\left(\bigcup_{n=1}^{\infty} E_n\right) = \sum_{n=1}^{\infty} m(E_n)$.

Measurable and Simple Functions

Let x be an extended real-valued function on $\mathbb{R}$, that is, $x : \mathbb{R} \to \mathbb{R} \cup \{\infty, -\infty\}$, where $-\infty < t < \infty$ for all $t \in \mathbb{R}$. Then x is called **(Lebesgue) measurable** if $x^{-1}(E)$ is a measurable subset of $\mathbb{R}$ for every open subset E of $\mathbb{R}$, and if the subsets $x^{-1}(\infty)$ and $x^{-1}(-\infty)$ of $\mathbb{R}$ are also measurable. Clearly, a real-valued continuous function on $\mathbb{R}$ is measurable. Let x and y be measurable functions on $\mathbb{R}$. We say that x is equal to y **almost everywhere** (which will be abbreviated as a.e.) if

$$m(\{t \in \mathbb{R} : x(t) \neq y(t)\}) = 0.$$

If $x^{-1}(\infty) \cap y^{-1}(-\infty) = \emptyset = x^{-1}(-\infty) \cap y^{-1}(\infty)$, then $x + y$ is a measurable function. Also, $x\,y$ is a measurable function (with the conventions $\infty \cdot 0 := 0$ and $-\infty \cdot 0 := 0$). Further, the functions $\max\{x, y\}$, $\min\{x, y\}$ and $|x|$ are measurable. Also, if (x_n) is a sequence of measurable functions such that $x_n(t) \to x(t)$ for each $t \in \mathbb{R}$, then x is measurable. Thus, a pointwise limit of measurable functions is measurable, just as we have seen in Sect. 1.3 that a uniform limit of continuous functions is continuous. Conversely, we shall now show that every measurable function is a pointwise limit of a sequence of functions of a special kind, just as every continuous function on $[a, b]$ is a uniform limit of polynomials (Theorem 1.16).

For a subset E of $\mathbb{R}$, let c_E denote the **characteristic function** of E:

$$c_E(t) := \begin{cases} 1, & \text{if } t \in E, \\ 0, & \text{if } t \in \mathbb{R} \setminus E. \end{cases}$$

A **simple function** is a function from $\mathbb{R}$ to $\mathbb{R}$ which takes only a finite number of values. Let s be a simple function. If $r_1, \ldots, r_m$ are the distinct values of s, then $s = \sum_{j=1}^{m} r_j c_{E_j}$, where $E_j := \{t \in \mathbb{R} : s(t) = r_j\}$ for $j = 1, \ldots, m$. Further, s is measurable if and only if the sets $E_1, \ldots, E_m$ are measurable.

For a function $x : \mathbb{R} \to \{r \in \mathbb{R} : r \geq 0\} \cup \{\infty\}$, and $n \in \mathbb{N}$, define a function $s_n : \mathbb{R} \to \{r \in \mathbb{R} : r \geq 0\}$ by

$$s_n(t) := \begin{cases} (j-1)/2^n, & \text{if } (j-1)/2^n \leq x(t) < j/2^n \text{ for } j = 1, \ldots, n2^n, \\ n, & \text{if } x(t) \geq n. \end{cases}$$

Then s_n is a simple function for each $n \in \mathbb{N}$, $0 \leq s_1(t) \leq s_2(t) \leq \cdots \leq x(t)$, and $s_n(t) \to x(t)$ for each $t \in \mathbb{R}$. If x is bounded, then the sequence (s_n) converges uniformly to x on $\mathbb{R}$.

If $x : \mathbb{R} \to \mathbb{R} \cup \{\infty, -\infty\}$, then by considering $x := x^+ - x^-$, where $x^+ := \max\{x, 0\}$ and $x^- := -\min\{x, 0\}$, we see that there is a sequence of simple functions converging to x at every point of $\mathbb{R}$. Note that if x is measurable, then each of the above simple functions is measurable.

For a nonnegative simple measurable function $s := \sum_{j=1}^{m} r_j c_{E_j}$, we define

$$\int_{\mathbb{R}} s \, dm := \sum_{j=1}^{m} r_j m(E_j),$$

and for a nonnegative measurable function x on $\mathbb{R}$, we define its integral by

$$\int_{\mathbb{R}} x \, dm := \sup \left\{ \int_{\mathbb{R}} s \, dm : 0 \leq s \leq x, \text{ and } s \text{ is a simple measurable function} \right\}.$$

In this case, the integral of x is equal to 0 if and only if $x = 0$ a.e. on $\mathbb{R}$.

Let now x be an extended real-valued measurable function on $\mathbb{R}$. If at least one of the integrals of x^+ and x^- is finite, then the **Lebesgue integral** of x with respect to the Lebesgue measure m is defined by

$$\int_{\mathbb{R}} x \, dm := \int_{\mathbb{R}} x^+ \, dm - \int_{\mathbb{R}} x^- \, dm.$$

A complex-valued function x defined on $\mathbb{R}$ is called **(Lebesgue) measurable** if its real part $\mathrm{Re}\, x$ and its imaginary part $\mathrm{Im}\, x$ are both measurable. In that case, we define

$$\int_{\mathbb{R}} x\, dm := \int_{\mathbb{R}} \mathrm{Re}\, x\, dm + i \int_{\mathbb{R}} \mathrm{Im}\, x\, dm,$$

whenever the Lebesgue integrals of $\mathrm{Re}\, x$ and $\mathrm{Im}\, x$ are well-defined. Then

$$\left| \int_{\mathbb{R}} x\, dm \right| \leq \int_{\mathbb{R}} |x|\, dm.$$

By a **measurable function** on $\mathbb{R}$, we shall mean either an extended real-valued measurable function or a complex-valued measurable function. If x is a measurable function on $\mathbb{R}$ and the Lebesgue integral of $|x|$ is finite, then we say that x is an **integrable** function on $\mathbb{R}$.

Suppose a function x is defined only on a closed and bounded subinterval $[a, b]$ of $\mathbb{R}$. We define the measurability of x, the existence of the Lebesgue integral of x, and the integrability of x on $[a, b]$ analogously. Such an integral will be denoted by

$$\int_{[a,\, b]} x\, dm \quad \text{or} \quad \int_a^b x(t)\, dm(t).$$

The following results concerning the convergence of a sequence of Lebesgue integrals of measurable functions are important.

Theorem 1.18 *Let $E := \mathbb{R}$ or $[a, b]$. Suppose (x_n) is a pointwise convergent sequence of measurable functions on E, and x is its pointwise limit.*

(i) **(Monotone convergence theorem)** *If $0 \leq x_1(t) \leq x_2(t) \leq \cdots$ for $t \in E$, then x is a nonnegative measurable function on E, and*

$$\int_E x_n\, dm \to \int_E x\, dm.$$

(ii) **(Dominated convergence theorem)** *If there is an integrable function y on E such that $|x_n(t)| \leq y(t)$ for $n \in \mathbb{N}$ and $t \in E$, then x is an integrable function on E, and*

$$\int_E x_n\, dm \to \int_E x\, dm.$$

(iii) **(Bounded convergence theorem)** *If there is $\alpha > 0$ such that $|x_n(t)| \leq \alpha$ for $n \in \mathbb{N}$ and $t \in [a, b]$, then x is an integrable function on $[a, b]$, and*

$$\int_{[a,b]} x_n\, dm \to \int_{[a,b]} x\, dm.$$

Proof We refer the reader to [26, 1.26 and 1.34] for proofs of parts (i) and (ii). Part (iii) follows from part (ii) by letting $y(t) := \alpha$ for $t \in [a, b]$. □

Corollary 1.19 *Let $E := \mathbb{R}$ or $[a, b]$, and let x and y be integrable functions on E. Then $x + y$ is integrable on E, and*

$$\int_E (x + y)\, dm = \int_E x\, dm + \int_E y\, dm.$$

Proof First let x and y be extended real-valued functions on E. Since x and y are integrable, $m(\{t \in E : |x(t)| = \infty\}) = m(\{t \in E : |y(t)| = \infty\}) = 0$. If x and y are simple measurable functions, then the desired result is obvious. For the general case, we write $x := x^+ - x^-$, $y := y^+ - y^-$, and approximate x^+, x^-, y^+ and y^- by nondecreasing sequences of simple measurable functions. The monotone convergence theorem (Theorem 1.18(i)) gives the desired result. If x and y are complex-valued functions, then we write $x := \operatorname{Re} x + i \operatorname{Im} x$, $y := \operatorname{Re} y + i \operatorname{Im} y$ to obtain the desired result. □

The above corollary allows us to conclude that the set X of all integrable functions on $E := \mathbb{R}$ or $[a, b]$ is a linear space over $\mathbb{K}$, and if we let $F(x) := \int_E x\, dm$ for $x \in X$, then F is a linear functional on X.

Although we shall usually restrict ourselves to the Lebesgue measure on $\mathbb{R}$, we shall refer to the Lebesgue measure on $\mathbb{R}^2$ on a few occasions. It generalizes the idea of area of a rectangle, just as the Lebesgue measure on $\mathbb{R}$ generalizes the idea of length of an interval. The following result will be useful.

Theorem 1.20 (Fubini and Tonelli) *Let $m \times m$ denote the Lebesgue measure on $\mathbb{R}^2$, and let $k(\cdot\,, \cdot) : [a, b] \times [c, d] \to \mathbb{K}$ be a measurable function. Suppose $k(\cdot\,, \cdot)$ is either integrable or nonnegative on $[a, b] \times [c, d]$. Then for almost every $s \in [a, b]$, the integral of $k(s\,, \cdot)$ on $[c, d]$ is well-defined, and for almost every $t \in [a, b]$, the integral of $k(\cdot\,, t)$ on $[a, b]$ is well-defined. Further, the two iterated integrals exist and satisfy*

$$\int_{[a,\,b]} \left(\int_{[c,\,d]} k(s, t)\, dm(t) \right) dm(s) = I = \int_{[c,\,d]} \left(\int_{[a,\,b]} k(s, t)\, dm(s) \right) dm(t),$$

where I is the Lebesgue integral of $k(\cdot\,, \cdot)$ on $[a, b] \times [c, d]$.

Proof We refer the reader to [24, pp. 307–309]. □

Calculus with Lebesgue Measure

We now state some basic results which relate Lebesgue integration with Riemann integration, and with differentiation.

Theorem 1.21 *A bounded function $x : [a, b] \to \mathbb{K}$ is Riemann integrable on $[a, b]$ if and only if the set of discontinuities of x on $[a, b]$ is of measure zero. In that case, x is Lebesgue integrable on $[a, b]$, and*

$$\int_{[a,b]} x \, dm = \int_a^b x(t) \, dt.$$

Proof We refer the reader to [25, Theorem 11.33]. □

We say that a function $x : [a, b] \to \mathbb{R}$ is **differentiable** on $[a, b]$ if the derivative of x exists for every $t \in (a, b)$, the right derivative of x exists at a, and the left derivative of x exists at b. We say that a function $x : [a, b] \to \mathbb{C}$ is **differentiable** on $[a, b]$ if the functions $\operatorname{Re} x$ and $\operatorname{Im} x$ are differentiable on $[a, b]$, and then we define $x' := (\operatorname{Re} x)' + i(\operatorname{Im} x)'$. Further, a function $x : [a, b] \to \mathbb{K}$ is called **continuously differentiable** if x is differentiable on $[a, b]$, and x' is continuous on $[a, b]$. The following well-known result relates Riemann integration and differentiation.

Theorem 1.22 (Fundamental theorem of calculus for Riemann integration) *A function $x : [a, b] \to \mathbb{K}$ is continuously differentiable on $[a, b]$ if and only if there is a continuous function y on $[a, b]$ such that*

$$x(t) = x(a) + \int_a^t y(s) \, ds \quad \textit{for all } t \in [a, b].$$

In that case, $y = x'$ on $[a, b]$.

Proof We refer the reader to [25, 6.20 and 6.21]. □

In order to formulate a similar result for Lebesgue integration, we need a stronger version of continuity. A function $x : [a, b] \to \mathbb{K}$ is called **absolutely continuous** on $[a, b]$ if for every $\epsilon > 0$, there is $\delta > 0$ such that

$$\sum_{j=1}^{n} |x(t_j) - x(s_j)| < \epsilon$$

whenever $a \le s_1 < t_1 \le s_2 < t_2 \le \cdots \le s_n < t_n \le b$ and $\sum_{j=1}^n (t_j - s_j) < \delta$.

Theorem 1.23 (Fundamental theorem of calculus for Lebesgue integration) *A function $x : [a, b] \to \mathbb{K}$ is absolutely continuous on $[a, b]$ if and only if there is a Lebesgue integrable function y on $[a, b]$ such that*

$$x(t) = x(a) + \int_{[a,t]} y \, dm \quad \textit{for all } t \in [a, b].$$

In that case, x is differentiable a.e. on $[a, b]$, and $y = x'$ a.e. on $[a, b]$.

Proof We refer the reader to [24, pp. 103-107]. □

Corollary 1.24 *Let* $x : [a, b] \to \mathbb{K}$ *be absolutely continuous on* $[a, b]$. *Then*

$$(b-a)\Big(|x(a)| - \int_{[a,\,b]} |x'|dm\Big) \le \int_a^b |x(t)|dt \le (b-a)\Big(|x(a)| + \int_{[a,\,b]} |x'|dm\Big).$$

Proof By Theorem 1.23, x' exists a.e. on $[a, b]$, x' is Lebesgue integrable on $[a, b]$, and $x(t) = x(a) + \int_{[a,\,t]} x'dm$ for $t \in [a, b]$. Hence

$$|x(t)| \le |x(a)| + \int_{[a,\,b]} |x'|dm \quad \text{and} \quad |x(a)| \le |x(t)| + \int_{[a,\,b]} |x'|dm.$$

Integrating over $[a, b]$, we obtain the desired inequalities. □

The inequalities in the above corollary will be useful in treating the Sobolev space of order (1,1) on $[a, b]$ in Example 2.24(v).

We now introduce a related property of functions. A $\mathbb{K}$-valued function x on $[a, b]$ is said to be of **bounded variation** if the set

$$E := \Big\{\sum_{j=1}^{n} |x(t_j) - x(t_{j-1})| : a = t_0 < t_1 < \cdots < t_{n-1} < t_n = b\Big\}$$

is bounded. If x is of bounded variation, then $V(x) := \sup E$ is called the **total variation** of x.

A continuous function on $[a, b]$ need not be of bounded variation. For example, let $x(0) := 0$ and $x(t) := t\,\sin(1/t)$ for t in $(0, 1]$. Then x is continuous on $[0, 1]$, but it is not of bounded variation. Also, a function of bounded variation on $[a, b]$ need not be continuous. For example, the characteristic function of the set $[0, 1/2]$ is of bounded variation on $[0, 1]$, but it is not continuous. If x is of bounded variation on $[a, b]$, then the derivative $x'(t)$ exists for almost every $t \in [a, b]$, and x' is Lebesgue integrable on $[a, b]$. (See [24], pp. 96-100.)

Further, an absolutely continuous function on $[a, b]$ is continuous, and it is of bounded variation. But a continuous function of bounded variation need not be absolutely continuous on $[a, b]$. For example, Cantor's ternary function is continuous and it is of bounded variation on $[0, 1]$, but it is not absolutely continuous. Also, a differentiable function on $[a, b]$ need not be absolutely continuous. For example, let $x(0) := 0$ and $x(t) := t^2 \sin(1/t^2)$ for $t \in (0, 1]$. Then x is differentiable on $[0, 1]$, but it is not absolutely continuous since its derivative is not Lebesgue integrable on $[0, 1]$. On the other hand, if a function is continuous on $[a, b]$, differentiable on (a, b) and its derivative is bounded on (a, b), then it is absolutely continuous on $[a, b]$. This follows from the mean value theorem.

L^p Spaces, $p \in \{1, 2, \infty\}$

Let $E := \mathbb{R}$ or $[a, b]$. Let us consider the set of all integrable functions on E. Thus a function x defined on E belongs to this set if and only if x is measurable and

$\int_E |x|dm < \infty$. We would like to define a suitable metric on this set. Suppose x, y and z are integrable functions on E. By Corollary 1.19, the functions $x - y$, $y - z$, $z - y$ are integrable on E. Since $|x(t) - y(t)| \leq |x(t) - z(t)| + |z(t) - y(t)|$ for all $t \in E$, we obtain

$$\int_E |x - y|dm \leq \int_E |x - z|dm + \int_E |z - y|dm.$$

Also,

$$\int_E |x - y|dm = \int_E |y - x|dm \quad \text{and} \quad \int_E |x - y|dm \geq 0.$$

Further, $\int_E |x - y|dm = 0$ if and only if $x = y$ a.e. on E.

Hence we may identify measurable functions x and y which are almost everywhere equal on E. Let us write $x \sim y$ if $x = y$ a.e. on E. It is easy to see that $\sim$ is an equivalence relation on the set of all measurable functions on E. For simplicity, we shall denote the equivalence class of a measurable function x on E by x itself and treat the equivalence class as if it is a function on E.

If x is integrable on E, and if $x \sim y$, then we see that y is integrable on E and $\int_E |x|dm = \int_E |y|dm$. Let us then consider the set $L^1(E)$ of all equivalence classes of integrable functions on E. It is easy to see that $L^1(E)$ is a linear space over $\mathbb{K}$. For $x, y \in L^1(E)$, define

$$d_1(x, y) := \int_E |x - y|dm.$$

It is now clear that d_1 is a metric on $L^1(E)$.

Next, let us consider the set of all **square-integrable functions** on E, that is, a function x defined on E belongs to this set if and only if x is measurable and $\int_E |x|^2 dm < \infty$. As above, let $L^2(E)$ denote the set of all equivalence classes of square-integrable functions on E. Before defining a metric on $L^2(E)$, we prove two inequalities involving nonnegative measurable functions.

Lemma 1.25 *Let $E := \mathbb{R}$, or let E be a closed and bounded subinterval of $\mathbb{R}$, and let $a, b : E \to \mathbb{R}$ be nonnegative measurable functions. Then*

(i) **(Schwarz inequality for functions)**

$$\int_E a(t)b(t)dm(t) \leq \left(\int_E a(t)^2 dm(t)\right)^{1/2} \left(\int_E b(t)^2 dm(t)\right)^{1/2},$$

(ii) **(Minkowski inequality for functions)**

$$\left(\int_E \big(a(t) + b(t)\big)^2 dm(t)\right)^{1/2} \leq \left(\int_E a(t)^2 dm(t)\right)^{1/2} + \left(\int_E b(t)^2 dm(t)\right)^{1/2}.$$

Proof In the proofs of the Schwarz inequality and the Minkowski inequality of Lemma 1.4, replace a_j and b_j, where $j = 1, \ldots, n$, by $a(t)$ and $b(t)$, where $t \in E$, and replace the summation from j to n by the Lebesgue integration over E. Then the inequalities in (i) and (ii) above follow. □

It follows from part (ii) of the above lemma that $L^2(E)$ is a linear space over $\mathbb{K}$. For $x, y \in L^2(E)$, define

$$d_2(x, y) := \left(\int_E |x - y|^2 dm \right)^{1/2}.$$

Now for $x, y, z \in L^2(E)$, let $a(t) := |x(t) - z(t)|$ and $b(t) := |z(t) - y(t)|$ for $t \in E$ in the Minkowski inequality proved above. Since

$$\left(\int_E |x(t) - y(t)|^2 dm(t) \right)^{1/2} \leq \left(\int_E (|x(t) - z(t)| + |z(t) - y(t)|)^2 dm(t) \right)^{1/2},$$

we see that $d_2(x, y) \leq d_2(x, z) + d_2(z, y)$. Hence d_2 is a metric on $L^2(E)$.

Finally, let us consider the set of all bounded measurable functions on E. If x and y are bounded measurable functions on E, then $\sup\{|x(t) - y(t)| : t \in E\} = 0$ if and only if $x(t) = y(t)$ for every $t \in E$. Thus if $x \sim y$ but $x \neq y$, then $\sup\{|x(t) - y(t)| : t \in E\} \neq 0$. This leads us to enlarge the set of all bounded measurable functions on E as follows. A measurable function x on E is called **essentially bounded** on E if there is $\alpha > 0$ such that $m(\{t \in E : |x(t)| > \alpha\}) = 0$, and then α is called an **essential bound** for $|x|$ on E. The **essential supremum** of $|x|$ on E is defined by

$$\text{ess sup}_E|x| := \inf\{\alpha : \alpha \text{ is an essential bound for } |x| \text{ on } E\}.$$

As an example of an essentially bounded function on $[0, 1]$ which is not bounded on $[0, 1]$, let $x(t) := n$ if $x := 1/n$ for some $n \in \mathbb{N}$, and let $x(t) := 0$ otherwise. Note that $\text{ess sup}_E|x| = 0$, although x is unbounded on $[0, 1]$.

Let x be an essentially bounded function on E. Since $\text{ess sup}_E|x| + (1/n)$ is an essential bound for $|x|$ on E for each $n \in \mathbb{N}$, $\text{ess sup}_E|x|$ is itself an essential bound for $|x|$ on E. Also, since $\{t \in E : x(t) \neq 0\} = \bigcup_{n=1}^{\infty}\{t \in E : |x(t)| > 1/n\}$, we see that $\text{ess sup}_E|x| = 0$ if and only if $x = 0$ a.e. on E. Let y be a measurable function on E such that $y \sim x$. Then $\alpha > 0$ is an essential bound for $|x|$ if and only if it is an essential bound for $|y|$. It follows that y is essentially bounded on E and $\text{ess sup}_E|y| = \text{ess sup}_E|x|$.

If x, y are essentially bounded functions on E, then $\text{ess sup}_E|x| + \text{ess sup}_E|y|$ is an essential bound for $|x + y|$ on E, since

$$\{t \in E : |(x + y)(t)| > \alpha + \beta\} \subset \{t \in E : |x(t)| > \alpha\} \cup \{t \in E : |y(t)| > \beta\}$$

for all $\alpha > 0$ and $\beta > 0$, and so

$$\text{ess sup}_E|x + y| \leq \text{ess sup}_E|x| + \text{ess sup}_E|y|.$$

Let $L^\infty(E)$ denote the set of all equivalence classes of essentially bounded functions on E. The above inequality shows that $L^\infty(E)$ is a linear space over $\mathbb{K}$. For $x, y \in L^\infty(E)$, define

$$d_\infty(x, y) := \operatorname{ess\,sup}_E|x - y|.$$

It follows that d_∞ is a metric on $L^\infty(E)$.

We note the following.

(i) $x_n \to x$ in $L^1(E)$ means $\int_E |x_n(t) - x(t)|dm(t) \to 0$. This convergence is known as **convergence in the mean** on E.
(ii) $x_n \to x$ in $L^2(E)$ means $\int_E |x_n(t) - x(t)|^2 dm(t) \to 0$. This convergence is known as the **mean square convergence** on E.
(iii) $x_n \to x$ in $L^\infty(E)$ means $\operatorname{ess\,sup}_E|x_n - x| \to 0$, that is, for every $\epsilon > 0$, there is $n_0 \in \mathbb{N}$ such that for every $n \geq n_0$, $|x_n(t) - x(t)| < \epsilon$ for almost every $t \in E$. This convergence is known as the **essentially uniform convergence** on E.

We now consider approximation of an integrable function, or of a square integrable function, or of an essentially bounded function defined on $[a, b]$ by sequences of functions of a special kind.

Proposition 1.26 *Let $E := [a, b]$, where $-\infty < a < b < \infty$.*

(i) *The set of all simple measurable functions on E is dense in $L^1(E)$, in $L^2(E)$ and in $L^\infty(E)$.*
(ii) *The set of all continuous functions on E is dense in $L^1(E)$ and in $L^2(E)$.*
(iii) *The set of all step functions on E is dense in $L^1(E)$ and in $L^2(E)$.*
(iv) *The metric spaces $L^1(E)$ and $L^2(E)$ are separable, but the metric space $L^\infty(E)$ is not separable.*

Proof (i) Let $x : E \to \{r \in \mathbb{R} : r \geq 0\} \cup \{\infty\}$ be a measurable function. As we have seen before, there is a sequence (s_n) of simple measurable functions on E such that $0 \leq s_1(t) \leq s_2(t) \leq \cdots \leq x(t)$, and $s_n(t) \to x(t)$ for all $t \in E$. Let $p \in \{1, 2\}$, and $x \in L^p(E)$. Since $0 \leq (x - s_n)^p \leq x^p$, and since x^p is integrable, the dominated convergence Theorem (1.18(ii)) shows that

$$d_p(x, s_n) = \left(\int_E |x - s_n|^p \, dm\right)^{1/p} \to 0.$$

If $x \in L^\infty(E)$, then $d_\infty(x, s_n) = \operatorname{ess\,sup}_E|s_n - x| \to 0$.

In general, if x is an extended real-valued function, then we consider $x := x^+ - x^-$, and if x is a complex-valued p-integrable function, then we consider $x := \operatorname{Re} x + i \operatorname{Im} x$, and obtain the desired conclusions.

(ii) Let $p \in \{1, 2\}$. A continuous function on E is bounded, and so it is in $L^p(E)$. Consider a closed subset F of E, and for $n \in \mathbb{N}$, let

$$x_n(t) := \frac{1}{1 + n\, d(t, F)}, \quad t \in E.$$

Since $t \longmapsto d(t, F)$ is a continuous function on E, each x_n is continuous on E. Also, $x_n(t) = 1$ for all $t \in F$, whereas $x_n(t) \to 0$ as $n \to \infty$ for all $t \in E \setminus F$. If c_F is the characteristic function of F, then $(c_F - x_n)(t) \to 0$ for $t \in E$. Since $|x_n(t)| \le 1$ for $t \in E$, by the bounded convergence Theorem (1.18(iii)),

$$d_p(c_F, x_n) = \left(\int_E |c_F - x_n|^p \, dm \right)^{1/p} \to 0.$$

Next, let G a measurable subset of E and c_G its characteristic function. Let $\epsilon > 0$. By the 'inner regularity' of the Lebesgue measure on E, there is a closed subset F of G such that $m(G \setminus F) < \epsilon$.[3] Hence $d_p(c_G, c_F) < \epsilon^{1/p}$. We thus approximate c_G, and so every simple measurable function on E, by a sequence of continuous functions on E. The result now follows from (i) above.

(iii) Let $p \in \{1, 2\}$. Consider a continuous function x on $[a, b]$. Since $E := [a, b]$ is compact, x is uniformly continuous on E. Given $\epsilon > 0$, there is $n \in \mathbb{N}$ such that $|x(s) - x(t)| < \epsilon$ whenever $s, t \in [a, b]$ and $|s - t| < (b - a)/n$. Let $t_0 := a$ and $t_j := a + j(b - a)/n$ for $j = 1, \ldots, n$. Define a step function y on $[a, b]$ by $y(t_0) := x(t_0)$, and $y(t) := x(t_{j-1})$ for $t_{j-1} < t \le t_j$. Then

$$d_p(x, y) = \left(\int_{[a,b]} |x(t) - y(t)|^p \, dm(t) \right)^{1/p} \le \left(\sum_{j=1}^{n} \frac{\epsilon^p (b - a)}{n} \right)^{1/p} = \epsilon (b - a)^{1/p}.$$

We can thus approximate each continuous function on E by a sequence of step functions. The result now follows from (ii) above.

(iv) Let $p \in \{1, 2\}$. A step function on E can be approximated by step functions which have steps at rational numbers in $[a, b]$, and whose real part and imaginary part take only rational values. Since such step functions are countable, it follows that $L^p([a, b])$ is separable by (iii) above.

To see that $X := L^\infty([a, b])$ is not separable, let $t \in [a, b]$, and denote the characteristic function of the interval $[a, t]$ by c_t. Then $d_\infty(c_s, c_t) = 1$ if $s \ne t$. Let $S := \{c_t : t \in [a, b]\}$, and let $\{x_1, x_2, \ldots\}$ be a countable subset of X. Then each $U(x_n, 1/2)$ contains at most one element of S. Since S is uncountable, there is $x \in S$ such that $x \notin U(x_n, 1/2)$ for any $n \in \mathbb{N}$. As $\{x_1, x_2, \ldots\} \cap U(x, 1/2) = \emptyset$, the set $\{x_1, x_2, \ldots\}$ cannot be dense in X. □

Fourier Series

Let x be a complex-valued integrable function on $[-\pi, \pi]$. For $k \in \mathbb{Z}$, the kth **Fourier series** of x is defined by

$$\hat{x}(k) := \frac{1}{2\pi} \int_{-\pi}^{\pi} x(t) e^{-ikt} \, dm(t).$$

[3] This can be seen as follows. By the definition of the Lebesgue measure of $E \setminus G$, there are open intervals $I_1, I_2, \ldots$ such that $E \setminus G \subset \bigcup_{n=1}^\infty I_n$ and $\sum_{n=1}^\infty \ell(I_n) < m(E \setminus G) + \epsilon$. Let $J := \bigcup_{n=1}^\infty I_n$ and $F := E \cap (\mathbb{R} \setminus J)$.

The series $\sum_{k=-\infty}^{\infty} \hat{x}(k)e^{ikt}$ is called the **Fourier series** of x. For $n = 0, 1, 2, \ldots$, consider the nth partial sum

$$s_n(x)(t) := \sum_{k=-n}^{n} \hat{x}(k)e^{ikt}, \quad t \in [-\pi, \pi],$$

of the Fourier series of x. Whether the sequence $(s_n(x))$ converges in some sense, and if so, whether its limit is equal to x are the central questions in the theory of Fourier series. In 1876, du Bois-Reymond showed that even if x is a continuous function on $[-\pi, \pi]$ satisfying $x(-\pi) = x(\pi)$, the sequence $(s_n(x))$ may not converge pointwise. In 1926, Kolmogorov gave an example of an integrable function x on $[-\pi, \pi]$ such that the sequence $(s_n(x)(t))$ diverges for each $t \in [-\pi, \pi]$. Further, the sequence $(d_1(s_n(x), x))$ may not converge to 0 for every integrable function x on $[-\pi, \pi]$.

A useful approach in this situation is to consider the arithmetic means of the partial sums of the Fourier series of an integrable function x. Thus for an integrable function x on $[-\pi, \pi]$, and for $m \in \mathbb{N}$, let

$$a_m(x) := \frac{1}{m}\big(s_0(x) + s_1(x) + \cdots + s_{m-1}(x)\big).$$

Theorem 1.27 (Féjer, 1904) *Let x be a complex-valued continuous function on $[-\pi, \pi]$ such that $x(\pi) = x(-\pi)$. Then the sequence $(a_m(x))$ converges uniformly to x on $[-\pi, \pi]$.*

Proof Let $X := \{x \in C([-\pi, \pi]) : x(\pi) = x(-\pi)\}$. For $x \in X$ and $m \in \mathbb{N}$, let $F_m(x)$ denote the mth arithmetic mean $a_m(x)$. Then F_m is a linear map from the linear space X to itself. It can be proved, by considering the integral representation of $a_m(x)$ in terms of the mth 'Féjer kernel', that $F_m(x) \geq 0$ if $x \geq 0$. Let $x_0(t) := 1$, $x_1(t) := \cos t$ and $x_2(t) := \sin t$ for $t \in [-\pi, \pi]$. Since the Fourier series of x_0, x_1 and x_2 are given by $1, \cos t = (e^{it} + e^{-it})/2$ and $\sin t = (e^{it} - e^{-it})/2i$, respectively, we see that

$$F_m(x_0)(t) = \frac{1 + \overset{m \text{ times}}{\cdots} + 1}{m} = 1,$$

$$F_m(x_1)(t) = \frac{0 + \cos t + \overset{m-1 \text{ times}}{\cdots} + \cos t}{m} = \frac{(m-1)\cos t}{m},$$

$$F_m(x_2)(t) = \frac{0 + \sin t + \overset{m-1 \text{ times}}{\cdots} + \sin t}{m} = \frac{(m-1)\sin t}{m}$$

for all $m \in \mathbb{N}$ and $t \in [-\pi, \pi]$. Hence for $j = 0, 1, 2$, the sequence $(F_m(x_j))$ converges uniformly to x_j on $[-\pi, \pi]$. These properties of arithmetic means of the partial sums of a Fourier series enable us to show that for every $x \in X$, the sequence $(F_m(x))$ converges uniformly to x on $[-\pi, \pi]$. For details, see [19, p. 38] or [2, p. 363]. For another proof, see [17, p. 17–18]. □

If x is an integrable function on $[-\pi, \pi]$, then $d_1(a_m(x), x) \to 0$. (See [17, pp. 18–19].) We shall show in Example 2.32(ii) that if x is square-integrable on $[-\pi, \pi]$, then $d_2(s_m(x), x) \to 0$. This result is the main motivation for our study of an orthonormal basis for a Hilbert space, as we shall point out in Sect. 2.4.

We now consider some interesting properties of the Fourier coefficients of an integrable function on $[-\pi, \pi]$.

Theorem 1.28 *Let x be an integrable function on $[-\pi, \pi]$. Then*

(i) **(Riemann–Lebesgue lemma, 1903)** *$\hat{x}(n) \to 0$ as $n \to \pm\infty$.*
(ii) *If $\hat{x}(n) = 0$ for all $n \in \mathbb{Z}$, then $x(t) = 0$ for almost all $t \in [-\pi, \pi]$.*

Proof (i) First suppose that there are $k_1, \ldots, k_m \in \mathbb{C}$ and distinct integers $j_1, \ldots, j_m$ such that $x(t) := k_1 e^{ij_1 t} + \cdots + k_m e^{ij_m t}$ for all $t \in [-\pi, \pi]$. Then $\hat{x}(j_1) = k_1, \ldots, \hat{x}(j_m) = k_m$, and $\hat{x}(n) = 0$ for all other $n \in \mathbb{Z}$. In particular, $\hat{x}(n) = 0$ for all $n \in \mathbb{Z}$ such that $|n| > \max\{|j_1|, \ldots, |j_m|\}$. Thus the desired result holds. Next, suppose that x is a continuous function on $[-\pi, \pi]$ with $x(\pi) = x(-\pi)$. Since the result holds for each arithmetic mean of the partial sums of the Fourier series of x, it follows from Theorem 1.27 of Féjer that the result holds for x as well. If x is an integrable function on $[-\pi, \pi]$, then a slight modification of the proof of Proposition 1.26(ii) shows that there is a sequence (x_m) of continuous functions on $[-\pi, \pi]$ with $x_m(\pi) = x_m(-\pi)$ for each $m \in \mathbb{N}$ such that $d_1(x, x_m) \to 0$ as $m \to \infty$, and so the result holds for x.

(ii) Suppose $\hat{x}(n) = 0$ for all $n \in \mathbb{Z}$. Define $y(s) := \int_{-\pi}^{s} x(t)dm(t)$ for s in $[-\pi, \pi]$. Then $y(\pi) = 2\pi\hat{x}(0) = 0$ and $y(-\pi) = 0$. By the fundamental theorem of calculus for Lebesgue integration Theorem 1.23, y is absolutely continuous on $[-\pi, \pi]$, and $y' = x$ a. e. on $[-\pi, \pi]$. Now for each nonzero $n \in \mathbb{Z}$,

$$\hat{y}(n) = \frac{1}{2\pi}\int_{-\pi}^{\pi} y(s)e^{-ins}\,dm(s) = \frac{1}{2\pi}\left[y(s)\frac{e^{-ins}}{-in}\right]_{-\pi}^{\pi} - \frac{1}{2\pi}\int_{-\pi}^{\pi} y'(s)\frac{e^{-ins}}{-in}\,dm(s).$$

Since $y(\pi) = y(-\pi)$ and $e^{-in\pi} = e^{in\pi}$, we see that

$$\hat{y}(n) = \frac{1}{2\pi in}\int_{-\pi}^{\pi} x(s)e^{-ins}\,dm(s) = \frac{\hat{x}(n)}{in} = 0 \quad \text{for all nonzero } n \in \mathbb{Z}.$$

Hence each arithmetic mean of the partial sums of the Fourier series of y is equal to $\hat{y}(0)$. By Theorem 1.27 of Féjer, we see that $y(t) = \hat{y}(0)$ for all $t \in [-\pi, \pi]$. Hence $x = y' = 0$ a.e. on $[-\pi, \pi]$. □

Corollary 1.29 (Inversion theorem) *Suppose x is an integrable function on $[-\pi, \pi]$ such that $\sum_{n=-\infty}^{\infty} |\hat{x}(n)| < \infty$. Then $x(t) = \sum_{n=-\infty}^{\infty} \hat{x}(n)e^{int}$ for almost all t in $[-\pi, \pi]$.*

Proof By the M-test of Weierstrass, the series $\sum_{n=-\infty}^{\infty} \hat{x}(n)e^{int}$ converges uniformly to, say, $y(t)$ for $t \in [-\pi, \pi]$. By Proposition 1.15, y is a continuous function on $[-\pi, \pi]$. Interchanging summation and integration, we obtain

$$\hat{y}(n) = \frac{1}{2\pi}\int_{-\pi}^{\pi} y(t)e^{-int}\,dm(t) = \sum_{k=-\infty}^{\infty}\frac{\hat{x}(k)}{2\pi}\int_{-\pi}^{\pi} e^{i(k-n)t}\,dm(t) = \hat{x}(n),\ \ n \in \mathbb{Z}.$$

Thus, all the Fourier coefficients of the integrable function $x - y$ are equal to zero. By Theorem 1.28(ii), $x(t) - y(t) = 0$, that is, $x(t) = y(t) = \sum_{n=-\infty}^{\infty} \hat{x}(n)e^{int}$ a.e. on $[-\pi, \pi]$. □

A parallel development for integrable and square-integrable functions on $\mathbb{R}$ is indicated in Example 4.32(ii).

Chapter 2
Basic Framework

In this chapter, we introduce the structure of a normed space. It involves the superposition of a metric structure on a linear space by means of a norm. Also, we introduce the structure of an inner product space and show that an inner product induces a special kind of norm. We consider the concept of orthogonality in the context of an inner product space. Our study of functional analysis will take place within these basic structures. In the last two sections, we investigate complete normed spaces (which are known as Banach spaces) as well as complete inner product spaces (which are known as Hilbert spaces). We consider many examples of Banach spaces and Hilbert spaces.

2.1 Normed Spaces

Let X be a linear space over $\mathbb{K}$. A map $p : X \to \mathbb{R}$ is a **seminorm** on X if

(i) $p(x+y) \le p(x) + p(y)$ for all $x, y \in X$, and
(ii) $p(kx) = |k|p(x)$ for all $x \in X$ and $k \in \mathbb{K}$.

If p is a seminorm on X, then $p(0) = p(0 \cdot 0) = |0|p(0) = 0$, and

$$p(x) = \frac{1}{2}\big(p(x) + p(x)\big) = \frac{1}{2}\big(p(x) + p(-x)\big) \ge \frac{1}{2}\, p(x - x) = 0 \quad \text{for all } x \in X.$$

If p is a seminorm on X, and if in addition,

(iii) $p(x) = 0$ only for $x = 0$,

then p is called a **norm** on X. A norm is usually denoted by the symbol $\|\cdot\|$. Thus a norm on X is a function $\|\cdot\| : X \to \mathbb{R}$ such that

(i) $\|x+y\| \le \|x\| + \|y\|$ for all $x, y \in X$,
(ii) $\|kx\| = |k|\|x\|$ for all $x \in X$ and $k \in \mathbb{K}$, and
(iii) $\|x\| \ge 0$ for all $x \in X$ and $\|x\| = 0$ if and only if $x = 0$.

© Springer Science+Business Media Singapore 2016, corrected publication 2023

B.V. Limaye, *Linear Functional Analysis for Scientists and Engineers*,
https://doi.org/10.1007/978-981-10-0972-3_2

A **normed space** $(X, \|\cdot\|)$ over $\mathbb{K}$ is a linear space X over $\mathbb{K}$ along with a norm $\|\cdot\|$ on it. For x and y in X, let $d(x, y) := \|x - y\|$. It is easy to see that d is a metric on X. Since $|\, \|x\| - \|y\| \,| \le \|x - y\|$ for all x and y in X, the function $\|\cdot\|$ is uniformly continuous on X. Further, if $x_n \to x$, $y_n \to y$ in X and $k_n \to k$ in $\mathbb{K}$, then $x_n + y_n \to x + y$ and $k_n x_n \to kx$ in X. This says that the operations of addition and scalar multiplication are continuous. As a consequence, if Y is a subspace of X, then its closure $\overline{Y}$ is also a subspace of X. Clearly, a norm $\|\cdot\|$ on X induces a norm on Y, and on $\overline{Y}$.

Let X be a normed space. For $x \in X$ and $r > 0$, consider the **open ball**

$$U(x, r) := \{y \in X : \|x - y\| < r\}$$

about x of radius r. If $y \in X$ and $\|x - y\| = r$, then $y_n := y + (x - y)/n$ is in $U(x, r)$ for all $n \in \mathbb{N}$, and $y_n \to y$. It follows that the closure of $U(x, r)$ is the **closed ball** $\overline{U}(x, r) := \{y \in X : \|x - y\| \le r\}$ about x of radius r.

The set $U(0, 1) = \{x \in X : \|x\| < 1\}$ is called the **open unit ball** of X, the set $\overline{U}(0, 1) = \{x \in X : \|x\| \le 1\}$ is called the **closed unit ball** of X and the set $\{x \in X : \|x\| = 1\}$ is called the **unit sphere** of X.

Examples 2.1 (i) **Euclidean Spaces**: Let $n \in \mathbb{N}$. For $x := (x(1), \ldots, x(n))$ in $\mathbb{K}^n$, define

$$\|x\|_1 := |x(1)| + \cdots + |x(n)|,$$

$$\|x\|_2 := \left(|x(1)|^2 + \cdots + |x(n)|^2\right)^{1/2},$$

$$\|x\|_\infty := \max\{|x(1)|, \ldots, |x(n)|\}.$$

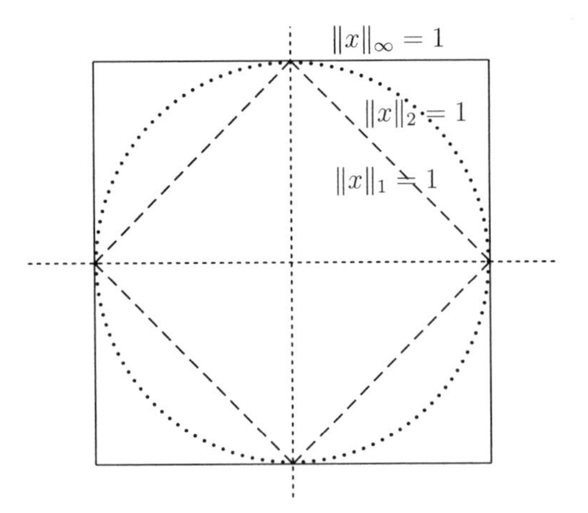

Fig. 2.1 Unit spheres of $\mathbb{R}^2$ with the norms $\|\cdot\|_1$, $\|\cdot\|_2$ and $\|\cdot\|_\infty$

It is easy to see that $\|\cdot\|_1$ and $\|\cdot\|_\infty$ are norms on $\mathbb{K}^n$. To show that $\|\cdot\|_2$ is also a norm on $\mathbb{K}^n$, consider $x := (x(1), \ldots, x(n))$ and $y := (y(1), \ldots, y(n))$ in $\mathbb{K}^n$, and let $a_j := |x(j)|$ and $b_j := |y(j)|$ for $j = 1, \ldots, n$ in the Minkowski inequality for numbers (Lemma 1.4(ii)). Then

$$\|x + y\|_2 = \Big(\sum_{j=1}^{n} |x(j) + y(j)|^2\Big)^{1/2} \le \Big(\sum_{j=1}^{n} \big(|x(j)| + |y(j)|\big)^2\Big)^{1/2} \le \|x\|_2 + \|y\|_2.$$

It follows that $\|\cdot\|_2$ is a norm on $\mathbb{K}^n$. We observe that the norms $\|\cdot\|_1$, $\|\cdot\|_2$ and $\|\cdot\|_\infty$ induce, respectively, the metrics d_1, d_2 and d_∞ on $\mathbb{K}^n$ introduced in Example 1.5(i). See Fig. 2.1 for the case $\mathbb{K} := \mathbb{R}$ and $n := 2$.

(ii) **Sequence spaces**: Consider the metric spaces ℓ^1, ℓ^2 and ℓ^∞ introduced in Example 1.5(ii). Define

$$\|x\|_1 := \sum_{j=1}^{\infty} |x(j)| \text{ for } x \in \ell^1,$$

$$\|x\|_2 := \Big(\sum_{j=1}^{\infty} |x(j)|^2\Big)^{1/2} \text{ for } x \in \ell^2,$$

$$\|x\|_\infty := \sup\{|x(j)| : j \in \mathbb{N}\} \text{ for } x \in \ell^\infty.$$

It is easy to see that ℓ^1 is a linear space over $\mathbb{K}$ and $\|\cdot\|_1$ is a norm on it, and that ℓ^∞ is a linear space over $\mathbb{K}$ and $\|\cdot\|_\infty$ is a norm on it. As for ℓ^2, let $x, y \in \ell^2$, $a_j := |x(j)|, b_j := |y(j)|$ for $j \in \mathbb{N}$, and let $n \to \infty$ in the Minkowski inequality for numbers (Lemma 1.4(ii)). Then

$$\Big(\sum_{j=1}^{\infty} |x(j) + y(j)|^2\Big)^{1/2} \le \Big(\sum_{j=1}^{\infty} \big(|x(j)| + |y(j)|\big)^2\Big)^{1/2} \le \|x\|_2 + \|y\|_2.$$

Hence $x + y \in \ell^2$ and $\|x + y\|_2 \le \|x\|_2 + \|y\|_2$. It follows that ℓ^2 is a linear space over $\mathbb{K}$, and that $\|\cdot\|_2$ is a norm on it.

We observe that the norm $\|\cdot\|_1$ induces the metric d_1 on ℓ^1, the norm $\|\cdot\|_2$ induces the metric d_2 on ℓ^2 and the norm $\|\cdot\|_\infty$ induces the metric d_∞ on ℓ^∞ introduced in Example 1.5(ii).

Note that if $x \in \ell^1$, then $x \in \ell^2$ and $\|x\|_2 \le \|x\|_1$, since $|x(1)|^2 + \cdots + |x(n)|^2 \le (|x(1)| + \cdots + |x(n)|)^2$ for all $n \in \mathbb{N}$. Also, if we let $x(j) := 1/j$ for $j \in \mathbb{N}$, then $x \in \ell^2$, but $x \notin \ell^1$. Further, if $x \in \ell^2$, then $x \in \ell^\infty$ and $\|x\|_\infty \le \|x\|_2$, since $|x(j)| \le \|x\|_2$ for all $j \in \mathbb{N}$. Also, if we let $x(j) := (-1)^j$ for $j \in \mathbb{N}$, then $x \in \ell^\infty$, but $x \notin \ell^2$.

Let us consider the following subspaces of ℓ^∞:

$$c := \{x \in \ell^\infty : \text{the sequence } (x(j)) \text{ converges in } \mathbb{K}\},$$

$$c_0 := \{x \in c : \text{the sequence } (x(j)) \text{ converges to } 0 \text{ in } \mathbb{K}\},$$

$$c_{00} := \{x \in c_0 : \text{there is } j_x \in \mathbb{N} \text{ such that } x(j) = 0 \text{ for all } j \ge j_x\}.$$

We may identify $x := (x(1), \ldots, x(n)) \in \mathbb{K}^n$ with $(x(1), \ldots, x(n), 0) \in \mathbb{K}^{n+1}$, which in turn can be identified with $(x(1), \ldots, x(n), 0, 0, \ldots) \in c_{00}$. Let us denote these identifications by the symbol $\hookrightarrow$. It is then easy to see that

$$\{0\} \subset \mathbb{K} \hookrightarrow \mathbb{K}^2 \hookrightarrow \cdots \hookrightarrow \mathbb{K}^n \hookrightarrow \mathbb{K}^{n+1} \hookrightarrow \cdots \hookrightarrow c_{00} \subset \ell^1 \subset \ell^2 \subset c_0 \subset c \subset \ell^\infty.$$

(iii) **Function spaces**: Consider the linear space $B(T)$ of all $\mathbb{K}$-valued functions defined on a nonempty set T introduced in Example 1.5(iii). For $x \in B(T)$, let

$$\|x\|_\infty := \sup\{|x(t)| : t \in T\}.$$

It is easy to see that $\|\cdot\|_\infty$ is a norm on $B(T)$. It is known as the **sup norm** on $B(T)$. We observe that the norm $\|\cdot\|_\infty$ induces the metric d_∞ on $B(T)$ defined in Example 1.5(iii).

Let T be a metric space. Consider the following subspaces of $B(T)$:

$$\begin{aligned} C(T) &:= \{x \in B(T) : x \text{ is continuous on } T\}, \\ C_0(T) &:= \{x \in C(T) : \text{for every } \epsilon > 0, \text{ there is a compact subset } T_\epsilon \text{ of } T \\ &\qquad \text{such that } |x(t)| < \epsilon \text{ for all } t \in T \setminus T_\epsilon\}, \\ C_{00}(T) &:= \{x \in C_0(T) : \text{there is a compact subset } T_0 \text{ of } T \text{ such that} \\ &\qquad x(t) = 0 \text{ for all } t \in T \setminus T_0\}. \end{aligned}$$

An element of $C_0(T)$ is known as a continuous function **vanishing at infinity**, and an element of $C_{00}(T)$ is known as a continuous function with **compact support**.[1] Clearly, $C_{00}(T) \subset C_0(T) \subset C(T)$. If T is a compact metric space, then $C_{00}(T) = C_0(T) = C(T)$. On the other hand, let $T := \mathbb{R}$ with the usual metric. If $x_0(t) := e^{-|t|}$ and $x_1(t) = 1$ for $t \in T$, then x_0 is in $C_0(T)$, but not in $C_{00}(T)$, while x_1 is in $C(T)$ but not in $C_0(T)$.

Let $T := [a, b]$. For $x \in C([a, b])$, define

$$\|x\|_1 := \int_a^b |x(t)|dt \quad \text{and} \quad \|x\|_2 := \Big(\int_a^b |x(t)|^2 dt\Big)^{1/2}.$$

It can be seen that $\|\cdot\|_1$ and $\|\cdot\|_2$ are norms on $C([a, b])$.

(iv) $\boldsymbol{L^p}$ **spaces**, $p \in \{1, 2, \infty\}$: Let m denote the Lebesgue measure on $\mathbb{R}$. Let $E := \mathbb{R}$, or let E be a closed and bounded interval in $\mathbb{R}$. Consider the spaces $L^1(E)$, $L^2(E)$ and $L^\infty(E)$ introduced in Sect. 1.4. Define

$$\begin{aligned} \|x\|_1 &:= \int_E |x(t)|dm(t) \text{ for } x \in L^1(E), \\ \|x\|_2 &:= \Big(\int_E |x(t)|^2 dm(t)\Big)^{1/2} \text{ for } x \in L^2(E), \\ \|x\|_\infty &:= \text{ess sup}_E |x| \text{ for } x \in L^\infty(E). \end{aligned}$$

[1] The space $C_{00}(T)$ is often denoted by $C_c(T)$.

It is easy to see that $L^1(E)$ is a linear space over $\mathbb{K}$ and $\|\cdot\|_1$ is a norm on it, and that $L^\infty(E)$ is a linear space over $\mathbb{K}$ and $\|\cdot\|_\infty$ is a norm on it. As for $L^2(E)$, let $x, y \in L^2(E)$, $a(t) := |x(t)|$ and $b(t) := |y(t)|$, $t \in E$, in the Minkowski inequality for functions (Lemma 1.25(ii)), and obtain

$$\left(\int_E |x(t) + y(t)|^2 dm(t)\right)^{1/2} \le \left(\int_E \left(|x(t)| + |y(t)|\right)^2 dm(t)\right)^{1/2} \le \|x\|_2 + \|y\|_2.$$

Hence $x + y \in L^2(E)$ and $\|x + y\|_2 \le \|x\|_2 + \|y\|_2$. It follows that $L^2(E)$ is a linear space, and that $\|\cdot\|_2$ is a norm on $L^2(E)$.

We observe that the norm $\|\cdot\|_1$ induces the metric d_1 on $L^1(E)$, the norm $\|\cdot\|_2$ induces the metric d_2 on $L^2(E)$ and the norm $\|\cdot\|_\infty$ induces the metric d_∞ on $L^\infty(E)$ introduced in Sect. 1.4.

Let E be a closed and bounded interval in $\mathbb{R}$. Then $m(E)$ is the length of E. Note that if $x \in L^\infty(E)$, then $x \in L^2(E)$ and $\|x\|_2 \le \sqrt{m(E)}\|x\|_\infty$, since $|x(t)| \le \|x\|_\infty$ for almost all $t \in E$. Also, if $x \in L^2(E)$, then $x \in L^1(E)$ and $\|x\|_1 \le \sqrt{m(E)}\|x\|_2$. This follows by letting $a(t) := |x(t)|$ and $b(t) := 1$ in the Schwarz inequality for functions (Lemma 1.25(i)). Thus we obtain $L^\infty(E) \subset L^2(E) \subset L^1(E)$, as opposed to $\ell^1 \subset \ell^2 \subset \ell^\infty$. ◊

Remark 2.2 Let X and Y be linear spaces, and let $F : X \to Y$ be a linear map. Suppose $\|\cdot\|_Y$ is a norm on Y. For $x \in X$, define $p(x) := \|F(x)\|_Y$. It is easy to see that p is a seminorm on X. It is in fact a norm on X if and only if the map F is one-one.

On the other hand, suppose $p : X \to \mathbb{R}$ is a seminorm on X, and let $Z := \{x \in X : p(x) = 0\}$. Then Z is a subspace of X. Suppose $x_1, x_2 \in X$ satisfy $x_1 + Z = x_2 + Z$, that is, $p(x_1 - x_2) = 0$. Then $p(x_1) \le p(x_1 - x_2) + p(x_2) = p(x_2)$. Similarly, $p(x_2) \le p(x_1)$, and so $p(x_1) = p(x_2)$. For $x + Z$ in X/Z, let $q(x + Z) := p(x)$. The function $q : X/Z \to \mathbb{R}$ is clearly a seminorm on X/Z. Also, if $x \in X$ and $q(x + Z) = 0$, then $x \in Z$, that is, $x + Z = 0$ in X/Z. Thus q is in fact a norm on X/Z. The norms on $L^1(E)$, $L^2(E)$ and $L^\infty(E)$ introduced in Example 2.1(iv) are of this kind. ◊

Comparability and Equivalence of Norms

Suppose $\|\cdot\|$ and $\|\cdot\|'$ are norms on a linear space X. We say that the norm $\|\cdot\|$ is **stronger** than the norm $\|\cdot\|'$ if $\|x_n\|' \to 0$ whenever $\|x_n\| \to 0$. The norms $\|\cdot\|$ and $\|\cdot\|'$ on X are called **comparable** if one of them is stronger than the other, and they are called **equivalent** if each is stronger than the other.

Proposition 2.3 *Let $\|\cdot\|$ and $\|\cdot\|'$ be norms on a linear space X. The norm $\|\cdot\|$ is stronger than the norm $\|\cdot\|'$ if and only if there is $\alpha > 0$ such that $\|x\|' \le \alpha\|x\|$ for all $x \in X$.*

Further, the norms $\|\cdot\|$ and $\|\cdot\|'$ are equivalent if and only if there are $\alpha > 0$ and $\beta > 0$ such that $\beta\|x\| \le \|x\|' \le \alpha\|x\|$ for all $x \in X$.

Proof Suppose $\|x\|' \le \alpha\|x\|$ for all $x \in X$, and let (x_n) be a sequence in X such that $\|x_n\| \to 0$. Clearly, $\|x_n\|' \to 0$. Thus the norm $\|\cdot\|$ is stronger than the norm $\|\cdot\|'$.

Conversely, suppose the norm $\|\cdot\|$ is stronger than the norm $\|\cdot\|'$. If there is no $\alpha > 0$ such that $\|x\|' \le \alpha\|x\|$ for all $x \in X$, then for every $n \in \mathbb{N}$, there is nonzero $x_n \in X$ such that $\|x_n\|' > n\|x_n\|$. Let $y_n := x_n/n\|x_n\|$ for $n \in \mathbb{N}$. Then $\|y_n\| = 1/n \to 0$, but $\|y_n\|' \not\to 0$ since $\|y_n\|' > 1$ for all $n \in \mathbb{N}$.

The statement about the equivalence of the two norms follows easily. □

Examples 2.4 (i) The norms $\|\cdot\|_1$, $\|\cdot\|_2$ and $\|\cdot\|_\infty$ on $\mathbb{K}^n$ are equivalent, since $\|x\|_\infty \le \|x\|_2 \le \|x\|_1$ and $\|x\|_1 \le \sqrt{n}\|x\|_2 \le n\|x\|_\infty$ for all $x \in \mathbb{K}^n$. We shall show that any two norms on $\mathbb{K}^n$ are equivalent in Lemma 2.8(ii).

(ii) On ℓ^1, the norm $\|\cdot\|_1$ is stronger than the norm $\|\cdot\|_2$ since $\|x\|_2 \le \|x\|_1$ for all $x \in \ell^1$. But these two norms on ℓ^1 are not equivalent since $\|x_n\|_1 = 1$ and $\|x_n\|_2 = 1/\sqrt{n}$, where $x_n := (e_1 + \cdots + e_n)/n$ for $n \in \mathbb{N}$. Similarly, on ℓ^2, the norm $\|\cdot\|_2$ is stronger than the norm $\|\cdot\|_\infty$ since $\|x\|_\infty \le \|x\|_2$ for all $x \in \ell^2$. But these two norms on ℓ^2 are not equivalent since $\|\sqrt{n}\, x_n\|_2 = 1$ and $\|\sqrt{n}\, x_n\|_\infty = 1/\sqrt{n}$ for $n \in \mathbb{N}$. Further, $\|x\|_2 \le \sqrt{\|x\|_1\|x\|_\infty}$ for all $x \in \ell^1$.

On the other hand, on $L^\infty([0, 1])$, the norm $\|\cdot\|_\infty$ is stronger than the norm $\|\cdot\|_2$ since $\|x\|_2 \le \|x\|_\infty$ for all $x \in L^\infty([0, 1])$. Similarly, on $L^2([0, 1])$ the norm $\|\cdot\|_2$ is stronger than the norm $\|\cdot\|_1$ since $\|x\|_1 \le \|x\|_2$ for all $x \in L^2([0, 1])$. (See also Exercise 2.5.) Further, $\|x\|_2 \le \sqrt{\|x\|_1\|x\|_\infty}$ for all $x \in L^\infty([0, 1])$.

(iii) For $x := (x(1), x(2), x(3), \ldots) \in c_{00}$, let

$$\|x\| := |x(1) - x(2) - x(3) - \cdots| + \sup\{|x(2)|, |x(3)|, \ldots\}$$

Then $\|\cdot\|$ is a norm on c_{00}. The norms $\|\cdot\|_\infty$ and $\|\cdot\|$ on c_{00} are not comparable since $\|e_2 + \cdots + e_{n+1}\|_\infty = 1 = \|ne_1 + e_2 + \cdots + e_{n+1}\|$, while $\|e_2 + \cdots + e_{n+1}\| = n + 1$ and $\|ne_1 + e_2 + \cdots + e_{n+1}\|_\infty = n$ for all $n \in \mathbb{N}$. ◊

Quotients and Products of Normed Spaces

Let X be a normed space. We have already seen that a subspace of X and its closure in X are normed spaces. We consider two other ways in which a normed space gives rise to new normed spaces.

Proposition 2.5 (i) *Let Y be a closed subspace of a normed space X. For $x + Y$ in the quotient space X/Y, let*

$$|||x + Y||| := \inf\{\|x + y\| : y \in Y\}.$$

Then $|||\cdot|||$ is a norm on X/Y, called the **quotient norm**.

A sequence $(x_n + Y)$ converges to $x + Y$ in X/Y if and only if there is a sequence (y_n) in Y such that $(x_n + y_n)$ converges to x in X.

(ii) *Let $m \in \mathbb{N}$ and for $j = 1, \ldots, m$, let $\|\cdot\|_j$ be a norm on a linear space X_j. For $x := (x(1), \ldots, x(m))$ in the product space $X := X_1 \times \cdots \times X_m$, let*

$$\|x\| := \|x(1)\|_1 + \cdots + \|x(m)\|_m,$$

Then $\|\cdot\|$ *is a norm on* X. *It is called a* **product norm**.
A sequence (x_n) *converges to* x *in* X *if and only if* $(x_n(j))$ *converges to* $x(j)$ *in* X_j *for every* $j = 1, \ldots, m$.

Proof (i) For $x \in X$, $|||x + Y||| = \inf\{\|x - y\| : y \in Y\} = d(x, Y)$. Let x_1 and x_2 be in X, and let $\epsilon > 0$. There are y_1 and y_2 in Y such that

$$\|x_1 + y_1\| < \inf\{\|x_1 + y\| : y \in Y\} + \frac{\epsilon}{2} = |||x_1 + Y||| + \frac{\epsilon}{2},$$
$$\|x_2 + y_2\| < \inf\{\|x_2 + y\| : y \in Y\} + \frac{\epsilon}{2} = |||x_2 + Y||| + \frac{\epsilon}{2}.$$

Then

$$\|x_1 + x_2 + y_1 + y_2\| \le \|x_1 + y_1\| + \|x_2 + y_2\| \le |||x_1 + Y||| + |||x_2 + Y||| + \epsilon.$$

Since $y_1 + y_2 \in Y$, we see that

$$|||(x_1 + x_2) + Y||| \le \|x_1 + x_2 + y_1 + y_2\| \le |||x_1 + Y||| + |||x_2 + Y||| + \epsilon.$$

It follows that $|||(x_1 + Y) + (x_2 + Y)||| \le |||x_1 + Y||| + |||x_2 + Y|||$.

Next, let $x \in X$ and $k \in \mathbb{K}$. By considering the cases $k = 0$ and $k \neq 0$, we see that $|||k(x + Y)||| = |k|\, |||x + Y|||$. Thus $|||\cdot|||$ is a seminorm on X/Y.

If $|||x + Y||| = 0$, then there is a sequence (y_n) in Y such that $x + y_n \to 0$, that is, $y_n \to -x$. Since the subspace Y is closed, $-x \in Y$, that is, $x + Y = Y$, which is the zero element of X/Y. Hence $|||\cdot|||$ is in fact a norm on X/Y.

Let $(x_n + Y)$ be a sequence in X/Y. Suppose there is sequence (y_n) in Y such that $x_n + y_n \to x$ in X. Then

$$|||(x_n + Y) - (x + Y)||| = |||(x_n - x) + Y||| \le \|x_n - x + y_n\|$$

for each $n \in \mathbb{N}$. Thus $x_n + Y \to x + Y$ in X/Y. Conversely, assume that $x_n + Y \to x + Y$ in X/Y. Since

$$|||(x_n + Y) - (x + Y)||| = \inf\{\|x_n - x + y\| : y \in Y\},$$

there is $y_n \in Y$ such that $\|x_n - x + y_n\| < |||(x_n + Y) - (x + Y)||| + 1/n$ for each $n \in \mathbb{N}$. It follows that $x_n - x + y_n \to 0$, that is, $x_n + y_n \to x$ in X.

(ii) It is easy to see that $\|\cdot\|$ is a norm on $X := X_1 \times \cdots \times X_m$. Let $x := (x(1), \ldots, x(m)) \in X$, and $x_n := (x_n(1), \ldots, x_n(m)) \in X$ for $n \in \mathbb{N}$. Since

$$\|x_n(j) - x(j)\|_j \le \|x_n - x\| = \|x_n(1) - x(1)\|_1 + \cdots + \|x_n(m) - x(m)\|_m$$

for every $j = 1, \ldots, m$, we see that $x_n \to x$ in X if and only if $x_n(j) \to x(j)$ in X_j for every $j = 1, \ldots, m$. □

Remarks 2.6 (i) We note that the quotient norm $|||x + Y|||$ of $x + Y$ defined in Proposition 2.5(i) is the distance $d(x, Y)$ from $x \in X$ to the closed subspace Y of X. It follows that $d(x_1 + x_2, Y) \le d(x_1, Y) + d(x_2, Y)$ and $d(k\,x_1, Y) = |k| d(x_1, Y)$ for $x_1, x_2 \in X$ and $k \in \mathbb{K}$. The linear map $Q : X \to X/Y$ given by $Q(x) := x + Y$ for $x \in X$, is called the **quotient map** on X/Y.

(ii) Proposition 2.5(ii) remains valid if we define a **product norm** by

$$\|x\| := \left(\|x(1)\|_1^2 + \cdots + \|x(m)\|_m^2\right)^{1/2} \text{ or } \|x\| := \max\{\|x(1)\|_1, \ldots, \|x(m)\|_m\}$$

for $x := (x(1), \ldots, x(m)) \in X := X_1 \times \cdots \times X_m$. In the former case, for $x := (x(1), \ldots, x(m))$ and $y := (y(1), \ldots, y(m))$ in X, we use the Minkowski inequality for numbers (Lemma 1.4(ii)) with $a_j := \|x(j)\|_j$ and $b_j := \|y(j)\|_j$ for $j = 1, \ldots, m$ to prove $\|x + y\| \le \|x\| + \|y\|$. The latter case is straightforward. As in Example 2.4(i), the product norms on $X_1 \times \cdots \times X_m$ considered here are equivalent. The linear map $P_j : X \to X_j$ given by $P_j(x) := x_j$ for $x := (x_1, \ldots, x_m)$, is called the **projection map** on the jth component X_j of X, $j = 1, \ldots, m$. ◊

Finite Dimensionality of a Normed Space

We shall give a criterion for the finite dimensionality of a normed space X in terms of the compactness of the closed unit ball of X. To obtain this criterion, we need two preliminary results which are also of independent interest.

Lemma 2.7 (F. Riesz, 1918) *Let X be a normed space, and Y be a closed subspace of X such that $Y \ne X$. Then for every $r \in (0, 1)$, there is $x_r \in X$ such that*

$$\|x_r\| = 1 \quad \text{and} \quad r < d(x_r, Y) \le 1.$$

Proof Let $x \in X \setminus Y$. Since Y is a closed subset of X, we see that $d(x, Y) = \inf\{\|x - y\| : y \in Y\} > 0$. Also, since $d(x, Y) < d(x, Y)/r$, there is $y_0 \in Y$ such that $0 < d(x, Y) \le \|x - y_0\| < d(x, Y)/r$. (See Fig. 2.2.) Hence

$$r < \frac{d(x, Y)}{\|x - y_0\|} = \frac{|||x + Y|||}{\|x - y_0\|} = \frac{|||(x - y_0) + Y|||}{\|x - y_0\|} = \left|\left|\left| \frac{x - y_0}{\|x - y_0\|} + Y \right|\right|\right|.$$

Let $x_r := (x - y_0)/\|x - y_0\|$. Then $r < d(x_r, Y) \le \|x_r\| = 1$. □

The Riesz lemma says that if Y is a closed proper subspace of a normed space X, then there is a point on the unit sphere of X whose distance from Y is as close to 1 as we please. If Y is finite dimensional, then there is in fact $x_1 \in X$ such that $\|x_1\| = 1 = d(x_1, Y)$. (See Exercise 2.39.)

Lemma 2.8 *Every finite dimensional normed space is complete. Suppose Y is a finite dimensional normed space, and let $\{y_1, \ldots, y_m\}$ be a basis for Y. Let (x_n)*

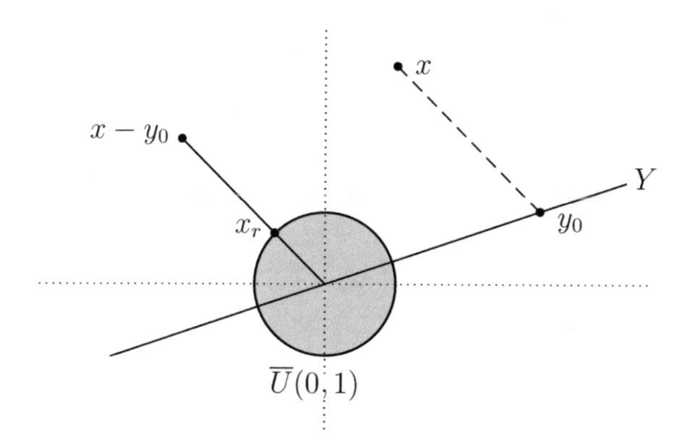

Fig. 2.2 Choice of x_r in the proof of the Riesz lemma

be a sequence in Y, and let $x_n := k_{n,1}y_1 + \cdots + k_{n,m}y_m$ for each $n \in \mathbb{N}$, where $k_{n,1}, \ldots, k_{n,m} \in \mathbb{K}$. Then

(i) *The sequence (x_n) is bounded in Y if and only if the sequence $(k_{n,j})$ is bounded in $\mathbb{K}$ for each $j = 1, \ldots, m$.*
(ii) *$x_n \to x := k_1y_1 + \cdots + k_my_m$ in Y if and only if $k_{n,j} \to k_j$ in $\mathbb{K}$ for each $j = 1, \ldots, m$. In particular, any two norms on Y are equivalent.*

Proof To prove the completeness of a finite dimensional normed space Y, we use mathematical induction on the dimension m of Y. Suppose $\|\cdot\|$ is a norm on Y. Let $m := 1$. Then there is nonzero $y_1 \in Y$ and $Y = \{ky_1 : k \in \mathbb{K}\}$. Consider a Cauchy sequence (x_n) in Y. Then $x_n := k_ny_1$, where $k_n \in \mathbb{K}$ for $n \in \mathbb{N}$, and

$$\|x_n - x_p\| = |k_n - k_p|\,\|y_1\| \quad \text{for all } n, p \in \mathbb{N}.$$

Since $\|y_1\| > 0$, (k_n) is a Cauchy sequence in $\mathbb{K}$. Since $\mathbb{K}$ is complete, there is $k \in \mathbb{K}$ such that $k_n \to k$. Then $x_n \to k\,y_1$ in Y by the continuity of scalar multiplication. Thus Y is complete. Let $m \geq 2$, and assume now that every $m-1$ dimensional normed space is complete. Let Y be an m dimensional normed space, and let $y_1, \ldots, y_m$ and (x_n) be as in the statement of this lemma. Fix $j \in \{1, \ldots, m\}$, and define

$$Y_j := \text{span}\,\{y_i : i = 1, \ldots, m \text{ and } i \neq j\}.$$

Then the $m-1$ dimensional subspace Y_j is complete by the inductive assumption. In particular, Y_j is closed in X. Since $y_j \notin Y_j$, $|||y_j + Y_j||| > 0$. Suppose (x_n) is a Cauchy sequence. Then for $n, p \in \mathbb{N}$,

$$\|x_n - x_p\| = \|(k_{n,1} - k_{p,1})y_1 + \cdots + (k_{n,m} - k_{p,m})y_m\|$$
$$\geq |||(k_{n,j} - k_{p,j})y_j + Y_j||| = |k_{n,j} - k_{p,j}|\, |||y_j + Y_j|||.$$

Hence $(k_{n,j})$ is a Cauchy sequence in $\mathbb{K}$. Let $k_{n,j} \to k_j$ in $\mathbb{K}$. Since this holds for each $j = 1, \ldots, m$, the continuity of addition and scalar multiplication in Y shows that $x_n \to k_1 y_1 + \cdots + k_m y_m \in Y$. Thus Y is complete.

(i) Suppose for each $j = 1, \ldots, m$, $(k_{n,j})$ is bounded in $\mathbb{K}$, and so there is $\alpha_j > 0$ such that $|k_{n,j}| \leq \alpha_j$ for all $n \in \mathbb{N}$. Then

$$\|x_n\| \leq \alpha_1 \|y_1\| + \cdots + \alpha_m \|y_m\|$$

for all $n \in \mathbb{N}$. Thus (x_n) is bounded in Y. Conversely, suppose (x_n) is bounded in Y. For each $j = 1, \ldots, m$,

$$\|x_n\| = \|k_{n,1} y_1 + \cdots + k_{n,m} y_m\| \geq |||k_{n,j} y_j + Y_j||| = |k_{n,j}|\, |||y_j + Y_j|||.$$

Thus $(k_{n,j})$ is bounded for each $j = 1, \ldots, m$.

(ii) Suppose $k_{n,j} \to k_j$ for each $j = 1, \ldots, m$, and let $x := k_1 y_1 + \cdots + k_m y_m$. Then we have already seen that $x_n \to x$ in Y. Conversely, suppose $x_n \to x$ in Y. Then

$$\|x_n - x\| = \|(k_{n,1} - k_1)y_1 + \cdots + (k_{n,m} - k_m)y_m\|$$
$$\geq |||(k_{n,j} - k_j)y_j + Y_j||| = |k_{n,j} - k_j|\, |||y_j + Y_j|||$$

for each $j = 1, \ldots, m$. Hence $k_{n,j} \to k_j$ for each $j = 1, \ldots, m$.

Let $\|\cdot\|'$ be another norm on Y. As above, $\|x_n - x\|' \to 0$ if and only if $k_{n,j} \to k_j$ for each $j = 1, \ldots, m$ if and only if $\|x_n - x\| \to 0$. Hence the norm $\|\cdot\|'$ is equivalent to the norm $\|\cdot\|$. □

Remark 2.9 Since every complete subset Y of a metric space X is closed in X, it follows from Lemma 2.8 that every finite dimensional subspace Y of a normed space X is closed in X. On the other hand, an infinite dimensional subspace Y of a normed space X may not be closed in X. For example, let $X := \ell^\infty$ and $Y := c_{00}$. For $n \in \mathbb{N}$, let $x_n := (1, 1/2, \ldots, 1/n, 0, 0, \ldots) \in c_{00}$ for $n \in \mathbb{N}$, and let $x := (1, 1/2, 1/3, \ldots) \in \ell^\infty$. Then $x_n \to x$ in ℓ^∞, since $\|x_n - x\|_\infty = 1/(n+1) \to 0$, but $x \notin Y$. Thus Y is not closed in X. (In fact, the closure c_{00} is c_0. See Exercise 2.3.) Next, let $x_n := (0, \ldots, 0, n, 0, 0, \ldots) \in c_{00}$ for $n \in \mathbb{N}$. Then $|x_n(j)| \leq j$ for all $n, j \in \mathbb{N}$. Hence the sequence $(x_n(j))$ is bounded for each $j \in \mathbb{N}$, but the sequence (x_n) is not bounded in c_{00} since $\|x_n\|_\infty = n$ for $n \in \mathbb{N}$. Finally, $e_n(j) \to 0$ for each $j \in \mathbb{N}$, but $e_n \not\to 0$ in c_{00}, since $\|e_n\|_\infty = 1$ for all $n \in \mathbb{N}$. Thus the conclusions of Lemma 2.8 may not hold in an infinite dimensional normed space. ◊

We are now in a position to prove a criterion for the finite dimensionality of a subspace of a normed space.

Theorem 2.10 *A normed space X is finite dimensional if and only if the closed unit ball of X is compact.*

Proof Let us denote the closed unit ball $\overline{U}(0, 1)$ of X by E.

Suppose X is finite dimensional. To prove that E is compact, consider a sequence (x_n) in E. Let $\{y_1, \ldots, y_m\}$ be a basis for X, and for $n \in \mathbb{N}$, let

$$x_n := k_{n,1}y_1 + \cdots + k_{n,m}y_m, \qquad \text{where } k_{n,1}, \ldots, k_{n,m} \in \mathbb{K}.$$

Since E is bounded in X, the sequence $(k_{n,j})$ is bounded in $\mathbb{K}$ for each $j = 1, \ldots, m$ by Lemma 2.8(i). By the Bolzano–Weierstrass theorem for $\mathbb{K}$, and by passing to a subsequence of a subsequence several times, we may find $n_1 < n_2 < \cdots$ in $\mathbb{N}$ such that $(k_{n_p,j})$ converges in $\mathbb{K}$ for each $j = 1, \ldots, m$. Again, by Lemma 2.8(ii), the subsequence (x_{n_p}) converges in X. Since the subsequence (x_{n_p}) is in E, and E is closed, we see that (x_{n_p}) converges in E.

Conversely, suppose X is not finite dimensional. Then there is an infinite linearly independent subset $\{y_1, y_2, \ldots\}$ of X. For $n \in \mathbb{N}$, let $Y_n := \text{span}\,\{y_1, \ldots, y_n\}$. Being finite dimensional, Y_n is complete by Lemma 2.8. Hence Y_n is a closed subspace of Y_{n+1}. Also, $Y_n \neq Y_{n+1}$, since the set $\{y_1, \ldots, y_{n+1}\}$ is linearly independent. By the Riesz lemma (Lemma 2.7), there is $x_n \in Y_{n+1}$ such that

$$\|x_n\| = 1 \quad \text{and} \quad d(x_n, Y_n) \geq \frac{1}{2} \quad \text{for each } n \in \mathbb{N}.$$

Then (x_n) is a sequence in E, but it has no convergent subsequence, since $\|x_n - x_p\| \geq 1/2$ for all $n \neq p$ in $\mathbb{N}$. Hence E cannot be compact. □

Remarks 2.11 Let X be a normed space. If X is finite dimensional, then the proof of Theorem 2.10 given above shows that every closed and bounded subset of X is compact since the only fact about $\overline{U}(0, 1)$ we have used is that it is closed and bounded. On the other hand, if X is infinite dimensional, then our proof shows that the unit sphere S of X is not compact; in fact, we have found, by using the Riesz lemma (Lemma 2.7), a sequence (x_n) in S which has no Cauchy subsequence, so that S is not even totally bounded (Proposition 1.8). Theorem 2.10 extends the classical Heine–Borel theorem stated in Corollary 1.13(i). It says that the classical version holds in every finite dimensional normed space, and it does not hold in any infinite dimensional normed space.

Suppose a normed space X is not complete. Then the closed unit ball of X is not complete, and hence it is not compact by Theorem 1.11. In this case, we need not use the Riesz lemma as we have done above. ◊

2.2 Inner Product Spaces

We now define a new structure on a linear space, which allows us to introduce the concept of orthogonality. It is a generalization of the structure induced by the **dot product** of two vectors $x := (x(1), \ldots, x(n))$ and $y := (y(1), \ldots, y(n))$ in $\mathbb{R}^n$ defined by $x \cdot y := x(1)y(1) + \cdots + x(n)y(n)$.

Let X be a linear space over $\mathbb{K}$, where $\mathbb{K} := \mathbb{R}$ or $\mathbb{K} := \mathbb{C}$. We shall write our definitions and results assuming that $\mathbb{K} := \mathbb{C}$; the corresponding results for $\mathbb{K} := \mathbb{R}$ will be obtained by dropping the bar over $k \in \mathbb{C}$ (as in $\overline{k}$) and all terms involving i (as in $k_1 + ik_2$).

An **inner product** on X is a function $\langle \cdot\,, \cdot \rangle : X \times X \to \mathbb{K}$ which is

(i) **linear in the first variable**, that is, $\langle x + y, z \rangle = \langle x, z \rangle + \langle y, z \rangle$ and $\langle k\,x, y \rangle = k\langle x, y \rangle$ for all $x, y, z \in X$ and $k \in \mathbb{K}$,
(ii) **conjugate-symmetric**, that is, $\langle y, x \rangle = \overline{\langle x, y \rangle}$ for all $x, y \in X$, and
(iii) **positive-definite**, that is, $\langle x, x \rangle \geq 0$ for all $x \in X$, and $\langle x, x \rangle = 0$ only for $x = 0$.

It follows from (i) and (ii) that $\langle \cdot\,, \cdot \rangle$ is **conjugate-linear in the second variable**, that is, $\langle x, y + z \rangle = \langle x, y \rangle + \langle x, z \rangle$ and $\langle x, k\,y \rangle = \overline{k}\langle x, y \rangle$ for all $x, y, z \in X$ and $k \in \mathbb{K}$. Also, $\langle 0, y \rangle = 0$ for all $y \in X$, since $\langle 0, y \rangle = \langle 0 + 0, y \rangle = \langle 0, y \rangle + \langle 0, y \rangle = 2\langle 0, y \rangle$.

An **inner product space** $(X, \langle \cdot\,, \cdot \rangle)$ over $\mathbb{K}$ is a linear space X over $\mathbb{K}$ along with an inner product $\langle \cdot\,, \cdot \rangle$ on it.

Examples 2.12 (i) Let $n \in \mathbb{N}$ and $X := \mathbb{K}^n$. Define

$$\langle x, y \rangle := \sum_{j=1}^{n} x(j)\overline{y(j)} \quad \text{for } x := (x(1), \ldots, x(n)),\ y := (y(1), \ldots, y(n)) \in \mathbb{K}^n.$$

It is easy to see that $\langle \cdot\,, \cdot \rangle$ is an inner product on $\mathbb{K}^n$. This inner product will be called the **usual inner product** on $\mathbb{K}^n$.

Let $w(1), \ldots, w(n)$ be positive real numbers, and define

$$\langle x, y \rangle_w := \sum_{j=1}^{n} w(j)x(j)\overline{y(j)} \quad \text{for } x, y \in \mathbb{K}^n.$$

Then $\langle \cdot\,, \cdot \rangle_w$ is an inner product on $\mathbb{K}^n$. If $w(1) = \cdots = w(n) = 1$, then $\langle \cdot\,, \cdot \rangle_w$ is the usual inner product on $\mathbb{K}^n$. We give an example of a function on $\mathbb{K}^4 \times \mathbb{K}^4$ which is linear in the first variable and conjugate-symmetric, but is not an inner product on $\mathbb{K}^4$. For $x := (x(1), \ldots, x(4))$, $y := (y(1), \ldots, y(4)) \in \mathbb{K}^4$, let

$$\langle x, y \rangle_M = x(1)\overline{y(1)} + x(2)\overline{y(2)} + x(3)\overline{y(3)} - x(4)\overline{y(4)}.$$

The linear space $\mathbb{R}^4$ with the function $\langle \cdot\,, \cdot \rangle_M : \mathbb{R}^4 \times \mathbb{R}^4 \to \mathbb{R}$ is called the **Minkowski space**. A vector $x \in \mathbb{R}^4$ is called a **space-like vector** if $\langle x, x \rangle_M > 0$, a **light-like vector** if $\langle x, x \rangle_M = 0$ and a **time-like vector** if $\langle x, x \rangle_M < 0$. This space plays a fundamental role in the theory of relativity.

(ii) Let $X := \ell^2$. Define

$$\langle x, y \rangle := \sum_{j=1}^{\infty} x(j)\overline{y(j)} \quad \text{for } x := (x(1), x(2), \ldots),\ y := (y(1), y(2), \ldots) \in \ell^2.$$

Here the series $\sum_{j=1}^{\infty} x(j)\overline{y(j)}$ converges absolutely in $\mathbb{K}$ since

$$\sum_{j=1}^{\infty} |x(j)\overline{y(j)}| \le \Big(\sum_{j=1}^{\infty} |x(j)|^2\Big)^{1/2} \Big(\sum_{j=1}^{\infty} |y(j)|^2\Big)^{1/2}.$$

This follows by letting $n \to \infty$ in the Schwarz inequality for numbers (Lemma 1.4(i)). It is now easy to see that $\langle \cdot\,, \cdot \rangle$ is an inner product on ℓ^2.

(iii) Let $X := L^2(E)$, where $E := \mathbb{R}$ or $[a, b]$. Define

$$\langle x, y \rangle := \int_E x(t)\overline{y}(t)dm(t) \quad \text{for } x, y \in L^2(E).$$

Here the function $x\overline{y}$ is integrable on E since

$$\int_E |x(t)\overline{y}(t)|dm(t) \le \Big(\int_E |x(t)|^2 dm(t)\Big)^{1/2} \Big(\int_E |y(t)|^2 dm(t)\Big)^{1/2}.$$

This follows by the Schwarz inequality for functions (Lemma 1.25(i)). It is now easy to see that $\langle \cdot\,, \cdot \rangle$ is an inner product on $L^2(E)$. ◊

We note that the Schwarz inequalities proved in Lemma 1.4(i) and in Lemma 1.25(i) have enabled us to show that the inner products on ℓ^2 and $L^2(E)$ introduced in the examples (ii) and (iii) above are well defined. Conversely, we show that similar inequalities hold in any inner product space X. Let us remark that if $x, u \in X$ and $\|u\| = 1$, then $\langle x, u \rangle u$ can be regarded as the component of the vector x in the direction of the vector u of norm 1.

Proposition 2.13 *Let $\langle \cdot\,, \cdot \rangle$ be an inner product on a linear space X, and let $x, y \in X$. Then*

(i) **(Schwarz inequality)**

$$|\langle x, y \rangle| \le \langle x, x \rangle^{1/2} \langle y, y \rangle^{1/2}.$$

(ii) **(Minkowski inequality)**

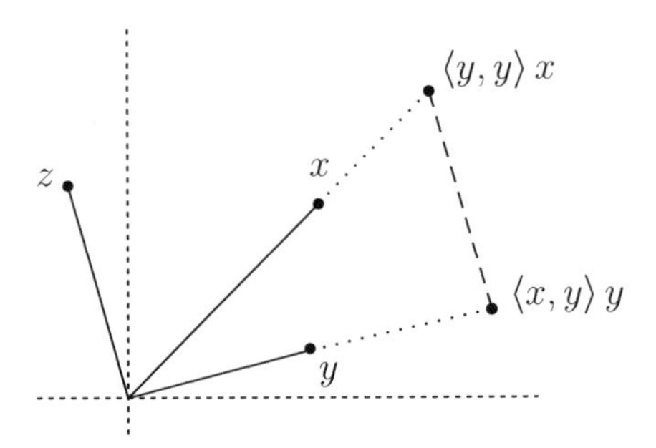

Fig. 2.3 Choice of z in the proof of the Schwarz inequality

$$\langle x+y, x+y\rangle^{1/2} \le \langle x, x\rangle^{1/2} + \langle y, y\rangle^{1/2}.$$

Proof (i) Let $z := \langle y, y\rangle x - \langle x, y\rangle y$. (See Fig. 2.3.) Then $\langle z, y\rangle = 0$ and

$$\begin{aligned} 0 \le \langle z, z\rangle &= \langle y, y\rangle^2 \langle x, x\rangle - \langle y, y\rangle \overline{\langle x, y\rangle} \langle x, y\rangle \\ &\quad - \langle x, y\rangle \langle y, y\rangle \langle y, x\rangle + \langle x, y\rangle \overline{\langle x, y\rangle} \langle y, y\rangle \\ &= \langle y, y\rangle \left(\langle x, x\rangle \langle y, y\rangle - |\langle x, y\rangle|^2 \right). \end{aligned}$$

If $\langle y, y\rangle > 0$, then it follows that $\langle x, x\rangle\langle y, y\rangle - |\langle x, y\rangle|^2 \ge 0$. If $\langle y, y\rangle = 0$, then $y = 0$, and so $|\langle x, y\rangle|^2 = 0 = \langle x, x\rangle\langle y, y\rangle$. This proves the Schwarz inequality.

(ii) By the Schwarz inequality in (i) above,

$$\begin{aligned} \langle x+y, x+y\rangle &= \langle x, x\rangle + \langle x, y\rangle + \langle y, x\rangle + \langle y, y\rangle \\ &= \langle x, x\rangle + 2\mathrm{Re}\, \langle x, y\rangle + \langle y, y\rangle \\ &\le \langle x, x\rangle + 2|\langle x, y\rangle| + \langle y, y\rangle \\ &\le \langle x, x\rangle + 2\langle x, x\rangle^{1/2} \langle y, y\rangle^{1/2} + \langle y, y\rangle \\ &= \left(\langle x, x\rangle^{1/2} + \langle y, y\rangle^{1/2} \right)^2. \end{aligned}$$

This proves the Minkowski inequality. □

Necessary and sufficient conditions for equality in (i) and (ii) above are given in Exercise 2.11.

Now we are in a position to show that an inner product on a linear space induces a norm, and hence a metric, on it.

Proposition 2.14 *Let $\langle \cdot\,, \cdot\rangle$ be an inner product on a linear space X. For $x \in X$, define $\|x\| := \langle x, x\rangle^{1/2}$. Then*

$$|\langle x, y\rangle| \le \|x\|\, \|y\| \quad \textit{for all } x, y \in X,$$

and the function $\|\cdot\| : X \to \mathbb{R}$ is a norm on X.

Also, if $\|x_n - x\| \to 0$ *and* $\|y_n - y\| \to 0$*, then* $\langle x_n, y_n\rangle \to \langle x, y\rangle$*, that is, the function* $\langle\cdot\,,\cdot\rangle$ *is continuous on* $X \times X$.

Proof Let $x, y \in X$. Then the Schwarz inequality (Proposition 2.13(i)) says

$$|\langle x, y\rangle| \le \langle x, x\rangle^{1/2}\langle y, y\rangle^{1/2} = \|x\|\,\|y\|.$$

Further, the Minkowski inequality (Proposition 2.13(ii)) says

$$\|x + y\| = \langle x + y, x + y\rangle^{1/2} \le \langle x, x\rangle^{1/2} + \langle y, y\rangle^{1/2} = \|x\| + \|y\|.$$

Also, for $k \in \mathbb{K}$, $\|kx\|^2 = \langle kx, kx\rangle = k\bar{k}\langle x, x\rangle = |k|^2\|x\|^2$, that is, $\|kx\| = |k|\,\|x\|$. Finally, $\|x\| = \langle x, x\rangle^{1/2} \ge 0$, and if $\|x\| = 0$, that is, $\langle x, x\rangle = 0$, then $x = 0$ by the positive-definiteness of $\langle\cdot\,,\cdot\rangle$. Hence $\|\cdot\|$ is a norm on X.

Next, let $\|x_n - x\| \to 0$ and $\|y_n - y\| \to 0$. Then

$$\begin{aligned}|\langle x_n, y_n\rangle - \langle x, y\rangle| &\le |\langle x_n, y_n\rangle - \langle x_n, y\rangle| + |\langle x_n, y\rangle - \langle x, y\rangle|\\ &= |\langle x_n, y_n - y\rangle| + |\langle x_n - x, y\rangle|\\ &\le \|x_n\|\,\|y_n - y\| + \|x_n - x\|\,\|y\|,\end{aligned}$$

again by the Schwarz inequality (Proposition 2.13(i)). Since $(\|x_n\|)$ is a bounded sequence, we see that $\langle x_n, y_n\rangle \to \langle x, y\rangle$. □

The usual inner product on $\mathbb{K}^n$ induces the norm $\|\cdot\|_2$ given in Example 2.1(i), while the inner products on ℓ^2 and $L^2(E)$ given in Example 2.12(ii) and (iii) induce the norms introduced in Example 2.1(ii) and (iii).

Remark 2.15 The norm $\|\cdot\|$ induced on a linear space X by an inner product $\langle\cdot\,,\cdot\rangle$ satisfies the **parallelogram law**, that is,

$$\|x + y\|^2 + \|x - y\|^2 = 2\left(\|x\|^2 + \|y\|^2\right) \quad \text{for all } x, y \in X.$$

This can be seen by writing both sides in terms of inner products. Conversely, suppose a norm $\|\cdot\|$ on a linear space X satisfies the parallelogram law. Define

$$\langle x, y\rangle := \frac{1}{4}\left(\|x + y\|^2 - \|x - y\|^2 + i\|x + iy\|^2 - i\|x - iy\|^2\right) \quad \text{for all } x, y \in X.$$

Jordan and von Neumann showed (1935) that $\langle\cdot\,,\cdot\rangle$ is an inner product on the linear space X. Clearly, it satisfies $\langle x, x\rangle^{1/2} = \|x\|$, $x \in X$. The parallelogram law thus distinguishes inner product spaces among all normed spaces. ◊

Orthogonality

From elementary geometry, we know that two vectors $x := (x(1), x(2))$ and $y := (y(1), y(2))$ in $\mathbb{R}^2$ are perpendicular if their dot product is equal to zero. This fact prompts us to introduce the following concept. Let $\langle \cdot\,,\cdot \rangle$ be an inner product on a linear space X over $\mathbb{K}$. Elements $x \neq y$ in X are called **orthogonal** if $\langle x, y \rangle = 0$, and then we write $x \perp y$. Let E be a subset of X. Then E is called **orthogonal** if $x \perp y$ for all $x \neq y$ in E. Further, let $E^\perp := \{y \in X : y \perp x \text{ for all } x \in E\}$. By the linearity and the continuity of the inner product in the first variable, $E^\perp$ is a closed subspace of X.

Proposition 2.16 *Let X be an inner product space, and let $\|\cdot\|$ be the induced norm on X. Consider an orthogonal subset E of X.*

(i) **(Pythagoras)** *Suppose $x_1, \ldots, x_n \in E$. Then*

$$\|x_1 + \cdots + x_n\|^2 = \|x_1\|^2 + \cdots + \|x_n\|^2.$$

(ii) *If $0 \notin E$, then E is linearly independent.*

Proof (i) Since $\langle x_j, x_k \rangle = 0$ for all $j \neq k$, we obtain

$$\begin{aligned}\|x_1 + \cdots + x_n\|^2 &= \langle x_1 + \cdots + x_n, x_1 + \cdots + x_n \rangle \\ &= \sum_{j=1}^{n} \langle x_j, x_j \rangle = \|x_1\|^2 + \cdots + \|x_n\|^2.\end{aligned}$$

(ii) Suppose $0 \notin E$. Let $x_1, \ldots, x_n$ be distinct (nonzero) elements of E, and let $k_1 x_1 + \cdots + k_n x_n = 0$, where $k_1, \ldots, k_n \in \mathbb{K}$. Fix $j \in \{1, \ldots, n\}$. Then $\langle k_1, x \rangle_1 + \cdots + k_n x_n x_j = k_j \langle x_j, x_j \rangle$. But $\langle k_1 x_1 + \cdots + k_n x_n, x_j \rangle = \langle 0, x_j \rangle = 0$. Since $\langle x_j, x_j \rangle \neq 0$, we obtain $k_j = 0$. Thus E is linearly independent. □

A linearly independent subset of an inner product space X need not be orthogonal. For example, if $X := \mathbb{R}^2$ with the usual inner product, then $E := \{(1,0), (1,1)\}$ is linearly independent, but not orthogonal.

If a subset E of an inner product space X is orthogonal and if $\|x\| = 1$ for all $x \in E$, then E is called **orthonormal**. If E is an orthogonal subset of X and if $0 \notin E$, then $E_1 := \{x/\|x\| : x \in E\}$ is an orthonormal subset of X. If E is an orthonormal subset of X, then $\|x - y\|^2 = \langle x - y, x - y \rangle = \langle x, x \rangle + \langle y, y \rangle = 2$, that is, $\|x - y\| = \sqrt{2}$ for all $x \neq y$ in E.

We now give a procedure for converting a countable linearly independent subset $\{x_1, x_2, \ldots\}$ of an inner product space X into an orthonormal subset of X while retaining the span of the elements $x_1, x_2, \ldots$ at each step. Let $n \geq 2$. The main idea is to write $x_n := y_n + z_n$, where y_n is a linear combination of $x_1, \ldots, x_{n-1}$, and $z_n \perp x_j$ for $j = 1, \ldots, n-1$.

Theorem 2.17 **(Gram–Schmidt orthonormalization)** *Let $\{x_1, x_2, \ldots\}$ be a countable linearly independent subset of an inner product space X. Define $y_1 := 0$, $z_1 := x_1$, $u_1 := z_1/\|z_1\|$, and for $n \geq 2$,*

$$y_n := \langle x_n, u_1 \rangle u_1 + \cdots + \langle x_n, u_{n-1} \rangle u_{n-1}, \quad z_n := x_n - y_n, \quad u_n := \frac{z_n}{\|z_n\|}.$$

Then $\{u_1, u_2, \ldots\}$ *is an orthonormal subset of* X, *and*

$$\text{span}\,\{u_1, \ldots, u_n\} = \text{span}\,\{x_1, \ldots, x_n\} \quad \textit{for all } n \in \mathbb{N}.$$

Proof Since $\{x_1\}$ is a linearly independent subset of X, we see that $z_1 = x_1 \neq 0$, $\|u_1\| = \|z_1\|/\|z_1\| = 1$, and $\text{span}\,\{u_1\} = \text{span}\,\{x_1\}$.

Let $n \in \mathbb{N}$. Suppose we have proved that $\{u_1, \ldots, u_n\}$ is an orthonormal set satisfying $\text{span}\,\{u_1, \ldots, u_n\} = \text{span}\,\{x_1, \ldots, x_n\}$. Note that

$$y_{n+1} = \langle x_{n+1}, u_1\rangle u_1 + \cdots + \langle x_{n+1}, u_n\rangle u_n \quad \text{and} \quad z_{n+1} = x_{n+1} - y_{n+1}.$$

For $j = 1, \ldots, n$, $\langle y_{n+1}, u_j\rangle = \langle x_{n+1}, u_j\rangle$ since $\langle u_k, u_j\rangle = 0$ for all $k \neq j$ and $\langle u_j, u_j\rangle = 1$. Also, since $\{x_1, \ldots, x_{n+1}\}$ is a linearly independent subset of X, we see that $x_{n+1} \notin \text{span}\,\{x_1, \ldots, x_n\} = \text{span}\,\{u_1, \ldots, u_n\}$. Hence $z_{n+1} \neq 0$. Since $u_{n+1} = z_{n+1}/\|z_{n+1}\|$, we obtain $\|u_{n+1}\| = 1$. Further,

$$\langle z_{n+1}, u_j\rangle = \langle x_{n+1}, u_j\rangle - \langle y_{n+1}, u_j\rangle = 0 \quad \text{for } j = 1, \ldots, n.$$

Now $\langle u_{n+1}, u_j\rangle = \langle z_{n+1}, u_j\rangle/\|z_{n+1}\| = 0$ for $j = 1, \ldots, n$. Thus $\{u_1, \ldots, u_{n+1}\}$ is an orthonormal subset of X. Also,

$$\text{span}\,\{u_1, \ldots, u_{n+1}\} = \text{span}\,\{x_1, \ldots, x_n, u_{n+1}\} = \text{span}\,\{x_1, \ldots, x_{n+1}\}.$$

By mathematical induction on n, the proof is complete. □

Examples 2.18 (i) Let $X := \ell^2$ with the usual inner product. For $n \in \mathbb{N}$, let $x_n := e_1 + \cdots + e_n$. Then $\{x_1, x_2, \ldots\}$ is a denumerable linearly independent subset of X. As in the Gram–Schmidt orthonormalization (Theorem 2.17), we obtain $y_1 = 0$, $y_n = e_1 + \cdots + e_{n-1}$ for all $n \geq 2$, and so $u_n = z_n/\|z_n\|_2 = (x_n - y_n)/\|z_n\|_2 = e_n$ for all $n \in \mathbb{N}$.

(ii) **(Legendre polynomials and Hermite polynomials)** Let $X := L^2([-1, 1])$ with the usual inner product. Let $x_0(t) := 1$, and for $n \in \mathbb{N}$, let $x_n(t) := t^n$, $t \in [-1, 1]$. Then $\{x_0, x_1, x_2, \ldots\}$ is a denumerable linearly independent subset of X. As in the Gram–Schmidt orthonormalization, we obtain $\{p_0, p_1, p_2, \ldots\}$. Here p_n is a real-valued polynomial of degree n for $n = 0, 1, 2, \ldots$. These polynomials are called the **Legendre polynomials**. We calculate the first three Legendre polynomials to illustrate this procedure.

Let $z_0(t) := x_0(t) = 1$ for $t \in [-1, 1]$, and so $\|z_0\|_2^2 = \int_{-1}^{1} ds = 2$. Hence $p_0 = z_0/\|z_0\| = 1/\sqrt{2}$. Next, let

$$y_1(t) := \langle x_1, p_0\rangle p_0(t) = \left(\int_{-1}^{1} \frac{s}{\sqrt{2}}\, ds\right) \frac{1}{\sqrt{2}} = 0 \quad \text{for } t \in [-1, 1].$$

Then $z_1(t) := x_1(t) - y_1(t) = t$ for $t \in [-1, 1]$, and $\|z_1\|_2^2 = \int_{-1}^{1} s^2 ds = 2/3$. Hence $p_1(t) = z_1(t)/\|z_1\|_2 = \sqrt{3}\,t/\sqrt{2}$ for $t \in [-1, 1]$. Further, let

$$\begin{aligned} y_2(t) &:= \langle x_2, p_0\rangle p_0(t) + \langle x_2, p_1\rangle p_1(t) \\ &= \left(\int_{-1}^{1} \frac{s^2}{\sqrt{2}} ds\right) \frac{1}{\sqrt{2}} + \left(\int_{-1}^{1} \frac{\sqrt{3}}{\sqrt{2}} s^3\, ds\right) \frac{\sqrt{3}}{\sqrt{2}}\, t \\ &= \frac{1}{3} \quad \text{for } t \in [-1, 1]. \end{aligned}$$

Then $z_2(t) := x_2(t) - y_2(t) = t^2 - (1/3), t \in [-1, 1]$, and $\|z_2\|_2^2 = \int_{-1}^{1}(s^2 - \frac{1}{3})^2 ds$ $= 8/45$. Hence $p_2(t) = z_2(t)/\|z_2\| = (\sqrt{10}/4)(3t^2 - 1)$ for $t \in [-1, 1]$.

Next, let $X := L^2(\mathbb{R})$ with the usual inner product. Let $w(t) := e^{-t^2/2}$ for $t \in \mathbb{R}$. Then $w \in L^2(\mathbb{R})$. Let $x_0(t) := w(t)$, and for $n \in \mathbb{N}$, let $x_n(t) := t^n w(t)$ for $t \in \mathbb{R}$. Clearly, $\{x_0, x_1, x_2, \ldots\}$ is a denumerable linearly independent subset of X. The Gram–Schmidt orthonormalization yields $\{u_0, u_1, u_2, \ldots\}$, where $u_n := q_n w$, q_n being a real-valued polynomial of degree n for $n = 0, 1, 2, \ldots$. These polynomials are called the **Hermite polynomials**. $\diamondsuit$

We now begin a deeper study of orthonormal sets. While dealing with countable orthonormal sets, we use the notation $\sum_n$, which stands for a sum over $n \in \mathbb{N}$, or over $n \in \{1, \ldots, n_0\}$, where n_0 is a fixed positive integer. We begin with a generalization of the Schwarz inequality stated in Proposition 2.13(i).

Proposition 2.19 (Bessel inequality) *Let $\{u_1, u_2, \ldots\}$ be a countable orthonormal set in an inner product space X, and let $x \in X$. Then*

$$\sum_n |\langle x, u_n\rangle|^2 \le \|x\|^2,$$

where equality holds if and only if $x = \sum_n \langle x, u_n\rangle u_n$.

Proof For $m \in \mathbb{N}$, let $s_m := \sum_{n=1}^{m} \langle x, u_n\rangle u_n$. Then

$$\|x - s_m\|^2 = \langle x - s_m, x - s_m\rangle = \langle x, x\rangle - \langle x, s_m\rangle - \langle s_m, x\rangle + \langle s_m, s_m\rangle.$$

It is easy to see that $\langle x, s_m\rangle = \langle s_m, x\rangle = \sum_{n=1}^{m} |\langle x, u_n\rangle|^2$. Also,

$$\langle s_m, s_m\rangle = \sum_{n=1}^{m} |\langle x, u_n\rangle|^2$$

since $\{u_1, \ldots, u_m\}$ is an orthonormal set. Hence

$$0 \le \|x - s_m\|^2 = \|x\|^2 - \sum_{n=1}^{m} |\langle x, u_n\rangle|^2.$$

Clearly, $\sum_{n=1}^{m} |\langle x, u_n\rangle|^2 = \|x\|^2$ if and only if $x = s_m = \sum_{n=1}^{m} \langle x, u_n\rangle u_n$.

In case the set $\{u_1, u_2, \ldots\}$ is infinite, we let $m \to \infty$ in the above inequality, and obtain

$$\sum_{n=1}^{\infty} |\langle x, u_n\rangle|^2 \leq \|x\|^2.$$

It is clear that equality holds here, that is, $\sum_{n=1}^{m} |\langle x, u_n\rangle|^2 \to \|x\|^2$ as $m \to \infty$ if and only if $\|x - s_m\|^2 \to 0$, that is,

$$x = \lim_{m\to\infty} s_m = \sum_{n=1}^{\infty} \langle x, u_n\rangle u_n,$$

as desired. □

Let $x, y \in X$ with $y \neq 0$, and let $u := y/\|y\|$. By the Bessel inequality, we obtain $|\langle x, u\rangle| \leq \|x\|$, that is, $|\langle x, y\rangle| \leq \|x\|\,\|y\|$, which is the Schwarz inequality stated in Proposition 2.13(i).

Corollary 2.20 *Let $\{u_\alpha : \alpha \in A\}$ be an orthonormal subset of an inner product space X. Let $x \in X$ and $E_x := \{u_\alpha : \alpha \in A$ and $\langle x, u_\alpha\rangle \neq 0\}$. Then E_x is a countable subset of X. If $E_x := \{u_1, u_2, \ldots\}$, then $(\langle x, u_n\rangle) \in \ell^2$, and consequently $\langle x, u_n\rangle \to 0$.*

Proof Fix $j \in \mathbb{N}$ and let $E_j := \{u_\alpha : \alpha \in A$ and $\|x\| < j|\langle x, u_\alpha\rangle|\}$. Suppose E_j contains m_j distinct elements $u_{\alpha_1}, \ldots, u_{\alpha_{m_j}}$. Then

$$m_j\|x\|^2 < j^2 \sum_{n=1}^{m_j} |\langle x, u_{\alpha_n}\rangle|^2 \leq j^2\|x\|^2$$

by the Bessel inequality (Proposition 2.19). This shows that $m_j < j^2$. Thus the set E_j contains less than j^2 elements. Since $E_x = \bigcup_{j=1}^{\infty} E_j$, we see that E_x is a countable subset of X. Also, if $E_x := \{u_1, u_2, \ldots\}$, then

$$\sum_{n} |\langle x, u_n\rangle|^2 \leq \|x\|^2 < \infty,$$

again by the Bessel inequality. Hence the sequence $(\langle x, u_n\rangle) \in \ell^2$, and so its nth term $\langle x, u_n\rangle$ tends to zero. □

Remarks 2.21 (i) For an example of an uncountable orthonormal set, see Exercise 2.21.

(ii) Let $\{u_1, u_2, \ldots\}$ be a denumerable orthonormal subset of an inner product space X. If $x \in X$, and we let $k_n := \langle x, u_n\rangle$, then Corollary 2.20 says that (k_n) is a sequence in ℓ^2. Conversely, if (k_n) is a sequence in ℓ^2, there must exist $x \in X$

such that $\langle x, u_n \rangle = k_n$ for all $n \in \mathbb{N}$? In general, the answer is in the negative. For example, let $X := c_{00}$ with the inner product given by $\langle x, y \rangle := \sum_{j=1}^{\infty} x(j)\overline{y(j)}$ for $x, y \in X$. Let $u_n := e_n$ and let $k_n := 1/n$ for $n \in \mathbb{N}$. Clearly, (k_n) is a sequence in ℓ^2, but there is no $x \in c_{00}$ such that $\langle x, e_n \rangle = k_n$, that is, $x(n) = 1/n$, for all $n \in \mathbb{N}$. We shall show that if the inner product space X is complete, then the answer to our question is in the affirmative. (See Theorem 2.29, the Riesz–Fischer theorem.)

(iii) We give an example of a denumerable orthonormal subset for which a strict inequality may hold in the Bessel inequality. Let $X := \ell^2$, $v_n := e_{2n}$ and let $x_0 := (1, 1/2, 1/3, \ldots) \in X$. Then

$$\sum_{n}^{\infty} |\langle x_0, v_n \rangle|^2 = \sum_{n=1}^{\infty} \frac{1}{(2n)^2} = \frac{1}{4} \sum_{n=1}^{\infty} \frac{1}{n^2} < \sum_{n=1}^{\infty} \frac{1}{n^2} = \|x_0\|^2.$$

Thus the Bessel inequality for $x_0 \in \ell^2$ is strict. ◊

Let $\{u_\alpha : \alpha \in A\}$ be an orthonormal subset of an inner product space X. For $x \in X$, let $E_x := \{u_\alpha : \alpha \in A \text{ and } \langle x, u_\alpha \rangle \neq 0\} = \{u_1, u_2, \ldots\}$, say. How can we ensure that equality holds in the Bessel inequality for every $x \in X$? Remark 2.21(iii) shows that $\{u_\alpha : \alpha \in A\}$ may not be large enough. Hence we may attempt to enlarge it as much as possible. This leads us to the following concept. For simplicity, we shall write $\{u_\alpha\}$ for the set $\{u_\alpha : \alpha \in A\}$.

A **maximal orthonormal subset** of an inner product space X is an orthonormal subset $\{u_\alpha\}$ of X such that if $\{u_\alpha\}$ is contained in an orthonormal subset E of X, then in fact $E = \{u_\alpha\}$. Suppose $\{u_\alpha\}$ and E are orthonormal subsets of X such that $\{u_\alpha\} \subset E$. Since $E \subset \{u_\alpha\} \cup \{u_\alpha\}^\perp$, it follows that $\{u_\alpha\}$ is a maximal orthonormal subset of X if and only if $\{u_\alpha\}^\perp = \{0\}$. For example, $\{e_n : n \in \mathbb{N}\}$ is a maximal orthonormal subset of ℓ^2 since $\{e_n : n \in \mathbb{N}\}^\perp = \{x \in \ell^2 : x(n) = 0 \text{ for all } n \in \mathbb{N}\} = \{0\}$.

Proposition 2.22 *Let X be an inner product space over $\mathbb{K}$. Suppose E_0 is an orthonormal subset of X. Then there is a maximal orthonormal subset $\{u_\alpha\}$ of X containing E_0.*

If X is nonzero, then there is a maximal orthonormal subset of X.

Proof Let $\mathcal{E} := \{E : E_0 \subset E \subset X \text{ and } E \text{ is orthonormal}\}$. Then $\mathcal{E}$ is nonempty since $E_0 \in \mathcal{E}$. The inclusion relation $\subset$ is a partial order on $\mathcal{E}$. Also, for a totally ordered subfamily $\mathcal{F}$ of $\mathcal{E}$, let E denote the union of all $F \in \mathcal{F}$. Clearly, $E_0 \subset E \subset X$. Let us show that E is orthonormal. It is clear that $\|x\| = 1$ for all $x \in E$. Let $x_1, x_2 \in E$ with $x_1 \neq x_2$. Then there are $F_1, F_2 \in \mathcal{F}$ such that $x_1 \in F_1$ and $x_2 \in F_2$. Since $\mathcal{F}$ is totally ordered, either $F_1 \subset F_2$ or $F_2 \subset F_1$. Hence either $x_1, x_2 \in F_1$, or $x_1, x_2 \in F_2$. Since F_1 and F_2 are both orthogonal subsets of X, we see that x_1 and x_2 are orthogonal. Thus E is an upper bound for $\mathcal{F}$ in $\mathcal{E}$. By the Zorn lemma stated in Sect. 1.1, $\mathcal{E}$ contains a maximal element $\{u_\alpha\}$. Clearly $\{u_\alpha\}$ is a maximal orthonormal subset of X containing E_0.

If $X \neq \{0\}$, consider $x_0 \in X$ such that $\|x_0\| = 1$. Letting $E_0 := \{x_0\}$, we conclude that there is a maximal orthonormal subset of X. □

The above proposition and its proof should be compared with Proposition 1.1 and Corollary 1.2.

Let us now return to the question posed earlier. Let $\{u_\alpha\}$ be a maximal orthonormal subset of an inner product space X. If $x \in X$, and the set $\{u_\alpha : \langle x, u_\alpha\rangle \neq 0\}$ is countable, say, $\{u_1, u_2, \ldots\}$, then must the equality $\sum_n |\langle x, u_n\rangle|^2 = \|x\|^2$ hold? The answer is still in the negative, as the following example shows. Let $x_0 := (1, 1/2, 1/3, \ldots) \in \ell^2$, and let X denote the span of $\{x_0, e_2, e_3, \ldots\}$ in ℓ^2. Then X is an inner product space, and the subset $\{e_2, e_3, \ldots\}$ of X is orthonormal. Let $x \in X$ be orthogonal to $\{e_2, e_3, \ldots\}$. There is $m \geq 2$ such that $x = k_0x_0 + k_2e_2 + \cdots + k_me_m$ with $k_0, k_2, \ldots, k_m$ in $\mathbb{K}$. Since $x \perp e_{m+1}$, we obtain $k_0 = 0$, and since $x \perp e_n$, we obtain $k_n = 0$ for $n = 2, \ldots, m$. Hence $x = 0$. Thus $\{e_2, e_3, \ldots\}^\perp = \{0\}$, and so $\{e_2, e_3, \ldots\}$ is a maximal othonormal subset of X. However,

$$\sum_{n=2}^{\infty} |\langle x_0, e_n\rangle|^2 = \sum_{n=2}^{\infty} \frac{1}{n^2} < \sum_{n=1}^{\infty} \frac{1}{n^2} = \|x_0\|^2.$$

Thus the Bessel inequality for $x_0 \in X$ is strict. However, if the inner product space X is complete, and $\{u_\alpha\}$ is a maximal orthonormal subset of X, then we shall show that the answer to our question is in the affirmative. (See the Parseval formula stated in Theorem 2.31(v).) We shall also find several necessary and sufficient conditions for an orthonormal subset of a complete inner product space to be maximal.

2.3 Banach Spaces

Banach recognized the importance of the completeness of a normed space and wrote a series of path breaking papers in the late 1920s. In his epoch-making book [3] written in 1932, he modestly refers to complete normed spaces as spaces of type B, since they were already being named after him.

A normed space $(X, \|\cdot\|)$ over $\mathbb{K}$ is called a **Banach space** if X is complete in the metric $d(x, y) := \|x - y\|$ for $x, y \in X$. Before giving several examples of Banach spaces, we prove a necessary and sufficient condition for a normed space X to be a Banach space by utilizing the additive structure of X. Let (x_n) be a sequence in X. For $m \in \mathbb{N}$, let $s_m := \sum_{n=1}^{m} x_n$. We say that the series $\sum_{n=1}^{\infty} x_n$ is **summable** or **convergent** in X if the sequence (s_m) of its **partial sums** converges in X. If (s_m) converges to s in X, then we write $s := \sum_{n=1}^{\infty} x_n$, and say that s is the **sum** of the series. A series $\sum_{n=1}^{\infty} x_n$ is said to be **absolutely summable** if the series $\sum_{n=1}^{\infty} \|x_n\|$ of nonnegative terms is summable. Recall that every absolutely summable series of terms in $\mathbb{K}$ is summable in $\mathbb{K}$. In fact, we have the following characterization of a Banach space in terms of summable series.

Theorem 2.23 *A normed space X is a Banach space if and only if every absolutely summable series of terms in X is summable in X.*

Proof Let X be a Banach space, (x_n) be a sequence in X, and suppose the series $\sum_{n=1}^{\infty} x_n$ is absolutely summable. Let (s_m) and (t_m) denote the sequences of the partial sums of the series $\sum_{n=1}^{\infty} x_n$ and $\sum_{n=1}^{\infty} \|x_n\|$, respectively. Then for all $m > p$ in $\mathbb{N}$,

$$\|s_m - s_p\| = \|x_{p+1} + \cdots + x_m\| \le \|x_{p+1}\| + \cdots + \|x_m\| = t_m - t_p = |t_m - t_p|.$$

Since (t_m) is a Cauchy sequence in $\mathbb{R}$, it follows that (s_m) is a Cauchy sequence in X. Since X is a complete metric space, the sequence (s_m) converges in X, that is, the series $\sum_{n=1}^{\infty} x_n$ is summable in X.

Conversely, suppose every absolutely summable series of terms in X is summable in X. Let (s_m) be a Cauchy sequence in X. Let $m_1 \in \mathbb{N}$ be such that $\|s_m - s_{m_1}\| < 1$ for all $m \ge m_1$. Find $m_2, m_3, \ldots$ inductively such that $m_n < m_{n+1}$ and $\|s_m - s_{m_n}\| < 1/n^2$ for all $m \ge m_n$. Let $x_n := s_{m_{n+1}} - s_{m_n}$ for $n \in \mathbb{N}$. Since $\sum_{n=1}^{\infty} \|x_n\| < \sum_{n=1}^{\infty} 1/n^2$, the series $\sum_{n=1}^{\infty} x_n$ is absolutely summable. By assumption, it is summable in X. Let $s \in X$ be its sum. Since

$$s_{m_{n+1}} - s_{m_1} = (s_{m_{n+1}} - s_{m_n}) + \cdots + (s_{m_2} - s_{m_1}) = x_n + \cdots + x_1 \text{ for } n \in \mathbb{N},$$

the subsequence (s_{m_n}) of the sequence (s_m) converges to $s + s_{m_1}$ in X. Hence the Cauchy sequence (s_m) itself converges in X. So X is a Banach space. □

We shall use the above result on several occasions in this section.

Examples 2.24 Since a subset of a complete metric space X is complete if and only if it is closed in X, it follows that a subspace Y of a Banach space X is itself a Banach space if and only if Y is closed in X.

(i) We have already seen in Lemma 2.8 that if a normed space X is finite dimensional, then it is complete, that is, it is a Banach space. In particular, the space $\mathbb{K}^n$ along with any of the norms $\|\cdot\|_1$, $\|\cdot\|_2$, $\|\cdot\|_\infty$ is a Banach space.

(ii) Let $X := \ell^p$, where $p \in \{1, 2, \infty\}$. We show that X is a Banach space. Let (x_n) be a Cauchy sequence in X, and let $\epsilon > 0$. There is $n_0 \in \mathbb{N}$ such that $\|x_n - x_m\|_p < \epsilon$ for all $n, m \ge n_0$. Fix $j \in \mathbb{N}$. Since $|x_n(j) - x_m(j)| \le \|x_n - x_m\|_p$ for all $n, m \in \mathbb{N}$, we see that $(x_n(j))$ is a Cauchy sequence in $\mathbb{K}$. Since $\mathbb{K}$ is complete, there is $k_j \in \mathbb{K}$ such that $x_n(j) \to k_j$ in $\mathbb{K}$. Define $x := (k_1, k_2, \ldots)$. We claim that $x \in X$ and $x_n \to x$ in X.

First let $p \in \{1, 2\}$. Let $i \in \mathbb{N}$. Keep $n \ge n_0$ fixed, and let $m \to \infty$ in the inequality

$$\left(\sum_{j=1}^{i} |x_n(j) - x_m(j)|^p\right)^{1/p} < \epsilon, \quad \text{where } m \ge n_0.$$

Then $\sum_{j=1}^{i} |x_n(j) - x(j)|^p \le \epsilon^p$ for all $n \ge n_0$ and for all $i \in \mathbb{N}$. As a result, $\sum_{j=1}^{\infty} |x_n(j) - x(j)|^p \le \epsilon^p$ for all $n \ge n_0$. In particular, $\sum_{j=1}^{\infty} |x_{n_0}(j) - x(j)|^p \le \epsilon^p$. Thus $x_{n_0} - x$, and hence $x = (x - x_{n_0}) + x_{n_0}$ are in ℓ^p, and $\|x_n - x\|_p \le \epsilon$ for all $n \ge n_0$. This shows that $x_n \to x$ in ℓ^p.

Next, let $p := \infty$. Let $j \in \mathbb{N}$. Keep $n \geq n_0$ fixed, and let $m \to \infty$ in the inequality

$$|x_n(j) - x_m(j)| < \epsilon, \quad \text{where } m \geq n_0.$$

Then $|x_n(j) - x(j)| \leq \epsilon$ for all $n \geq n_0$ and $j \in \mathbb{N}$. In particular, $|x_{n_0}(j) - x(j)| \leq \epsilon$ for all $j \in \mathbb{N}$. Thus $x_{n_0} - x$ and hence $x = (x - x_{n_0}) + x_{n_0}$ are in ℓ^∞, and $\|x_n - x\|_\infty \leq \epsilon$ for all $n \geq n_0$. This shows that $x_n \to x$ in ℓ^∞.

We have seen in Remark 2.9 that the subspace c_{00} of ℓ^∞ is not closed in ℓ^∞. Hence c_{00} is not a Banach space.

(iii) Let T be a nonempty set, and let $B(T)$ denote the set of all $\mathbb{K}$-valued bounded functions on T along with the sup norm $\|\cdot\|_\infty$. If we replace $j \in \mathbb{N}$ by $t \in T$ in the proof of the completeness of ℓ^∞ given in (ii) above, we obtain a proof of the completeness of $B(T)$. Thus $B(T)$ is a Banach space.

Let now T be a metric space. Since convergence in $B(T)$ is the uniform convergence on T, and since a uniform limit of a sequence of continuous functions is a continuous function by Proposition 1.15, we see that $C(T)$ is a closed subspace of $B(T)$. Hence $C(T)$ is a Banach space. However, $C_{00}(T)$ may not be a closed subspace of $C(T)$. For example, let $T := [0, \infty)$, and let $x(t) := e^{-t}$ for $t \in T$. For $n \in \mathbb{N}$, define

$$x_n(t) := \begin{cases} e^{-t} & \text{if } t \in [0, n], \\ e^{-n}(n + 1 - t) & \text{if } n < t < n + 1, \\ 0 & \text{if } t \geq n + 1. \end{cases}$$

Then $x \in C([0, \infty))$, $x_n \in C_{00}([0, \infty))$ for all $n \in \mathbb{N}$ and $x_n \to x$ in $C([0, \infty))$, but $x \notin C_{00}([0, \infty))$. Hence $C_{00}([0, \infty))$ is not a Banach space.

Now let $T := [a, b]$, and let us consider the subspace of $C([a, b])$ consisting of all $\mathbb{K}$-valued **continuously differentiable** functions on $[a, b]$, namely,

$$C^1([a, b]) := \{x \in C([a, b]) : x \text{ is differentiable on } [a, b] \text{ and } x' \in C([a, b])\}.$$

Since $C^1([a, b])$ contains all polynomial functions, Theorem 1.16 of Weierstrass shows that $C^1([a, b])$ is dense in $C([a, b])$. Also, $C^1([a, b]) \neq C([a, b])$. Hence $C^1([a, b])$ is not closed in $C([a, b])$, and so it is not a Banach space.

Now for $x \in C^1([a, b])$, let $\|x\|_{1,\infty} := \max\{\|x\|_\infty, \|x'\|_\infty\}$. Clearly, $\|\cdot\|_{1,\infty}$ is a norm on $C^1([a, b])$. We show that $(C^1([a, b]), \|\cdot\|_{1,\infty})$ is a Banach space. Let (x_n) be a Cauchy sequence in $C^1([a, b])$. Since (x_n) and (x_n') are Cauchy sequences in the Banach space $(C([a, b]), \|\cdot\|_\infty)$, there are $x, y \in C([a, b])$ such that $\|x_n - x\|_\infty \to 0$ and $\|x_n' - y\|_\infty \to 0$. By a well-known result in Real Analysis ([25, Theorem 7.17]), x is differentiable on $[a, b]$ and $x' = y$. Thus $x_n \to x$ in $C^1([a, b])$. In Exercise 2.29(i), subspaces of k times continuously differentiable functions on $[a, b]$ are treated, where $k \in \mathbb{N}$.

(iv) Let $E := \mathbb{R}$ or $[a, b]$, and let $X := L^p(E)$, where $p \in \{1, 2, \infty\}$. We show that X is a Banach space.

First let $p := 1, 2$. Recalling Theorem 2.23, we consider an absolutely summable series $\sum_{n=1}^{\infty} x_n$ of terms in $L^p(E)$, and show that it is summable in $L^p(E)$. Let $\alpha := \sum_{n=1}^{\infty} \|x_n\|_p$. For each $m \in \mathbb{N}$, define

$$y_m(t) := \sum_{n=1}^{m} |x_n(t)|, \ t \in E.$$

By the Minkowski inequality for functions, $\|y_m\|_p \le \sum_{n=1}^{m} \|x_n\|_p \le \alpha$, and so $\int_E y_m(t)^p dm(t) \le \alpha^p$ for all $m \in \mathbb{N}$. For $t \in E$, $(y_m(t))$ is a monotonically increasing sequence in $[0, \infty)$, and so $y_m(t) \to y(t)$. Now y is a measurable function on E, and by the monotone convergence theorem (Theorem 1.18(i)),

$$\int_E y_m(t)^p dm(t) \to \int_E y(t)^p dm(t), \quad \text{so that} \int_E y(t)^p dm(t) \le \alpha^p.$$

Thus $y \in L^p(E)$. Hence the set $A := \{t \in E : y(t) = \infty\}$ has Lebesgue measure zero. For each $t \in E \setminus A$, we see that $\sum_{n=1}^{\infty} x_n(t)$ is an absolutely summable series of terms in $\mathbb{R}$, and so it is summable in $\mathbb{R}$. Fix $m \in \mathbb{N}$. Define

$$s_m(t) := \sum_{n=1}^{m} x_n(t) \quad \text{and} \quad s(t) := \sum_{n=1}^{\infty} x_n(t) \quad \text{for } t \in E \setminus A,$$

and $s_m(t) := 0$, $s(t) := 0$ for $t \in A$. Then s_m and s are measurable functions, $|s_m(t)| \le y_m(t) \le y(t)$ and $|s(t)| \le y(t)$ for $t \in E$. Hence $s \in L^p(E)$, and

$$|(s_m - s)(t)|^p \le (|s_m(t)| + |s(t)|)^p \le 2^p y(t)^p \quad \text{for } t \in E.$$

Since $s_m(t) - s(t) \to 0$ as $m \to \infty$ for all $t \in E$, and since y^p is integrable on E, we obtain $\|s_m - s\|_p \to 0$ by the dominated convergence theorem (Theorem 1.18(ii)). Thus the series $\sum_{n=1}^{\infty} x_n$ is summable in $L^p(E)$.

Next, let $p := \infty$. Let (x_n) be a Cauchy sequence in $L^\infty(E)$. For each $i \in \mathbb{N}$, there is $n_i \in \mathbb{N}$ such that $\|x_n - x_m\|_\infty < 1/i$ for all $m, n \ge n_i$, and so there is a subset A_i of E having Lebesgue measure zero such that $|x_n(t) - x_m(t)| < 1/i$ for all $t \in E \setminus A_i$. Define $A := \bigcup_{i=1}^{\infty} A_i$. Then A has Lebesgue measure zero, and the sequence $(x_n(t))$ is Cauchy for each $t \in E \setminus A$. Since $\mathbb{K}$ is complete, there is $k_t \in \mathbb{K}$ such that $x_n(t) \to k_t$ for $t \in E \setminus A$. Define $x(t) := k_t$ for $t \in E \setminus A$ and $x(t) := 0$ for $t \in A$. Then x is a measurable function on E. Keep $n \ge n_i$ fixed, and let $m \to \infty$, in the inequality $\|x_n - x_m\|_\infty < 1/i$ which holds for all $m, n \ge n_i$, and obtain

$$|x_n(t) - x(t)| \le \frac{1}{i} \quad \text{for all } t \in E \setminus A.$$

In particular, $|x_{n_i}(t) - x(t)| \leq 1/i$ for all $t \in E \setminus A$. Thus $x_{n_i} - x$ and hence $x = (x - x_{n_i}) + x_{n_i}$ are in $L^\infty(E)$, and $\|x_n - x\|_\infty \leq 1/i$ for all $n \geq n_i$. This shows that $x_n \to x$ in $L^\infty(E)$.

Consider the space $C([0, 2])$ consisting of all $\mathbb{K}$-valued continuous functions defined on $[0, 2]$. Clearly, $C([0, 2])$ can be considered as a closed subspace of $L^\infty([0, 2])$. But $C([0, 2])$, considered as a subspace of $L^1([0, 2])$, is not closed in $L^1([0, 2])$.

To see this, it suffices to consider

$$x(t) := \begin{cases} 0 & \text{if } t \in [0, 1), \\ 1 & \text{if } t \in [1, 2] \end{cases} \quad \text{and} \quad x_n(t) := \begin{cases} 0 & \text{if } t \in [0, (n-1)/n], \\ n(t-1)+1 & \text{if } t \in ((n-1)/n, 1), \\ 1 & \text{if } t \in [1, 2]. \end{cases}$$

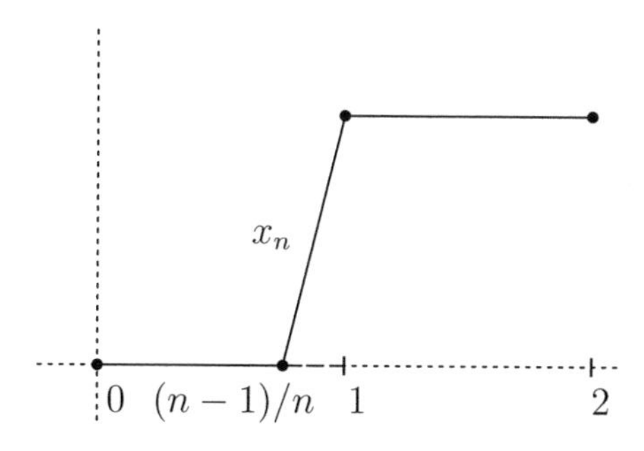

Fig. 2.4 Graph of $x_n \in C([0, 2])$

(See Fig. 2.4.) Note that $x \in L^1([0, 2])$, $x_n \in C([0, 2])$ and $\|x_n - x\|_1 = 1/2n$ for each $n \in \mathbb{N}$, but $x \notin C([0, 2])$. Hence $C([0, 2])$ with the norm $\|\cdot\|_1$ is not a Banach space. Similarly, since $\|x_n - x\|_2^2 = 1/3n$ for each $n \in \mathbb{N}$, $C([0, 2])$ with the norm $\|\cdot\|_2$ is not a Banach space.

(v) Let X denote the linear space of all $\mathbb{K}$-valued absolutely continuous functions on a $[a, b]$. By the fundamental theorem of calculus for Lebesgue integration (Theorem 1.23), every $x \in X$ is differentiable a.e. on $[a, b]$, and its derivative x' is integrable on $[a, b]$. For $x \in X$, define $\|x\|_{1,1} := \|x\|_1 + \|x'\|_1$. Clearly, $\|\cdot\|_{1,1}$ is a norm on X. We show that X is a Banach space. Let (x_n) be a Cauchy sequence in X. Since (x_n) and (x_n') are Cauchy sequences in the Banach space $L^1([a, b])$, there are $x, y \in L^1([a, b])$ such that $\|x_n - x\|_1 \to 0$ and $\|x_n' - y\|_1 \to 0$. By Corollary 1.24,

$$|x_n(a) - x_m(a)| \leq \frac{1}{b-a}\|x_n - x_m\|_1 + \|x_n' - x_m'\|_1 \quad \text{for all } n, m \in \mathbb{N}.$$

It follows that $(x_n(a))$ is a Cauchy sequence in $\mathbb{K}$. Let $k_0 \in \mathbb{K}$ be such that $x_n(a) \to k_0$. Define

$$z(t) := k_0 + \int_{[a,t]} y\,dm \quad \text{for } t \in [a, b].$$

Again, by Theorem 1.23, z is an absolutely continuous function, and $z' = y$ a.e. on $[a, b]$. By Corollary 1.24,

$$\|x_n - z\|_1 \le (b-a)\big(|x_n(a) - z(a)| + \|x_n' - z'\|_1\big) = (b-a)\big(|x_n(a) - k_0| + \|x_n' - y\|_1\big)$$

for $n \in \mathbb{N}$. Hence $\|x_n - z\|_1 \to 0$, $x = z$ a.e. on $[a, b]$, and $x' = z' = y$ a.e. on $[a, b]$. Thus we see that if (x_n) is a sequence of absolutely continuous functions on $[a, b]$ such that the sequences (x_n) and (x_n') are convergent in $L^1([a, b])$, then there is an absolutely continuous function x on $[a, b]$ such that $\|x_n - x\|_1 \to 0$ and $\|x_n' - x'\|_1 \to 0$, that is, $\|x_n - x\|_{1,1} \to 0$, as desired. This is an analogue of a well-known result in Real Analysis about uniform convergence and differentiation. The space X is known as the **Sobolev space** of order (1,1) on $[a, b]$, and it is denoted by $W^{1,1}([a, b])$. In Exercise 2.29(ii), Sobolev spaces of order $(k, 1)$ on $[a, b]$ are treated, where $k \in \mathbb{N}$. ◊

We now consider the completeness of quotient spaces and product spaces.

Theorem 2.25 (i) *Let X be a normed space, and let Y be a closed subspace of X. Then X is a Banach space if and only if Y is a Banach space in the induced norm and X/Y is a Banach space in the quotient norm.*

(ii) *Let $X_1, \ldots, X_m$ be normed spaces, and let $X := X_1 \times \cdots \times X_m$. Then X is a Banach space in the product norm introduced in Proposition 2.5(ii) if and only if $X_1, \ldots, X_m$ are Banach spaces.*

Proof (i) Suppose X is a Banach space. Then Y is a Banach space since it is closed in X. To show X/Y is a Banach space, consider a series $\sum_{n=1}^{\infty}(x_n + Y)$ of terms in X/Y which is absolutely summable. By the definition of the quotient norm $|||\cdot|||$, there is $y_n \in Y$ such that

$$\|x_n + y_n\| < |||x_n + Y||| + \frac{1}{n^2} \quad \text{for each } n \in \mathbb{N}.$$

Now the series $\sum_{n=1}^{\infty}(x_n + y_n)$ of terms in X is absolutely summable. Since X is a Banach space, the series $\sum_{n=1}^{\infty}(x_n + y_n)$ is summable in X by Theorem 2.23. Let $s \in X$ be its sum, that is, let $\sum_{n=1}^{m}(x_n + y_n) \to s$ as $m \to \infty$. By Proposition 2.5(i), we obtain

$$\sum_{n=1}^{m}(x_n + Y) = \sum_{n=1}^{m}(x_n + y_n + Y) = \Big(\sum_{n=1}^{m} x_n + y_n\Big) + Y \to s + Y \quad \text{as } m \to \infty,$$

that is, $\sum_{n=1}^{\infty}(x_n + Y) = s + Y \in X/Y$. It follows, again by Theorem 2.23, that X/Y is a Banach space.

Conversely, suppose Y and X/Y are Banach spaces. Consider a Cauchy sequence (x_n) in X. Since

$$|||(x_n + Y) - (x_m + Y)||| = |||(x_n - x_m) + Y||| \le \|x_n - x_m\| \quad \text{for all } n, m \in \mathbb{N},$$

we see that $(x_n + Y)$ is a Cauchy sequence in X/Y. Let $x_n + Y \to x + Y$ in X/Y. By Proposition 2.5(i), there is a sequence (y_n) in Y such that $x_n + y_n \to x$ in X. In particular, $(x_n + y_n)$ is a Cauchy sequence in X. Since $y_n - y_m = (x_n + y_n) - (x_m + y_m) - (x_n - x_m)$, we obtain

$$\|y_n - y_m\| \le \|(x_n + y_n) - (x_m + y_m)\| + \|x_n - x_m\| \quad \text{for all } n, m \in \mathbb{N}.$$

It follows that (y_n) is a Cauchy sequence in Y. Let $y_n \to y$ in Y. Then $x_n = (x_n + y_n) - y_n \to x - y$ in X. This shows that X is a Banach space.

(ii) Recall that the product norm on $X := X_1 \times \cdots \times X_m$ introduced in Proposition 2.5(ii) is given by $\|x\| := \|x(1)\|_1 + \cdots + \|x(m)\|_m$, $x := (x(1), \ldots, x(m)) \in X$.

Suppose $X_1, \ldots, X_m$ are Banach spaces. Let (x_n) be a Cauchy sequence in X. Fix $j \in \{1, \ldots, m\}$. Then $(x_n(j))$ is a Cauchy sequence in X_j. Since X_j is complete, there is $x(j) \in X_j$ with $x_n(j) \to x(j)$ in X_j. Then $x_n \to x := (x(1), \ldots, x(m))$ in X by Proposition 2.5(ii). Hence X is a Banach space.

Conversely, suppose X is a Banach space. Fix $j \in \{1, \ldots, m\}$. Let $(x_n(j))$ be a Cauchy sequence in X_j, and define

$$x_n := \big(0, \ldots, 0, x_n(j), 0, \ldots, 0\big) \quad \text{for } n \in \mathbb{N}.$$

Clearly, (x_n) is a Cauchy sequence in X. Let $x_n \to x$ in X. Then $x_n(j) \to x(j)$ in X_j again by Proposition 2.5(ii). Hence X_j is a Banach space. □

Before concluding this section, we prove a striking result regarding the number of elements in a (Hamel) basis for a Banach space. It will lead us to modify the concept of a basis for a Banach space.

Theorem 2.26 *A Banach space cannot have a denumerable (Hamel) basis.*

Proof Assume for a moment that a Banach space X has a denumerable (Hamel) basis $\{x_1, x_2, \ldots\}$. Let $m \in \mathbb{N}$, and define

$$Y_m := \operatorname{span}\{x_1, \ldots, x_m\}.$$

By Lemma 2.8, the subspace Y_m of X is complete, and hence it is closed in X, that is, $D_m := X \setminus Y_m$ is open in X. Also, since the set $\{x_1, \ldots, x_{m+1}\}$ is linearly independent, $x_{m+1} \notin Y_m$. Thus $D_m \neq \emptyset$. In fact, D_m is dense in X. To see this, consider $y \in X \setminus D_m$, that is, $y \in Y_m$. Let $n \in \mathbb{N}$ and $z_n := y + (x_{m+1}/n)$. (See Fig. 2.5.) Then $z_n \notin Y_m$ since $y \in Y_m$, $x_{m+1} \notin Y_m$ and Y_m is a subspace of X. Thus (z_n) is a sequence in D_m, and clearly $z_n \to y$.

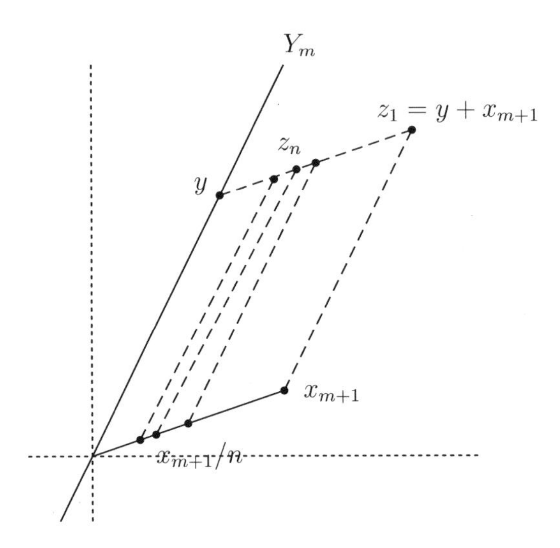

Fig. 2.5 Choice of the sequence (z_n) in the proof of Theorem 2.26

Since X is complete, the intersection $\bigcap_{m=1}^{\infty} D_m$ of dense open subsets of X is dense in X by Theorem 1.10 of Baire. On the other hand, if $x \in X$, then there is $m \in \mathbb{N}$ such that $x \in Y_m$ since span $\{x_1, x_2, \ldots\} = X$, and so $x \notin D_m$. Thus $\bigcap_{m=1}^{\infty} D_m = \emptyset$. But $\emptyset$ cannot be dense in X. □

The above result says that if a linear space X has a denumerable basis, then no norm on X makes it a Banach space. For example, there is no norm on c_{00} which makes it a Banach space. In other words, a Banach space X has either a finite basis or an uncountable basis. Thus any basis for an infinite dimensional Banach space is unwieldy! Let us then relax the requirement of a (Hamel) basis that every element of a linear space must be a *finite* linear combination of the basis elements, and admit denumerable linear combinations of elements of a Banach space X. This leads us to the following concept.

Let X be a nonzero Banach space. A countable subset $\{x_1, x_2, \ldots\}$ of X is called a **Schauder basis** for X if $\|x_n\| = 1$ for each $n \in \mathbb{N}$, and if for every $x \in X$, there are unique $k_1, k_2, \ldots$ in $\mathbb{K}$ such that $x = \sum_n k_n x_n$.

Examples 2.27 (i) Suppose $(X, \|\cdot\|)$ is a finite dimensional normed space, and $\{x_1, \ldots, x_n\}$ is a (Hamel) basis for X. Then X is a Banach space (Lemma 2.8), and $\{x_1/\|x_1\|, \ldots, x_n/\|x_n\|\}$ is a Schauder basis for X. In particular, the standard basis for $\mathbb{K}^n$ is a Schauder basis for $(\mathbb{K}^n, \|\cdot\|_p)$, where $p \in \{1, 2, \infty\}$.

(ii) Let $X := \ell^p$, where $p \in \{1, 2\}$. Clearly, $\{e_1, e_2, \ldots\}$ is not a (Hamel) basis for X. However, it is a Schauder basis for X. This can be seen as follows. For $x := (x(1), x(2), \ldots) \in X$, and for $m \in \mathbb{N}$, let

$$s_m(x) := (x(1), \ldots, x(m), 0, 0, \ldots) = \sum_{n=1}^{m} x(n)e_n.$$

Then $\|s_m(x) - x\|_p^p = \sum_{n=m+1}^{\infty} |x(n)|^p \to 0$ as $m \to \infty$. This says $x = \sum_{n=1}^{\infty} x(n)e_n$. Further, suppose $x \in X$ and there are $k_1, k_2, \ldots$ in $\mathbb{K}$ such that $x = \sum_{n=1}^{\infty} k_n e_n$. Then clearly $k_n = x(n)$ for each $n \in \mathbb{N}$. The Schauder basis $\{e_1, e_2, \ldots\}$ is known as the **standard Schauder basis** for X.

(iii) A Schauder basis for $C([0, 1])$ can be constructed as follows. For $t \in \mathbb{R}$, let $y_0(t) := t,\ y_1(t) := 1 - t$,

$$y_2(t) := \begin{cases} 2t & \text{if } 0 \le t \le 1/2, \\ 2 - 2t & \text{if } 1/2 < t \le 1, \\ 0 & \text{if } t < 0 \text{ or } t > 1, \end{cases}$$

and define $y_{2^n+j}(t) := y_2(2^n t - j + 1)$ for $n \in \mathbb{N}$ and $j = 1, \ldots, 2^n$. Let x_n denote the restriction of y_n to $[0, 1]$ for $n = 0, 1, 2, \ldots$. Then $\{x_0, x_1, x_2, \ldots\}$ can be shown to be a Schauder basis for $C([0, 1])$. Observe that each x_n is a nonnegative piecewise linear continuous function defined on $\mathbb{R}$; for $n \ge 2$, x_n is known as a **saw-tooth function** because of its shape.

(iv) Let $p \in \{1, 2\}$. A well-known Schauder basis for $L^p([0, 1])$ consists of the functions $\{x_1, x_2, \ldots\}$ defined as follows. For $t \in [0, 1]$, let $x_1(t) := 1$,

$$x_2(t) := \begin{cases} 1 & \text{if } 0 \le t \le 1/2, \\ -1 & \text{if } 1/2 < t \le 1, \end{cases}$$

and for $n \in \mathbb{N}$ and $j = 1, \ldots, 2^n$, define

$$x_{2^n+j}(t) := \begin{cases} 2^{n/p} & \text{if } (2j-2)/2^{n+1} \le t \le (2j-1)/2^{n+1}, \\ -2^{n/p} & \text{if } (2j-1)/2^{n+1} < t \le 2j/2^{n+1}, \\ 0 & \text{otherwise.} \end{cases}$$

Observe that each x_n is a step function defined on $[0, 1]$. The set $\{x_1, x_2, x_3, \ldots\}$ is known as a **Haar system**.

(v) It will follow from Theorem 2.31 of Sect. 2.4 that if the norm on a nonzero Banach space X is induced by an inner product, and if $\{u_1, u_2, \ldots\}$ is a maximal orthonormal subset of X, then it is a Schauder basis for X. (Compare Remark 2.34(iv).) In particular, the Haar system defined in (iv) above is a Schauder basis for $L^2([0, 1])$ of this kind. $\diamond$

If a nonzero Banach space X has a Schauder basis, then X must be separable. To see this, let $\{x_1, x_2, \ldots\}$ be a Schauder basis for X. Then

$$\{k_1 x_1 + \cdots + k_m x_m : m \in \mathbb{N},\ k_j \in \mathbb{K} \text{ with } \operatorname{Re} k_j, \operatorname{Im} k_j \in \mathbb{Q} \text{ for } j = 1, \ldots, m\}$$

is a countable dense subset of X. All classical separable Banach spaces are known to have Schauder bases. Whether *every* nonzero separable Banach space has a Schauder basis was an open question for a long time. It was settled in the negative by Enflo ([9]) in 1973 by finding a closed subspace of c_0 having no Schauder basis.

2.4 Hilbert Spaces

Let X be an inner product space over $\mathbb{K}$. We have seen that the inner product induces a norm, and hence a metric, on X (Proposition 2.14). If X is complete in this metric, then X is called a **Hilbert space**. Equivalently, if X is Banach space with a norm that is induced by an inner product on X, then X is called a Hilbert space.

Examples 2.28 Let G be a subspace of a Hilbert space H. It is clear that G is itself a Hilbert space if and only if G is closed in H. From our considerations in Examples 2.1, 2.12 and 2.24, we obtain the following.

(i) $\mathbb{K}^n$ with the usual inner product is a Hilbert space, and so is every subspace of $\mathbb{K}^n$. The induced norm is denoted by $\|\cdot\|_2$.

(ii) ℓ^2 with the usual inner product is a Hilbert space. The induced norm is denoted by $\|\cdot\|_2$. The subspace c_{00} of ℓ^2 is not a Hilbert space.

(iii) $L^2(E)$, where $E := \mathbb{R}$ or $[a, b]$, with the usual inner product is a Hilbert space. The induced norm is denoted by $\|\cdot\|_2$. The subspace $C(E)$ of $L^2(E)$ is not a Hilbert space. (Compare Example 2.24(iv).)

(iv) Let H denote the linear space of all absolutely continuous functions on $[a, b]$ whose derivatives are in $L^2([a, b])$. For $x, y \in H$, define

$$\langle x, y \rangle_{1,2} := \int_a^b x(t)\overline{y}(t)dt + \int_a^b x'(t)\overline{y}'(t)dm(t).$$

Here the functions $x\overline{y}$ and $x'\overline{y}'$ are integrable on $[a, b]$ by the Schwarz inequality for functions (Lemma 1.25(i)). It is easy to see that $\langle \cdot, \cdot \rangle$ is an inner product on H. The induced norm on H is given by

$$\|x\|_{1,2} := \left(\|x\|_2^2 + \|x'\|_2^2\right)^{1/2} \quad \text{for } x \in H.$$

We show that H is a Hilbert space. Let (x_n) be a Cauchy sequence in H. Since (x_n) and (x_n') are Cauchy sequences in the Banach space $L^2([a, b])$, they converge in $L^2([a, b])$, and so in $L^1([a, b])$. We have seen in Example 2.24(v) that there is an absolutely continuous function x on $[a, b]$ such that $\|x_n - x\|_1 \to 0$ and $\|x_n' - x'\|_1 \to 0$. It follows that $\|x_n - x\|_2 \to 0$, x' is in $L^2([a, b])$, and $\|x_n' - x'\|_2 \to 0$, that is, $\|x_n - x\|_{1,2} \to 0$, as desired. The space H is known as the **Sobolev space** of order (1,2) on $[a, b]$, and it is denoted by $W^{1,2}([a, b])$. In Exercise 2.29(iii), Sobolev spaces of order $(k, 2)$ on $[a, b]$ are treated, where $k \in \mathbb{N}$.

The subspace $C^1([-1, 1])$ of H consisting of all continuously differentiable functions on $[-1, 1]$ is not closed in H. To see this, it suffices to consider

$$x(t) := |t| \quad \text{for } t \in [-1, 1] \text{ and } x_n(t) := \begin{cases} -t & \text{if } t \in [-1, -1/n], \\ (n^2t^2 + 1)/2n & \text{if } t \in (-1/n, 1/n), \\ t & \text{if } t \in [1/n, 1]. \end{cases}$$

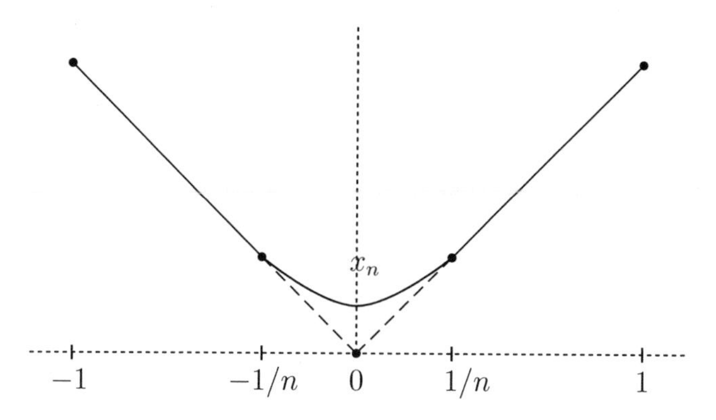

Fig. 2.6 Graph of $x_n \in C^1([-1, 1])$

(See Fig. 2.6.) Now $x \in H$, but $x \notin C^1([-1, 1])$. Also, $x_n \in C^1([-1, 1])$ and $\|x_n - x\|_{1,2}^2 = \|x_n - x\|_2^2 + \|x_n' - x'\|_2^2 = (1/10n^3) + (2/3n) \to 0$, since

$$x'(t) = \begin{cases} -1 & \text{if } t \in [-1, 0), \\ 1 & \text{if } t \in (0, 1], \end{cases} \quad \text{while} \quad x_n'(t) = \begin{cases} -1 & \text{if } t \in [-1, -1/n], \\ nt & \text{if } t \in (-1/n, 1/n), \\ 1 & \text{if } t \in [1/n, 1]. \end{cases}$$

Hence $C^1([-1, 1])$ with the induced norm $\|\cdot\|_{1,2}$ is not a Hilbert space. $\diamond$

Exercises 2.28 and 2.40 treat product and quotient spaces of Hilbert spaces.

In Sect. 2.2, we had introduced the concept of orthogonality of a subset of an inner product space X and had begun an in-depth study of orthonormal subsets of X. In Remark 2.21(ii), we had asked whether a series $\sum_n k_n u_n$, where $\{u_1, u_2, \ldots\}$ is a denumerable orthonormal set in X and (k_n) is a sequence in ℓ^2, would always converge in X. Also, after Proposition 2.22, we had asked whether a maximal orthonormal subset of an inner product space would ensure equality in the Bessel inequality. We shall now answer these questions in the affirmative if X is complete, that is, if it is a Hilbert space.

Theorem 2.29 (Riesz–Fischer theorem, 1907) *Let H be a nonzero Hilbert space, let $\{u_1, u_2, \ldots\}$ be a countable orthonormal set in H and let $k_1, k_2, \ldots$ be in $\mathbb{K}$. The series $\sum_n k_n u_n$ of terms in H is summable in H if and only if $\sum_n |k_n|^2 < \infty$. In that case, if $y := \sum_n k_n u_n$, then*

$$\langle y, u_n \rangle = k_n \textit{ for all } n \in \mathbb{N} \quad \textit{and} \quad \|y\|^2 = \sum_n |k_n|^2.$$

Proof Suppose $\sum_n |k_n|^2 < \infty$. Define $s_m := \sum_{n=1}^m k_n u_n$ for $m \in \mathbb{N}$. Then by the orthonormality of the set $\{u_1, u_2, \ldots\}$, we obtain

$$\|s_m - s_p\|^2 = \langle s_m - s_p, s_m - s_p\rangle = \sum_{n=p+1}^{m} |k_n|^2 \quad \text{for all } m > p \text{ in } \mathbb{N}.$$

It follows that (s_m) is a Cauchy sequence in H, and hence it converges in H, that is, the series $\sum_n k_n u_n$ is summable in H.

Conversely, suppose the series $\sum_n k_n u_n$ of terms in H is summable in H, and its sum is $y := \sum_n k_n u_n$. By the continuity and the linearity of the inner product $\langle \cdot\,, \cdot\rangle$ in the first variable, we see that $\langle y, u_n\rangle = k_n$ for all $n \in \mathbb{N}$. Also, by the equality criterion in the Bessel inequality (Proposition 2.19), we obtain $\|y\|^2 = \sum_n |\langle y, u_n\rangle|^2 = \sum_n |k_n|^2$. □

Corollary 2.30 *Let H be a nonzero Hilbert space and $\{u_\alpha\}$ be an orthonormal set in H. Let $x \in H$, and $\{u_\alpha : \langle x, u_\alpha\rangle \neq 0\} := \{u_1, u_2, \ldots\}$. Then $\sum_n \langle x, u_n\rangle u_n$ is summable in H. If y is its sum, then $(x - y) \perp u_\alpha$ for all α.*

Proof Let $k_n := \langle x, u_n\rangle$ for $n \in \mathbb{N}$. By the Bessel inequality, $\sum_n |k_n|^2 = \sum_n |\langle x, u_n\rangle|^2 \leq \|x\|^2 < \infty$. The Riesz–Fischer theorem (Theorem 2.29) shows that $\sum_n \langle x, u_n\rangle u_n$ is summable in H. Let $y := \sum_n \langle x, u_n\rangle u_n$. Then $\langle y, u_n\rangle = \langle x, u_n\rangle$ for all $n \in \mathbb{N}$. Also, if $u_\alpha \notin \{u_1, u_2, \ldots\}$, then $\langle x, u_\alpha\rangle = 0$, and also $\langle y, u_\alpha\rangle = 0$ since $u_\alpha \perp u_n$ for all $n \in \mathbb{N}$. Thus $\langle x, u_\alpha\rangle = \langle y, u_\alpha\rangle$ for all α. □

Theorem 2.31 *Let H be a nonzero Hilbert space, and let $\{u_\alpha\}$ be an orthonormal subset of H. Then the following conditions are equivalent.*

(i) span $\{u_\alpha\}$ *is dense in H.*
(ii) $\{u_\alpha\}^\perp = \{0\}$.
(iii) $\{u_\alpha\}$ *is a maximal orthonormal subset of H.*
(iv) **(Fourier expansion)** *For $x \in H$, and $\{u_1, u_2, \ldots\} := \{u_\alpha : \langle x, u_\alpha\rangle \neq 0\}$,*

$$x = \sum_n \langle x, u_n\rangle u_n.$$

(v) **(Parseval formula)** *For $x \in H$, and $\{u_1, u_2, \ldots\} := \{u_\alpha : \langle x, u_\alpha\rangle \neq 0\}$,*

$$\|x\|^2 = \sum_n |\langle x, u_n\rangle|^2.$$

Proof (i) $\Longrightarrow$ (ii): Let $x \in \{u_\alpha\}^\perp$. By (i), there is a sequence (x_m) in span $\{u_\alpha\}$ such that $x_m \to x$. Then $\langle x, x_m\rangle = 0$ for all $m \in \mathbb{N}$ and $\langle x_m, x\rangle \to \langle x, x\rangle$. Hence $\langle x, x\rangle = 0$, that is, $x = 0$.

(ii) $\Longleftrightarrow$ (iii): We have seen this right after defining a maximal orthonormal set in Sect. 2.2.

(iii) $\Longrightarrow$ (iv): Let $x \in H$, and let $\{u_1, u_2, \ldots\} := \{u_\alpha : \langle x, u_\alpha\rangle \neq 0\}$. By Corollary 2.30 of the Riesz–Fischer theorem, there is $y \in H$ such that $y = \sum_n \langle x, u_n\rangle u_n$ and $\langle x - y, u_\alpha\rangle = 0$ for all α, that is, $(x - y) \in \{u_\alpha\}^\perp$. By condition (ii), $x - y = 0$, that is, $x = y = \sum_n \langle x, u_n\rangle u_n$.

(iv) $\Longleftrightarrow$ (v): This is the criterion for equality to hold in the Bessel inequality proved in Proposition 2.19.

(iv) $\Longrightarrow$ (i): Let $x \in H$, and let $\{u_1, u_2, \ldots\} := \{u_\alpha : \langle x, u_\alpha \rangle \neq 0\}$. By (iv), $x = \sum_n \langle x, u_n \rangle u_n$, that is, $s_m \to x$, where $s_m := \sum_{n=1}^m \langle x, u_n \rangle u_n \in \operatorname{span}\{u_\alpha\}$ for each $m \in \mathbb{N}$. Thus $\operatorname{span}\{u_\alpha\}$ is dense in H. □

Let H be a nonzero Hilbert space. In the above proof, the completeness of H is used only in the implication '(iii) $\Longrightarrow$ (iv)' via the Riesz–Fischer theorem. It follows that the implications '(i) $\Longrightarrow$ (ii) $\Longleftrightarrow$ (iii)' and the implications '(iv) $\Longleftrightarrow$ (v) $\Longrightarrow$ (i)' hold in any inner product space. In fact, the implication '(i) $\Longrightarrow$ (iv)' also holds in any inner product space. (See Exercise 4.19(i).)

Conditions (i) and (ii) of the above result, therefore, can be used to check whether a given orthonormal subset of an inner product space X is in fact a maximal orthonormal subset of X. On the other hand, if a maximal orthonormal subset of H is presented to us, then the equivalent conditions (iv) and (v) yield useful information about each $x \in H$. An orthonormal subset of H satisfying one of the equivalent conditions given in Theorem 2.31 is called an **orthonormal basis** for H. Proposition 2.22 ensures that H has an orthonormal basis; in fact every orthonormal subset of H can be extended to an orthonormal basis for H. By Theorem 2.31, a countable orthonormal basis for H is just an orthogonal Schauder basis for H.

Examples 2.32 (i) Let $H := \ell^2$. Then the orthonormal subset $\{e_1, e_2, \ldots\}$ of ℓ^2 is an orthonormal basis for ℓ^2 since $\{e_1, e_2, \ldots\}^\perp = \{0\}$.

Next, let $u_n := (e_{2n-1} + e_{2n})/\sqrt{2}$ and $v_n := (e_{2n-1} - e_{2n})/\sqrt{2}$ for $n \in \mathbb{N}$, and let $E := \{u_1, v_1, u_2, v_2, \ldots\}$. Clearly, E is an orthonormal subset of ℓ^2. Also, if $x \in E^\perp$, then $x(2n-1) + x(2n) = 0$ and $x(2n-1) - x(2n) = 0$ for all $n \in \mathbb{N}$, so that $x(n) = 0$ for all $n \in \mathbb{N}$, that is, $E^\perp = \{0\}$. Hence E is also an orthonormal basis for ℓ^2.

(ii) Let $\mathbb{K} := \mathbb{C}$ and $H := L^2([-\pi, \pi])$. For $k \in \mathbb{Z}$, let

$$u_k(t) := \frac{e^{ikt}}{\sqrt{2\pi}}, \quad t \in [-\pi, \pi].$$

It is easy to check that $\{u_k : k \in \mathbb{Z}\}$ is an orthonormal subset of $L^2([-\pi, \pi])$. For $x \in L^2([-\pi, \pi])$,

$$\langle x, u_k \rangle = \frac{1}{\sqrt{2\pi}} \int_{-\pi}^{\pi} x(t) e^{-ikt}\, dm(t) = \sqrt{2\pi}\hat{x}(k),$$

where $\hat{x}(k)$ is the kth Fourier coefficient of x introduced in Sect. 1.4. We proved in Theorem 1.28(ii) that if $x \in L^1([-\pi, \pi])$ and $\hat{x}(k) = 0$ for all $k \in \mathbb{Z}$, then $x = 0$ a.e. on $[-\pi, \pi]$. Thus if $x \in L^2([-\pi, \pi])$ and $\langle x, u_k \rangle = 0$ for all $k \in \mathbb{Z}$, then x is the zero element of $L^2([-\pi, \pi])$, that is, $\{u_k : k \in \mathbb{Z}\}^\perp = \{0\}$. Hence $\{u_k : k \in \mathbb{Z}\}$ is an orthonormal basis for $L^2([-\pi, \pi])$. This is one of the most useful orthonormal bases, well known from the classical times. In fact, much of the Hilbert space theory is modelled after this example.

The result $\langle x, u_k \rangle \to 0$ proved in Corollary 2.20 simply restates the Riemann–Lebesgue lemma (Theorem 1.28(i)) for square-integrable functions.

The Riesz–Fischer theorem (Theorem 2.29) implies that if (c_k) is a square-summable sequence of complex numbers, then there is a square-integrable function x on $[-\pi, \pi]$ whose kth Fourier coefficient is equal to c_k for all $k \in \mathbb{Z}$.

The Fourier expansion of Theorem 2.31 says that for every $x \in L^2([-\pi, \pi])$,

$$x = \sqrt{2\pi} \sum_{k=-\infty}^{\infty} \hat{x}(k) u_k,$$

that is, the sequence (s_n) of partial sums of the Fourier series of a square-integrable function x on $[-\pi, \pi]$ converges to the function in the mean square:

$$\int_{-\pi}^{\pi} |x(t) - s_n(t)|^2 \, dm(t) \to 0, \quad \text{where } s_n := \sqrt{2\pi} \sum_{k=-n}^{n} \hat{x}(k) u_k.$$

The Parseval formula of Theorem 2.31 says that for every $x \in L^2([-\pi, \pi])$,

$$\frac{1}{2\pi} \int_{-\pi}^{\pi} |x(t)|^2 \, dm(t) = \sum_{k=-\infty}^{\infty} |\hat{x}(k)|^2.$$

(iii) Let $H := L^2([-1, 1])$. For $n = 0, 1, 2, \ldots$, let u_n denote the Legendre polynomial of degree n introduced in Example 2.18(ii). Since $u_0, u_1, u_2, \ldots$ are obtained by the Gram–Schmidt orthonormalization, $\{u_0, u_1, u_2, \ldots\}$ is an orthonormal subset of $L^2([-1, 1])$. Since span $\{u_0, \ldots, u_n\}$ is the set of all polynomials of degree at most n for each $n \in \mathbb{N}$, span $\{u_0, u_1, u_2, \ldots\}$ is the set of all polynomials on $[-1, 1]$. Let $x \in L^2([-1, 1])$ and $\epsilon > 0$. Since the set of all continuous functions on $[-1, 1]$ is a dense subset of $L^2([-1, 1])$ (Proposition 1.26(ii)), there is $y \in C([-1, 1])$ such that $\|x - y\|_2 < \epsilon$. By Theorem 1.16 of Weierstrass, there is $p \in$ span $\{u_0, u_1, u_2, \ldots\}$ such that $|y(t) - p(t)| < \epsilon$ for all $t \in [-1, 1]$. Hence

$$\|y - p\|_2^2 = \int_{-1}^{1} |y(t) - p(t)|^2 \, dt \leq 2\epsilon^2.$$

Thus $\|x - p\|_2 \leq \|x - y\|_2 + \|y - p\|_2 < (1 + \sqrt{2})\epsilon$. This shows that span $\{u_0, u_1, u_2, \ldots\}$ is dense in $L^2([-1, 1])$. Hence $\{u_0, u_1, u_2, \ldots\}$ is an orthonormal basis for $L^2([-1, 1])$ by condition (i) of Theorem 2.31.

Next, for $n = 0, 1, 2, \ldots$, let $u_n := p_n w$, where p_n denotes the nth Hermite polynomial introduced in Example 2.18(ii), and $w(t) := e^{-t^2/2}$ for $t \in \mathbb{R}$. Since $u_0, u_1, u_2, \ldots$ are obtained by the Gram–Schmidt orthonormalization, $\{u_0, u_1, u_2, \ldots\}$ is an orthonormal subset of $L^2(\mathbb{R})$. In fact, it is an orthonormal basis for $L^2(\mathbb{R})$. We refer the reader to [5, pp. 95–97] for an ingenious proof of this fact due to von Neumann. $\Diamond$

We now prove an interesting result which shows that $\mathbb{K}^n$ and ℓ^2 serve as models for separable Hilbert spaces. Its proof utilizes the entire machinery we have built so far to study orthonormal sets.

Theorem 2.33 *Let H be a nonzero Hilbert space over $\mathbb{K}$ with inner product $\langle \cdot\,, \cdot \rangle_H$. Then the following conditions are equivalent.*

(i) *H is separable.*
(ii) *H has a countable orthonormal basis.*
(iii) *There is a linear map F from H onto $\mathbb{K}^n$, or from H onto ℓ^2, such that $\langle F(x), F(y)\rangle = \langle x, y\rangle_H$ for all $x, y \in H$, where $\langle \cdot\,, \cdot\rangle$ denotes the usual inner product on $\mathbb{K}^n$, or on ℓ^2.*

Proof (i) $\Longrightarrow$ (ii): Let $\{y_1, y_2, \ldots\}$ be a countable dense subset of H. Let x_1 be the first nonzero element among $y_1, y_2, \ldots$. Next, let x_2 be the first element among $y_2, y_3, \ldots$ which does not belong to span $\{x_1\}$. After defining $x_1, \ldots, x_n$ in a similar manner, let x_{n+1} be the first element among $y_{n+1}, y_{n+2}, \ldots$ which does not belong to span $\{x_1, \ldots, x_n\}$ for $n \in \mathbb{N}$. Then the set $\{x_1, x_2, \ldots\}$ is linearly independent and span $\{x_1, x_2, \ldots\} =$ span $\{y_1, y_2, \ldots\}$.

By the Gram–Schmidt orthonormalization (Theorem 2.17), we can find an orthonormal set $\{u_1, u_2, \ldots\}$ such that span $\{u_1, u_2, \ldots\} =$ span $\{x_1, x_2, \ldots\}$. It follows that span $\{u_1, u_2, \ldots\}$ is dense in H. Hence $\{u_1, u_2, \ldots\}$ is a countable orthonormal basis for H by condition (i) of Theorem 2.31.

(ii) $\Longrightarrow$ (iii): Let $\{u_1, u_2, \ldots\}$ be a countable orthonormal basis for H. Define $F(x) := (\langle x, u_1\rangle_H, \langle x, u_2\rangle_H, \ldots)$ for $x \in H$. If $\{u_1, u_2, \ldots\}$ is a finite set having n elements, then F is a map from H to $\mathbb{K}^n$. If $\{u_1, u_2, \ldots\}$ is a denumerable set, then F is a map from H to ℓ^2 since $\sum_{n=1}^{\infty} |\langle x, u_n\rangle_H|^2 \le \|x\|_H^2$ by the Bessel inequality. The linearity of the inner product $\langle \cdot\,, \cdot\rangle_H$ in the first variable shows the map F is linear. Let $x, y \in H$. By condition (iv) of Theorem 2.31, we obtain $x = \sum_n \langle x, u_n\rangle_H u_n$ and $y = \sum_n \langle y, u_n\rangle_H u_n$. The continuity of the inner product $\langle \cdot\,, \cdot\rangle_H$ in both variables (Proposition 2.14), its linearity in the first variable, its conjugate linearity in the second variable and the orthogonality of the set $\{u_1, u_2, \ldots\}$ show that

$$\langle x, y\rangle_H = \sum_n \langle x, u_n\rangle_H \overline{\langle y, u_n\rangle_H} = \langle F(x), F(y)\rangle.$$

Finally, F maps H onto $\mathbb{K}^n$ or onto ℓ^2 by Theorem 2.29 of Riesz and Fischer.

(iii) $\Longrightarrow$ (i): We have seen in Example 1.7(i) and (ii) that $\mathbb{K}^n$ and ℓ^2 are separable. Also, $\|x\|_H^2 = \langle x, x\rangle_H = \langle F(x), F(x)\rangle = \|F(x)\|_2^2$ for all $x \in H$, so that $\|x - y\|_H = \|F(x) - F(y)\|_2$ for all $x, y \in H$, that is, F is an isometry from H onto $\mathbb{K}^n$, or onto ℓ^2. Hence H is separable. $\square$

Remarks 2.34 (i) The equality

$$\langle x, y\rangle_H = \sum_n \langle x, u_n\rangle_H \overline{\langle y, u_n\rangle_H} \quad \text{for } x, y \in H$$

used in the proof of '(ii) $\Longrightarrow$ (iii)' given above is called the **Parseval identity**.

(ii) Let H and G be Hilbert spaces with inner products $\langle \cdot\,, \cdot \rangle_H$ and $\langle \cdot\,, \cdot \rangle_G$ respectively. If F is a linear map from H onto G such that $\langle F(x), F(y)\rangle_G = \langle x, y\rangle_H$ for all $x, y \in H$, then F is called a **Hilbert space isomorphism**. Clearly, a Hilbert space isomorphism is an isometry since $\|F(x)\|_G^2 = \langle F(x), F(x)\rangle_G = \langle x, x\rangle_H = \|x\|_H^2$ for all $x \in H$.

The map F in condition (iii) of Theorem 2.33 is an example of a Hilbert space isomorphism. Here is another example. Let $\mathbb{K} := \mathbb{C}$ and $H := L^2([-\pi, \pi])$. Consider the orthonormal basis $\{u_k : k \in \mathbb{Z}\}$ for $L^2([-\pi, \pi])$ given in Example 2.32(ii). Let G denote the linear space of all doubly infinite square-summable sequences $(\ldots, y(-2), y(-1), y(0), y(1), y(2), \ldots)$ in $\mathbb{C}$ along with the inner product

$$\langle y_1, y_2\rangle := \sum_{k\in\mathbb{Z}} y_1(k)\overline{y_2(k)} \quad \text{for } y_1, y_2 \in G.$$

Then G is a Hilbert space. As before, let $\hat{x}(k)$ be the kth Fourier coefficient of $x \in L^2([-\pi, \pi])$ for each $k \in \mathbb{Z}$. The map $F : L^2([-\pi, \pi]) \to G$ defined by

$$F(x) := \sqrt{2\pi}(\ldots, \hat{x}(-2), \hat{x}(-1), \hat{x}(0), \hat{x}(1), \hat{x}(2), \ldots), \quad x \in L^2([-\pi, \pi]),$$

is a Hilbert space isomorphism. Since the index set $\mathbb{Z}$ is denumerable, there is a Hilbert space isomorphism from G onto ℓ^2 as well.

(iii) It follows from condition (iv) of Theorem 2.31 that a countable orthonormal basis for H is, in particular, a Schauder basis for H. Hence by Theorem 2.33, every separable Hilbert space has a Schauder basis. Note that a separable Banach space need not have a Schauder basis, as we have pointed out at the end of Sect. 2.3. $\Diamond$

Let X be a normed space, and let Y be a closed subspace of X. Suppose there is a closed subspace Z of X such that $X = Y + Z$ and $Y \cap Z = \{0\}$. Then we write $X = Y \oplus Z$, and say that X is the **direct sum** of the subspaces Y and Z. In this case, for every $x \in X$, there are unique $y \in Y$ and $z \in Z$ such that $x = y + z$. To see the uniqueness of $y \in Y$ and $z \in Z$, let $x = y_0 + z_0$, where $y_0 \in Y$ and $z_0 \in Z$. Then $y - y_0 = z_0 - z$, where $(y - y_0) \in Y$ and $(z_0 - z) \in Z$. Since $Y \cap Z = \{0\}$, we obtain $y_0 = y$ and $z_0 = z$.

We shall now show that for every closed subspace G of a Hilbert space H, there is a closed subspace $\widetilde{G}$ of H such that $H = G \oplus \widetilde{G}$. In fact, $\widetilde{G}$ can be so chosen that $y \perp z$ for every $y \in G$ and $z \in \widetilde{G}$.

Theorem 2.35 (Projection theorem) *Let H be a Hilbert space, and let G be a closed subspace of H. Then $H = G \oplus G^\perp$. Consequently, $G^{\perp\perp} = G$.*

Proof If $G := \{0\}$, then $G^\perp = H$ and there is nothing to prove. Let then $G \neq \{0\}$. Since G is a closed subspace of H, it is a Hilbert space. Let $\{v_\alpha\}$ be an orthonormal basis for G. Let $x \in H$. By Corollary 2.20, $\{v_\alpha : \langle x, v_\alpha\rangle \neq 0\}$ is a countable orthonormal subset of H, say $\{v_1, v_2, \ldots\}$. By Corollary 2.30 of the Riesz–Fischer theorem, the series $\sum_n \langle x, v_n\rangle v_n$ is summable in H, and if $y := \sum_n \langle x, v_n\rangle v_n$, then

$(x - y) \perp v_\alpha$ for each α. Since $\{v_\alpha\}$ is an orthonormal basis for G, condition (i) of Theorem 2.31 shows that span $\{v_\alpha\}$ is dense in G. Hence $(x - y) \in G^\perp$. Also, since each $v_n \in G$, and since G is a closed subspace, it follows that $y \in G$. Let $z := x - y$. Then $x = y + z$, where $y \in G$ and $z \in G^\perp$. Thus $H = G + G^\perp$. Since $G \cap G^\perp = \{0\}$, we obtain $H = G \oplus G^\perp$.

Next, we prove that $G^{\perp\perp} = G$. Let $y \in G$. Then $\langle y, z\rangle = 0$ for every $z \in G^\perp$, that is, $y \in (G^\perp)^\perp = G^{\perp\perp}$. Conversely, let $x \in G^{\perp\perp}$. Then there are $y \in G$ and $z \in G^\perp$ such that $x = y + z$. By what we have just proved, $y \in G^{\perp\perp}$. Hence $z = (x - y) \in G^{\perp\perp}$. Thus $z \in G^\perp \cap G^{\perp\perp}$. But $G^\perp \cap G^{\perp\perp} = \{0\}$, and we obtain $x = y$, showing that $x \in G$. □

We remark that the projection theorem does not hold in any incomplete inner product space. (See Exercise 4.18.)

Let G be a closed subspace of H. Then the closed subspace $G^\perp$ is called the **orthogonal complement** of G in H. Let $x \in H$. If $x = y + z$, where $y \in G$ and $z \in G^\perp$, then y is called the **orthogonal projection** of x on G. If G has a countable orthonormal basis $\{v_1, v_2, \ldots\}$, then the orthogonal projection of $x \in H$ on G is given by $y := \sum_n \langle x, v_n\rangle v_n$. We have already come across such orthogonal projections in the following context. Consider a countable linearly independent subset $\{x_1, x_2, \ldots\}$ of H. Let $G_1 := \{0\}$, and for $n \geq 2$, let $G_n :=$ span $\{x_1, \ldots, x_{n-1}\}$. Being finite dimensional, the subspace G_n is closed in H. In the Gram–Schmidt orthonormalization procedure given in Theorem 2.17, we have obtained $x_n = y_n + z_n$, where $y_n \in G_n$ and $z_n \in G_n^\perp$, that is, y_n is the orthogonal projection of x_n on G_n for each $n \in \mathbb{N}$.

Exercises

2.1. Let p be a seminorm on a linear space X, and let $U := \{x \in X : p(x) < 1\}$. Then the set U is **convex** (that is, $(1 - t)x + ty \in U$ whenever $x, y \in U$ and $t \in (0, 1)$), **absorbing** (that is, for every $x \in X$, there is $r > 0$ such that $(x/r) \in U$) and **balanced** (that is, $kx \in U$ whenever $x \in U$ and $k \in \mathbb{K}$ with $|k| \leq 1$). (Compare Lemma 4.8.)

2.2. Let $m \geq 2$, and let $p_1, \ldots, p_m$ be seminorms on a linear space X. Define

$$p(x) := \max\{p_1(x), \ldots, p_m(x)\} \quad \text{and} \quad q(x) := \min\{p_1(x), \ldots, p_m(x)\}$$

for $x \in X$. If one of $p_1, \ldots, p_m$ is a norm, then p is a norm on X. However, q may not be a seminorm even if each of $p_1, \ldots, p_m$ is a norm on X.

2.3. The closure of c_{00} in $(\ell^1, \|\cdot\|_1)$ is ℓ^1, the closure of c_{00} in $(\ell^2, \|\cdot\|_2)$ is ℓ^2 and the closure of c_{00} in $(\ell^\infty, \|\cdot\|_\infty)$ is c_0.

2.4. The inclusions $L^\infty([0, 1]) \subset L^2([0, 1]) \subset L^1([0, 1])$ are proper, but there is no inclusion relation among the normed spaces $L^1(\mathbb{R})$, $L^2(\mathbb{R})$, $L^\infty(\mathbb{R})$.

2.5. Let $X := C([0, 1])$, and let $\|\cdot\|_p$ be the induced norm on X as a subspace of $L^p([0, 1])$, $p \in \{1, 2, \infty\}$. Then the norm $\|\cdot\|_\infty$ is stronger than the norm $\|\cdot\|_2$, and the norm $\|\cdot\|_2$ is stronger than the norm $\|\cdot\|_1$, but any two of these norms are not equivalent.

2.6. Let $a < b$, and $X := C^1([a, b])$. For $x \in X$, let $\|x\|' := \max\{|x(a)|, \|x'\|_\infty\}$. The norm $\|\cdot\|'$ on X is equivalent to the norm $\|\cdot\|_{1,\infty}$ on X, which is stronger than but not equivalent to the norm $\|\cdot\|_\infty$ on X.

2.7. Let $(X, \|\cdot\|)$ be a normed space, Y be a linear space, and let F be a linear map from X onto Y. Define $q(y) := \inf\{\|x\| : x \in X \text{ and } F(x) = y\}$ for $y \in Y$. If $x \in X$ and $F(x) = y$, then $q(y) = \inf\{\|x + z\| : z \in Z(F)\}$. Consequently, q is a seminorm on Y. In fact, q is a norm on Y if and only if $Z(F)$ is a closed subset of X.

2.8. Suppose E is a compact subset of a normed space X. If there are $x_0 \in X$ and $r > 0$ such that $U(x_0, r) \subset E$, then X is finite dimensional.

2.9. Let $E := \{x \in \ell^2 : \sum_{j=1}^{\infty} |x(j)|^2 \le 1\}$. Then E is closed and bounded, but not compact. The **Hilbert cube** $C := \{x \in \ell^2 : j|x(j)| \le 1 \text{ for all } j \in \mathbb{N}\}$ is compact.

2.10. A normed space $(X, \|\cdot\|)$ is called **strictly convex** if $\|(x + y)/2\| < 1$ whenever $x, y \in X$, $\|x\| = 1 = \|y\|$ and $x \neq y$. If a norm $\|\cdot\|$ on a linear space X is induced by an inner product, then $(X, \|\cdot\|)$ is strictly convex.
If $p := 2$, then the normed spaces $(\ell^p, \|\cdot\|_p)$ and $(L^p([0, 1]), \|\cdot\|_p)$ are strictly convex, but if $p \in \{1, \infty\}$, then they are not strictly convex, and as a result, the norm $\|\cdot\|_p$ is not induced by any inner product.

2.11. Let X be an inner product space, and let $x, y \in X$. Then equality holds in the Schwarz inequality, that is, $|\langle x, y\rangle| = \|x\|\|y\|$ if and only if $\langle y, y\rangle x = \langle x, y\rangle y$, and equality holds in the triangle inequality, that is, $\|x + y\| = \|x\| + \|y\|$ if and only if $\|y\|x = \|x\|y$.

2.12. **(Parallelepiped law)** Let X be an inner product space. Then

$$\|x + y + z\|^2 + \|x + y - z\|^2 + \|x - y + z\|^2 + \|x - y - z\|^2 = 4(\|x\|^2 + \|y\|^2 + \|z\|^2)$$

for all $x, y, z \in X$, that is, the sum of the squares of the lengths of the diagonals of a parallelepiped equals the sum of the squares of the lengths of its edges. If $z := 0$, then we obtain the parallelogram law (Remark 2.15).

2.13. Let X be a linear space over $\mathbb{K}$. Suppose a map $\langle\cdot\,,\cdot\rangle : X \times X \to \mathbb{K}$ is linear in the first variable, conjugate-symmetric, and it satisfies $\langle x, x\rangle \ge 0$ for all $x \in X$. Then $Z := \{x \in X : \langle x, x\rangle = 0\}$ is a subspace of X. If we let $\langle\langle x + Z, y + Z\rangle\rangle := \langle x, y\rangle$ for $x + Z,\ y + Z$ in X/Z, then $\langle\langle\cdot\,,\cdot\rangle\rangle$ defines an inner product on X/Z. In particular, $|\langle x, y\rangle| \le \langle x, x\rangle^{1/2}\langle y, y\rangle^{1/2}$ for all $x, y \in X$. (Compare Remark 2.2 and Exercise 4.37.)

2.14. **(Polarization identity)** Let $(X, \langle\cdot\,,\cdot\rangle)$ be an inner product space. Then

$$4\langle x, y\rangle = \langle x + y, x + y\rangle - \langle x - y, x - y\rangle + i\langle x + iy, x + iy\rangle - i\langle x - iy, x - iy\rangle$$

for all $x, y \in X$. (Note: The inner product $\langle \cdot\,, \cdot\rangle$ is determined by the 'diagonal' subset $\{\langle x, x\rangle : x \in X\}$ of $X \times X$. Compare Exercise 4.40.)

2.15. Let $\langle \cdot\,, \cdot\rangle$ be an inner product on a linear space X. For nonzero $x, y \in X$, define the **angle** between x and y by

$$\theta_{x,y} := \arccos \frac{\mathrm{Re}\, \langle x, y\rangle}{\sqrt{\langle x, x\rangle\langle y, y\rangle}},$$

where $\arccos : [-1, 1] \to [0, \pi]$. Then $\theta_{x,y}$ satisfies the **law of cosines**

$$\|x\|^2 + \|y\|^2 - 2\|x\|\,\|y\| \cos \theta_{x,y} = \|x - y\|^2.$$

In particular, $\theta_{x,y} = \pi/2$ if and only if $\|x\|^2 + \|y\|^2 = \|x - y\|^2$.

2.16. Let X denote the linear space of all $m \times n$ matrices with entries in $\mathbb{K}$. For $M := [k_{i,j}]$ and $N := [\ell_{i,j}]$ in X, define $\langle M, N\rangle := \sum_{i=1}^{m} \sum_{j=1}^{n} k_{i,j}\overline{\ell_{i,j}}$. Then $\langle \cdot\,, \cdot\rangle$ is an inner product on X. The induced norm

$$\|M\|_F := \left(\sum_{i=1}^{m} \sum_{j=1}^{n} |k_{i,j}|^2 \right)^{1/2}$$

is known as the **Frobenius norm** of the matrix M. If $m = n$, and I_n denotes the $n \times n$ identity matrix, then $\|I_n\|_F = \sqrt{n}$.

2.17. Let $w := (w(1), w(2), \ldots) \in \ell^\infty$ be such that $w(j) > 0$ for all $j \in \mathbb{N}$, and define $\langle x, y\rangle_w := \sum_{j=1}^{\infty} w(j)x(j)\overline{y(j)}$ for $x, y \in \ell^2$. Then $\langle \cdot\,, \cdot\rangle_w$ is an inner product on ℓ^2. Also, the norm $\|\cdot\|_2$ on ℓ^2 is stronger than the norm $\|\cdot\|_w$ induced by the inner product $\langle \cdot\,, \cdot\rangle_w$. Let $v := (1/w(1), 1/w(2), \ldots)$. The norms $\|\cdot\|_2$ and $\|\cdot\|_w$ are equivalent if and only if $v \in \ell^\infty$, and then

$$\frac{1}{\|v\|_\infty} \|x\|_2^2 \le \|x\|_w^2 \le \|w\|_\infty \|x\|_2^2 \quad \text{for all } x \in \ell^2.$$

2.18. **(Helmert basis)** Let $m \ge 2$, and consider the usual inner product on $\mathbb{K}^m$. If $x_1 := e_1 + \cdots + e_m$, and $x_n := e_1 - e_n$ for $n = 2, \ldots, m$, then the Gram–Schmidt procedure yields the basis $\{u_1, \ldots, u_m\}$ for $\mathbb{K}^m$, where $u_1 := (e_1 + \cdots + e_m)/\sqrt{m}$, and $u_n := \big(e_1 + \cdots + e_{n-1} - (n-1)e_n\big)/\sqrt{(n-1)n}$ for $n = 2, \ldots, m$. (Note: This basis is useful in Multivariate Statistics.)

2.19. **(QR factorization)** Let A be an $m \times n$ matrix such that the n columns of A form a linearly independent subset of $\mathbb{K}^m$. Then there is a unique $m \times n$ matrix Q, whose columns form an orthonormal subset of $\mathbb{K}^m$, and there is a unique $n \times n$ matrix R which is upper triangular and has positive diagonal entries, such that $A = QR$. The result also holds for an infinite matrix A whose columns form a linearly independent subset of ℓ^2.

2.20. Let $x \in C([-1, 1])$, and suppose that x is not a polynomial. If $m \in \mathbb{N}$ and $p_0, p_1, \ldots, p_m$ are the Legendre polynomials of degrees $0, 1, \ldots, m$, then

$$\sum_{n=0}^{m} \left| \int_{-1}^{1} x(t)p_n(t)dt \right|^2 < \int_{-1}^{1} |x(t)|^2 dt.$$

2.21. **(Trigonometric polynomials on $\mathbb{R}$)** For $r \in \mathbb{R}$, let $u_r(t) := e^{irt}$, $t \in \mathbb{R}$. Let $\mathbb{K} := \mathbb{C}$, and let X be the subspace of $C(\mathbb{R})$ spanned by $\{u_r : r \in \mathbb{R}\}$. For $p, q \in X$, define

$$\langle p, q \rangle := \lim_{T \to \infty} \frac{1}{2T} \int_{-T}^{T} p(t)\overline{q(t)}dt.$$

Then $\langle \cdot\,, \cdot \rangle$ is an inner product on X, and $\{u_r : r \in \mathbb{R}\}$ is an uncountable orthonormal subset of X.

2.22. Let X and Y be linear spaces, and let $\langle \cdot\,, \cdot \rangle_Y$ be an inner product on Y. Suppose $F : X \to Y$ is a linear map. For $x_1, x_2 \in X$, define $\langle x_1, x_2 \rangle_X := \langle F(x_1), F(x_2) \rangle_Y$. Then

(i) $\langle \cdot\,, \cdot \rangle_X$ is an inner product on X if and only if the map F is one-one. (Compare Remark 2.2.)

(ii) Suppose the map F is one-one, and let $\{u_\alpha\}$ be an orthonormal subset of X. Then $\{F(u_\alpha)\}$ is an orthonormal subset of Y.

(iii) Suppose the map F is one-one and onto, and let $\{u_\alpha\}$ be a maximal orthonormal subset of X. Then $\{F(u_\alpha)\}$ is a maximal orthonormal subset of Y.

2.23. Let $\|\cdot\|$ and $\|\cdot\|'$ be equivalent norms on a linear space X. Then $(X, \|\cdot\|)$ is a Banach space if and only if $(X, \|\cdot\|')$ is a Banach space.

2.24. $(c_0, \|\cdot\|_\infty)$ and $(c, \|\cdot\|_\infty)$ are Banach spaces. Also, if T is a metric space, then $(C_0(T), \|\cdot\|_\infty)$ is a Banach space.

2.25. Let $\|\cdot\|$ be a norm on the linear space X consisting of all polynomials defined on $[a, b]$ with coefficients in $\mathbb{K}$. Then there is a sequence (p_n) in X such that $\sum_{n=1}^{\infty} \|p_n\| < \infty$, but $\sum_{n=1}^{\infty} p_n$ does not converge in X.

2.26. Let X be an inner product space.

(i) Every finite dimensional subspace of X is complete. (Note: This can be proved without using Lemma 2.8.)

(ii) The closed unit ball of X is compact if and only if X is finite dimensional. (Note: This can be proved without using Theorem 2.10.)

(iii) If X is complete, then it cannot have a denumerable (Hamel) basis. (Note: This can be proved without using Theorem 2.26.)

2.27. $\{e_1, e_2, \ldots\}$ is a Schauder basis for c_0. Also, $\{e_0, e_1, e_2, \ldots\}$, is a Schauder basis for c, where $e_0 := (1, 1, \ldots)$. If $u_n := (e_{2n-1} + e_{2n})/2$ and $v_n := (e_{2n-1} - e_{2n})/2$ for $n \in \mathbb{N}$, then $\{u_1, v_1, u_2, v_2, \ldots\}$ is a Schauder basis for ℓ^1. On the other hand, neither ℓ^∞ nor $L^\infty([a, b])$ has a Schauder basis.

2.28. For $j \in \mathbb{N}$, let $(X_j, \langle \cdot\,, \cdot \rangle_j)$ be an inner product space over $\mathbb{K}$. Let $X := \big\{(x(1), x(2), \ldots) : x(j) \in X_j$ for all $j \in \mathbb{N}$ and $\sum_{j=1}^{\infty} \langle x(j), x(j) \rangle_j < \infty\big\}$.

Define $\langle x, y \rangle := \sum_{j=1}^{\infty} \langle x(j), y(j) \rangle_j$ for $x, y \in X$. Then X is a linear space over $\mathbb{K}$ with componentwise addition and scalar multiplication, and $\langle \cdot , \cdot \rangle$ is an inner product on X. Further, X is a Hilbert space if and only if X_j is a Hilbert space for each $j \in \mathbb{N}$. (Note: $X = \ell^2$ if $X_j := \mathbb{K}$ for each $j \in \mathbb{N}$.)

2.29. (i) Let $C^k([a, b])$ denote the linear space of all k times differentiable functions on $[a, b]$ whose kth derivatives are continuous on $[a, b]$. For $x \in C^k([a, b])$, let $\|x\|_{k,\infty} := \max\{\|x\|_\infty, \|x'\|_\infty, \ldots, \|x^{(k)}\|_\infty\}$. Then $C^k([a, b])$ is a Banach space.

(ii) Let $W^{k,1}([a, b])$ denote the linear space of all $k - 1$ times differentiable functions on $[a, b]$ whose $(k - 1)$th derivatives are absolutely continuous on $[a, b]$. For $x \in W^{k,1}([a, b])$, let $\|x\|_{k,1} := \sum_{j=0}^{k} \|x^{(j)}\|_1$. Then the **Sobolev space** $W^{k,1}([a, b])$ of order $(k, 1)$ is a Banach space.

(iii) Let $W^{k,2}([a, b])$ denote the linear space of all functions in $W^{k,1}([a, b])$ whose kth derivatives are in $L^2([a, b])$. For x and y in $W^{k,2}([a, b])$, let

$$\langle x, y \rangle_{k,2} := \sum_{j=0}^{k} \int_a^b x^{(j)}(t)\overline{y}^{(j)}(t)dm(t).$$

Then the **Sobolev space** $W^{k,2}([a, b])$ of order $(k, 2)$ is a Hilbert space.

2.30. Let (x_n) be a sequence in a Hilbert space H such that the set $\{x_n : n \in \mathbb{N}\}$ is orthogonal. Then the following conditions are equivalent.

(i) The series $\sum_{n=1}^{\infty} x_n$ is summable in H.
(ii) There is $s \in H$ such that $\langle s, x_n \rangle = \|x_n\|^2$ for all $n \in \mathbb{N}$.
(iii) $\sum_{n=1}^{\infty} \|x_n\|^2 < \infty$.

2.31. For $n \in \mathbb{N}$, let $u_n := (e_{3n-2} + e_{3n-1} + e_{3n})/\sqrt{3}$, $v_n := (e_{3n-2} - e_{3n-1})/\sqrt{2}$ and $w_n := (e_{3n-2} + e_{3n-1} - 2e_{3n})/\sqrt{6}$. Then $\{u_1, v_1, w_1, u_2, v_2, w_2, \ldots\}$ is an orthonormal basis for ℓ^2.

2.32. Let $\mathbb{K} := \mathbb{C}$ and $\omega := e^{2\pi i/3}$. Let $u_n := (e_{3n-2} + e_{3n-1} + e_{3n})/\sqrt{3}$, $v_n := (e_{3n-2} + \omega\, e_{3n-1} + \omega^2 e_{3n})/\sqrt{3}$ and $w_n := (e_{3n-2} + \omega^2 e_{3n-1} + \omega\, e_{3n})/\sqrt{3}$ for $n \in \mathbb{N}$. Then $\{u_1, v_1, w_1, u_2, v_2, w_2, \ldots\}$ is an orthonormal basis for ℓ^2.

2.33. Let $H := L^2([-\pi, \pi])$. Let $u_0(t) := 1/\sqrt{2\pi}$, and for $n \in \mathbb{N}$, let $u_n(t) := \cos nt/\sqrt{\pi}$, $v_n(t) := \sin nt/\sqrt{\pi}$, $t \in [-\pi, \pi]$. Then $\{u_0, u_1, v_1, u_2, v_2, \ldots\}$ is an orthonormal basis for H. Let $x \in H$. If $2\pi a_0 = \int_{-\pi}^{\pi} x(t)dm(t)$, and $\pi a_n = \int_{-\pi}^{\pi} x(t) \cos nt\, dm(t)$, $\pi b_n = \int_{-\pi}^{\pi} x(t) \sin nt\, dm(t)$ for $n \in \mathbb{N}$, then $x(t) = a_0 + \sum_{n=1}^{\infty} \big(a_n \cos nt + b_n \sin nt\big)$, where the series converges in the mean square, and

$$\frac{1}{\pi} \int_{-\pi}^{\pi} |x(t)|^2 dm(t) = 2|a_0|^2 + \sum_{n=1}^{\infty} \left(|a_n|^2 + |b_n|^2\right).$$

2.34. Let $H := L^2([0, 1])$. Let $u_0(t) := 1$, and for $n \in \mathbb{N}$, let $u_n(t) := \sqrt{2} \cos n\pi t$, $v_n(t) := \sqrt{2} \sin n\pi t$, $t \in [0, 1]$. Then both $\{u_0, u_1, u_2, \ldots\}$ and $\{v_1, v_2, \ldots\}$

are orthonormal bases for H. Let $x \in H$. If $a_0 := \int_0^1 x(t)dm(t)$, and $a_n := 2\int_0^1 x(t)\cos n\pi t\, dm(t)$, $b_n := 2\int_0^1 x(t)\sin n\pi t\, dm(t)$ for $n \in \mathbb{N}$, then $x(t) = a_0 + \sum_{n=1}^{\infty} a_n \cos n\pi t$, and $x(t) = \sum_{n=1}^{\infty} b_n \sin n\pi t$, where both series converge in the mean square.

2.35. Let $H := L^2([-1, 1])$. Let $x_0(t) := 1$, and for $n \in \mathbb{N}$, let $x_n(t) := t^{3n}$ for t in $[-1, 1]$. Suppose $v_0, v_1, v_2, \ldots$ are obtained by the orthonormalization of the linearly independent subset $\{x_0, x_1, x_2, \ldots\}$ of H. Then $\{v_0, v_1, v_2, \ldots\}$ is an orthonormal basis for H. Also, $v_0(t) := 1/\sqrt{2}$, $v_1(t) := \sqrt{7}\,t^3/\sqrt{2}$, and $v_2(t) := \sqrt{13}(7t^6 - 1)/6\sqrt{2}$ for $t \in [-1, 1]$.

2.36. Let $\{u_\alpha : \alpha \in A\}$ be an orthonormal basis for a Hilbert space H, and let $\{v_\beta : \beta \in B\}$ be an orthonormal basis for a Hilbert space G. Suppose ϕ is a one-one function from A onto B. For $x := \sum_\alpha \langle x, u_\alpha\rangle u_\alpha \in H$, define $F(x) := \sum_\alpha \langle x, u_\alpha\rangle v_{\phi(\alpha)} \in G$, where $\sum_\alpha$ denotes a countable sum. Then F is a Hilbert space isomorphism from H onto G.

2.37. Let E be a subset of a Hilbert space H. Then $E^{\perp\perp}$ is the closure of span E.

2.38. Let G be a closed subspace of a Hilbert space H. For $x \in H$, let y be the orthogonal projection of x on G. Then y is the unique **best approximation** of x from G, that is, y is the unique element of G such that $\|x - y\| = d(x, G)$.

2.39. (i) Let Y be a finite dimensional proper subspace of a normed space X. Then there is $x_1 \in X$ with $\|x_1\| = 1 = d(x_1, Y)$.

(ii) Let G be a closed proper subspace of a Hilbert space H. Then there is $x_1 \in H$ such that $\|x_1\| = 1 = d(x_1, G)$.

(Compare Lemma 2.7 of Riesz.)

2.40. Let H be a Hilbert space with an inner product $\langle\cdot\,,\cdot\rangle$, and let G be a closed subspace of H. For $x_1 + G$ and $x_2 + G$ in H/G, define $\langle\langle x_1 + G, x_2 + G\rangle\rangle := \langle x_1 - y_1, x_2 - y_2\rangle$, where y_1 and y_2 are the orthogonal projections of x_1 and x_2 on G respectively. Then $\langle\langle\cdot\,,\cdot\rangle\rangle$ is an inner product on H/G, and it induces the quotient norm on H/G. Further, H/G is a Hilbert space.

Chapter 3
Bounded Linear Maps

Having described the basic framework of a normed space in the previous chapter, we study the continuity of linear maps between normed spaces in this chapter. The notion of the operator norm of a continuous linear map is important in this context. We give many examples of continuous linear maps which include matrix transformations and Fredholm integral maps, and attempt to find their operator norms. Four major results are proved in the second and the third section: the uniform boundedness principle, the closed graph theorem, the bounded inverse theorem and the open mapping theorem. These are easily deduced from a theorem of Zabreiko which states that a countably subadditive seminorm on a Banach space is continuous. We give several applications of these major results. In the last section of this chapter, we introduce compact linear maps. They provide a useful generalization of finite rank continuous maps.

3.1 Continuity of a Linear Map

In Sect. 2.1, we have defined a seminorm and a norm on a linear space. Let F be a linear map from a normed space X to a normed space Y, and consider the seminorm p on X defined by $p(x) := \|F(x)\|$, $x \in X$, mentioned in Remark 2.2. We shall see that F is continuous if and only if p is continuous. With this in mind, let us undertake a study of the continuity of a seminorm on a normed space. This will also be useful in Sect. 3.2.

Lemma 3.1 *Let X be a normed space with norm $\|\cdot\|$, and let p be a seminorm on X. Then p is continuous on X if and only if there is $\alpha > 0$ such that $p(x) \leq \alpha\|x\|$ for all $x \in X$.*

Proof Suppose there is $\alpha > 0$ such that $p(x) \leq \alpha\|x\|$ for all $x \in X$. Let $x_n \to x$ in X. Then for all $n \in \mathbb{N}$,

$$p(x_n) - p(x) \leq p(x_n - x) \quad \text{and} \quad p(x) - p(x_n) \leq p(x - x_n),$$

© Springer Science+Business Media Singapore 2016, corrected publication 2023

B.V. Limaye, *Linear Functional Analysis for Scientists and Engineers*,
https://doi.org/10.1007/978-981-10-0972-3_3

and so $|p(x_n) - p(x)| \le p(x_n - x) \le \alpha\|x_n - x\|$, which tends to 0. Thus p is continuous on X.

Conversely, suppose there is no $\alpha > 0$ such that $p(x) \le \alpha\|x\|$ for all $x \in X$. Then for each $n \in \mathbb{N}$, there is $x_n \in X$ such that $p(x_n) > n\|x_n\|$. Let $y_n := x_n/n\|x_n\|$ for $n \in \mathbb{N}$. Then $\|y_n\| = 1/n \to 0$, but $p(y_n) \not\to 0$ since $p(y_n) > 1$ for all $n \in \mathbb{N}$. Thus p is not continuous at $0 \in X$. (Compare the proof of Proposition 2.3.) □

Proposition 3.2 *Let $(X, \|\cdot\|_X)$ and $(Y, \|\cdot\|_Y)$ be normed spaces, and let $F : X \to Y$ be a linear map. The following conditions are equivalent.*

(i) *The linear map F is continuous on X.*
(ii) *The seminorm p defined by $p(x) := \|F(x)\|_Y$, $x \in X$, is continuous on X.*
(iii) *There is $\alpha > 0$ such that $\|F(x)\|_Y \le \alpha\|x\|_X$ for all $x \in X$.*
(iv) *F is bounded on the closed unit ball $\overline{U}_X(0, 1)$ of X.*

Proof (i) $\Longrightarrow$ (ii): Let F be continuous. If $x_n \to x$ in X, then $F(x_n) \to F(x)$, and so $p(x_n) = \|F(x_n)\| \to \|F(x)\| = p(x)$. Hence p is continuous.

(ii) $\Longrightarrow$ (iii): This follows from Lemma 3.1.

(iii) $\Longrightarrow$ (i): If $x_n \to x$ in X, then

$$\|F(x_n) - F(x)\|_Y = \|F(x_n - x)\|_Y \le \alpha\|x_n - x\|_X \to 0.$$

Hence F is continuous.

(iii) $\Longleftrightarrow$ (iv): This follows by noting that $\|F(x)\|_Y \le \alpha\|x\|_X$ for all $x \in X$ if and only if $\|F(x)\|_Y \le \alpha$ for all $x \in \overline{U}_X(0, 1)$. □

Remark 3.3 Let X and Y be normed spaces. When the context is clear, we shall often drop the suffixes in the notation $\|\cdot\|_X$ and $\|\cdot\|_Y$, and simply write $\|\cdot\|$ for both. Let $F : X \to Y$ be a linear map. If F is continuous at $0 \in X$, then linearity of F shows that F is continuous at every $x \in X$. Also, if F is continuous on X, then the condition (iii) of Proposition 3.2 shows that F is uniformly continuous on X.

In view of condition (iv) of Proposition 3.2, a continuous linear map on X is known as a **bounded linear map** on X. It should be kept in mind that the boundedness of a linear map refers to its boundedness on the closed unit ball of X, and not on X. In fact, the only linear map which is bounded on X is the map which sends every element of X to zero.

The set of all bounded linear maps from a normed space X to a normed space Y will be denoted by $BL(X, Y)$. Clearly, $BL(X, Y)$ is a linear space over $\mathbb{K}$. Let $F \in BL(X, Y)$. If Z is a normed space over $\mathbb{K}$, and $G \in BL(Y, Z)$, then the composition $G \circ F \in BL(X, Z)$. We also write GF for $G \circ F$. If $Y := X$, we denote $BL(X, X)$ simply by $BL(X)$. If $F, G \in BL(X)$, then $F \circ G$, $G \circ F \in BL(X)$. It follows that $BL(X)$ is an algebra over $\mathbb{K}$. An element of $BL(X)$ is called a **bounded operator** on X. ◊

Corollary 3.4 *Let X and Y be normed spaces, and $F : X \to Y$ be a linear map. Let $Z := Z(F)$, and for $x + Z \in X/Z$, let $\widetilde{F}(x + Z) := F(x)$. Then $F \in BL(X, Y)$ if and only if Z is closed in X and $\widetilde{F} \in BL(X/Z, Y)$.*

Proof Recall that the quotient norm on X/Z is defined by $|||x + Z||| := \inf\{\|x + z\| : z \in Z\}$ for $x + Z \in X/Z$ if the subspace Z is closed in X. (See Proposition 2.5.) We also note that the map $\widetilde{F} : X/Z \to Y$ is well defined.

Suppose Z is closed in X, and $\widetilde{F} \in BL(X/Z, Y)$. By Proposition 3.2, there is $\alpha > 0$ such that $\|\widetilde{F}(x + Z)\| \leq \alpha|||x + Z|||$ for all $x + Z \in X/Z$. Then

$$\|F(x)\| = \|\widetilde{F}(x + Z)\| \leq \alpha|||x + Z||| \leq \alpha\|x\| \quad \text{for } x \in X.$$

Thus $F \in BL(X, Y)$ again by Proposition 3.2.

Conversely, suppose $F \in BL(X, Y)$. To show that Z is closed in X, let (z_n) be a sequence in Z such that $z_n \to x$ in X. Since $F(z_n) = 0$ for all $n \in \mathbb{N}$, and $F(z_n) \to F(x)$, it follows that $F(x) = 0$. Now by Proposition 3.2, there is $\alpha > 0$ such that $\|F(x)\| \leq \alpha\|x\|$ for all $x \in X$. Let $x \in X$. Then

$$\|\widetilde{F}(x + Z)\| = \|\widetilde{F}(x + z + Z)\| = \|F(x + z)\| \leq \alpha\|x + z\| \quad \text{for all } z \in Z,$$

and so $\|\widetilde{F}(x + Z)\| \leq \alpha \inf\{\|x + z\| : z \in Z\} = \alpha|||x + Z|||$. This shows that $\widetilde{F} \in BL(X/Z, Y)$, again by Proposition 3.2. □

Recall that a function F from a metric space X to a metric space Y is called a homeomorphism if it is one-one and continuous, and if $F^{-1} : R(F) \to X$ is also continuous. We shall now characterize a linear homeomorphism from a normed space X to a normed space Y. For this purpose, we introduce an additional terminology. We have called a linear map 'bounded' if there is $\alpha > 0$ such that $\|F(x)\| \leq \alpha\|x\|$ for all $x \in X$. On the other hand, a linear map $F : X \to Y$ is called **bounded below** if there is $\beta > 0$ such that $\beta\|x\| \leq \|F(x)\|$ for all $x \in X$.

Proposition 3.5 *Let X and Y be normed spaces, and $F : X \to Y$ be linear. Then F is a homeomorphism if and only if $F \in BL(X, Y)$ and F is bounded below, that is, there are α, $\beta > 0$ such that*

$$\beta\|x\| \leq \|F(x)\| \leq \alpha\|x\| \quad \textit{for all } x \in X.$$

Proof First note that if a map $F : X \to Y$ is linear and one-one, then the inverse map $F^{-1} : R(F) \to X$ is also linear.

By Proposition 3.2, $F \in BL(X, Y)$ if and only if there is $\alpha > 0$ such that $\|F(x)\| \leq \alpha\|x\|$ for all $x \in X$. Also, if F is one-one, then $F^{-1} \in BL(R(F), X)$ if and only if there is $\gamma > 0$ such that $\|F^{-1}(y)\| \leq \gamma\|y\|$ for all $y \in R(F)$, that is, $\|x\| \leq \gamma\|F(x)\|$ for all $x \in X$. Letting $\beta := 1/\gamma$, we see that F is a homeomorphism if and only if there are α, $\beta > 0$ such that $\beta\|x\| \leq \|F(x)\| \leq \alpha\|x\|$ for all $x \in X$. □

Proposition 3.6 *Let X and Y be normed spaces, and $F : X \to Y$ be linear.*

(i) *Suppose X is finite dimensional. Then F is continuous.*
(ii) *Suppose the map F is of finite rank. Then F is continuous if and only if the zero space $Z(F)$ of F is closed in X.*

Proof (i) Let $\dim X := m$, and $\{y_1, \ldots, y_m\}$ be a basis for X. Let (x_n) be a sequence in X, and for each $n \in \mathbb{N}$, let $x_n := k_{n,1}y_1 + \cdots + k_{n,m}y_m$, where $k_{n,1}, \ldots, k_{n,m} \in \mathbb{K}$. Also, let $x := k_1y_1 + \cdots + k_my_m \in X$, where $k_1, \ldots, k_n \in \mathbb{K}$. If $x_n \to x$, then $k_{n,j} \to k_j$ in $\mathbb{K}$ for each $j = 1, \ldots, m$, and so $F(x_n) = k_{n,1}F(y_1) + \cdots + k_{n,m}F(y_m) \to k_1F(y_1) + \cdots + k_mF(y_m) = F(x)$ by Lemma 2.8(ii). Hence F is continuous on X. (Compare Exercise 3.2(i).)

(ii) Let $Z := Z(F)$. If F is continuous, then Z is closed in X by Corollary 3.4. Conversely, suppose Z is closed in X. Since $R(F)$ is finite dimensional, so is X/Z. To see this, consider $y_1, \ldots, y_m \in Y$ such that $R(F) = \text{span}\,\{y_1, \ldots, y_m\}$, and let $x_j \in X$ be such that $F(x_j) = y_j$ for $j = 1, \ldots, m$. If $x \in X$, then there are $k_1, \ldots, k_m \in \mathbb{K}$ such that $F(x) = k_1y_1 + \cdots + k_my_m$, and so $(x - k_1x_1 - \cdots - k_mx_m) \in Z$, that is, $x + Z = k_1(x_1 + Z) + \cdots + k_m(x_m + Z)$. Thus $X/Z = \text{span}\,\{x_1 + Z, \ldots, x_m + Z\}$. Since X/Z is finite dimensional, the linear map $\widetilde{F} : X/Z \to Y$ defined by $\widetilde{F}(x + Z) := F(x)$, $x + Z \in X/Z$, is continuous by (i) above. Corollary 3.4 shows that F is continuous. □

Corollary 3.7 *Let X and Y be normed spaces. Suppose X is finite dimensional, and $F : X \to Y$ is one-one and linear. Then F is a homeomorphism.*

Proof The map F is continuous by Proposition 3.6(i). Since $R(F)$ is also finite dimensional, the map $F^{-1} : R(F) \to X$ is continuous as well. Hence F is a homeomorphism. □

Remarks 3.8 (i) Suppose there is a linear homeomorphism F from a normed space X onto a normed space Y. Proposition 3.5 shows that (x_n) is a Cauchy sequence in X if and only if $(F(x_n))$ is a Cauchy sequence in Y, and (x_n) converges in X if and only if $(F(x_n))$ converges in Y. Hence X is a Banach space if and only if Y is a Banach space. This result may be contrasted with our comment in Sect. 1.3 that a complete metric space can be homeomorphic to an incomplete metric space. For example, $\mathbb{R}$ is homeomorphic to $(-1, 1)$.

(ii) Let Y be a normed space of dimension m, and consider a norm on $\mathbb{K}^m$. If $\{y_1, \ldots, y_m\}$ is a basis for Y, then the map $F : \mathbb{K}^m \to Y$ given by $F\big((k_1, \ldots, k_m)\big) := k_1y_1 + \cdots + k_my_m$ is linear, one-one and onto. By Corollary 3.7, F is a linear homeomorphism from $\mathbb{K}^m$ onto Y.

Let us ask a related question. Suppose $m \geq 2$. Is there a linear isometry from $(\mathbb{K}^m, \|\cdot\|_p)$ onto $(\mathbb{K}^m, \|\cdot\|_r)$, where $p, r \in \{1, 2, \infty\}$ and $p \neq r$. If $m = 2$ and $\mathbb{K} := \mathbb{R}$, then the linear map $F : \mathbb{R}^2 \to \mathbb{R}^2$ given by $F\big(x(1), x(2)\big) := \big(x(1) + x(2), x(1) - x(2)\big)$ is an isometry from $(\mathbb{R}^2, \|\cdot\|_1)$ onto $(\mathbb{R}^2, \|\cdot\|_\infty)$, since $|a| + |b| = \max\{|a + b|, |a - b|\}$ for all $a, b \in \mathbb{R}$. Except for this case, our question has a negative answer. For example, assume that F is a linear isometry from $(\mathbb{K}^m, \|\cdot\|_1)$ to $(\mathbb{K}^m, \|\cdot\|_2)$. Since $\|e\|_1 = \|e_2\|_1 = 1$ and $\|e_1 + e_2\|_1 = 2$, we must have $\|F(e_1)\|_2 = \|F(e_2)\|_2 = 1$ and $\|F(e_1) + F(e_2)\|_2 = \|F(e_1 + e_2)\|_2 = 2$, but then $\|F(e_1) - F(e_2)\|_2^2 = 2\|F(e_1)\|_2^2 + 2\|F(e_2)\|_2^2 - \|F(e_1) + F(e_2)\|_2^2 = 0$, and so $F(e_1) = F(e_2)$, a contradiction. The general case is proved in [27]. ◊

Examples 3.9 (i) Let X and Y be finite dimensional normed spaces of dimensions n and m, and let $\{x_1, \ldots, x_n\}$ and $\{y_1, \ldots, y_m\}$ be bases for X and Y respectively. We have seen in Example 1.3(i) that every $m \times n$ matrix $M := [k_{i,j}]$ defines a linear map F from X to Y with respect to these bases. Then F is continuous. (In Exercise 3.2(i), we give a constant $\alpha \geq 0$ such that $\|F(x)\| \leq \alpha\|x\|$ for all $x \in X$, in terms of $\|F(x_j)\|$ and the distance of x_j from the span of $\{x_i : i = 1, \ldots, n \text{ and } i \neq j\}$ for $j = 1, \ldots, n$.) Conversely, it is easy to see that every linear map from X to Y is of this form.

(ii) Let $f(x) := \sum_{j=1}^{\infty} jx(j)$ for $x \in c_{00}$. Then f is a linear map from c_{00} to $\mathbb{K}$. Suppose $\|\cdot\|$ is norm on c_{00} such that $\|e_n\| \leq \alpha$ for some $\alpha > 0$ and all $n \in \mathbb{N}$. (The norms $\|\cdot\|_1, \|\cdot\|_2, \|\cdot\|_\infty$ satisfy this condition with $\alpha = 1$.) Then f is discontinuous, since $|f(e_n/\alpha)| = n/\alpha$ for $n \in \mathbb{N}$, and so f is not bounded on the closed unit ball of c_{00}. Note that the normed space c_{00} is infinite dimensional. In fact, if X is any infinite dimensional normed space, then there is a discontinuous linear map from X to $\mathbb{K}$. See Exercise 3.1.

(iii) Let $X := C^1([0, 1])$ and $Y := C([0, 1])$, both with the sup norm $\|\cdot\|_\infty$. For $x \in X$, let $F(x) := x'$, the derivative of x. Then F is a linear map from X to Y, but it is not continuous. To see this, let $x_n(t) := t^n$, $t \in [0, 1]$, for $n \in \mathbb{N}$. Then $\|x_n\|_\infty = 1$, while $\|F(x_n)\|_\infty = \|x_n'\|_\infty = n$ for all $n \in \mathbb{N}$, and so F is not bounded on the closed unit ball of X. Note that $Z(F)$ consists of the set of all constant functions on [0,1], which is closed in X. By the fundamental theorem of calculus for Riemann integration (Theorem 1.22), $R(F) = Y$, which is infinite dimensional. This shows that the assumption 'F is of finite rank' in Proposition 3.6(ii) cannot be omitted.

For $x \in X$, let $f(x) := x'(1)$. Again, f is a linear map from X to $\mathbb{K}$, but it is not continuous. This follows by considering the sequence (x_n) in X mentioned above. By Proposition 3.6, $Z(f)$ is not closed.

On the other hand, if we define $g : Y \to \mathbb{K}$ by $g(y) := \int_0^1 y(t)dt$ for $y \in Y$, then g is linear, and it is continuous since $|g(y)| \leq \|y\|_\infty$ for all $y \in Y$. ◊

We now introduce a norm on the linear space $BL(X, Y)$.

Theorem 3.10 *Let X and Y be normed spaces. For $F \in BL(X, Y)$, let*

$$\|F\| := \sup\{\|F(x)\| : x \in X \text{ and } \|x\| \leq 1\}.$$

(i) **(Basic inequality)** *Let $F \in BL(X, Y)$. Then $\|F(x)\| \leq \|F\|\,\|x\|$ for all $x \in X$. In fact,*

$$\|F\| = \inf\{\alpha > 0 : \|F(x)\| \leq \alpha\|x\| \text{ for all } x \in X\}.$$

(ii) *$\|\cdot\|$ is a norm on $BL(X, Y)$, called the* **operator norm**. *Further, if Z is a normed space, and $G \in BL(Y, Z)$, then $G \circ F \in BL(X, Z)$, and $\|G \circ F\| \leq \|G\|\,\|F\|$.*

(iii) *Let X and Y be normed spaces, and let $F \in BL(X, Y)$. Let $Z := Z(F)$, and for $x + Z \in X/Z$, define $\widetilde{F}(x + Z) := F(x)$. Then $\widetilde{F} \in BL(X/Z, Y)$ and $\|\widetilde{F}\| = \|F\|$.*

Proof Consider the subset $E := \{\|F(x)\| : x \in X \text{ and } \|x\| \le 1\}$ of $\mathbb{R}$. Since F belongs to $BL(X, Y)$, it is bounded on the closed unit ball of X by Proposition 3.2. Thus the set E is bounded, and so $\|F\|$ is well defined.

(i) Let $x \in X$. If $x = 0$, then $F(0) = 0$. If $x \neq 0$, then the norm of $x_1 := x/\|x\|$ is equal to 1. Since $\|F\|$ is an upper bound of the set E, $\|F(x_1)\| \le \|F\|$, that is, $\|F(x)\| \le \|F\|\,\|x\|$.

Let $\alpha_0 := \inf\{\alpha \in \mathbb{R} : \|F(x)\| \le \alpha\|x\| \text{ for all } x \in X\}$. By (i) above, $\alpha_0 \le \|F\|$. Also, if $\alpha \in \mathbb{R}$ and $\|F(x)\| \le \alpha\|x\|$ for all $x \in X$, then in particular, $\|F(x)\| \le \alpha$ for all $x \in X$ with $\|x\| \le 1$, that is, α is an upper bound of the set E. Hence $\|F\| \le \alpha$, and consequently, $\|F\| \le \alpha_0$.

(ii) Let $F, G \in BL(X, Y)$. Then for all $x \in X$,

$$\|(F + G)(x)\| = \|F(x) + G(x)\| \le \|F(x)\| + \|G(x)\| \le (\|F\| + \|G\|)\|x\|$$

by the basic inequality stated in (i) above. Taking supremum over all $x \in X$ with $\|x\| \le 1$, we obtain $\|F + G\| \le \|F\| + \|G\|$. Next, let $k \in \mathbb{K}$. Then

$$\|(kF)(x)\| = \|kF(x)\| = |k|\,\|F(x)\| \quad \text{for all } x \in X.$$

Again, taking supremum over all $x \in X$ with $\|x\| \le 1$, we obtain $\|kF\| = |k|\|F\|$. It is clear that $\|F\| \ge 0$, and if $\|F\| = 0$, then $F = 0$. Thus $\|\cdot\|$ is a norm on $BL(X, Y)$.

Let Z be a normed space, and let $G \in BL(Y, Z)$. Clearly, $G \circ F$ belongs to $BL(X, Z)$. Also, by (i) above, $\|(G \circ F)(x)\| \le \|G\|\,\|F(x)\| \le \|G\|\,\|F\|\,\|x\|$ for all $x \in X$. Hence $\|G \circ F\| \le \|G\|\,\|F\|$.

(iii) By Corollary 3.4, $\widetilde{F} \in BL(X/Z, Y)$. Also, as in the proof of Corollary 3.4, we see that for $\alpha > 0$, $\|\widetilde{F}(x + Z)\| \le \alpha|||x + Z|||$ for all $x + Z \in X/Z$ if and only if $\|F(x)\| \le \alpha\|x\|$ for all $x \in X$. Hence $\|\widetilde{F}\| = \|F\|$. □

Remarks 3.11 (i) If $F \in BL(X, Y)$, then its operator norm $\|F\|$ should really be denoted by $\|F\|_{BL(X,Y)}$. But we shall use the simpler notation $\|F\|$.

(ii) If either $X := \{0\}$ or $Y := \{0\}$, then clearly $BL(X, Y) = \{0\}$. Conversely, if $X \neq \{0\}$ and $Y \neq \{0\}$, must $BL(X, Y) \neq \{0\}$? The answer is affirmative. (See Remark 4.7(i).)

Let $X := \mathbb{K}$. Given $y \in Y$, consider the map $F_y : \mathbb{K} \to Y$ defined by $F_y(k) := k\,y$ for $k \in \mathbb{K}$. Then $F_y \in BL(\mathbb{K}, Y)$ and $\|F_y\| = \|y\|$. In fact, the function $\Phi : Y \to BL(\mathbb{K}, Y)$ defined by $\Phi(y) := F_y$, $y \in Y$, is a linear isometry from Y onto $BL(\mathbb{K}, Y)$. On the other hand, let $Y := \mathbb{K}$. We shall study the normed space $B(X, \mathbb{K})$ extensively in Sect. 4.2.

(iii) Let $X \neq \{0\}$ and $F \in BL(X, Y)$. It is easy to see that $\|F\| = \sup\{\|F(x)\| : x \text{ in } X \text{ and } \|x\| = 1\}$. The computation of the operator norm $\|F\|$ of $F \in BL(X, Y)$ involves a constrained optimization problem, namely,

$$\text{'Maximize } \|F(x)\|, \text{ subject to } \|x\| = 1,\ x \in X\text{'}.$$

If X is finite dimensional, then the unit sphere of X is compact, and the function $x \longmapsto \|F(x)\|$ is continuous on it, so that there is $x_0 \in X$ such that $\|x_0\| = 1$ and $\|F(x_0)\| = \|F\|$. If X is infinite dimensional, there may not be any $x \in X$ such that $\|x\| = 1$ and $\|F(x)\| = \|F\|$. But for every $\epsilon > 0$, there is $x_\epsilon \in X$ such that $\|x_\epsilon\| = 1$ and $\|F\| < \|F(x_\epsilon)\| + \epsilon$. For example, define $f : \ell^1 \to \mathbb{K}$ by $f(x) := \sum_{j=1}^{\infty} j\, x(j)/(j+1)$ for $x := (x(1), x(2), \ldots)$ in ℓ^1. Then $f(e_j) = j/(j+1)$ for each $j \in \mathbb{N}$ and $|f(x)| < \sum_{j=1}^{\infty} |x(j)| = \|x\|_1$ for every nonzero $x \in \ell^1$. Hence $\|f\| = 1$, and there is no $x \in \ell^1$ such that $\|x\|_1 = 1$ and $\|f\| = |f(x)|$. But if $\epsilon > 0$ and $j_0 \in \mathbb{N}$ is such that $j_0 > (1 - \epsilon)/\epsilon$, then $\|e_{j_0}\|_1 = 1$ and $\|f\| < |f(e_{j_0})| + \epsilon$. (See also Exercise 3.5.) $\diamond$

Calculation of $\|F\|$ can be very difficult. We may accomplish it only in some special cases. For example, if $Y := X$, and I denotes the identity operator on X, then $\|I\| = 1$. Often one has to be satisfied with an upper bound for $\|F\|$. We now give a number of important examples.

If $p = 1$, let $q := \infty$; if $p = 2$, let $q := 2$; and if $p = \infty$, let $q := 1$, so that $(1/p) + (1/q) = 1$, that is, let p and q be **conjugate exponents**. The **signum function** sgn : $\mathbb{K} \to \mathbb{K}$ is defined as follows:

$$\operatorname{sgn} k := \begin{cases} \overline{k}/|k| & \text{if } k \neq 0, \\ 0 & \text{if } k = 0. \end{cases}$$

Then $(\operatorname{sgn} k)k = |k|$ for $k \in \mathbb{K}$. Also, if T is a set, and $x : T \to \mathbb{K}$, then we let $(\operatorname{sgn} x)(t) := \operatorname{sgn} x(t)$. The signum function will be useful in what follows.

Example 3.12 For $p \in \{1, 2, \infty\}$, let q be the conjugate exponent. If $x \in \ell^p$ and $y \in \ell^q$, then $\sum_{j=1}^{\infty} |x(j)y(j)| \leq \|x\|_p \|y\|_q$. If $p \in \{1, \infty\}$, then this is obvious, and if $p = 2$, then this follows from the Schwarz inequality for numbers stated in Lemma 1.4(i). Hence we see that $\sum_{j=1}^{\infty} x(j)y(j)$ is convergent in $\mathbb{K}$.

Fix $y := (y(1), y(2), \ldots) \in \ell^q$, and define

$$f_y(x) := \sum_{j=1}^{\infty} x(j)y(j) \quad \text{for } x := (x(1), x(2), \ldots) \in \ell^p.$$

Clearly, $f_y \in BL(\ell^p, \mathbb{K})$, and $\|f_y\| \leq \|y\|_q$. We show that $\|f_y\| = \|y\|_q$.

Let $p := 1$. Then $|y(j)| = |f_y(e_j)| \leq \|f_y\|$, where $e_j \in \ell^1$ and $\|e_j\|_1 = 1$ for every $j \in \mathbb{N}$. Hence $\|f_y\| \geq \|y\|_\infty$.

Next, let $p := \infty$. Define $x := (\operatorname{sgn} y(1), \operatorname{sgn} y(2), \ldots) \in \ell^\infty$. Now $\|x\|_\infty \leq 1$, and $\|f_y\| \geq |f_y(x)| = \sum_{j=1}^{\infty} |y(j)| = \|y\|_1$. Hence $\|f_y\| \geq \|y\|_1$.

Finally, let $p := 2$. Define $x := \big(\overline{y(1)}, \overline{y(2)}, \ldots\big) \in \ell^2$. Then $\|x\|_2 = \|y\|_2$, and $|f_y(x)| = \sum_{j=1}^{\infty} |y(j)|^2 = \|y\|_2^2$. Hence $\|f_y\| \geq \|y\|_2$.

Thus in all cases, $\|f_y\| = \|y\|_q$, as desired. ◊

Example 3.13 For $p \in \{1, 2, \infty\}$, let q be the conjugate exponent. Let us denote $L^1([a, b])$ by L^p. If $x \in L^p$ and $y \in L^q$, then $\int_{[a,b]} |xy| dm \leq \|x\|_p \|y\|_q$. If $p \in \{1, \infty\}$, then this is obvious, and if $p = 2$, then this follows from the Schwarz inequality for functions stated in Lemma 1.25(i). Hence $\int_{[a,b]} xy\, dm$ is a well-defined element in $\mathbb{K}$.

Fix $y \in L^q$, and define

$$f_y(x) := \int_{[a,b]} xy\, dm \quad \text{for } x \in L^p.$$

Clearly, $f_y \in BL(L^p, \mathbb{K})$, and $\|f_y\| \leq \|y\|_q$. We show that $\|f_y\| = \|y\|_q$.

Let $p := 1$. To show $\|y\|_\infty \leq \|f_y\|$, that is, $|y(t)| \leq \|f_y\|$ for almost all $t \in [a, b]$, consider $E := \{t \in [a, b] : |y(t)| > \|f_y\|\}$, and $E_n := \{t \in [a, b] : |y(t)| > \|f_y\| + (1/n)\}$ for $n \in \mathbb{N}$. Since E is the union of $E_1, E_2, \ldots$, we obtain $m(E) \leq \sum_{n=1}^{\infty} m(E_n)$ by the countable subadditivity of the Lebesgue measure m. For $n \in \mathbb{N}$, let c_n denote the characteristic function of E_n, and let $x_n := (\operatorname{sgn} y)c_n$. Then $x_n y = |y|c_n$, $|x_n| = c_n$ and $\|x_n\|_1 = m(E_n)$. Hence

$$\left(\|f_y\| + \frac{1}{n}\right) m(E_n) \leq \int_{E_n} |y| dm = \int_{[a,b]} x_n y\, dm = f_y(x_n) \leq \|f_y\|\, m(E_n),$$

and so, $m(E_n) = 0$ for $n \in \mathbb{N}$, showing that $m(E) = 0$, that is, $\|f_y\| \geq \|y\|_\infty$.

Next, let $p := \infty$. Define $x := \operatorname{sgn} y \in L^\infty$. Now $\|x\|_\infty \leq 1$, and $\|f_y\| \geq |f_y(x)| = \int_{[a,b]} |y| dm = \|y\|_1$. Hence $\|f_y\| \geq \|y\|_1$.

Finally, let $p := 2$. Define $x := \overline{y} \in L^2$. Then $\|x\|_2 = \|y\|_2$, and $|f_y(x)| = \int_{[a,b]} |y|^2 dm = \|y\|_2^2$. Hence $\|f_y\| \geq \|y\|_2$.

Thus in all cases, $\|f_y\| = \|y\|_q$, as desired. ◊

Examples 3.14 Let X and Y denote sequence spaces considered in Example 2.1(ii). Consider an infinite matrix $M := [k_{i,j}]$ having $k_{i,j} \in \mathbb{K}$ as the element in the ith row and the jth column for $i, j \in \mathbb{N}$. Suppose for every $x := (x(1), x(2), \ldots) \in X$ and for every $i \in \mathbb{N}$, the series $\sum_{j=1}^{\infty} k_{i,j} x(j)$ is summable in $\mathbb{K}$, and if $y(i) := \sum_{j=1}^{\infty} k_{i,j} x(j)$ for $i \in \mathbb{N}$, then suppose $y := (y(1), y(2), \ldots) \in Y$. Let us define $F(x) := y$ for $x \in X$. It is easy to see that F is a linear map from X to Y, and for each $j \in \mathbb{N}$, $F(e_j) = (k_{1,j}, k_{2,j}, \ldots)$, the jth column of M. In this case, we say that the matrix M **defines** the map F from X to Y, and the map is known as a **matrix transformation**. This definition is motivated, as in the finite dimensional case treated in Sect. 1.2, by the matrix multiplication

$$\begin{bmatrix} k_{1,1} & \ldots & k_{1,j} & \ldots & \ldots \\ \vdots & & \vdots & & \\ k_{i,1} & \ldots & k_{i,j} & \ldots & \ldots \\ \vdots & & \vdots & & \\ \vdots & & \vdots & & \end{bmatrix} \begin{bmatrix} x(1) \\ \vdots \\ x(j) \\ \vdots \\ \vdots \end{bmatrix} = \begin{bmatrix} \sum_j k_{1,j} x(j) \\ \vdots \\ \sum_j k_{i,j} x(j) \\ \vdots \\ \vdots \end{bmatrix}.$$

If $M := [k_{i,j}]$ with $k_{i,j} \in \mathbb{K}$ for $i, j \in \mathbb{N}$, then let $|M|$ denote the matrix having $|k_{i,j}|$ as the element in the ith row and the jth column for $i, j \in \mathbb{N}$.

(i) Let $X = Y := \ell^1$. Suppose the set of all column sums of the matrix $|M|$ is a bounded subset of $\mathbb{R}$. Define

$$\alpha_1(j) := \sum_{i=1}^{\infty} |k_{i,j}| \text{ for } j \in \mathbb{N} \quad \text{and} \quad \alpha_1 := \sup\{\alpha_1(j) : j \in \mathbb{N}\}.$$

We show that M defines $F \in BL(\ell^1)$ and $\|F\| = \alpha_1$. Let $x \in \ell^1$. For $i \in \mathbb{N}$,

$$\sum_{j=1}^{\infty} |k_{i,j} x(j)| \leq \sup\{|k_{i,j}| : j \in \mathbb{N}\} \sum_{j=1}^{\infty} |x(j)| \leq \alpha_1 \|x\|_1,$$

and hence the series $\sum_{j=1}^{\infty} k_{i,j} x(j)$ is summable in $\mathbb{K}$ to, say, $y(i)$. Now

$$\|y\|_1 = \sum_{i=1}^{\infty} |y(i)| = \sum_{i=1}^{\infty} \Big| \sum_{j=1}^{\infty} k_{i,j} x(j) \Big| \leq \sum_{j=1}^{\infty} \sum_{i=1}^{\infty} |k_{i,j}|\, |x(j)| \leq \alpha_1 \|x\|_1.$$

Hence $y \in \ell^1$. Define $F(x) := y$. Thus M defines $F \in BL(\ell^1)$ and $\|F\| \leq \alpha_1$. Further, $\|e_j\|_1 = 1$ and $F(e_j) = (k_{1,j}, k_{2,j}, \ldots) \in \ell^1$ for $j \in \mathbb{N}$, and so

$$\|F\| \geq \sup\{\|F(e_j)\|_1 : j \in \mathbb{N}\} = \sup\{\alpha_1(j) : j \in \mathbb{N}\} = \alpha_1.$$

Thus $\|F\| = \alpha_1$.

Conversely, if a matrix M defines a bounded operator F on ℓ^1, then $\alpha_1 = \sup\{\|F(e_j)\|_1 : j \in \mathbb{N}\} \leq \|F\|$. In fact, we shall show that if M defines a map F from ℓ^1 to itself, then $F \in BL(\ell^1)$. (See Proposition 3.30.)

For example, let $k_{i,j} = (-1)^{i+j}/i^2 j$ for $i, j \in \mathbb{N}$. Then $\alpha_1(j) = \sum_{i=1}^{\infty} 1/i^2 j = \pi^2/6j$ for each $j \in \mathbb{N}$, and so $\alpha_1 = \pi^2/6$. Hence the infinite matrix $M := [k_{i,j}]$ defines a bounded operator F on ℓ^1, and $\|F\| = \pi^2/6$. On the other hand, let $k_{i,j} = (-1)^{i+j} j/i^2$ for $i, j \in \mathbb{N}$. Then $\alpha_1(j) = \sum_{i=1}^{\infty} j/i^2 = \pi^2 j/6$ for each $j \in \mathbb{N}$, and so the set $\{\alpha_1(j) : j \in \mathbb{N}\}$ is not bounded. Hence the infinite matrix $M := [k_{i,j}]$ does not define a bounded operator on ℓ^1.

(ii) Let $X = Y := \ell^\infty$. Suppose the set of all row sums of the matrix $|M|$ is a bounded subset of $\mathbb{R}$. Define

$$\beta_1(i) := \sum_{j=1}^{\infty} |k_{i,j}| \text{ for } i \in \mathbb{N} \quad \text{and} \quad \beta_1 := \sup\{\beta_1(i) : i \in \mathbb{N}\}.$$

We show that M defines $F \in BL(\ell^\infty)$ and $\|F\| = \beta_1$. Let $x \in \ell^\infty$. For $i \in \mathbb{N}$,

$$\sum_{j=1}^{\infty} |k_{i,j}x(j)| \leq \beta_1(i)\|x\|_\infty,$$

and hence the series $\sum_{j=1}^{\infty} k_{i,j}x(j)$ is summable in $\mathbb{K}$ to, say, $y(i)$. Now

$$\|y\|_\infty = \sup\{|y(i)| : i \in \mathbb{N}\} \leq \sup\{\beta_1(i)\|x\|_\infty : i \in \mathbb{N}\} = \beta_1\|x\|_\infty.$$

Hence $y \in \ell^\infty$. Define $F(x) := y$. Thus M defines $F \in BL(\ell^\infty)$ and $\|F\| \leq \beta_1$. Further, for each $i \in \mathbb{N}$, let $x_i := (\operatorname{sgn} k_{i,1}, \operatorname{sgn} k_{i,2}, \ldots)$. Then $\|x_i\|_\infty \leq 1$ and $F(x_i)(i) = \sum_{j=1}^{\infty} k_{i,j}(\operatorname{sgn} k_{i,j}) = \sum_{j=1}^{\infty} |k_{i,j}| = \beta_1(i)$ for each $i \in \mathbb{N}$, and so

$$\|F\| \geq \sup\{\|F(x_i)\|_\infty : i \in \mathbb{N}\} \geq \sup\{\beta_1(i) : i \in \mathbb{N}\} = \beta_1.$$

Thus $\|F\| = \beta_1$.

Conversely, if a matrix M defines a bounded operator F on ℓ^∞, then $\beta_1 = \sup\{|F(x_i)(i)| : i \in \mathbb{N}\} \leq \|F\|$. In fact, we shall show that if M defines a map F from ℓ^∞ to itself, then $F \in BL(\ell^\infty)$. (See Proposition 3.30.)

For example, let $k_{i,j} = (-1)^{i+j}/ij^2$ for $i, j \in \mathbb{N}$. Then $\beta_1(i) = \sum_{j=1}^{\infty} 1/ij^2 = \pi^2/6i$ for each $i \in \mathbb{N}$, and so $\beta_1 = \pi^2/6$. Hence the infinite matrix $M := [k_{i,j}]$ defines a bounded operator F on ℓ^∞, and $\|F\| = \pi^2/6$. On the other hand, let $k_{i,j} = (-1)^{i+j}i/j^2$ for $i, j \in \mathbb{N}$. Then $\beta_1(i) = \sum_{j=1}^{\infty} i/j^2 = \pi^2 i/6$ for each $i \in \mathbb{N}$, and so the set $\{\beta_1(i) : j \in \mathbb{N}\}$ is not bounded. Hence the infinite matrix $M := [k_{i,j}]$ does not define a bounded operator on ℓ^∞.

(iii) Let $X = Y := \ell^2$. Suppose the set of all column sums and the set of all row sums of the matrix $|M|$ are bounded, and define α_1 and β_1 as in (i) and (ii) above. We show that M defines $F \in BL(\ell^2)$ and $\|F\| \leq \sqrt{\alpha_1\beta_1}$.

Let $x \in \ell^2$. Then $x \in \ell^\infty$, and so for each $i \in \mathbb{N}$, the series $\sum_{j=1}^{\infty} k_{i,j}x(j)$ is summable to, say $y(i)$, as in (ii) above. Now

$$\|y\|_2^2 = \sum_{i=1}^{\infty} |y(i)|^2 = \sum_{i=1}^{\infty} \Big|\sum_{j=1}^{\infty} k_{i,j}x(j)\Big|^2 \leq \sum_{i=1}^{\infty} \Big(\sum_{j=1}^{\infty} |k_{i,j}x(j)|\Big)^2.$$

Writing $|k_{i,j}x(j)| = |k_{i,j}|^{1/2}\big(|k_{i,j}|^{1/2}|x(j)|\big)$ for $i, j \in \mathbb{N}$, and letting $n \to \infty$ in the Schwarz inequality for numbers (Lemma 1.4(i)), we obtain

$$\|y\|_2^2 \le \sum_{i=1}^{\infty}\Big(\sum_{j=1}^{\infty}|k_{i,j}|\Big)\Big(\sum_{j=1}^{\infty}|k_{i,j}|\,|x(j)|^2\Big) \le \beta_1 \sum_{j=1}^{\infty}\sum_{i=1}^{\infty}|k_{i,j}|\,|x(j)|^2,$$

which is less than or equal to $\beta_1\alpha_1\|x\|_2^2$. Hence $y \in \ell^2$. Define $F(x) := y$. Thus M defines $F \in BL(\ell^2)$ and $\|F\| \le \sqrt{\alpha_1\beta_1}$.

For example, let $k_{i,j} = (-1)^{i+j}/i^2j^2$ for $i, j \in \mathbb{N}$. Then $\alpha_1 = \pi^2/6 = \beta_1$. Hence the infinite matrix $M := [k_{i,j}]$ defines a bounded operator F on ℓ^2, and $\|F\| \le \pi^2/6$.

Alternatively, suppose

$$\gamma_{2,2} := \Big(\sum_{i=1}^{\infty}\sum_{j=1}^{\infty}|k_{i,j}|^2\Big)^{1/2} < \infty.$$

Let $x \in \ell^2$. For $i \in \mathbb{N}$, we obtain

$$\sum_{j=1}^{\infty}|k_{i,j}x(j)| \le \Big(\sum_{j=1}^{\infty}|k_{i,j}|^2\Big)^{1/2}\Big(\sum_{j=1}^{\infty}|x(j)|^2\Big)^{1/2} \le \gamma_{2,2}\|x\|_2 < \infty,$$

by letting $n \to \infty$ in the Schwarz inequality for numbers (Lemma 1.4(i)), and hence the series $\sum_{j=1}^{\infty} k_{i,j}x(j)$ is summable in $\mathbb{K}$ to, say $y(i)$. Now

$$\|y\|_2^2 \le \sum_{i=1}^{\infty}\Big(\sum_{j=1}^{\infty}|k_{i,j}|^2\Big)\Big(\sum_{j=1}^{\infty}|x(j)|^2\Big) \le \gamma_{2,2}^2\|x\|_2^2.$$

Hence $y \in \ell^2$. Define $F(x) := y$. Thus M defines $F \in BL(\ell^2)$ and $\|F\| \le \gamma_{2,2}$.

For example, let $k_{i,j} = (-1)^{i+j}/ij$, $i, j \in \mathbb{N}$. Then $\gamma_{2,2}^2 = \sum_{i=1}^{\infty}\sum_{j=1}^{\infty} 1/i^2j^2 = \sum_{i=1}^{\infty} 1/i^2 \sum_{j=1}^{\infty} 1/j^2 = \pi^4/36$. Hence the infinite matrix $M := [k_{i,j}]$ defines a bounded operator F on ℓ^2, and $\|F\| \le \pi^2/6$. In fact, if we let $x(j) := (-1)^j/j$ for $j \in \mathbb{N}$, then $x \in \ell^2$, and $\|x\|_2 = \pi/\sqrt{6}$, whereas $\|F(x)\|_2 = \pi^3/6\sqrt{6}$. Hence $\|F\| = \pi^2/6$.

We give examples to show that the conditions given above for an infinite matrix to define a map in $BL(\ell^2)$ are not necessary. Let

$$M := \begin{bmatrix} 1 & 0 & 0 & \cdots \\ 0 & 1 & 0 & \cdots \\ 0 & 0 & 1 & \cdots \\ \vdots & \vdots & \vdots & \end{bmatrix} \quad \text{and} \quad N := \begin{bmatrix} 1 & 1/2 & 1/3 & \cdots \\ 1/2 & 0 & 0 & \cdots \\ 1/3 & 0 & 0 & \cdots \\ \vdots & \vdots & \vdots & \end{bmatrix}.$$

The column sums as well as the row sums of $|M|$ are bounded, but $\gamma_{2,2} = \infty$ for M, while $\gamma_{2,2}^2 = 1 + 2\sum_{j=2}^{\infty}(1/j^2) < \infty$ for N, but the first column sum and the first row sum of $|N|$ are infinite. Since M as well as N defines a map in $BL(\ell^2)$, so does $M + N$, although $\alpha_1 = \beta_1 = \gamma_{2,2} = \infty$ for $M + N$.

In the special case where all entries of the infinite matrix M, except possibly those in the first row, are equal to 0, we obtain

$$\alpha_1 = \sup\{|k_{1,j}| : j \in \mathbb{N}\}, \quad \beta_1 = \sum_{j=1}^{\infty} |k_{1,j}| \quad \text{and} \quad \gamma_{2,2} = \Big(\sum_{j=1}^{\infty} |k_{1,j}|^2\Big)^{1/2}.$$

These correspond to the norms of the linear functional considered in Example 3.12 when $p = 1$, $p = \infty$ and $p = 2$ respectively. Here $\gamma_{2,2} \le \sqrt{\alpha_1 \beta_1}$. ◊

Examples 3.15 Arguing as in Example 3.14(i), (ii) and (iii), we obtain the following results for linear maps defined by finite matrices. Let $X := \mathbb{K}^n$, $Y := \mathbb{K}^m$, and let $M := [k_{i,j}]$ be an $m \times n$ matrix.

(i) Consider the norm $\|\cdot\|_1$ on both $\mathbb{K}^n$ and $\mathbb{K}^m$, and let

$$\alpha_1 := \max\Big\{\sum_{i=1}^{m} |k_{i,j}| : j = 1, \ldots, n\Big\}.$$

Then M defines $F \in BL(\mathbb{K}^n, \mathbb{K}^m)$, and $\|F\| = \alpha_1$. Thus there is j_0 in $\{1, \ldots, m\}$ such that $\|F\| = \sum_{i=1}^{m} |k_{i,j_0}|$.

(ii) Consider the norm $\|\cdot\|_\infty$ on both $\mathbb{K}^n$ and $\mathbb{K}^m$, and let

$$\beta_1 := \max\Big\{\sum_{j=1}^{n} |k_{i,j}| : i = 1, \ldots, m\Big\}.$$

Then M defines $F \in BL(\mathbb{K}^n, \mathbb{K}^m)$, and $\|F\| = \beta_1$. Thus there is i_0 in $\{1, \ldots, m\}$ such that $\|F\| = \sum_{j=1}^{n} |k_{i_0,j}|$.

(iii) Consider the norm $\|\cdot\|_2$ on both $\mathbb{K}^n$ and $\mathbb{K}^m$, and let

$$\gamma_{2,2} := \Big(\sum_{i=1}^{m}\sum_{j=1}^{n} |k_{i,j}|^2\Big)^{1/2}, \quad \text{called the } \textbf{Frobenius norm} \text{ of } M.$$

Then M defines $F \in BL(\mathbb{K}^n, \mathbb{K}^m)$, and $\|F\| \le \min\{\sqrt{\alpha_1\beta_1}, \gamma_{2,2}\}$. Here a strict inequality can hold as the following simple example shows. Let $m = n := 2$ and $M := \begin{bmatrix} 1 & 1 \\ 1 & 0 \end{bmatrix}$. Then $F(x(1), x(2)) = (x(1) + x(2), x(1))$ for all $(x(1), x(2)) \in \mathbb{K}^2$. Let $(x(1), x(2)) \in \mathbb{K}^2$ with $|x(1)|^2 + |x(2)|^2 = 1$. Then

$$\begin{aligned} |x(1) + x(2)|^2 + |x(1)|^2 &\le (|x(1)| + |x(2)|)^2 + |x(1)|^2 \\ &= 1 + 2|x(1)|\,|x(2)| + |x(1)|^2 \\ &= 1 + 2|x(1)|\sqrt{1 - |x(1)|^2} + |x(1)|^2. \end{aligned}$$

For $s \in [0, 1]$, let $\phi(s) := 1 + 2s\sqrt{1 - s^2} + s^2$. To find the maximum of the function ϕ, we compare its values at the boundary points 0 and 1 of the interval $[0, 1]$, and at its critical points $\big((5 + \sqrt{5})/10\big)^{1/2}$ and $\big((5 - \sqrt{5})/10\big)^{1/2}$. On considering $x(1) := \big((5 + \sqrt{5})/10\big)^{1/2}$ and $x(2) := \big((5 - \sqrt{5})/10\big)^{1/2}$, we obtain

$$\|F\|^2 := \sup\big\{|x(1) + x(2)|^2 + |x(1)|^2 : |x(1)|^2 + |x(2)|^2 = 1\big\} = \frac{3 + \sqrt{5}}{2}.$$

Hence $\|F\| = (1 + \sqrt{5})/2$, which is less than $\sqrt{\alpha_1 \beta_1} = \sqrt{2 \cdot 2} = 2$, and also less than $\gamma_{2,2} = \sqrt{3}$. This example illustrates the difficulty in solving the optimization problem involved in finding $\|F\|$. (See Remark 3.11(iii).) ◊

Examples 3.16 As 'continuous' analogues of bounded linear maps defined by finite or infinite matrices, we consider bounded linear maps defined by kernels. Let X and Y denote function spaces considered in Example 2.1(iii) with $T := [a, b]$. Consider a measurable function $k(\cdot\,, \cdot)$ on $[a, b] \times [a, b]$. Suppose for every $x \in X$, and for almost every $s \in [a, b]$, the integral $\int_a^b k(s, t)x(t)dm(t)$ exists, and if we let $y(s) := \int_a^b k(s, t)x(t)dm(t)$, $s \in [a, b]$, then $y \in Y$. Let us define $F(x) := y$ for $x \in X$. It is easy to see that F is a linear map from X to Y. In this case, we say that the **kernel** $k(\cdot\,, \cdot)$ **defines** F, and F is called a **Fredholm integral map** from X to Y. Note that the discrete variables i and j in Example 3.14 are here replaced by the continuous variables s and t, and summation is replaced by integration. In what follows, we shall denote $L^p([a, b])$ by L^p for $p = 1, 2, \infty$.

(i) Let $X = Y := L^1$. Suppose the function $t \longmapsto \int_a^b |k(s, t)|dm(s)$ is essentially bounded on $[a, b]$, and let

$$\alpha_1 := \text{ess sup}_{[a,b]} \int_a^b |k(s, \cdot)|dm(s).$$

Then $F \in BL(L^1)$ and $\|F\| = \alpha_1$.

(ii) Let $X = Y := L^\infty$. Suppose the function $s \longmapsto \int_a^b |k(s, t)|dm(t)$ is essentially bounded on $[a, b]$, and let

$$\beta_1 := \text{ess sup}_{[a,b]} \int_a^b |k(\cdot\,, t)|dm(t).$$

Then $F \in BL(L^\infty)$ and $\|F\| = \beta_1$.

(iii) Let $X = Y := L^2$. Suppose the functions $t \longmapsto \int_a^b |k(s, t)|dm(s)$ and $s \longmapsto \int_a^b |k(s, t)|dm(t)$ are essentially bounded on $[a, b]$, and define α_1 and β_1 as in (i) and (ii) above. Then $F \in BL(L^2)$ and $\|F\| \le \sqrt{\alpha_1 \beta_1}$.

Alternatively, suppose

$$\gamma_{2,2} := \left(\int_a^b \int_a^b |k(s,t)|^2 dm(t) dm(s) \right)^{1/2} < \infty.$$

For $x \in L^2$ and $s \in [a,b]$,

$$|F(x)(s)| \le \int_a^b |k(s,t)|\,|x(t)|\,dm(t) \le \left(\int_a^b |k(s,t)|^2\,dm(t) \right)^{1/2} \|x\|_2$$

by the Schwarz inequality for functions (Lemma 1.25(ii)), and hence

$$\|F(x)\|_2^2 = \int_a^b |F(x)(s)|^2\,dm(s) \le \left(\int_a^b \int_a^b |k(s,t)|^2\,dm(t)\,dm(s) \right) \|x\|_2^2.$$

Hence $F \in BL(L^2)$ and $\|F\| \le \gamma_{2,2}$.

The proofs of the results for L^p are similar to the proofs for ℓ^p treated in Example 3.14, but additional care must be taken to justify the measurability of functions involved therein. These difficulties do not arise if the linear space $L^p([a,b])$ is repaced by the linear space $C([a,b])$, and the kernel is continuous on $[a,b] \times [a,b]$. See Exercise 3.11 in this regard. Also, see [15] for a detailed account of the case $L^2([a,b])$ mentioned in (iii) above. ◊

In this section, we have studied the normed space $BL(X,Y)$ of all bounded linear maps from a normed space X to a normed space Y. Before concluding this study, we mention two results about $BL(X,Y)$ when Y is a Banach space.

Proposition 3.17 *Let X be a normed space, and Y a Banach space. Then*

(i) *$BL(X,Y)$ is a Banach space.*
(ii) *Suppose X_0 is a dense subspace of X, and $F_0 \in BL(X_0,Y)$. Then there is a unique F in $BL(X,Y)$ such that $F(x) = F_0(x)$ for all $x \in X_0$. Also, $\|F\| = \|F_0\|$.*

Proof (i) Let (F_n) be a Cauchy sequence in $BL(X,Y)$. Let $\epsilon > 0$. There is $n_0 \in \mathbb{N}$ such that for all $n, m \ge n_0$,

$$\|F_n(x) - F_m(x)\| \le \|F_n - F_m\|\,\|x\| < \epsilon\,\|x\|, \quad x \in X.$$

Fix $x \in X$. Now $(F_n(x))$ is a Cauchy sequence in Y, and so there is $y \in Y$ such that $F_n(x) \to y$. Define $F(x) := y$. Clearly, the map $F : X \to Y$ is linear. Fix $n \ge n_0$, and let $m \to \infty$ in the inequality $\|F_n(x) - F_m(x)\| < \epsilon\,\|x\|$. It follows that $\|(F_n - F)(x)\| \le \epsilon\,\|x\|$ for all $x \in X$. In particular, $F_{n_0} - F \in BL(X,Y)$. Since $F = (F - F_{n_0}) + F_{n_0}$, we see that $F \in BL(X,Y)$. Also, $\|F_n - F\| \le \epsilon$ for all $n \ge n_0$.

Hence (F_n) converges to F in $BL(X, Y)$. Thus $BL(X, Y)$ is a Banach space.

(ii) Let $x \in X$. There is a sequence (x_n) in X_0 such that $x_n \to x$. Since

$$\|F_0(x_n) - F_0(x_m)\| \leq \|F_0\| \, \|x_n - x_m\|$$

for all $n, m \in \mathbb{N}$, we see that $(F_0(x_n))$ is a Cauchy sequence, and hence it converges to, say, y in the Banach space Y. If we let $F(x) := y$, then it is easy to check that $F : X \to Y$ is well defined and linear. Also, F is continuous and $\|F\| \leq \|F_0\|$ since

$$\|F(x)\| = \lim_{n\to\infty} \|F_0(x_n)\| \leq \|F_0\| \lim_{n\to\infty} \|x_n\| = \|F_0\| \, \|x\| \quad \text{for all } x \in X.$$

In fact, $\|F\| = \|F_0\|$ since $F(x) = F_0(x)$ for all $x \in X_0$. The uniqueness of F follows from the denseness of X_0 in X. □

The following converse of Proposition 3.17(i) holds. If $X \neq \{0\}$ and $BL(X, Y)$ is a Banach space, then so is Y. (See Exercise 4.5.)

3.2 Zabreiko Theorem and Uniform Boundedness

Let X be a linear space over $\mathbb{K}$, and let $p : X \to \mathbb{R}$ be a seminorm on X. It is easy to see that a seminorm p is **finitely subadditive**, that is,

$$p\left(\sum_{k=1}^{n} x_k\right) \leq \sum_{k=1}^{n} p(x_k) \quad \text{for all } x_1, \ldots, x_n \in X.$$

Let $\|\cdot\|$ be a norm on X, and consider the metric on X defined by $d(x, y) := \|x - y\|$, $x, y \in X$. Lemma 3.1 gives us a criterion for the seminorm p to be continuous on X with respect to this metric. We now describe an important consequence of the continuity of a seminorm. For this purpose, let us call a seminorm p on a normed space X **countably subadditive** if

$$p\left(\sum_{k=1}^{\infty} x_k\right) \leq \sum_{k=1}^{\infty} p(x_k) \quad \text{whenever the series } \sum_{k=1}^{\infty} x_k \text{ is summable in } X.$$

Since p is nonnegative on X, either the series $\sum_{k=1}^{\infty} p(x_k)$ converges in $\mathbb{R}$ or it diverges to ∞. Hence to prove the countable subadditivity of a seminorm, it is sufficient to verify the above condition only when $\sum_{k=1}^{\infty} p(x_k) < \infty$.

Lemma 3.18 *Let p be a seminorm on a normed space $(X, \|\cdot\|)$. If p is continuous on X, then p is countably subadditive.*

Proof Let $s := \sum_{k=1}^{\infty} x_k$ in X, and define $s_n := \sum_{k=1}^{n} x_k$ for $n \in \mathbb{N}$. Then $s_n \to s$. Since p is continuous at $s \in X$, we see that $p(s_n) \to p(s)$. Also, since p is finitely subadditive, we see that

$$p(s_n) \le \sum_{k=1}^{n} p(x_k) \le \sum_{k=1}^{\infty} p(x_k) \quad \text{for all } n \in \mathbb{N}.$$

Letting $n \to \infty$, we obtain $p(s) \le \sum_{k=1}^{\infty} p(x_k)$, as desired. □

Examples 3.19 Let $(X, \|\cdot\|_X)$ and $(Y, \|\cdot\|_Y)$ be normed spaces, and let F be a linear map from X to Y. Define $p(x) := \|F(x)\|_Y$ for $x \in X$. By Proposition 3.2, the seminorm p is continuous on X if and only if F is continuous on X. Thus if F is continuous on X, then p is countably subadditive by Lemma 3.18. On the other hand, if F is discontinuous on X, then p may or may not be countably subadditive, as the following examples show.

Let $X := C^1([0, 1])$ with the sup norm $\|\cdot\|_\infty$.

(i) Let $Y := C([0, 1])$ with the sup norm $\|\cdot\|_\infty$, and let $F : X \to Y$ be defined by $F(x) := x'$, the derivative of the function x. Then F is linear but discontinuous on X, as we have seen in Example 3.9(iii). Hence the seminorm p on X given by $p(x) := \|F(x)\| = \|x'\|_\infty$, $x \in X$, is discontinuous. We show that p is countably subadditive. Let $s := \sum_{k=1}^{\infty} x_k$ in X with $\sum_{k=1}^{\infty} p(x_k) = \sum_{k=1}^{\infty} \|x_k'\|_\infty < \infty$. Now $x_k' \in Y$ for each $k \in \mathbb{N}$, and Y is a Banach space. By Theorem 2.23, the absolutely summable series $\sum_{k=1}^{\infty} x_k'$ is summable in Y, that is, there is $y \in Y$ such that $y = \sum_{k=1}^{\infty} x_k'$. Define $s_n := \sum_{k=1}^{n} x_k$ for $n \in \mathbb{N}$. Then the sequence (s_n) of differentiable functions defined on $[0, 1]$ converges (uniformly) to the function $s \in C^1([0, 1])$, and the derived sequence (s_n'), where $s_n' = \sum_{k=1}^{n} x_k'$, $n \in \mathbb{N}$, converges uniformly to the function $y \in C([0, 1])$. By a well-known theorem in Real Analysis, $y = s'$. (See [25, Theorem 7.17].) Thus $s' = \sum_{k=1}^{\infty} x_k'$, and so $\|s'\|_\infty \le \sum_{k=1}^{\infty} \|x_k'\|_\infty$, that is, $p(s) \le \sum_{k=1}^{\infty} p(x_k)$. Hence the seminorm p is countably subadditive. This example shows that the converse of Lemma 3.18 does not hold.

(ii) Let $Y := \mathbb{K}$ with the norm given by $\|k\| := |k|$ for $k \in Y$, and let $f : X \to Y$ be defined by $f(x) := x'(1)$, the derivative of the function x at 1. Then f is linear, but discontinuous on X, as we have seen in Example 3.9(iii). Hence the seminorm p on X given by $p(x) = |f(x)| = |x'(1)|$, $x \in X$, is discontinuous. We show that p is not countably subadditive. Note that

$$t = \left(t - \frac{t^2}{2}\right) + \left(\frac{t^2}{2} - \frac{t^3}{3}\right) + \cdots = \sum_{k=1}^{\infty} \left(\frac{t^k}{k} - \frac{t^{k+1}}{k+1}\right),$$

where the series converges uniformly for $t \in [0, 1]$: Let $s(t) := t$ and $x_k(t) := (t^k/k) - \big(t^{k+1}/(k+1)\big)$ for $k \in \mathbb{N}$ and $t \in [0, 1]$. If $s_n := \sum_{k=1}^{n} x_k$ for $n \in \mathbb{N}$, then

$s_n(t) = t - \big(t^{n+1}/(n+1)\big)$ for $t \in [0, 1]$, and so $\|s_n - s\|_\infty \to 0$, that is, $s = \sum_{k=1}^{\infty} x_k$ in X. But $p(s) = |s'(1)| = 1$, while $p(x_k) = |x_k'(1)| = 1 - 1 = 0$ for all $k \in \mathbb{N}$, so that $\sum_{k=1}^{\infty} p(x_k) = 0$. Thus $\sum_{k=1}^{\infty} p(x_k) < p(s)$. ◊

We shall prove a theorem of Zabreiko, first given in [31], which says that the converse of Lemma 3.18 holds if X is a Banach space. For this purpose, we need a technical result about a seminorm defined on a normed space.

Lemma 3.20 *Let p be a seminorm on a normed space $(X, \|\cdot\|)$. For $\alpha > 0$, let $V_\alpha := \{x \in X : p(x) \le \alpha\}$, and let $\overline{V}_\alpha$ denote the closure of V_α in X. Suppose $a \in X$ and $r > 0$ are such that $\overline{U}(a, r) \subset \overline{V}_\alpha$. Then $\overline{U}(0, r) \subset \overline{V}_\alpha$, and in fact $\overline{U}(0, \delta r) \subset \overline{V}_{\delta\alpha}$ for every $\delta > 0$.*

Proof If $x, y \in \overline{V}_\alpha$, and $t \in (0, 1)$, then $(1-t)x + ty \in \overline{V}_\alpha$. To see this, let (x_n) and (y_n) be sequences in V_α such that $x_n \to x$ and $y_n \to y$. If we let $z_n := (1-t)x_n + ty_n$ for $n \in \mathbb{N}$, then $z_n \to (1-t)x + ty$, and each $z_n \in V_\alpha$ as $p(z_n) \le (1-t)p(x_n) + tp(y_n) \le \alpha$. Similarly, if $x \in \overline{V}_\alpha$ and $k \in \mathbb{K}$, then we see that $kx \in \overline{V}_{|k|\alpha}$.

Suppose $a \in X$ and $r > 0$ satisfy $\overline{U}(a, r) \subset \overline{V}_\alpha$. Let x belong to $\overline{U}(0, r)$. Then $x + a$ is in $\overline{U}(a, r) \subset \overline{V}_\alpha$. Also, $-x \in \overline{U}(0, r)$, and so $(-x + a) \in \overline{U}(a, r) \subset \overline{V}_\alpha$. Hence $(x - a)$ and $x = \frac{1}{2}\big((x + a) + (x - a)\big)$ are in $\overline{V}_\alpha$. Thus $\overline{U}(0, r) \subset \overline{V}_\alpha$. (See Fig. 3.1.) Next, let $\delta > 0$. If $x \in \overline{U}(0, \delta r)$, then $(x/\delta) \in \overline{U}(0, r) \subset \overline{V}_\alpha$, and so $x \in \overline{V}_{\delta\alpha}$. Thus $\overline{U}(0, \delta r) \subset \overline{V}_{\delta\alpha}$. □

Theorem 3.21 **(Zabreiko, 1969)** *Let p be a countably subadditive seminorm on a Banach space X. Then p is continuous on X.*

Proof For $n \in \mathbb{N}$, let $V_n := \{x \in X : p(x) \le n\}$. Then

$$X = \bigcup_{n=1}^{\infty} V_n = \bigcup_{n=1}^{\infty} \overline{V}_n, \quad \text{where } \overline{V}_n \text{ denotes the closure of } V_n \text{ in } X.$$

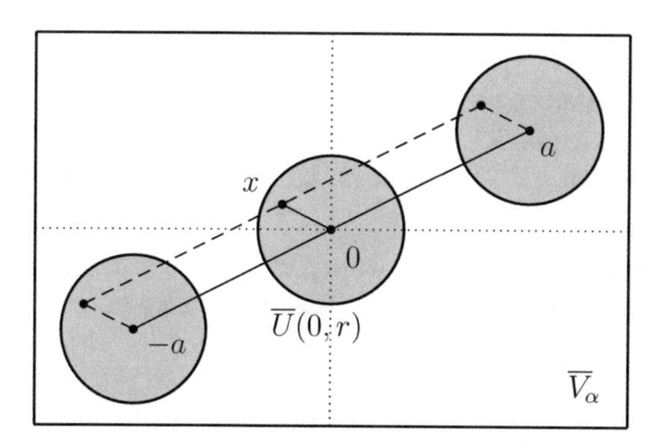

Fig. 3.1 Illustration of $\overline{U}(0, r) \subset \overline{V}_\alpha$ in the proof of Lemma 3.20

Hence $\bigcap_{n=1}^{\infty}(X \setminus \overline{V}_n) = \emptyset$. Now X is a complete metric space, and $X \setminus \overline{V}_n$ is an open subset of X for each $n \in \mathbb{N}$. Theorem 1.10 of Baire shows that one of the sets $X \setminus \overline{V}_n$ is not dense in X. Thus there is $m \in \mathbb{N}$, and there are $a \in X$ and $r > 0$ such that $\overline{U}(a, r) \subset \overline{V}_m$. We shall use Lemma 3.20 repeatedly to prove that $p(x) \leq m\|x\|/r$ for all $x \in X$.

Let $x \in X$, and let $\epsilon > 0$. We shall find $x_1, x_2, \ldots$ in X such that $x = \sum_{k=1}^{\infty} x_k$, $p(x_1) \leq m\|x\|/r$ and $p(x_k) \leq m\,\epsilon/2^{k-1}$ for $k \geq 2$, and then use the countable subadditivity of p.

Let $x \in X$. Define $\epsilon_0 := \|x\|/r$, and $\epsilon_k := \epsilon/2^k$ for $k \in \mathbb{N}$. Since $\|x\| = \epsilon_0 r$, we see that $x \in \overline{V}_{\epsilon_0 m}$ by Lemma 3.20 (with $\alpha := m$ and $\delta = \epsilon_0$), and so there is $x_1 \in X$ with $p(x_1) \leq \epsilon_0 m$ and $\|x - x_1\| \leq \epsilon_1 r$. Let $u_1 := x - x_1$, so that $x = x_1 + u_1$, where $\|u_1\| \leq \epsilon_1 r$. Lemma 3.20 (with $\alpha = m$ and $\delta = \epsilon_1$) shows that $u_1 \in \overline{V}_{\epsilon_1 m}$, and so there is $x_2 \in X$ with $p(x_2) \leq \epsilon_1 m$ and $\|u_1 - x_2\| \leq \epsilon_2 r$. Let $u_2 := u_1 - x_2$, so that $x = x_1 + u_1 = x_1 + x_2 + u_2$, where $\|u_2\| \leq \epsilon_2 r$. Continuing in this manner, we find $x_k, u_k \in X$ such that $p(x_k) \leq \epsilon_{k-1} m$ and $x = x_1 + x_2 + \cdots + x_k + u_k$, where $\|u_k\| \leq \epsilon_k r$ for $k \in \mathbb{N}$. Since $\epsilon_k = \epsilon/2^k \to 0$, it follows that $u_k \to 0$, that is, $x = \sum_{k=1}^{\infty} x_k$.

By the countable subadditivity of the seminorm p,

$$p(x) \leq \sum_{k=1}^{\infty} p(x_k) \leq m \sum_{k=1}^{\infty} \epsilon_{k-1} = m\left(\epsilon_0 + \sum_{k=1}^{\infty} \epsilon_k\right) = \frac{m\|x\|}{r} + m\,\epsilon.$$

Since this inequality holds for every $\epsilon > 0$, we obtain $p(x) \leq m\|x\|/r$ for all $x \in X$. Lemma 3.1 now shows that p is continuous on X. □

Corollary 3.22 (Equicontinuity of seminorms) *Let X be a Banach space, and let $\mathcal{P}$ be a set of continuous seminorms on X such that the set $\{p(x) : p \in \mathcal{P}\}$ is bounded for each $x \in X$. Then there is $\alpha > 0$ such that $p(x) \leq \alpha\|x\|$ for all $p \in \mathcal{P}$ and $x \in X$. Consequently, $\mathcal{P}$ is equicontinuous.*

Proof Define $\wp(x) := \sup\{p(x) : p \in \mathcal{P}\}$ for $x \in X$. Since each $p \in \mathcal{P}$ is a seminorm on X, it is easy to see that $\wp$ is a seminorm on X. Let $x := \sum_{k=1}^{\infty} x_k$ in X. Since each $p \in \mathcal{P}$ is continuous, it is countably subadditive by Lemma 3.18, and so $p(x) \leq \sum_{k=1}^{\infty} p(x_k) \leq \sum_{k=1}^{\infty} \wp(x_k)$. Hence $\wp(x) \leq \sum_{k=1}^{\infty} \wp(x_k)$. Thus $\wp$ is a countably subadditive seminorm on X. By the Zabreiko Theorem (Theorem 3.21), $\wp$ is continuous, and by Lemma 3.1, there is $\alpha > 0$ with $\wp(x) \leq \alpha\|x\|$ for all $x \in X$, that is, $p(x) \leq \alpha\|x\|$ for all $p \in \mathcal{P}$ and for all $x \in X$. The equicontinuity of the set $\mathcal{P}$ follows easily. □

A family of continuous functions from a metric space to a metric space can be bounded at each point without being uniformly bounded on the metric space. For example, for $n \in \mathbb{N}$, define

$$x_n(t) := \begin{cases} n^2 t & \text{if } 0 \leq t \leq (1/n), \\ 1/t & \text{if } (1/n) < t \leq 1. \end{cases}$$

Then for each $n \in \mathbb{N}$, x_n is continuous on $[0, 1]$, $x_n(0) = 0$ and $|x_n(t)| \leq 1/t$ for each $t \in (0, 1]$, but $x_n(1/n) = n$. Hence the family $\{x_n : n = 1, 2, \ldots\}$ of continuous functions on $[0, 1]$ is bounded at each point of [0,1], but it is not uniformly bounded on [0,1]. We now show that if a family of continuous linear maps from a Banach space X to a normed space Y is bounded at each $x \in X$, then it is uniformly bounded on the closed unit ball of X, and hence it is uniformly bounded on each bounded subset of X. We shall deduce this result from Corollary 3.22.

Theorem 3.23 (Uniform boundedness principle) *Let $(X, \|\cdot\|)$ be a Banach space. For each s in an index set S, let $(Y_s, \|\cdot\|_s)$ be a normed space, and let $F_s \in BL(X, Y_s)$ be such that the set $\{\|F_s(x)\|_s : s \in S\}$ is bounded for each $x \in X$. Then the set $\{\|F_s\| : s \in S\}$ is bounded.*

Proof For $s \in S$, define $p_s : X \to \mathbb{R}$ by $p_s(x) := \|F_s(x)\|_s$ for $x \in X$. Since F_s is linear and continuous, we see that p_s is a continuous seminorm on X for each $s \in S$. Let $\mathcal{P} := \{p_s : s \in S\}$. For a fixed $x \in X$, the set $\{p(x) : p \in \mathcal{P}\} = \{\|F_s(x)\|_s : s \in S\}$ is bounded. By Corollary 3.22, there is $\alpha > 0$ such that

$$\|F_s(x)\|_s = p_s(x) \leq \alpha\|x\| \text{ and all } s \in S \text{ and all } x \in X.$$

This shows that $\|F_s\| \leq \alpha$ for all $s \in S$. □

Theorem 3.24 (Banach–Steinhaus, 1927) *Let X be a Banach space, Y be a normed space, and let (F_n) be a sequence in $BL(X, Y)$ such that $(F_n(x))$ converges in Y for every $x \in X$. Define $F(x) := \lim_{n\to\infty} F_n(x)$ for $x \in X$. Then the sequence $(\|F_n\|)$ is bounded, $F \in BL(X, Y)$, and $\|F\| \leq \sup_{n\in\mathbb{N}}\{\|F_n\|\}$.*

Proof It is clear that the map $F : X \to Y$ is linear. For each $x \in X$, the set $\{\|F_n(x)\| : n \in \mathbb{N}\}$ is bounded because $(F_n(x))$ is a convergent sequence in Y. By the uniform boundedness principle (Theorem 3.23), the set $\{\|F_n\| : n \in \mathbb{N}\}$ is bounded. Let $\alpha := \sup\{\|F_n\| : n \in \mathbb{N}\}$. Then $\|F(x)\| = \lim_{n\to\infty} \|F_n(x)\| \leq \alpha\|x\|$ for all $x \in X$. This shows that F is continuous and $\|F\| \leq \alpha$. □

Remark 3.25 The hypothesis of completeness of X cannot be dropped from the Banach–Steinhaus theorem (Theorem 3.24) and from the result on the equicontinuity of seminorms (Corollary 3.22). For example, let $X := c_{00}$ with the norm $\|\cdot\|_\infty$, and define

$$f_n(x) := \sum_{j=1}^{n} x(j) \quad \text{and} \quad f(x) := \sum_{j=1}^{\infty} x(j) \quad \text{for } x \in X.$$

Then each f_n is a continuous linear functional on X. In fact, $\|f_n\| = n$ for each $n \in \mathbb{N}$. Now fix $x \in X$, and find m_x such that $x(j) = 0$ for all $j > m_x$. Then $f_n(x) = f(x)$ for all $n \geq m_x$. Hence $f_n(x) \to f(x)$ for every $x \in X$. However, the linear functional f is not continuous on X, since $f(e_1 + \cdots + e_n) = n$, where $\|e_1 + \cdots + e_n\|_\infty = 1$ for all $n \in \mathbb{N}$. Also, the set $\{\|f_n\| : n \in \mathbb{N}\}$ is unbounded. Similarly, for $n \in \mathbb{N}$ and $x \in X$, let $p_n(x) := \sum_{j=1}^{n} |x(j)|$, and so $p_n(x) \leq n\|x\|_\infty$. Hence each p_n is a continuous seminorm on X. Also, for $x \in X$, the set $\{p_n(x) : n \in \mathbb{N}\}$ is bounded above by $m_x \|x\|_\infty$. But the seminorm given by $\wp(x) := \sup\{p_n(x) : n \in \mathbb{N}\} = \sum_{j=1}^{\infty} |x(j)|$, $x \in X$, is not continuous on X. Note that the linear space c_{00} is not complete with respect to any norm. ◊

Corollary 3.26 *Let $p, r \in \{1, 2, \infty\}$, and let q satisfy $(1/p) + (1/q) = 1$. Suppose an infinite matrix M defines a map F from ℓ^p to ℓ^r. Then every row of M belongs to ℓ^q. In fact, the rows of M form a bounded subset of ℓ^q.*

Proof Let $M := [k_{i,j}]$, where $k_{i,j}$ is the element in the ith row and the jth column. Fix $i \in \mathbb{N}$. For $n \in \mathbb{N}$, define $f_{i,n} : \ell^p \to \mathbb{K}$ by

$$f_{i,n}(x) := \sum_{j=1}^{n} k_{i,j} x(j), \quad x \in \ell^p.$$

Considering $y_{i,n} := (k_{i,1}, \ldots, k_{i,n}, 0, 0, \ldots) \in \ell^q$ in Example 3.12, it follows that $f_{i,n}$ is in $BL(\ell^p, \mathbb{K})$ and $\|f_{i,n}\| = \|y_{i,n}\|_q = \|(k_{i,1}, \ldots, k_{i,n}, 0, 0, \ldots)\|_q$. Since M defines a linear map from ℓ^p to ℓ^r, the series $\sum_{j=1}^{\infty} k_{i,j} x(j)$ converges in $\mathbb{K}$, that is, $f_{i,n}(x) \to \sum_{j=1}^{\infty} k_{i,j} x(j)$ as $n \to \infty$ for every $x \in \ell^p$. Also, ℓ^p is a Banach space (Example 2.24(ii)). By the Banach–Steinhaus theorem (Theorem 3.24), there is a positive number α_i such that $\|(k_{i,1}, \ldots, k_{i,n}, 0, 0, \ldots)\|_q = \|f_{i,n}\| \leq \alpha_i$ for all $n \in \mathbb{N}$, that is, the ith row $(k_{i,1}, k_{i,2}, \ldots)$ of M belongs to ℓ^q. Now define

$$f_i(x) := \sum_{j=1}^{\infty} k_{i,j} x(j), \quad x \in \ell^p.$$

Considering $y_i := (k_{i,1}, k_{i,2}, \ldots) \in \ell^q$ in Example 3.12, it follows that f_i is in $BL(\ell^p, \mathbb{K})$ and $\|f_i\| = \|y_i\|_q = \|(k_{i,1}, k_{i,2}, \ldots)\|_q$. Note that $F(x)(i) = f_i(x)$ for all $x \in \ell^p$. Further, if we fix $x \in \ell^p$, then

$$|f_i(x)| = |F(x)(i)| \leq \|F(x)\|_r \quad \text{for all } i \in \mathbb{N}.$$

By the uniform boundedness principle (Theorem 3.23), the set $\{\|f_i\| : i \in \mathbb{N}\}$ is bounded, that is, the rows of the matrix M are bounded in ℓ^q. □

See Corollary 3.31 for a similar result for the columns of an infinite matrix.

Quadrature Formulæ

Let $X := C([a, b])$ with the sup norm, and define $Q : X \to \mathbb{K}$ by

$$Q(x) := \int_a^b x(t)dt, \quad x \in X.$$

It is clear that the functional Q is linear. Also, $|Q(x)| \leq (b - a)\|x\|_\infty$ for all $x \in X$, and $Q(1) = b - a$. Hence Q is continuous, and $\|Q\| = b - a$. Although it is of great importance to find the exact value of the linear functional Q at a given continuous function x, it is seldom possible to do so. The Riemann sum $\sum_{j=1}^{m}(s_j - s_{j-1})x(t_j)$, where $a := s_0 < s_1 < \cdots < s_{m-1} < s_m := b$ and $s_{j-1} \le t_j \le s_j$ for $j = 1, \ldots, m$, approximates $Q(x)$ if we make the mesh $\max\{|s_j - s_{j-1}| : j = 1, \ldots, m\}$ of the partition small. A variety of sums of this type are used to calculate $Q(x)$ approximately.

A **quadrature formula** Q_m is a function on X defined by

$$Q_m(x) := \sum_{j=1}^{m} w_j x(t_j),$$

where the **nodes** $t_1, \ldots, t_m$ satisfy $a \le t_1 < \cdots < t_m \le b$ and the **weights** $w_1, \ldots, w_m$ are in $\mathbb{K}$. Clearly, Q_m is a linear functional, and since

$$|Q_m(x)| \le \left(\sum_{j=1}^{m} |w_j|\right) \|x\|_\infty \quad \text{for all } x \in X,$$

we see that $\|Q_m\| \le \sum_{j=1}^{m} |w_j|$. In fact, define $x \in X$ by $x(t_j) := \operatorname{sgn} w_j$ for $j = 1, \ldots, m$, and

$$x(t) := \begin{cases} x(t_1) & \text{if } a \le t < t_1, \\ c_j(t - t_{j-1}) + d_j & \text{if } t_{j-1} \le t < t_j \text{ for } j = 2, \ldots, m, \\ x(t_m) & \text{if } t_m \le t \le b, \end{cases}$$

where $c_j := \big(x(t_j) - x(t_{j-1})\big)/(t_j - t_{j-1})$ and $d_j := x(t_{j-1})$ for $j = 2, \ldots, m$. Then $\|x\|_\infty \le \max\{|x(t_1)|, \ldots, |x(t_m)|\} \le 1$ and $Q_m(x) = \sum_{j=1}^{m} |w_j|$. It follows that

$$\|Q_m\| = \sum_{j=1}^{m} |w_j|.$$

A sequence (Q_n) of quadrature formulæ is said to be **convergent** if $Q_n(x)$ converges to $Q(x)$ for every $x \in C([a, b])$. The following result gives a set of necessary and sufficient conditions for a sequence of quadrature formulæ to be convergent.

Theorem 3.27 **(Polya, 1933)** *For $n \in \mathbb{N}$ and $x \in C([a, b])$, let*

$$Q_n(x) := \sum_{j=1}^{m_n} w_{n,j} x(t_{n,j}).$$

Then the sequence (Q_n) of quadrature formulæ is convergent if and only if

(i) *$Q_n(x) \to \int_a^b x(t)dt$ for every x in a subset E of $C([a, b])$ such that* span *E is dense in $C([a, b])$, and*

(ii) *there is* $\alpha > 0$ *such that* $\sum_{j=1}^{m_n} |w_{n,j}| \le \alpha$ *for all* $n \in \mathbb{N}$.

Proof Let $X := C([a, b])$ along with the sup norm $\|\cdot\|_\infty$, and for $x \in X$, let $Q(x) := \int_a^b x(t)dt$ as before. Suppose the sequence (Q_n) is convergent. Then condition (i) is satisfied with $E = C([a, b])$. Also, we have seen that $\|Q_n\| = \sum_{j=1}^{m_n} |w_{n,j}|$ for all $n \in \mathbb{N}$. Thus condition (ii) follows from the Banach–Steinhaus theorem.

Conversely, suppose conditions (i) and (ii) hold, and let $X_0 := \operatorname{span} E$. It is easy to see that $Q_n(x) \to Q(x)$ for every $x \in X_0$. Consider now $x \in X$ and $\epsilon > 0$. Since X_0 is dense in X, there is $x_0 \in X_0$ such that $\|x - x_0\|_\infty < \epsilon$. Also, since $Q_n(x_0) \to Q(x_0)$, there is $n_0 \in \mathbb{N}$ such that $|Q_n(x_0) - Q(x_0)| < \epsilon$ for all $n \ge n_0$. Further, $\|Q_n\| = \sum_{j=1}^{m_n} |w_{n,j}| \le \alpha$ for all $n \in \mathbb{N}$. Hence

$$\begin{aligned} |Q_n(x) - Q(x)| &\le |Q_n(x - x_0)| + |Q_n(x_0) - Q(x_0)| + |Q(x_0 - x)| \\ &\le \|Q_n\| \|x - x_0\| + |Q_n(x_0) - Q(x_0)| + \|Q\| \|x_0 - x\| \\ &\le (\alpha + 1 + b - a)\,\epsilon \end{aligned}$$

for all $n \ge n_0$. Thus $Q_n(x) \to Q(x)$, $x \in X$, that is, (Q_n) is convergent. □

As a special case, suppose the weights $w_{n,j}$ appearing in a sequence (Q_n) of quadrature formulæ are all nonnegative. Then condition (ii) of Theorem 3.27 is satisfied if the set E in condition (i) contains a nonzero constant function, and so $\sum_{j=1}^{m_n} w_{n,j} \to b - a$. The weights in the classical compound quadrature formulæ which use the mid-point rule, the trapezoidal rule, the Gauss two-point rule and the Simpson rule are nonnegative. (See [12, p. 339].)

3.3 Closed Graph and Open Mapping Theorems

Let (X, d_X) and (Y, d_Y) be metric spaces, and let $F : X \to Y$. We introduce a property of F which is, in general, weaker than continuity.

We say that F is a **closed map** if the conditions $x_n \to x$ in X and $F(x_n) \to y$ in Y imply that $y = F(x)$. Let us define the **graph** of F by

$$\operatorname{Gr}(F) := \{(x, F(x)) : x \in X\} \subset X \times Y.$$

Consider the metric d on $X \times Y$ defined by $d\big((x_1, y_1), (x_2, y_2)\big) := d_X(x_1, x_2) + d_Y(y_1, y_2)$ for $(x_1, y_1), (x_2, y_2) \in X \times Y$. Then $(x_n, F(x_n)) \to (x, y)$ in $X \times Y$ if and only if $x_n \to x$ in X and $F(x_n) \to y$ in Y. It follows that $F : X \to Y$ is a closed map if and only if $\operatorname{Gr}(F)$ is a closed subset of $X \times Y$.

Suppose F is continuous. Since the condition $x_n \to x$ in X implies that $F(x_n) \to F(x)$ in Y, we see that F is a closed map. However, a closed map may not be continuous in general. For example, define $F : \mathbb{R} \to \mathbb{R}$ by $F(t) := 1/t$ if $t \ne 0$ and

$F(0) := 0$. Clearly, F is not continuous at 0, but it is a closed map since its graph $\{(0, 0)\} \cup \{(t, 1/t) : t \in \mathbb{R}, t \neq 0\}$ is a closed subset of $\mathbb{R}^2$.

Let now X and Y be normed spaces, and let $F : X \to Y$ be a linear map. Then F is a closed map if and only if the conditions $x_n \to 0$ in X and $F(x_n) \to y$ in Y imply that $y = 0$. In particular, if F is a closed map, then its zero space $Z(F)$ is closed in X.[1] As a consequence, a closed linear map of finite rank is continuous by Proposition 3.6(ii).

In general, a closed linear map need not be continuous. For example, let $X := C^1([a, b])$ and $Y := C([a, b])$, both with the sup norm $\|\cdot\|_\infty$, and define $F(x) = x'$, the derivative of the function $x \in X$. We have seen in Example 3.9(iii) that F is not continuous. To see that F is a closed map, let (x_n) be a sequence in X such that $x_n \to 0$ and $F(x_n) \to y$ in Y, that is, (x_n) converges uniformly to 0 on $[a, b]$ and (x_n') converges uniformly to a continuous function y on $[a, b]$. As we have pointed out in Example 3.19(i), a well-known theorem in Real Analysis shows that $y = 0$. Thus F is a closed map.

We shall now use the Zabreiko theorem to prove the following major result.

Theorem 3.28 (Closed graph theorem) *Let X and Y be Banach spaces, and let $F : X \to Y$ be closed linear map. Then F is continuous.*

Proof Define a seminorm $p : X \to \mathbb{R}$ by $p(x) := \|F(x)\|$ for $x \in X$. We claim that p is countably subadditive. Let $x = \sum_{k=1}^{\infty} x_k \in X$ be such that $\sum_{k=1}^{\infty} p(x_k) = \sum_{k=1}^{\infty} \|F(x_k)\| < \infty$. Since Y is a Banach space, the absolutely summable series $\sum_{k=1}^{\infty} F(x_k)$ of terms in Y is summable in Y, that is, there is $y \in Y$ such that $y = \sum_{k=1}^{\infty} F(x_k)$. Define $s_n := \sum_{k=1}^{n} x_k$ for $n \in \mathbb{N}$. Then $s_n \to x$ in X and $F(s_n) = \sum_{k=1}^{n} F(x_k) \to y$ in Y. Since F is a closed map, $y = F(x)$, that is, $F(x) = \sum_{k=1}^{\infty} F(x_k)$. Hence

$$p(x) = \|F(x)\| \leq \sum_{k=1}^{\infty} \|F(x_k)\| = \sum_{k=1}^{\infty} p(x_k),$$

as claimed. Since X is a Banach space, p is a continuous seminorm by the Zabreiko Theorem (Theorem 3.21). By Proposition 3.2, F is continuous. □

Remark 3.29 Let X and Y be normed spaces. Let $F : X \to Y$ be a linear map, and define a seminorm on X by $p(x) := \|F(x)\|$ for $x \in X$. The above proof shows that if Y is a Banach space, and F is a closed map, then p is countably subadditive.

[1] Here is an example of a one-one linear map which is not closed. Let X and Y denote the linear spaces consisting of polynomial functions on $[0, 1]$ and on $[2, 3]$, respectively, along with the respective sup norms. If $x \in X$, and $x(t) := a_0 + a_1 t + \cdots + a_n t^n$ for $t \in [0, 1]$, then define $F(x)(t) := a_0 + a_1 t + \cdots + a_n t^n$ for $t \in [2, 3]$. Clearly, $F : X \to Y$ is one-one and linear. Consider a continuous function y defined on $[0, 3]$ as follows: $y(t) := 0$ if $t \in [0, 1]$, $y(t) := t - 1$ if $t \in (1, 2)$ and $y(t) := 1$ if $t \in [2, 3]$. By Theorem 1.16 of Weierstrass, there is a sequence (y_n) of polynomial functions on $[0, 3]$ which converges uniformly to y on $[0, 3]$. Let x_n denote the restriction of y_n to $[0, 1]$ for $n \in \mathbb{N}$. Then $x_n \to 0$ in X but $F(x_n) \to 1$ in Y. Hence F is not a closed map. Note that $Z(F) = \{0\}$ is a closed subspace of X.

Conversely, Zabreiko has shown in an email correspondence that if p is countably subadditive, then F is a closed map. ◊

Matrix Transformations

If an infinite matrix M defines a map from ℓ^p to ℓ^r, where $p, r \in \{1, 2, \infty\}$, then Corollary 3.26 says that the rows of M form a bounded subset of ℓ^q, where $(1/p) + (1/q) = 1$. Before obtaining a similar result for the columns of the matrix M, we prove a general result about matrix transformations.

Proposition 3.30 *Let $p, r \in \{1, 2, \infty\}$. Then every matrix transformation from ℓ^p to ℓ^r is continuous. Conversely, if $p \neq \infty$, then every continuous linear map from ℓ^p to ℓ^r is a matrix transformation.*

Proof Let an infinite matrix $M := [k_{i,j}]$ define a (linear) map F from ℓ^p to ℓ^r. We show that F is a closed map. Let $x_n \to x$ in ℓ^p and $F(x_n) \to y$ in ℓ^r. Fix $i \in \mathbb{N}$. Then $F(x_n)(i) = \sum_{j=1}^{\infty} k_{i,j} x_n(j) \to y(i)$. By Corollary 3.26, the ith row $y_i := (k_{i,1}, k_{i,2}, \ldots)$ of M is in ℓ^q, where $(1/p) + (1/q) = 1$, and so

$$\left| \sum_{j=1}^{\infty} k_{i,j} x_n(j) - \sum_{j=1}^{\infty} k_{i,j} x(j) \right| = \left| \sum_{j=1}^{\infty} k_{i,j} \big(x_n(j) - x(j) \big) \right| \leq \|y_i\|_q \|x_n - x\|_p$$

for all $n \in \mathbb{N}$, as in Example 3.12. Hence $\sum_{j=1}^{\infty} k_{i,j} x_n(j) \to \sum_{j=1}^{\infty} k_{i,j} x(j)$. This shows that $y(i) = \sum_{j=1}^{\infty} k_{i,j} x(j)$, that is, $y(i) = F(x)(i)$. Since this holds for every $i \in \mathbb{N}$, we obtain $y = F(x)$. Thus F is a closed map. Also, ℓ^p and ℓ^r are Banach spaces (Example 2.24(ii)). By the closed graph theorem (Theorem 3.28), F is continuous.

Conversely, let $p \neq \infty$, and consider $F \in BL(\ell^p, \ell^r)$. Let $x \in \ell^p$. Since $x = \sum_{j=1}^{\infty} x(j) e_j$, the continuity and the linearity of F shows that $F(x) = \sum_{j=1}^{\infty} x(j) F(e_j)$ belongs to ℓ^r. In particular, the series $\sum_{j=1}^{\infty} x(j) F(e_j)(i)$ converges to $F(x)(i)$ for every $i \in \mathbb{N}$. Let $k_{i,j} := F(e_j)(i)$ for $i, j \in \mathbb{N}$. It follows that the infinite matrix $M := [k_{i,j}]$ defines the map F. □

The above result shows that matrix transformations from ℓ^p to ℓ^r follow the finite dimensional pattern where every $m \times n$ matrix defines a continuous linear map from $\mathbb{K}^n$ to $\mathbb{K}^m$. However, this result should not lead the reader to think that every linear map from ℓ^p to ℓ^r is continuous. (See Exercise 3.1.) Also, if $p = \infty$, and $r \in \{1, 2, \infty\}$, then there is $F \in BL(\ell^p, \ell^r)$ which is not a matrix transformation. (See the footnote given in Example 4.18(iii).)

Corollary 3.31 *Let $p, r \in \{1, 2, \infty\}$, and let an infinite matrix M define a map F from ℓ^p to ℓ^r. Then the columns of M form a bounded subset of ℓ^r.*

Proof Let $M := [k_{i,j}]$, where $k_{i,j}$ is the element in the ith row and the jth column. Fix $j \in \mathbb{N}$. Since $F(e_j)(i) = k_{i,j}$ for all $i \in \mathbb{N}$, we see that $F(e_j)$ is the jth column

$(k_{1,j}, k_{2,j}, \ldots)$ of M. By Proposition 3.30, the linear map F is continuous. Hence $\|F(e_j)\|_r \leq \|F\| \, \|e_j\|_p = \|F\|$ by the basic inequality for the operator norm (Theorem 3.10(i)). Hence the desired result follows. □

Remark 3.32 Let M denote an infinite matrix. Combining our results in Example 3.14(i) and (ii) with Proposition 3.30, we obtain the following:

(i) M defines a map F from ℓ^1 to ℓ^1 if and only if the set $\{\alpha_1(j) : j \in \mathbb{N}\}$ of all column sums of $|M|$ is bounded. In that case, $F \in BL(\ell^1)$ and $\|F\| = \sup\{\alpha_1(j) : j \in \mathbb{N}\}$.

(ii) M defines a map F from ℓ^∞ to ℓ^∞ if and only if the set $\{\beta_1(i) : i \in \mathbb{N}\}$ of all row sums of $|M|$ is bounded. In that case, $F \in BL(\ell^\infty)$ and $\|F\| = \sup\{\beta_1(i) : i \in \mathbb{N}\}$.

Similar results for matrix transformations from ℓ^1 to ℓ^2, from ℓ^1 to ℓ^∞, and from ℓ^2 to ℓ^∞ are given in Exercise 3.31. For the remaining cases, it is not possible to give conditions in terms of the entries of M which are necessary as well as sufficient; some sufficient conditions are given in Exercise 3.36, and some necessary conditions follow from Corollaries 3.26 and 3.31. See also Exercise 3.33. We shall introduce a Hilbert–Schmidt operator on a Hilbert space in Exercise 3.40, and invite the reader to show that the condition $\gamma_{2,2}^2 := \sum_{i=1}^\infty \sum_{j=1}^\infty |k_{i,j}|^2 < \infty$ is necessary as well as sufficient for an infinite matrix $M := [k_{i,j}]$ to define such an operator on ℓ^2. ◊

Projection Operators

We consider an interesting consequence of the closed graph theorem. A linear map P from a linear space X to itself is called a **projection operator** on X if $P^2 = P$. Let P be a projection operator on X. Then $I - P$ is also a projection operator on X. Further, the range spaces and the zero spaces of P and $I - P$ are related as follows: $R(P) = Z(I - P)$ and $Z(P) = R(I - P)$. Since $x = P(x) + (I - P)(x)$ for all $x \in X$, and $Z(P) \cap Z(I - P) = \{0\}$, $X = R(P) \oplus Z(P)$. Conversely, let Y and Z be subspaces of X such that $X := Y \oplus Z$. Then for $x \in X$, there are unique $y \in Y$ and $z \in Z$ such that $x = y + z$. The map $P : X \to Y$ given by $P(x) := y$ is a projection operator on X with $R(P) = Y$ and $Z(P) = Z$. This map is called the projection operator **onto** Y **along** Z.

Proposition 3.33 *Let X be a normed space, and let $P : X \to X$ be a projection operator. Then P is a closed map if and only if its range space and its zero space are closed in X. If X is a Banach space, and the range space and the zero space of P are closed in X, then P is in fact continuous.*

Proof Let P be a closed map. Let $y_n \in R(P)$, $z_n \in Z(P)$ and $y_n \to y$, $z_n \to z$ in X. Then $P(y_n) = y_n \to y$ and $P(z_n) = 0 \to 0$ in X, so that $P(y) = y$ and $P(z) = 0$. Thus $y \in R(P)$ and $z \in Z(P)$, showing that the subspaces $R(P)$ and $Z(P)$ are closed in X.

Conversely, suppose $R(P)$ and $Z(P)$ are closed in X. Let $x_n \to 0$ and $P(x_n) \to y$ in X. Then $P(x_n) - x_n \to y$ in X. Since $P(x_n) \in R(P)$ for each $n \in \mathbb{N}$, and $R(P)$ is closed, we see that $y \in R(P)$. Also, since $P(x_n) - x_n$ belongs to $Z(P)$ for each $n \in \mathbb{N}$, and $Z(P)$ is closed, we see that y belongs to $Z(P)$. Thus $y \in R(P) \cap Z(P)$, that is, $y = 0$. Hence P is a closed map.

Let X be a Banach space. By the closed graph theorem (Theorem 3.28), a closed linear operator on X is continuous. The desired result follows. □

Let X be an inner product space. If P is a projection operator on X such that $R(P) \perp Z(P)$, that is, $y \perp z$ for all $y \in R(P)$ and all $z \in Z(P)$, then P is called an **orthogonal projection operator**. A projection operator on X which is not orthogonal is called an **oblique projection operator**. Figure 3.2 shows an oblique projection operator P onto Y along Z, and also the orthogonal projection operator Q onto Y (along $Y^\perp$) defined on $\mathbb{R}^2$.

Remark 3.34 Let us briefly discuss the existence of projection operators. Let X be linear space, and let Y be a subspace of X. Let $\{y_s\}$ be a (Hamel) basis for Y, and extend it to a basis $\{y_s\} \cup \{z_t\}$ for X (Proposition 1.1). If we let $Z := \text{span}\,\{z_t\}$, then clearly $X = Y \oplus Z$. Thus there *d*oes exist a projection operator onto Y (along Z). Now let X be a normed space, and let Y be a closed subspace of X. Does there exist a closed projection operator P on X such that $R(P) = Y$? By Proposition 3.33, such a projection operator exists if and only if there is a closed subspace Z of X such that $X = Y \oplus Z$. In that case, Z is called a **closed complement** of Y in X. If Y is a finite dimensional (and hence closed) subspace of X, then it has a closed complement in X. This can be seen by considering $Z := \{x \in X : f_j(x) = 0 \text{ for } j = 1, \ldots, m\}$, where $f_1, \ldots, f_m$ are as in Proposition 4.6(iii). On the other hand, it is known that c_0 has no closed complement in ℓ^∞, and $C([0, 1])$ has no closed complement in $B([0, 1])$. (See [10, 30].)

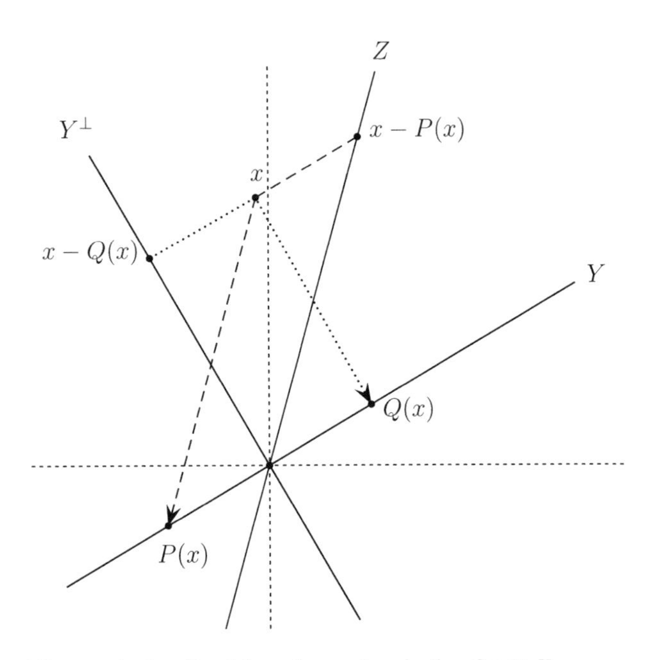

Fig. 3.2 An oblique projection P and the orthogonal projection Q onto Y

On the other hand, for every closed subspace G of a Hilbert space H, the orthogonal complement $G^{\perp}$ of H satisfies $H = G \oplus G^{\perp}$ by Theorem 2.35. In fact, if X is a Banach space such that for every closed subspace Y of X, there is a closed subspace Z of X with $X = Y \oplus Z$, then there is a linear homeomorphism from X onto a Hilbert space H. This result was proved by Lindenstrauss and Tzafriri ([22]) in 1974. $\diamondsuit$

Inverse Map

Let F be a one-one continuous map from a metric space X onto a metric space Y. In general, the inverse map F^{-1} from Y onto X may not be continuous. For example, let $X := [0, 2\pi)$, $Y := \{z \in \mathbb{C} : |z| = 1\}$, and $F(x) := e^{ix}$ for $x \in X$. Then F is one-one, continuous and onto, but F^{-1} is not continuous at 1, since $y_n := e^{2\pi i(n-1)/n} \to 1$, but $F^{-1}(y_n) = 2\pi(n-1)/n \to 2\pi$, whereas $F^{-1}(1) = 0$.

On the other hand, if F is a one-one closed map from a metric space X onto a metric space Y, then the inverse map F^{-1} from Y onto X is also closed. To see this, let $y_n \to y$ in Y and $F^{-1}(y_n) \to x$ in X. Let $x_n := F^{-1}(y_n)$ for $n \in \mathbb{N}$. Then $x_n \to x$ in X and $F(x_n) = y_n \to y$ in Y. Since F is a closed map, $y = F(x)$, that is, $F^{-1}(y) = x$, as desired. Based on this observation, we give a quick proof of the following important result by making use of the closed graph theorem.

Theorem 3.35 (Bounded inverse theorem) *Let X and Y be Banach spaces, and let $F \in BL(X, Y)$ be one-one and onto. Then $F^{-1} \in BL(Y, X)$.*

Proof We know F^{-1} is linear. Since F is continuous, it is a closed map. But then F^{-1} is a closed map from the Banach space Y to the Banach space X. By the closed graph theorem (Theorem 3.28), F^{-1} is continuous. $\square$

Here is a consequence of the above result for equivalence of norms.

Theorem 3.36 (Two-norm theorem) *Let $(X, \|\cdot\|)$ be a Banach space. Then a norm $\|\cdot\|'$ on the linear space X is equivalent to the norm $\|\cdot\|$ if and only if $(X, \|\cdot\|')$ is a Banach space and the norm $\|\cdot\|'$ is comparable to the norm $\|\cdot\|$.*

Proof Suppose the norm $\|\cdot\|'$ is equivalent to the norm $\|\cdot\|$. Then by Proposition 2.3, there are $\alpha > 0$ and $\beta > 0$ such that $\beta\|x\| \le \|x\|' \le \alpha\|x\|$ for all $x \in X$. Clearly, the norm $\|\cdot\|'$ is comparable to the norm $\|\cdot\|$. Also, if a sequence in X is Cauchy in $(X, \|\cdot\|')$, then it is Cauchy in $(X, \|\cdot\|)$, and if it convergent in $(X, \|\cdot\|)$, then it is convergent in $(X, \|\cdot\|')$. Since $(X, \|\cdot\|)$ is complete, it follows that $(X, \|\cdot\|')$ is also complete.

Conversely, assume that $(X, \|\cdot\|')$ is a Banach space, and the norms $\|\cdot\|$ and $\|\cdot\|'$ are comparable. We can assume, without loss of generality, that the norm $\|\cdot\|$ is stronger than the norm $\|\cdot\|'$. Again, by Proposition 2.3, there is some $\alpha > 0$ such that $\|x\|' \le \alpha\|x\|$ for all $x \in X$. Consider the identity map $I : (X, \|\cdot\|) \to (X, \|\cdot\|')$. Clearly, I is one-one, onto, linear and continuous. By the bounded inverse theorem (Theorem 3.35), $I^{-1} : (X, \|\cdot\|') \to (X, \|\cdot\|)$ is also continuous. By Proposition 3.2, there is $\gamma > 0$ such that $\|x\| \le \gamma\|x\|'$ for all $x \in X$. Letting $\beta = 1/\gamma$, we see that $\beta\|x\| \le \|x\|' \le \alpha\|x\|$ for all $x \in X$. Hence the norm $\|\cdot\|'$ is equivalent to the norm $\|\cdot\|'$. $\square$

Examples 3.37 The two-norm theorem says that two comparable complete norms on a linear space are in fact equivalent.

(i) Let $X := C([0, 2])$. The norms $\|\cdot\|_\infty$ and $\|\cdot\|_1$ are comparable, since $\|x\|_1 \le 2\|x\|_\infty$ for all $x \in X$. But $(X, \|\cdot\|_\infty)$ is a Banach space as we have seen in Example 2.24(iii), while $(X, \|\cdot\|_1)$ is not a Banach space, as we have seen in Example 2.24(iv). Hence the two norms are not equivalent.

(ii) Let $X := L^1([-\pi, \pi])$, and consider the usual norm $\|\cdot\|_1$ on X. Now $(X, \|\cdot\|_1)$ is a Banach space, as we have seen in Example 2.24(iv). Further, if $\|x_n - x\|_1 \to 0$, then it is easy to see that the kth Fourier coefficient $\hat{x}_n(k)$ of x_n tends to the kth Fourier coefficient $\hat{x}(k)$ of x for every $k \in \mathbb{Z}$.

Let $\|\cdot\|'$ be a complete norm on X such that $\hat{x}_n(k) \to \hat{x}(k)$ for every $k \in \mathbb{Z}$ whenever $\|x_n - x\|' \to 0$. We show that the identity map I from $(X, \|\cdot\|_1)$ to $(X, \|\cdot\|')$ is a closed map. Let $\|x_n\|_1 \to 0$ and $\|x_n - y\|' = \|I(x_n) - y\|' \to 0$. Then $\hat{y}(k) = \lim_{n\to\infty} \hat{x}_n(k) = 0$ for every $k \in \mathbb{Z}$. By Theorem 1.28(ii), $y = 0$ almost everywhere on $[-\pi, \pi]$, that is, $y = 0$. The closed graph theorem shows that the map I is continuous. Hence there is $\alpha > 0$ such that $\|x\|' \le \alpha\|x\|_1$ for all $x \in X$, that is, $\|\cdot\|_1$ is stronger than $\|\cdot\|'$. Hence the complete and comparable norms $\|\cdot\|_1$ and $\|\cdot\|'$ on X are equivalent. This gives a characterization of the usual norm on $L^1([-\pi, \pi])$ up to equivalence. For a similar characterization of the sup norm on $C([a, b])$, see Exercise 3.24 ◊

Open Map

We now prove a generalization of the bounded inverse theorem by making use of that result itself. Let X and Y be metric spaces, and let $F : X \to Y$. Recall that F is continuous if and only if $F^{-1}(E)$ is open in X for every open subset E of Y. We say that F is an **open map** if $F(E)$ is open in Y for every open subset E of X. If F is one-one and onto, then it follows that F is open if and only if F^{-1} is continuous. An important example of an open map is a nonconstant analytic function defined on a domain in $\mathbb{C}$.

Let X and Y be normed spaces, and let $F : X \to Y$ be an open linear map. Then $R(F)$ is a subspace of Y containing an $U_Y(0, \delta)$ for some $\delta > 0$, and so $R(F) = Y$. Thus F is onto.

We now give an important example of a linear map that is open.

Lemma 3.38 *Let X be a normed space, Z be a closed subspace of X, and let $Q : X \to X/Z$ be the quotient map. Then $Q(U_X) = U_{X/Z}$, where U_X and $U_{X/Z}$ are the open unit balls of X and X/Z respectively. Consequently, the map Q is continuous as well as open.*

Proof Let $x \in U_X$. Then $|||Q(x)||| \le \|x\| < 1$, so that $Q(x) \in U_{X/Z}$. Thus $Q(U_X) \subset U_{X/Z}$. Conversely, let $(x + Z) \in U_{X/Z}$, that is, $|||x + Z||| < 1$. By the definition of the quotient norm $|||\cdot|||$, there is $z \in Z$ such that $\|x + z\| < 1$. Then $(x + z) \in U_X$ and $Q(x + z) = Q(x) = x + Z$. Thus $Q(U_X) \supset U_{X/Z}$. Hence $Q(U_X) = U_{X/Z}$. Clearly, the map Q is continuous and $\|Q\| \le 1$.

Let E be an open subset of X, and let $x \in E$. Then there is $\delta > 0$ such that $U_X(x, \delta) \subset E$. Now by the linearity of the map Q, we obtain $U_{X/Z}(x+Z, \delta) = Q(U_X(x, \delta)) \subset Q(E)$. Hence Q is an open map. □

Theorem 3.39 (Open mapping theorem) *Let X and Y be Banach spaces, and let F be a continuous linear map from X onto Y. Then F is an open map.*

Proof Let $Z := Z(F)$. For $x + Z \in X/Z$, let $\widetilde{F}(x + Z) := F(x)$. It is easy to see that $\widetilde{F} : X/Z \to Y$ is well defined. Since $F \in BL(X, Y)$, we see that $\widetilde{F} \in BL(X/Z, Y)$ by Corollary 3.4. Also, since F maps X onto Y, $\widetilde{F}$ maps X/Z onto Y. Moreover, $\widetilde{F}$ is one-one.

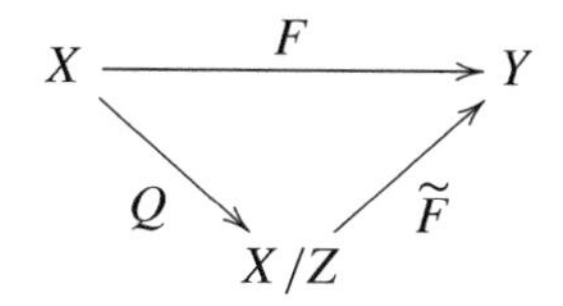

By the bounded inverse theorem (Theorem 3.35), $\widetilde{F}^{-1} : Y \to X/Z$ is continuous, that is, $\widetilde{F} : X/Z \to Y$ is an open map. Also, by Lemma 3.38, the quotient map $Q : X \to X/Z$ is open. Let E be an open subset of X. Then $Q(E)$ is an open subset of X/Z, and so $F(E) = \widetilde{F}(Q(E))$ is an open subset of Y. Thus F is an open map. □

Remark 3.40 Let X and Y be normed spaces, and let $F : X \to Y$ be a linear map. Suppose either X is finite dimensional or Y is finite dimensional and $Z(F)$ is closed in X. By Proposition 3.6, F is continuous. Suppose, in addition, F is onto. Then F is open as well. This follows from the proof of the open mapping theorem given above if we note that $Z(F)$ is closed in X, and $Y = R(F)$ is finite dimensional, and so the linear map $\widetilde{F}^{-1} : Y \to X/Z$ is continuous.

Now suppose X and Y are infinite dimensional. Then a linear map $F : X \to Y$ need not be open even if it is one-one, onto and continuous. For example, let f be a discontinuous linear functional on $(X, \|\cdot\|)$. (See Example 3.9(ii) and (iii), and Exercises 3.1 and 3.4.) Consider the norm on X defined by $\|x\|_f := \|x\| + |f(x)|$, $x \in X$. Then the identity operator I from $(X, \|\cdot\|_f)$ to $(X, \|\cdot\|)$ is obviously one-one, onto and continuous, but it is not open since its inverse is not continuous. We note that the domain space $(X, \|\cdot\|_f)$ of the map I is not complete. To see this, let (x_n) be a bounded sequence in X such that $0 < |f(x_n)| \to \infty$, and let $y_n := x_n/f(x_n)$ for $n \in \mathbb{N}$. Since $\|y_n\| \to 0$ and $f(y_n) = 1$ for all $n \in \mathbb{N}$, (y_n) is a nonconvergent Cauchy sequence in $(X, \|\cdot\|_f)$. (Compare Exercise 3.20(iii).) As an example of a situation where the image space of a linear map is not complete, let $X := C^1([a, b])$ with the norm defined by $\|x\|_{1,\infty} := \max\{\|x\|_\infty, \|x'\|_\infty\}$, $x \in X$. Then the identity operator I from $(X, \|\cdot\|_{1,\infty})$ to $(X, \|\cdot\|_\infty)$ is one-one, onto and continuous, but it is not open. This cannot happen if both the domain space and the image space are complete, according to the open mapping theorem. ◊

Before considering an application of the open mapping theorem for solutions of operator equations, we give a characterization of an open linear map.

Proposition 3.41 *Let X and Y be normed spaces, and let $F : X \to Y$ be a linear map. Then F is an open map if and only if there is $\beta > 0$ such that $U_Y(0, \beta) \subset F(U_X)$, or equivalently, there is $\gamma > 0$ such that every $y \in Y$ is equal to $F(x)$ for some $x \in X$ satisfying $\|x\| \le \gamma \|y\|$.*

Proof Let $x \in X$ and $r > 0$. Since F is linear, $y \in F(U_X(x, r))$ if and only if $y = F(x) + rv$, where $v \in F(U_X)$, that is, $F(U_X(x, r)) = F(x) + rF(U_X)$.

Suppose F is an open map. Since U_X is open in X, and $0 \in F(U_X)$, there is $\beta > 0$ such that $U_Y(0, \beta) \subset F(U_X)$. Conversely, suppose there is $\beta > 0$ such that $U_Y(0, \beta) \subset F(U_X)$. Let E be an open subset of X, and let $y \in F(E)$. Then there are $x \in X$ and $\delta > 0$ such that $y = f(x)$ and

$$U_Y(y; \delta\beta) = F(x) + \delta\, U_Y(0, \beta) \subset F(x) + \delta F(U_X) = F(U_X(x, \delta)) \subset F(E).$$

Hence F is an open map. (Compare the proof of Lemma 3.38.)

To prove the equivalence of the last two conditions in the statement of the proposition, let $\gamma := 1/\beta$ and $\beta := 1/\gamma$. □

The above result shows that a linear map from a normed space to a normed space is one-one and open if and only if it is bounded below and onto. (Compare Proposition 5.1.)

Operator Equations

Suppose X and Y are Banach spaces and $F \in BL(X, Y)$. Consider the **operator equation** $F(x) = y$. Suppose this equation has a solution in X for every $y \in Y$, that is, F maps X onto Y. By the open mapping theorem and Proposition 3.41, there exists $\gamma > 0$ such that for every $y \in Y$, this operator equation has a solution x in X such that $\|x\| \le \gamma \|y\|$. This estimate on the norm of a solution x in terms of the norm of the 'free term' y of the operator equation is important in many situations. (See Exercise 3.26.)

For example, suppose the above-mentioned operator equation has a unique solution $x \in X$ for every $y \in Y$. Often we are unable to compute the unique solution x for every $y \in Y$, but a computation of the unique solution is feasible for all y belonging to a specified dense subset D of Y. If $y \in Y \backslash D$, then one may find a sequence (y_n) in D such that $y_n \to y$, and a sequence (x_n) in X such that $F(x_n) = y_n$ for all $n \in \mathbb{N}$. Let us ask whether the sequence (x_n) will converge to the unique element $x \in X$ such that $F(x) = y$. The answer is in the affirmative. Since

$$F(x_n - x) = F(x_n) - F(x) = y_n - y \quad \text{for all } n \in \mathbb{N},$$

we obtain $\|x_n - x\| \le \gamma \|y_n - y\| \to 0$, and so $x_n \to x$. In this case, the element x_n can be called an **approximate solution** of the operator equation $F(x) = y$ for every large enough $n \in \mathbb{N}$. This also shows that the solution $x \in X$ corresponding to the 'free term' y depends continuously on y. We have thus established the validity of the **perturbation technique** used in the theory of operator equations. It consists of changing the 'free term' a little bit and allowing a small change in the exact solution.

We describe a concrete case. Let $m \in \mathbb{N}$, and consider an mth order nonhomogeneous linear ordinary differential equation with variable coefficients:

$$a_m(t)x^{(m)}(t) + \cdots + a_0(t)x(t) = y(t), \quad t \in [0, 1],$$

where $a_j \in C([0, 1])$ for $j = 1, \ldots, m$, and $a_m(t) \neq 0$ for every $t \in [0, 1]$. Also, consider the m initial conditions

$$x(0) = x'(0) = \cdots = x^{(m-1)}(0) = 0.$$

It is well known that for every $y \in C([0, 1])$, there is a unique solution of the above-mentioned differential equation satisfying the given m initial conditions. (See, for instance, [4, Theorem 8 of Chap. 6].)

Let $X := \{x \in C^{(m)}([0, 1]) : x(0) = \cdots = x^{(m-1)}(0) = 0\}$, and let $Y := C([0, 1])$. For $x \in X$, let

$$F(x) := a_m x^{(m)} + \cdots + a_0 x \in Y.$$

Then $F : X \to Y$ is linear, one-one and onto. For $x \in X$, let

$$\|x\|_{m,\infty} := \max\left\{\|x\|_\infty, \ldots, \|x^{(m)}\|_\infty\right\}.$$

It can be seen that $(X, \|\cdot\|_{m,\infty})$ is a Banach space. (See Exercise 2.29(i).) Also, $(Y, \|\cdot\|_\infty)$ is a Banach space. Further, $F \in BL(X, Y)$ since

$$\|F(x)\|_\infty \leq \max\{\|a_0\|_\infty, \ldots, \|a_m\|_\infty\}\, \|x\|_{m,\infty} \quad \text{for all } x \in X.$$

Hence there is $\gamma > 0$ such that $\|x\|_{m,\infty} \leq \gamma \|F(x)\|_\infty$ for all $x \in X$.

Let D denote the set of all polynomials on $[0, 1]$, and suppose that for every $y \in D$, we can compute the unique element $x \in X$ such that $F(x) = y$. Consider a continuous function y on $[0, 1]$ which is not a polynomial. Let

$$B_n(y)(t) := \sum_{k=0}^{n} y\left(\frac{k}{n}\right)\binom{n}{k} t^k (1-t)^{n-k} \quad \text{for } n \in \mathbb{N} \text{ and } t \in [0, 1].$$

Then $B_n(y) \in D$ for each $n \in \mathbb{N}$ and, as we have seen in the proof of Theorem 1.16, $\|B_n(y) - y\|_\infty \to 0$. For each $n \in \mathbb{N}$, find $x_n \in X$ such that

$$a_m x_n^{(m)} + \cdots + a_0 x_n = B_n(y).$$

Let $x \in X$ satisfy $F(x) = y$. Since $\|x_n - x\|_{m,\infty} \leq \gamma \|B_n(y) - y\|_\infty$ for all $n \in \mathbb{N}$, the sequence (x_n) converges in X to the unique $x \in X$ such that

$$a_m x^{(m)} + \cdots + a_0 x = y.$$

In other words, the approximate solution x_n of the initial value problem converges to the exact solution x uniformly, and also the jth derivative $x_n^{(j)}$ of x_n converges uniformly to the jth derivative $x^{(j)}$ of x on $[0, 1]$ for $j = 1, \ldots, m$.

3.4 Compact Linear Maps

In this section, we consider a natural and useful generalization of a continuous linear map of finite rank. We have seen in Proposition 3.2 that a linear map F from a normed space X to a normed space Y is continuous if and only if it is bounded on the closed unit ball of X, that is, the sequence $(F(x_n))$ is bounded in Y for every bounded sequence (x_n) in X. We introduce a property of a linear map which is, in general, stronger than continuity.

Let X and Y be normed spaces, and let $F : X \to Y$ be linear. Then F is called a **compact linear map** if for every bounded sequence (x_n) in X, the sequence $(F(x_n))$ has a subsequence which converges in Y. If F is not continuous, then there is a bounded sequence (x_n) in X such that $\|F(x_n)\| > n$ for all $n \in \mathbb{N}$, and so no subsequence of the sequence $(F(x_n))$ would converge in Y. This shows that every compact linear map is continuous. However, not every continuous linear map is compact. For example, if X is an infinite dimensional normed space, then the identity map $I : X \to X$ is clearly linear and continuous, but it is not compact since, by Theorem 2.10, there is a sequence (x_n) in the closed unit ball of X which has no convergent subsequence. If X is an infinite dimensional inner product space, then this can be seen more easily as follows. Let $\{u_1, u_2, \ldots\}$ be an infinite orthonormal set in X. Since $\|u_n\| = 1$ for all $n \in \mathbb{N}$, and $\|u_n - u_m\| = \sqrt{2}$ for all $n \neq m$ in $\mathbb{N}$, no subsequence of (u_n) converges in X. We now establish a relationship between compact linear maps and continuous linear maps of finite rank.

Theorem 3.42 *Let X and Y be normed spaces, and let $F : X \to Y$ be a linear map. If F is continuous and of finite rank, then F is a compact linear map and $R(F)$ is closed in Y. Conversely, if X and Y are Banach spaces, F is a compact linear map from X to Y, and if $R(F)$ is closed in Y, then F is continuous and of finite rank.*

Proof Suppose F is continuous and of finite rank. Let (x_n) be a bounded sequence in X, and let $y_n := F(x_n)$ for $n \in \mathbb{N}$. Since F is continuous, (y_n) is a bounded sequence in $R(F)$. Hence there is $\alpha > 0$ such that the sequence (y_n/α) is in the closed unit ball of $R(F)$, which is compact by Theorem 2.10. This shows that $(F(x_n))$ has a subsequence which converges in Y. Thus F is a compact map. Also, since the subspace $R(F)$ is finite dimensional, it is complete by Lemma 2.8. Hence $R(F)$ is closed in Y.

Conversely, suppose X and Y are Banach spaces, F is a compact linear map, and $R(F)$ is closed in Y. We have seen that F is continuous. Note that $R(F)$ is a Banach space, and F maps X onto $R(F)$. By Theorem 3.39, the map $F : X \to R(F)$ is an open map. By Proposition 3.41, there is $\gamma > 0$ such that for every $y \in R(F)$, there is $x \in X$ such that $y = F(x)$ and $\|x\| \leq \gamma\|y\|$. Let (y_n) be a sequence in $R(F)$ such that $\|y_n\| \leq 1$. Then there is $x_n \in X$ such that $y_n = F(x_n)$ for $n \in \mathbb{N}$ and $\|x_n\| \leq \gamma$. Since F is a compact map from X to Y, (y_n) has a subsequence (y_{n_k}) which converges in Y. Since $R(F)$ is closed in Y, (y_{n_k}) in fact converges in $R(F)$. Thus the closed unit ball of the normed space $R(F)$ is compact. It follows from Theorem 2.10 that $R(F)$ is finite dimensional, that is, F is of finite rank. □

The above result, along with Proposition 3.6(ii), shows that if F is a linear map of finite rank from a normed space X to a normed space Y such that $Z(F)$ is closed in X, then F is a compact linear map.

The set of all compact linear maps from a normed space X to a normed space Y is denoted by $CL(X, Y)$. It is a subset of $BL(X, Y)$, and in general, it is a proper subset of $BL(X, Y)$. If X or Y is finite dimensional, then $CL(X, Y) = BL(X, Y)$. If $Y := X$, we denote $CL(X, X)$ simply by $CL(X)$. An element of $CL(X)$ is called a **compact operator** on X.

Proposition 3.43 *Let X and Y be normed spaces. Then $CL(X, Y)$ is a subspace of $BL(X, Y)$. If Y is a Banach space, (F_n) is a sequence in $CL(X, Y)$, and $F \in BL(X, Y)$ is such that $\|F_n - F\| \to 0$, then $F \in CL(X, Y)$, that is, $CL(X, Y)$ is a closed subspace of $BL(X, Y)$.*

Proof Let $F, G \in CL(X, Y)$, and let (x_n) be a bounded sequence in X. There is a subsequence (x_{n_k}) of (x_n) such that $(F(x_{n_k}))$ converges in Y. Further, there is a subsequence $(x_{n_{k_j}})$ of (x_{n_k}) such that $(G(x_{n_{k_j}}))$ converges in Y. Then the subsequence $((F + G)(x_{n_{k_j}}))$ of $((F + G)(x_n))$ converges in Y. This shows that $(F + G) \in CL(X, Y)$. Similarly, $k\,F \in CL(X, Y)$ for every $k \in \mathbb{K}$. Thus $CL(X, Y)$ is a subspace of $BL(X, Y)$.

Let Y be a Banach space. Suppose (F_n) is a sequence in $CL(X, Y)$, and there is $F \in BL(X, Y)$ such that $\|F_n - F\| \to 0$. Let (x_j) be a bounded sequence in X. Then there is $\alpha > 0$ such that $\|x_j\| \leq \alpha$ for all $j \in \mathbb{N}$. Since F_1 is a compact linear map, (x_j) has a subsequence $(x_{1,j})$ such that $(F_1(x_{1,j}))$ converges in Y. Now $(x_{1,j})$ is a bounded sequence in X, and since F_2 is a compact linear map, $(x_{1,j})$ has a subsequence $(x_{2,j})$ such that $(F_2(x_{2,j}))$ converges in Y. Continuing this process, for each $n \in \mathbb{N}$, we find a subsequence $(x_{n+1,j})$ of $(x_{n,j})$ such that $(F_{n+1}(x_{n+1,j}))$ converges in Y. Define $u_j := x_{j,j}$ for $j \in \mathbb{N}$. The 'diagonal sequence' (u_j) is a subsequence of the given sequence (x_j). We show that the sequence $(F(u_j))$ converges in Y.

For each $n \in \mathbb{N}$, from the nth term onward, (u_j) is a subsequence of $(x_{n,j})$. Since $(F_n(x_{n,j}))$ converges in Y, we see that $(F_n(u_j))$ also converges in Y. Let $\epsilon > 0$. There is $n_0 \in \mathbb{N}$ such that $\|F_{n_0} - F\| < \epsilon$. Since $(F_{n_0}(u_j))$ converges in Y, it is a Cauchy sequence in Y. Hence there is $j_0 \in \mathbb{N}$ such that $\|F_{n_0}(u_j) - F_{n_0}(u_k)\| < \epsilon$ for all $j, k \geq j_0$, and so

$$\begin{aligned}\|F(u_j) - F(u_k)\| &= \|F(u_j) - F_{n_0}(u_j)\| + \|F_{n_0}(u_j) - F_{n_0}(u_k)\| \\ &\quad + \|F_{n_0}(u_k) - F(u_k)\| \\ &\leq \|F - F_{n_0}\| \, \|u_j\| + \|F_{n_0}(u_j) - F_{n_0}(u_k)\| + \|F_{n_0} - F\| \, \|u_k\| \\ &\leq \epsilon\,\alpha + \epsilon + \epsilon\,\alpha = (2\alpha + 1)\epsilon\end{aligned}$$

for all $j, k \geq j_0$. Thus $(F(u_j))$ is a Cauchy sequence in Y, and since Y is a Banach space, it converges in Y. This shows that $F \in CL(X, Y)$. □

If a continuous linear map from a normed space X to a normed space Y can be approximated by a sequence of compact linear maps in the operator norm, and if Y is complete, then by Proposition 3.43, the given continuous linear map is in fact a compact linear map. On the other hand, if (F_n) is a sequence in $CL(X, Y)$ and $F \in BL(X, Y)$ such that $F_n(x) \to F(x)$ in Y for each x in X, then F need not be in $CL(X, Y)$. For example, let $X := Y := \ell^2$, $F := I$, and $F_n(x) := (x(1), \ldots, x(n), 0, 0, \ldots)$ for $x := (x(1), x(2), \ldots) \in \ell^2$.

Remarks 3.44 (i) Suppose X, Y and Z are normed spaces. Let $F \in BL(X, Y)$ and $G \in BL(Y, Z)$. We have seen in Theorem 3.10(ii) that $G \circ F \in BL(X, Z)$. It is easy to see that if one of the maps F and G is a compact linear map, then so is $G \circ F$. In particular, if $Y = Z = X$, $F \in CL(X)$ and $G \in BL(X)$, then both $G \circ F$ and $F \circ G$ are in $CL(X)$. Hence the linear space $CL(X)$ of all compact operators on X is in fact a two-sided ideal of the algebra $BL(X)$ of all bounded operators on X. If X is a Banach space, then $CL(X)$ is a closed ideal of $BL(X)$.

(ii) We have seen in Theorem 3.42 that every continuous linear map of finite rank from a normed space X to a normed space Y is a compact linear map. Also, it follows from Proposition 3.43 that if Y is a Banach space, (F_n) is a sequence in $BL(X, Y)$, $F \in BL(X, Y)$ and $\|F_n - F\| \to 0$, and if each F_n is of finite rank, then F is a compact linear map. Quite often this is how the compactness of a linear map is established, as we shall see below. ♢

Examples 3.45 An $m \times n$ matrix defines a linear map F from $\mathbb{K}^n$ to $\mathbb{K}^m$. Since F is continuous and of finite rank, it is a compact map when $\mathbb{K}^n$ and $\mathbb{K}^m$ are equipped with any norms including the norms $\|\cdot\|_1$, $\|\cdot\|_2$, $\|\cdot\|_\infty$.

Let us now consider an infinite matrix $M := [k_{i,j}]$. Let $p, r \in \{1, 2, \infty\}$. We give below sufficient conditions for M to define a compact linear map from ℓ^p to itself. Sufficient conditions for M to define a compact linear map from ℓ^p to ℓ^r are given in Exercises 3.31 and 3.36. For necessary conditions, see Exercises 3.34 and 4.21. Conditions that are necessary as well as sufficient for a tridiagonal matrix to define a compact linear map from ℓ^p to itself are given in Exercise 3.35.

(i) Let $p := 1$. For $j \in \mathbb{N}$, consider the jth column sum $\alpha_1(j) := \sum_{i=1}^{\infty} |k_{i,j}|$ of the matrix $|M| := [|k_{i,j}|]$. If $\alpha_1(j) < \infty$ for each $j \in \mathbb{N}$, and if $\alpha_1(j) \to 0$, then M defines a compact linear map F from ℓ^1 to itself. This can be seen as follows. For $n \in \mathbb{N}$, let M_n denote the infinite matrix whose first n columns are the same as those of the matrix M, and the remaining columns are zero. Example 3.14(i) shows that the matrices M, M_n and $M - M_n$ define continuous linear maps F, F_n and $F - F_n$ from ℓ^1 to itself, and

$$\|F - F_n\| = \sup\{\alpha_1(j) : j = n+1, n+2, \ldots\} \quad \text{for each } n \in \mathbb{N}.$$

Now $\|F - F_n\| \to 0$ since $\alpha_1(j) \to 0$. Further, each F_n is of finite rank since

$$F_n(x)(i) = \sum_{j=1}^{n} k_{i,j}x(j) = \sum_{j=1}^{n} x(j)F(e_j)(i) \quad \text{for all } x \in \ell^1 \text{ and } i \in \mathbb{N},$$

that is, $F_n(x) = x(1)F(e_1) + \cdots + x(n)F(e_n)$ for all $x \in \ell^1$. Since ℓ^1 is a Banach space, F is a compact linear map by Proposition 3.43. However, the condition '$\alpha_1(j) \to 0$' does not follow from the compactness of F. To see this, consider the matrix $M := [k_{i,j}]$, where $k_{1,j} := 1$ and $k_{i,j} := 0$ if $i = 2, 3, \ldots$ and $j \in \mathbb{N}$. Then the matrix M defines $F : \ell^1 \to \ell^1$ given by $F(x) := \left(\sum_{j=1}^{\infty} x(j)\right)e_1$ for all $x \in \ell^1$. Since $F \in BL(\ell^1)$ and $R(F) = \text{span}\,\{e_1\}$, we see that $F \in CL(\ell^1)$.

(ii) Let $p := \infty$. For $i \in \mathbb{N}$, consider the ith row sum $\beta_1(i) := \sum_{j=1}^{\infty} |k_{i,j}|$ of the matrix $|M| := [|k_{i,j}|]$. If $\beta_1(i) < \infty$ for each $i \in \mathbb{N}$, and if $\beta_1(i) \to 0$, then M defines a compact linear map F from ℓ^∞ to itself. This can be seen as follows. For $n \in \mathbb{N}$, let M_n denote the infinite matrix whose first n rows are the same as those of the matrix M, and the remaining rows are zero. Example 3.14(ii) shows that the matrices M, M_n and $M - M_n$ define continuous linear maps F, F_n and $F - F_n$ from ℓ^∞ to itself, and

$$\|F - F_n\| = \sup\{\beta_1(i) : i = n+1, n+2, \ldots\} \quad \text{for each } n \in \mathbb{N}.$$

Now $\|F - F_n\| \to 0$ since $\beta_1(i) \to 0$. Further, each F_n is of finite rank since

$$F_n(x)(i) = \sum_{j=1}^{\infty} k_{i,j}x(j) = F(x)(i) \text{ if } 1 \le i \le n, \text{ and } F_n(x)(i) = 0 \text{ if } i > n,$$

that is, $F_n(x) = F(x)(1)\, e_1 + \cdots + F(x)(n)\, e_n$ for all $x \in \ell^\infty$. Since ℓ^∞ is a Banach space, F is a compact linear map by Proposition 3.43. However, the condition '$\beta_1(i) \to 0$' does not follow from the compactness of F. To see this, consider the matrix $M := [k_{i,j}]$, where $k_{i,1} := 1$ and $k_{i,j} := 0$ if $j = 2, 3, \ldots$ and $i \in \mathbb{N}$. Then the matrix M defines $F : \ell^\infty \to \ell^\infty$ given by $F(x) = x(1)(1, 1, \ldots)$ for all $x \in \ell^\infty$. Since $F \in BL(\ell^\infty)$ and $R(F) = \text{span}\,\{F(e_1)\}$, we see that $F \in CL(\ell^\infty)$.

(iii) Let $p := 2$. Suppose $\alpha_1(j) < \infty$ for each $j \in \mathbb{N}$, and $\beta_1(i) < \infty$ for each $i \in \mathbb{N}$. If one of the sequences $(\alpha_1(j))$ and $(\beta_1(i))$ is bounded and the other sequence tends to zero, then M defines a compact linear map F from ℓ^2 to itself. Assume first that the sequence $(\alpha_1(j))$ is bounded and $\beta_1(i) \to 0$. Consider the matrix M_n and the map F_n given in the case (ii): $p := \infty$ above. Then, as in Example 3.14(iii),

$$\|F - F_n\| \le \alpha_1^{1/2}\left(\sup\{\beta_1(i) : i = n+1, n+2, \ldots\}\right)^{1/2},$$

where $\alpha_1 := \sup\{\alpha_1(j) : j = 1, 2, \ldots\}$. Since ℓ^2 is a Banach space, we obtain the desired result by Proposition 3.43. A similar argument holds if $\alpha_1(j) \to 0$ and the sequence $(\beta_1(i))$ is bounded. However, these conditions do not follow from the compactness of the map F. To see this, consider the matrix $M := [k_{i,j}]$, where $k_{i,1} := 1/i$ if $i \in \mathbb{N}$, $k_{1,j} := 1/j$ if $j \in \mathbb{N}$ and $k_{i,j} := 0$ otherwise. This matrix satisfies the condition '$\gamma_{2,2} < \infty$' given below.

We have seen in Example 3.14(iii) that if

$$\gamma_{2,2} := \left(\sum_{i=1}^{\infty} \sum_{j=1}^{\infty} |k_{i,j}|^2 \right)^{1/2} < \infty,$$

then M defines a continuous linear map F from ℓ^2 to itself. In fact, the map F is compact. This follows by considering the matrix M_n and the map F_n given in the case $p := \infty$ above and noting, as in Example 3.14(iii), that

$$\|F - F_n\|^2 \leq \sum_{i=n+1}^{\infty} \left(\sum_{j=1}^{\infty} |k_{i,j}|^2 \right) \quad \text{for each } n \in \mathbb{N}.$$

Note that $\|F - F_n\| \to 0$ since $\gamma_{2,2} < \infty$, and each F_n is of finite rank. However, this condition does not follow from the compactness of the map F. To see this, consider the matrix $M := \operatorname{diag}(1, 1/\sqrt{2}, 1/\sqrt{3}, \ldots)$. For this matrix, $\alpha_1(j) \to 0$ and $\beta_1(i) \to 0$. ◊

Examples 3.46 As 'continuous' analogues of bounded linear maps defined by finite or infinite matrices, we have considered Fredholm integral maps defined by kernels in Example 3.16. First, suppose $k(\cdot\,, \cdot)$ is a continuous function on $[a, b] \times [a, b]$, and for $x \in L^1([a, b])$, let

$$F(x)(s) := \int_a^b k(s, t)x(t)\, dm(t), \quad s \in [a, b].$$

If $s_n \to s$ in $[a, b]$, then the dominated convergence theorem (Theorem 1.18(ii)) shows that $F(x)(s_n) \to F(x)(s)$. Hence $F(x)$ is a continuous function on $[a, b]$. Let (x_n) be a bounded sequence in $L^1([a, b])$. We prove that the sequence $(F(x_n))$ has a subsequence which converges uniformly on $[a, b]$. There are $\alpha > 0$ and $\beta > 0$ such that $\|x_n\|_1 \leq \alpha$ for all $n \in \mathbb{N}$ and $|k(s, t)| \leq \beta$ for all $s, t \in [a, b]$. Then $|F(x_n)(s)| \leq \beta\alpha$ for all $n \in \mathbb{N}$ and all $s \in [a, b]$. Thus the subset $E := \{F(x_n) : n \in \mathbb{N}\}$ of $C([a, b])$ is uniformly bounded. Also, the set E is equicontinuous on $[a, b]$. To see this, we note that

$$|F(x_n)(s) - F(x_n)(u)| \leq \int_a^b |k(s, t) - k(u, t)|\, |x_n(t)|\, dm(t)$$

for all $n \in \mathbb{N}$ and all $s, u \in [a, b]$. Let $\epsilon > 0$. By Proposition 1.14, the function $k(\cdot\,, \cdot)$ is uniformly continuous on $[a, b] \times [a, b]$. Hence there is $\delta > 0$ such that $|k(s, t) - k(u, t)| < \epsilon$ for all $s, u \in [a, b]$, satisfying $|s - u| < \delta$ and all $t \in [a, b]$. Thus $|F(x_n)(s) - F(x_n)(u)| \le \epsilon\alpha$ for all $n \in \mathbb{N}$ if $|s - u| < \delta$. By Theorem 1.17(ii) of Arzelà, the sequence $(F(x_n))$ has a subsequence which converges uniformly on $[a, b]$. Thus we have shown that the Fredholm integral map $F : L^1([a, b]) \to C([a, b])$ is a compact linear map. Also, $\|F\| \le \|k(\cdot\,, \cdot)\|_\infty$. For another way of proving the compactness of F, see Exercise 3.37.

Next, let X as well as Y denote one of the normed spaces $L^1([a, b])$, $L^2([a, b])$, $L^\infty([a, b])$ and $C([a, b])$. If $x \in X$, then $x \in L^1([a, b])$, and if (x_n) is a bounded sequence in X, then (x_n) is a bounded sequence in $L^1([a, b])$. On the other hand, if $y \in C([a, b])$, then $y \in Y$, and if (y_n) is a uniformly convergent sequence in $C([a, b])$, then (y_n) converges in Y. By what we have proved above, it follows that for every bounded sequence (x_n) in X, the sequence $(F(x_n))$ has a subsequence which converges in Y. Thus the Fredholm integral map F from X to Y is a compact linear map, provided the kernel $k(\cdot\,, \cdot)$ is continuous on $[a, b] \times [a, b]$.

What can we say if the kernel $k(\cdot\,, \cdot)$ is not continuous? Let $X := L^2([a, b])$, $Y := L^2([a, b])$, and let $k(\cdot\,, \cdot) \in L^2([a, b] \times [a, b])$. We show that the Fredholm integral operator F on $L^2([a, b])$ with kernel $k(\cdot\,, \cdot)$ is a compact operator. As in Example 3.16(iii), F is a bounded operator, and $\|F\| \le \gamma_{2,2} := \|k(\cdot\,, \cdot)\|_2$. Let $\{u_1, u_2, \ldots\}$ be a denumerable orthonormal basis for $L^2([a, b])$ consisting of continuous functions on $[a, b]$. (See Example 2.32(ii) and (iii).) For i, j in $\mathbb{N}$, let $w_{i,j}(s, t) := u_i(s)\overline{u_j}(t)$, $(s, t) \in [a, b] \times [a, b]$, and for $n \in \mathbb{N}$, let

$$k_n(s, t) := \sum_{i=1}^{n} \sum_{j=1}^{n} c_{i,j} w_{i,j}(s, t), \quad (s, t) \in [a, b] \times [a, b], \quad \text{where}$$

$$c_{i,j} := \int_a^b \int_a^b k(s, t)\overline{u_i}(s)u_j(t)dm(s)dm(t) \quad \text{for } i, j \in \mathbb{N}.$$

Then $k_n(\cdot\,, \cdot) \in C([a, b] \times [a, b])$ for $n \in \mathbb{N}$, and $\|k_n(\cdot\,, \cdot) - k(\cdot\,, \cdot)\|_2 \to 0$. In fact, $\{w_{i,j} : i, j \in \mathbb{N}\}$ is a denumerable orthonormal basis for $L^2([a, b] \times [a, b])$, and so $k(\cdot\,, \cdot)$ has a convergent Fourier expansion. For $n \in \mathbb{N}$ and $x \in L^2([a, b])$, let

$$F_n(x)(s) := \int_a^b k_n(s, t)x(t)\, dm(t), \quad s \in [a, b].$$

By what we have seen earlier, each F_n is a compact linear map from $L^2([a, b])$ to itself for each $n \in \mathbb{N}$. Also,

$$\|F_n - F\| \le \|k_n(\cdot\,, \cdot) - k(\cdot\,, \cdot)\|_2 \to 0.$$

Since $L^2([a,b])$ is a Banach space, Proposition 3.43 shows that F is also a compact linear map. Specific compact integral operators are considered in Example 5.4(iii) and (iv), Example 5.23(ii) and Exercise 5.19. ◊

Exercises

3.1 Let X be an infinite dimensional normed space. Then there is a discontinuous linear functional on X. Also, there is a one-one linear discontinuous map from X onto X. Further, if Y is a nonzero normed space, then there is a discontinuous linear map from X to Y.

3.2 Let X and Y be normed spaces, and let $F : X \to Y$ be a linear map.

(i) Suppose X is finite dimensional, and let $\{x_1, \ldots, x_n\}$ be a basis for X. Let $y_j := F(x_j)$, and let $X_j := \text{span}\,\{x_i : i = 1, \ldots, n,\ i \neq j\}$ for $j = 1, \ldots, n$. Define

$$\alpha := \sum_{j=1}^{n} \frac{\|y_j\|}{|||x_j + X_j|||}.$$

Then $\|F(x)\| \leq \alpha \|x\|$ for every $x \in X$. (Compare Proposition 3.6(i).)

(ii) Suppose $R(F)$ is finite dimensional, and let $\{y_1, \ldots, y_m\}$ be a basis for $R(F)$. Let $x_i \in X$ be such that $F(x_i) = y_i$, and let $Y_i := \text{span}\,\{y_j : j = 1, \ldots, m,\ j \neq i\}$ for $i = 1, \ldots, m$. Define

$$\gamma := \sum_{i=1}^{m} \frac{\|x_i\|}{|||y_i + Y_i|||}.$$

Then for every $y \in R(F)$, there is $x \in X$ such that $F(x) = y$ and $\|x\| \leq \gamma \|y\|$. (Compare Remark 3.40 and Proposition 3.41.)

3.3 Let f be a linear functional on a normed space X. If Y is a subspace of X such that $Z(f) \subset Y$ and $Y \neq Z(f)$, then $Y = X$. Also, f is continuous if and only if either $Z(f) = X$ or $Z(f)$ is not dense in X.

3.4 Let $r \in \mathbb{R}$. For $x \in c_{00}$, define $f_r(x) := \sum_{j=1}^{\infty} x(j)/j^r$. Then f_r is continuous on $(c_{00}, \|\cdot\|_1)$ if and only if $r \geq 0$, and then $\|f_r\| = 1$; f_r is continuous on $(c_{00}, \|\cdot\|_2)$ if and only if $r > 1/2$, and then $\|f_r\| = \left(\sum_{j=1}^{\infty} j^{-2r}\right)^{1/2}$; f_r is continuous on $(c_{00}, \|\cdot\|_\infty)$ if and only if $r > 1$, and then $\|f_r\| = \sum_{j=1}^{\infty} j^{-r}$.

3.5 For $x := (x(1), x(2), \ldots) \in \ell^1$, and $i \in \mathbb{N}$, define $y(i) := \left(\sum_{j=i}^{\infty} x(j)\right)/i^2$, and let $F(x) := (y(1), y(2), \ldots)$. Then $F \in BL(\ell^1)$ and $\|F\| = \pi^2/6$. Also, there is no $x \in \ell^1$ such that $\|x\|_1 = 1$ and $\|F(x)\|_1 = \|F\|$.

3.6 Let X be a normed space.

(i) Let $P \in BL(X)$ be a projection operator. Then $\|P\| = 0$ or $\|P\| \geq 1$. If in fact X is an inner product space, then P is an orthogonal projection operator if and only if $\|P\| = 0$ or $\|P\| = 1$.

(ii) Let Z be a closed subspace of X, and let Q denote the quotient map from X to X/Z. Then $Q \in BL(X, X/Z)$, and $\|Q\| = 0$ or $\|Q\| = 1$.

3.7 An infinite matrix M defines a map F from c_{00} to itself if and only if each column of M belongs to c_{00}. In this case, define α_1 and β_1 as in Example 3.14(i) and (ii). Then F is a continuous map from $(c_{00}, \|\cdot\|_1)$ to itself if and only if $\alpha_1 < \infty$, and then $\|F\| = \alpha_1$, while F is a continuous map from $(c_{00}, \|\cdot\|_\infty)$ to itself if and only if $\beta_1 < \infty$, and then $\|F\| = \beta_1$.

3.8 Let $\{u_j\}$ and $\{v_i\}$ be denumerable orthonormal sets in inner product spaces X and Y. Let $M := [k_{i,j}]$ be an infinite matrix. If for each $x \in X$ and for each $i \in \mathbb{N}$, the series $\sum_{j=1}^{\infty} k_{i,j}\langle x, u_j\rangle$ is summable in $\mathbb{K}$ with sum $f_i(x)$, and if for each $x \in X$, the series $\sum_{i=1}^{\infty} f_i(x)v_i$ is summable in Y, then we say that the matrix M defines a map from X to Y with respect to the given orthonormal subsets. In this case, if we let $F(x) := \sum_{i=1}^{\infty} f_i(x)v_i$, $x \in X$, then F is a linear map from X to Y. Suppose Y is a Hilbert space.

(i) **(Schur test)** Let $\alpha_1, \beta_1 < \infty$. Then M defines $F \in BL(X, Y)$, and $\|F\| \leq \sqrt{\alpha_1\beta_1}$, where α_1 and β_1 are defined in Example 3.14(i), (ii).

(ii) **(Hilbert–Schmidt test)** Let $\gamma_{2,2} < \infty$. Then M defines $F \in BL(X, Y)$, and $\|F\| \leq \gamma_{2,2}$, where $\gamma_{2,2}$ is defined in Example 3.14(iii).

3.9 Let $M := [k_{i,j}]$, where $k_{i,j} := (-1)^{i+j}/i^2j^2$ for $i, j \in \mathbb{N}$. Then $M := [k_{i,j}]$ defines a bounded operator F on ℓ^p for $p \in \{1, 2, \infty\}$. In fact, $\|F\| = \pi^2/6$ if $p \in \{1, \infty\}$, and $\|F\| = \pi^4/90$ if $p = 2$.

3.10 Let $X := C([a, b])$, and let $y \in X$. Define $f_y : X \to \mathbb{K}$ by

$$f_y(x) := \int_a^b x(t)y(t)dt \quad \text{for } x \in X.$$

(i) f_y is a continuous linear functional on $(X, \|\cdot\|_1)$ and $\|f_y\| = \|y\|_\infty$. In fact, if $t_0 \in (a, b)$, and for $t \in [a, b]$, we let $x_n(t) := n - n^2|t - t_0|$ when $|t - t_0| \leq 1/n$ and $x_n(t) := 0$ otherwise, then $x_n \in X$, $x_n \geq 0$, $\|x_n\|_1 \leq 1$ for $n \in \mathbb{N}$ and $f_y(x_n) \to y(t_0)$. (Compare x_n with a hat function.)

(ii) f_y is a continuous linear functional on $(X, \|\cdot\|_\infty)$, and $\|f_y\| = \|y\|_1$. In fact, if $x_n := n\bar{y}/(1 + n|y|)$ for $n \in \mathbb{N}$, then $x_n \in X$, $\|x_n\|_\infty \leq 1$ and $f_y(x_n) \to \|y\|_1$. (Compare x_n with the function $\operatorname{sgn} y$.)

(iii) f_y is a continuous linear functional on $(X, \|\cdot\|_2)$, and $\|f_y\| = \|y\|_2$.

(Compare Examples 3.12 and 3.13.)

3.11 Let $X := C([a, b])$ and $k(\cdot\,, \cdot) \in C([a, b] \times [a, b])$. Let

$$\alpha_1 := \sup_{t\in[a,b]} \int_a^b |k(s,t)|ds \quad \text{and} \quad \beta_1 := \sup_{s\in[a,b]} \int_a^b |k(s,t)|dt.$$

Define $F(x)(s) := \displaystyle\int_a^b k(s,t)x(t)dt$ for $x \in X$ and $s \in [a,b]$. Then

(i) F is a bounded operator on $(X, \|\cdot\|_1)$, and $\|F\| = \alpha_1$.

(ii) F is a bounded operator on $(X, \|\cdot\|_\infty)$, and $\|F\| = \beta_1$.

(iii) F is a bounded operator on $(X, \|\cdot\|_2)$, and $\|F\| \le \sqrt{\alpha_1\beta_1}$.

3.12 (i) Let F_1 and F_2 be operators on ℓ^2 defined by infinite matrices $M_1 := [k_1(i,j)]$ and $M_2 := [k_2(i,j)]$ satisfying $\sum_{i=1}^\infty \sum_{j=1}^\infty |k_1(i,j)|^2 < \infty$ and $\sum_{i=1}^\infty \sum_{j=1}^\infty |k_2(i,j)|^2 < \infty$ respectively. Then the operator $F := F_1 \circ F_2$ on ℓ^2 is defined by the infinite matrix $M := [k_{i,j}]$, where $k_{i,j} := \sum_{n=1}^\infty k_1(i,n)$ $k_2(n,j)$ for $i,j \in \mathbb{N}$, and

$$\|F\|^2 \le \sum_{i=1}^\infty \sum_{j=1}^\infty |k(i,j)|^2 \le \Big(\sum_{i=1}^\infty \sum_{j=1}^\infty |k_1(i,j)|^2\Big)\Big(\sum_{i=1}^\infty \sum_{j=1}^\infty |k_2(i,j)|^2\Big).$$

(ii) Let F_1 and F_2 be Fredholm integral operators on $L^2([a,b])$ with kernels $k_1(\cdot\,,\cdot)$ and $k_2(\cdot\,,\cdot)$ in $L^2([a,b]\times[a,b])$ respectively. Then $F := F_1 \circ F_2$ is a Fredholm integral operator on $L^2([a,b])$ with kernel

$$k(s,t) := \int_a^b k_1(s,u)k_2(u,t)\,dm(u) \quad \text{for } (s,t) \in [a,b]\times[a,b],$$

and $\|F\| \le \|k(\cdot\,,\cdot)\|_2 \le \|k_1(\cdot\,,\cdot)\|_2\|k_2(\cdot\,,\cdot)\|_2$.

3.13 Let $X := C^2([0,1])$ with the norm given by $\|x\|_{1,\infty} := \max\{\|x\|_\infty, \|x'\|_\infty\}$ for $x \in X$. Let $p(x) := \|x''\|_\infty$ for $x \in X$. Then p is discontinuous, but it is countably subadditive on X.

3.14 (i) Every seminorm on a finite dimensional normed space is continuous.

(ii) A seminorm p on a normed space X is called **lower semicontinuous** if $p(x) \le \lim_{n\to\infty} \inf\{p(x_m) : m \ge n\}$ whenever $x_n \to x$ in X. If p is a lower semicontinuous seminorm on a Banach X, then p is continuous.

3.15 Let X be a Banach space. Let $r \in \{1, 2, \infty\}$, and let $F : X \to \ell^r$ be a linear map. For $j \in \mathbb{N}$, let $f_j(x) := F(x)(j)$, $x \in X$. Then $F \in BL(X, \ell^r)$ if and only if $f_j \in BL(X, \mathbb{K})$ for each $j \in \mathbb{N}$.

3.16 Let X be a Banach space, and let Y be a normed space. If (F_n) is a sequence in $BL(X,Y)$ such that $(F_n(x))$ converges in Y for every $x \in X$, then (F_n) converges uniformly on each totally bounded subset of X.

3.17 Let X and Y be Banach spaces, and let (F_n) be a sequence in $BL(X,Y)$. Then there is $F \in BL(X,Y)$ such that $F_n(x) \to F(x)$ for each $x \in X$ if and only if

$(\|F_n\|)$ is a bounded sequence and $(F_n(x))$ is a Cauchy sequence in Y for each x in a subset of X whose span is dense in X.

3.18 Let X be a Banach space, and let (P_n) be a sequence of projection operators in $BL(X)$ such that $R(P_n) \subset R(P_{n+1})$ for all $n \in \mathbb{N}$. Then $P_n(x) \to x$ for each $x \in X$ if and only if $(\|P_n\|)$ is a bounded sequence and $\bigcup_{n=1}^{\infty} R(P_n)$ is dense in X.

3.19 Let (Q_n) be a sequence of quadrature formulæ as given in Theorem 3.27. Then (Q_n) is convergent if and only if

(i) $\sum_{j=1}^{m_n} w_{n,j} t_{n,j}^k \to \left(b^{k+1} - a^{k+1}\right)/(k+1)$ for $k = 0, 1, 2, \ldots$, and
(ii) the sequence $\left(\sum_{j=1}^{m_n} |w_{n,j}|\right)$ is bounded.

The condition (ii) above is redundant if $w_{n,j} \geq 0$ for all $n, j \in \mathbb{N}$.

3.20 **(Norm of the graph)** Let $(X, \|\cdot\|_X)$ and $(Y, \|\cdot\|_Y)$ be normed spaces, and let $F : X \to Y$ be a linear map. For $x \in X$, let $\|x\|_F := \|x\|_X + \|F(x)\|_Y$.

(i) The norms $\|\cdot\|_X$ and $\|\cdot\|_F$ on X are equivalent if and only if F is continuous.
(ii) Let $(X, \|\cdot\|_X)$ and $(Y, \|\cdot\|_Y)$ be Banach spaces, and let F be a closed map. Then $(X, \|\cdot\|_F)$ is a Banach space.
(iii) Let $(X, \|\cdot\|_F)$ be a Banach space. Then F is a closed map.
(iv) Let $(X, \|\cdot\|_X)$ and $(X, \|\cdot\|_F)$ be Banach spaces. Then F is continuous.

3.21 Let X_1 and X_2 be Banach spaces, and let Y be a normed space. Suppose maps $F_1 \in B(X_1, Y)$ and $F_2 \in BL(X_2, Y)$ are such that for every $x_1 \in X_1$, there is a unique $x_2 \in X_2$ satisfying $F_1(x_1) = F_2(x_2)$, and let us define $F(x_1) := x_2$. Then $F \in BL(X_1, X_2)$.

3.22 For $x := (x(1), x(2), \ldots)$ and $y := (y(1), y(2), \ldots)$, where $x(j), y(j) \in \mathbb{K}$, define $xy := (x(1)y(1), x(2)y(2), \ldots)$. Let $p, r \in \{1, 2, \infty\}$, and let q satisfy $(1/p) + (1/q) = 1$. Then

(i) $y \in \ell^q$ if and only if $xy \in \ell^1$ for all $x \in \ell^p$. In this case, if we define $F : \ell^p \to \ell^1$ by $F(x) := xy$, then $F \in BL(\ell^p, \ell^1)$ and $\|F\| = \|y\|_q$.
(ii) $y \in \ell^r$ if and only if $xy \in \ell^r$ for all $x \in \ell^\infty$. In this case, if we define $F : \ell^\infty \to \ell^r$ by $F(x) := xy$, then $F \in BL(\ell^\infty, \ell^r)$ and $\|F\| = \|y\|_r$.
(iii) $y \in \ell^\infty$ if and only if $xy \in \ell^r$ for all $x \in \ell^p$, where $p \leq r$. In this case, if we define $F : \ell^p \to \ell^r$ by $F(x) := xy$, then $F \in BL(\ell^p, \ell^r)$ and $\|F\| = \|y\|_\infty$.

3.23 Let $y : [a, b] \to \mathbb{K}$ be a measurable function. Let $p, r \in \{1, 2, \infty\}$, and let q satisfy $(1/p) + (1/q) = 1$. Denote $L^p([0, 1])$ by L^p. Then

(i) $y \in L^q$ if and only if $xy \in L^1$ for all $x \in L^p$. In this case, if we define $F : L^p \to L^1$ by $F(x) := xy$, then $F \in BL(L^p, L^1)$ and $\|F\| = \|y\|_q$.

(ii) $y \in L^r$ if and only if $xy \in L^r$ for all $x \in L^\infty$. In this case, if we define $F : L^\infty \to L^r$ by $F(x) := xy$, then $F \in BL(L^\infty, L^r)$ and $\|F\| = \|y\|_r$.

(iii) $y \in L^\infty$ if and only if $xy \in L^2$ for all $x \in L^2$. In this case, if we define $F : L^2 \to L^2$ by $F(x) := xy$, then $F \in BL(L^p)$ and $\|F\| = \|y\|_\infty$.

3.24 Let $\|\cdot\|$ be a complete norm on $C([a, b])$ such that $x_n(t) \to x(t)$ for every $t \in [a, b]$ whenever $\|x_n - x\| \to 0$. Then a sequence (x_n) in $C([a, b])$ converges uniformly to x on $[a, b]$ if and only if $\|x_n - x\| \to 0$.

3.25 Let X be a Banach space, and let Y, Z be closed subspaces of X such that $X = Y + Z$. Then there is $\gamma > 0$ such that for every $x \in X$, there are $y \in Y$ and $z \in Z$ satisfying $x = y + z$ and $\|y\| + \|z\| \le \gamma\|x\|$.

3.26 Let X and Y be Banach spaces, and let $F \in BL(X, Y)$ be onto. Suppose $y \in Y$ and $x \in X$ satisfy $F(x) = y$. If $y_n \to y$ in Y, then there is a sequence (x_n) in X such that $F(x_n) = y_n$ for all $n \in \mathbb{N}$ and $x_n \to x$ in X.

3.27 Let X and Y be normed spaces, and let $F : X \to Y$ be a linear map.

(i) **(Bounded inverse theorem)**Suppose F is continuous, one-one and onto. For $y \in Y$, define $q(y) := \|F^{-1}(y)\|$. If X is Banach space, then q is a countably subadditive seminorm on Y, and if Y is also a Banach space, then q is continuous, so that $F^{-1} \in BL(Y, X)$.

(ii) **(Open mapping theorem)** Suppose F is continuous and onto. For $y \in Y$, define $q(y) := \inf\{\|x\| : x \in X \text{ and } F(x) = y\}$. If X is a Banach space, then q is a countably subadditive seminorm on Y, and if Y is also a Banach space, then q is continuous, so that F is an open map.

(iii) **(Closed graph theorem)** Define $\Phi : \text{Gr}(F) \to X$ by $\Phi(x, F(x)) := x$ for $x \in X$. Then Φ is linear, continuous, one-one and onto. If X and Y are Banach spaces, and F is a closed map, then $F \in BL(X, Y)$.

3.28 Let $p, p', r, r' \in \{1, 2, \infty\}$, and let $L^p := L^p([0, 1])$. For F in $BL(\ell^p, \ell^r)$ as well as for F in $BL(L^p, L^r)$, let $\|F\|_{p,r}$ denote its operator norm.

(i) Let $p' \le p$ and $r \le r'$. Then $BL(\ell^p, \ell^r) \subset BL(\ell^{p'}, \ell^{r'})$, and $\|F\|_{p',r'} \le \|F\|_{p,r}$ for all $F \in BL(\ell^p, \ell^r)$. Also, $CL(\ell^p, \ell^r) \subset CL(\ell^{p'}, \ell^{r'})$. In particular, $BL(\ell^p) \subset BL(\ell^{p'}, \ell^{r'})$ and $CL(\ell^p) \subset CL(\ell^{p'}, \ell^{r'})$ if $p' \le p \le r'$.

(ii) Let $p' \ge p$ and $r \ge r'$. Then $BL(L^p, L^r) \subset BL(L^{p'}, L^{r'})$, and $\|F\|_{p',r'} \le \|F\|_{p,r}$ for all $BL(L^p, L^r)$. Also, $CL(L^p, L^r) \subset CL(L^{p'}, L^{r'})$. In particular, $BL(L^p) \subset BL(L^{p'}, L^{r'})$ and $CL(L^p) \subset CL(L^{p'}, L^{r'})$ if $r' \le p \le p'$.

3.29 Let X be an infinite dimensional normed space, and let $F \in CL(X)$. Define $G := k_0 I + k_1 F + \cdots + k_n F^n$, where $k_0, k_1, \ldots, k_n$ are in $\mathbb{K}$. Then $G \in CL(X)$ if and only if $k_0 = 0$.

3.30 Let X be a Banach space, and $P \in BL(X)$ be a projection operator. Then P belongs to $CL(X)$ if and only if P is of finite rank.

3.31 Let $M := [k_{i,j}]$ be an infinite matrix, let $p, r \in \{1, 2, \infty\}$, and let q be the conjugate exponent of p. For $j \in \mathbb{N}$, let $\alpha_r(j)$ denote the r-norm of the jth column of M, and for $i \in \mathbb{N}$, let $\beta_q(i)$ denote the q-norm of the ith row of M. Further, let $\alpha_r := \sup\{\alpha_r(j) : j \in \mathbb{N}\}$ and $\beta_q := \sup\{\beta_q(i) : i \in \mathbb{N}\}$.

(i) If M defines a map F from ℓ^p to ℓ^r, then $\max\{\alpha_r, \beta_q\} \le \|F\| < \infty$.

(ii) M defines a (linear) map F from ℓ^1 to ℓ^r if and only if $\{\alpha_r(j) : j \in \mathbb{N}\}$ is a bounded subset of $\mathbb{K}$. In this case, $F \in BL(\ell^1, \ell^r)$ and $\|F\| = \alpha_r$. If, in addition, $\alpha_r(j) \to 0$, then $F \in CL(\ell^1, \ell^r)$.

(iii) M defines a (linear) map F from ℓ^p to ℓ^∞ if and only if $\{\beta_q(i) : i \in \mathbb{N}\}$ is a bounded subset of $\mathbb{K}$. In this case, $F \in BL(\ell^p, \ell^\infty)$ and $\|F\| = \beta_q$. If, in addition, $\beta_q(i) \to 0$, then $F \in CL(\ell^p, \ell^\infty)$.

3.32 Let $M := [k_{i,j}]$ be an infinite matrix, and let $p, r \in \{1, 2, \infty\}$.

(i) Let $k_{i,j} := 1$ if $1 \le i \le j$ and $k_{i,j} := 0$ if $i > j \ge 1$. Then M defines a map F from ℓ^p to ℓ^r if and only if $p = 1$ and $r = \infty$. In this case, $F \in BL(\ell^1, \ell^\infty)$, and $\|F\| = 1$. But $F \notin CL(\ell^1, \ell^\infty)$.

(ii) Let $k_{i,j} := 1/i$ if $1 \le i \le j$ and $k_{i,j} := 0$ if $i > j \ge 1$. Then M defines a map from ℓ^p to ℓ^r if and only if $p = 1$ and $r \in \{2, \infty\}$. In this case, $F \in BL(\ell^1, \ell^r)$, and $\|F\| = \pi/\sqrt{6}$ if $r = 2$, while $\|F\| = 1$ if $r = \infty$. In fact, $F \in CL(\ell^1, \ell^r)$ for $r \in \{2, \infty\}$.

3.33 Let $M := [k_{i,j}]$ be an infinite matrix, and let $p, r \in \{1, 2, \infty\}$.

(i) The converse of Corollary 3.31 holds if and only if $p = 1$.

(ii) The converse of Corollary 3.26 holds if and only if $r = \infty$.

(iii) Suppose $p \ne 1$ and $r \ne \infty$. Then there is a matrix M such that $\{\alpha_r(j) : j \in \mathbb{N}\}$ and $\{\beta_q(i) : i \in \mathbb{N}\}$ are bounded subsets of $\mathbb{K}$, but M does not define a map from ℓ^p to ℓ^r.

3.34 Let $p \in \{2, \infty\}$ and $r \in \{1, 2, \infty\}$. If a matrix M defines a map belonging to $CL(\ell^p, \ell^r)$, then the sequence of the columns of M tends to 0 in ℓ^r. This result does not hold if $p = 1$. (Compare Exercises 4.16 (v) and 4.21.)

3.35 (i) **(Tridiagonal operator)** Let (a_j), (b_j), (c_j) be sequences in $\mathbb{K}$, and let $M := [k_{i,j}]$, where for $j \in \mathbb{N}$, $k_{j+1,j} := a_j$, $k_{j,j} := b_j$, $k_{j,j+1} := c_j$, and $k_{i,j} := 0$ if $i \notin \{j+1, j, j-1\}$. Then the tridiagonal matrix M defines $F \in BL(\ell^p)$ if and only if (a_j), (b_j), (c_j) are bounded sequences, and $F \in CL(\ell^p)$ if and only if $a_j \to 0$, $b_j \to 0$, $c_j \to 0$, where $p \in \{1, 2, \infty\}$.

(ii) **(Diagonal operator)** Let (k_j) be a sequence in $\mathbb{K}$, and let $M := \text{diag}(k_1, k_2, \ldots)$. The diagonal matrix M defines $F \in BL(\ell^p)$ if and only if (k_j) is a bounded sequence, and $F \in CL(\ell^p)$ if and only if $k_j \to 0$.

(iii) **(Weighted right shift operator)** Let (w_j) be a sequence in $\mathbb{K}$, and let $M := [k_{i,j}]$, where for $j \in \mathbb{N}$, $k_{j+1,j} := w_j$ and $k_{i,j} := 0$ if $i \ne j + 1$. Then the lower weighted-shift matrix M defines $F \in BL(\ell^p)$ if and only if (w_j) is a bounded sequence, and $F \in CL(\ell^p)$ if and only if $w_j \to 0$.

3.36 Let $M := [k_{i,j}]$ be an infinite matrix. Let $p \in \{2, \infty\}$, q be the conjugate exponent of p, and let $r \in \{1, 2\}$. Define $\gamma_{1,1} := \sum_{i=1}^{\infty} \beta_1(i)$, $\gamma_{1,2} := \left(\sum_{i=1}^{\infty} \beta_1(i)^2\right)^{1/2}$, $\gamma_{2,1} := \sum_{i=1}^{\infty} \beta_2(i)$, and $\gamma_{2,2} := \left(\sum_{i=1}^{\infty} \beta_2(i)^2\right)^{1/2}$.
If $\gamma_{q,r} < \infty$, then M defines $F \in CL(\ell^p, \ell^r)$ and $\|F\| \le \gamma_{q,r}$.
If $\gamma_{1,1} < \infty$, then in fact M defines $F \in CL(\ell^p, \ell^r)$ for all $p, r \in \{1, 2, \infty\}$.

3.37 Let $X := L^1([0, 1])$, $Y := C([0, 1])$, and $k(\cdot\,, \cdot) \in C([0, 1] \times [0, 1])$. Let F denote the Fredholm integral map from X to Y with kernel $k(\cdot\,, \cdot)$. For $n \in \mathbb{N}$, let F_n denote the Fredholm integral map from X to Y with kernel

$$k_n(s, t) := \sum_{i,j=0}^{n} k\left(\frac{i}{n}, \frac{j}{n}\right)\binom{n}{i}\binom{n}{j} s^i (1-s)^{n-i} t^j (1-t)^{n-j}, \quad s, t \in [0, 1],$$

called the nth **Bernstein polynomial in two variables**. Then each F_n belongs to $BL(X, Y)$, F_n is of finite rank, $\|F - F_n\| \to 0$, and $F \in CL(X, Y)$.

3.38 Let X and Y be normed spaces, and let $F : X \to Y$ be linear. Let $\overline{U}$ denote the closed unit ball of X. Then $F \in CL(X, Y)$ if and only if the closure of $F(\overline{U})$ is a compact subset of Y. If $F \in CL(X, Y)$, then $F(\overline{U})$ is a totally bounded subset of Y, and conversely, if Y is a Banach space and $F(\overline{U})$ is a totally bounded subset of Y, then $F \in CL(X, Y)$.

3.39 Let X be a Banach space, and let $F \in CL(X)$. Let (F_n) be a sequence in $BL(X)$ such that $F_n(x) \to F(x)$ for every $x \in X$. Then $\|(F_n - F)F\| \to 0$.
(Note: This result is important in operator approximation theory.)

3.40 Let H and G be Hilbert spaces, and let $A \in BL(H, G)$. If there is a countable orthonormal basis $\{u_1, u_2, \ldots\}$ for H such that $\sum_j \|A(u_j)\|^2 < \infty$, then A is called a **Hilbert–Schmidt map**. In this case, if $\{\tilde{u}_1, \tilde{u}_2, \ldots\}$ is any orthonormal basis for H, then $\sum_k \|A(\tilde{u}_k)\|^2 = \sum_j \|A(u_j)\|^2$.

(i) Every Hilbert–Schmidt map is a compact linear map.

(ii) Let $A \in BL(\ell^2)$. Then A is a Hilbert–Schmidt operator if and only if A is defined by a matrix $M := [k_{i,j}]$ satisfying $\gamma_{2,2} < \infty$.

(iii) Let $L^2 := L^2([a, b])$, and $A \in BL(L^2)$. Then A is a Hilbert–Schmidt operator if and only if A is a Fredholm integral operator defined by a kernel $k(\cdot\,, \cdot) \in L^2([a, b] \times [a, b])$.

Chapter 4
Dual Spaces, Transposes and Adjoints

In this chapter we develop a duality between a normed space X and the space X' consisting of all bounded linear functionals on X, known as the dual space of X. As a consequence of the Hahn–Banach extension theorem, we show that $X' \neq \{0\}$ if $X \neq \{0\}$. We also prove a companion result which is geometric in nature and is known as the Hahn–Banach separation theorem. We characterize duals of several well-known normed spaces. To a bounded linear map F from a normed space X to a normed space Y, we associate a bounded linear map F' from Y' to X', known as the transpose of F. To a bounded linear map A from a Hilbert space H to a Hilbert space G, we associate a bounded linear map A^* from G to H, known as the adjoint of A. We study maps that are 'well behaved' with respect to the adjoint operation. We also introduce the numerical range of a bounded linear map from a nonzero inner product space to itself. These considerations will be useful in studying the spectral theory in the next chapter.

4.1 Hahn–Banach Theorems

In Chap. 3, we have studied properties of continuous linear maps from one normed space to another. We also studied properties of some variants of continuous linear maps, namely, the closed linear maps and the compact linear maps. We have so far not wondered whether continuous linear maps always exist on infinite dimensional normed spaces. As a start, we may look for a continuous linear functional defined on a possibly finite dimensional subspace, and attempt to extend it to the entire normed space.

Let X be a normed space, let Y be a subspace of X and let g be a continuous linear functional defined on Y. There is certainly a linear extension of g to X. To see this, let $\{y_s\}$ be a (Hamel) basis for Y. By Corollary 1.2, there is a subset $\{z_t\}$ of X such that $\{y_s\} \cup \{z_t\}$ is a (Hamel) basis for X. Given $x \in X$, there are unique $y_{s_1}, \ldots, y_{s_m}, z_{t_1}, \ldots, z_{t_n}$ in X, and $k_1, \ldots, k_m, \ell_1, \ldots, \ell_n$ in $\mathbb{K}$ with $x = k_1 y_{s_1} + \cdots + k_m y_{s_m} + \ell_1 z_{t_1} + \cdots + \ell_n z_{t_n}$. Define $f(x) := k_1 g(y_{s_1}) + \cdots + k_m g(y_{s_m})$. Then

© Springer Science+Business Media Singapore 2016, corrected publication 2023

B.V. Limaye, *Linear Functional Analysis for Scientists and Engineers*,
https://doi.org/10.1007/978-981-10-0972-3_4

the functional f is linear on X and $f(y) = g(y)$ for all $y \in Y$. But there is no reason for f to be continuous on X. On the other hand, a continuous linear functional g on Y can be extended to a continuous linear functional $\tilde{g}$ on the closure $\overline{Y}$ of Y by Proposition 3.17(ii). Because $\overline{Y}$ is closed in X, the Tietze extension theorem mentioned in Sect. 1.3 gives a continuous functional f on X such that $f(y) = \tilde{g}(y)$ for all $y \in \overline{Y}$, and hence $f(y) = g(y)$ for all $y \in Y$. But there is no reason for f to be linear on X. Thus it is far from obvious how one may extend g to X both linearly and continuously.

In this section, we shall see how a continuous linear functional g defined on a subspace Y of a normed space X can be extended to a continuous linear functional f on X satisfying $\|f\| = \|g\|$. Later we shall briefly discuss the case of a continuous linear map from Y to a normed space Z. We first prove a preliminary result for a normed space over $\mathbb{R}$, and then show how to treat a normed space over $\mathbb{C}$ by regarding it as a linear space over $\mathbb{R}$.

Let X be a linear space over $\mathbb{K}$. A **sublinear functional** on X is a function $p : X \to \mathbb{R}$ such that

(i) $p(x + y) \leq p(x) + p(y)$ for all $x, y \in X$, and
(ii) $p(t\,x) = t\,p(x)$ for all $x \in X$ and $t \geq 0$.

Every seminorm on X as well as every linear functional on X is a sublinear functional on X, but there are other sublinear functionals as well. For example, let $X := \mathbb{R}$, and $\alpha, \beta \in \mathbb{R}$ be such that $\alpha \leq \beta$. Define $p(x) := \alpha x$ if $x < 0$ and $p(x) := \beta x$ if $x \geq 0$. Then p is a sublinear functional on $\mathbb{R}$. If $\beta \geq 0$, then the choice $\alpha := -\beta$ yields the seminorm $p(x) := \beta|x|$, $x \in \mathbb{R}$, while the choice $\alpha := \beta$ yields the linear functional $p(x) := \beta x$, $x \in \mathbb{R}$.

Lemma 4.1 *Let X be a linear space over $\mathbb{R}$, and let $p : X \to \mathbb{R}$ be a sublinear functional on X. Suppose Y is a subspace of X, and $g : Y \to \mathbb{R}$ is a linear functional on Y such that $g(y) \leq p(y)$ for all $y \in Y$. Then there is a linear functional $f : X \to \mathbb{R}$ such that $f(y) = g(y)$ for all $y \in Y$, and $f(x) \leq p(x)$ for all $x \in X$.*

Proof If $Y = X$, then there is nothing to prove. If $Y \neq X$, then there is $a \in X$ such that $a \notin Y$. Let $Y_1 := \operatorname{span}\{Y, a\}$. As a modest beginning, we first show that there is a linear functional $g_1 : Y_1 \to \mathbb{R}$ such that $g_1(y) = g(y)$ for all $y \in Y$, and $g_1(y_1) \leq p(y_1)$ for all $y_1 \in Y_1$. For $y, z \in Y$,

$$g(y) + g(z) = g(y + z) \leq p(y + z) = p(y - a + z + a) \leq p(y - a) + p(z + a),$$

that is, $g(y) - p(y - a) \leq p(z + a) - g(z)$. Let

$$\alpha := \sup\{g(y) - p(y - a) : y \in Y\} \quad \text{and} \quad \beta := \inf\{p(y + a) - g(y) : y \in Y\}.$$

Then $\alpha \leq \beta$. Let γ be a real number satisfying $\alpha \leq \gamma \leq \beta$, and define $g_1 : Y_1 \to \mathbb{R}$ by $g_1(y + t\,a) := g(y) + t\,\gamma$ for $y \in Y$ and $t \in \mathbb{R}$. Clearly, g_1 is linear and $g_1(y) = g(y)$ for all $y \in Y$. To show that $g_1(y_1) \leq p(y_1)$ for all $y_1 \in Y_1$, we proceed as follows.

Let $y_1 \in Y_1$. Then there are unique $y \in Y$ and $t \in \mathbb{R}$ such that $y_1 = y + t\,a$. If $t = 0$, then $y_1 = y$ and $g_1(y_1) = g(y) \le p(y) = p(y_1)$. If $t > 0$, then

$$g_1(y_1) = g(y) + t\,\gamma \le g(y) + t\,\beta \le g(y) + t\Big[p\Big(\frac{y}{t} + a\Big) - g\Big(\frac{y}{t}\Big)\Big] = p(y + t\,a) = p(y_1).$$

On the other hand, if $t < 0$, then we let $r := -t > 0$ and obtain

$$g_1(y_1) = g(y) - r\gamma \le g(y) - r\,\alpha \le g(y) - r\Big[g\Big(\frac{y}{r}\Big) - p\Big(\frac{y}{r} - a\Big)\Big] = p(y - r\,a) = p(y_1).$$

Thus $g_1(y_1) \le p(y_1)$ for all $y_1 \in Y_1$, as claimed.

Consider the set $\mathcal{Z}$ of all pairs (Z, h), where Z is a subspace of X containing Y, and $h : Z \to \mathbb{R}$ is a linear functional such that $h(y) = g(y)$ for all $y \in Y$, and $h(z) \le p(z)$ for all $z \in Z$. Then the set $\mathcal{Z}$ is nonempty, since $(Y, g) \in \mathcal{Z}$. Let us partially order $\mathcal{Z}$ by defining $(Z_1, h_1) \le (Z_2, h_2)$ if $Z_1 \subset Z_2$ and $h_2(z_1) = h_1(z_1)$ for all $z_1 \in Z_1$. It can be checked that every totally ordered subset of $\mathcal{Z}$ has an upper bound in $\mathcal{Z}$. By the Zorn lemma stated in Sect. 1.1, there is a maximal element (W, f) of $\mathcal{Z}$. We claim that $W = X$. Otherwise, there is $b \in X \setminus W$. Let $W_1 := \text{span}\,\{W, b\}$. By what we have proved earlier, there is a linear functional $f_1 : W_1 \to \mathbb{R}$ such that $f_1(w) = f(w)$ for all $w \in W$, and $f_1(w_1) \le p(w_1)$ for all $w_1 \in W_1$. Thus $(W_1, f_1) \in \mathcal{Z}$, $(W, f) \le (W_1, f_1)$, but $(W, f) \ne (W_1, f_1)$. This contradicts the maximality of (W, f) in $\mathcal{Z}$. Hence $W = X$, and so f is a linear functional on X such that $f(y) = g(y)$ for all $y \in Y$, and $f(x) \le p(x)$ for all $x \in X$. □

Remark 4.2 The first part of the proof of Lemma 4.1 involves choosing a suitable real number γ which lies between the supremum α of a subset of $\mathbb{R}$ and the infimum β of another subset of $\mathbb{R}$. This yields an extension g_1 of g to a larger subspace Y_1 of X. Whenever such choices are possible, one can extend the functional g_1 to a still larger subspace Y_2 of X containing Y_1 in a similar manner, and repeat this process. But there is no guarantee that we can thus extend g to all of X. Hence we resorted to transfinite induction by using the Zorn lemma and lost the constructive nature of our proof. ◊

Lemma 4.3 *Let X be a linear space over $\mathbb{C}$. Consider functions $f : X \to \mathbb{C}$ and $u : X \to \mathbb{R}$. Then f is a complex-linear functional on X and $u(x) = \text{Re}\, f(x)$ for all $x \in X$ if and only if u is a real-linear functional on X (regarded as a linear space over $\mathbb{R}$) and $f(x) = u(x) - iu(ix)$ for all $x \in X$.*

If, in addition, X is a normed space, and f and u are related as above, then f is a continuous complex-linear functional on X if and only if u is a continuous real-linear functional on X, and in this case $\|f\| = \|u\|$.

Proof Suppose f is complex-linear, and $u(x) := \text{Re}\, f(x)$ for $x \in X$. It is easy to see that u is real-linear. Also, for all $x \in X$,

$$f(x) = \text{Re}\, f(x) + i\,\text{Im}\, f(x) = u(x) - i\,\text{Re}\, if(x) = u(x) - i\,\text{Re}\, f(ix) = u(x) - iu(ix).$$

Conversely, suppose u is real-linear, and $f(x) := u(x) - iu(ix)$ for $x \in X$. Then $\mathrm{Re}\, f(x) = u(x)$ for all $x \in X$, and it is easy to see that f is real-linear. Also, since $f(ix) = u(ix) - iu(-x) = u(ix) + iu(x) = i\,[u(x) - iu(ix)] = if(x)$ for all $x \in X$, it follows that f is in fact complex-linear.

Let X be a normed space. Let $f : X \to \mathbb{C}$ be complex-linear, $u : X \to \mathbb{R}$ be real-linear, and let them be related as above. Then it is clear that f is continuous if and only if u is continuous. Further, $|u(x)| = |\mathrm{Re}\, f(x)| \le |f(x)|$ for all $x \in X$. Also, if $x \in X$, and we let $k := \mathrm{sgn}\, f(x)$, then $|f(x)| = kf(x) = f(kx) = \mathrm{Re}\, f(kx) = u(kx)$, and $\|kx\| \le \|x\|$. Hence $\|f\| = \|u\|$. □

Theorem 4.4 (Hahn–Banach extension theorem) *Let X be a normed space over $\mathbb{K}$. If Y is a subspace of X, and $g : Y \to \mathbb{K}$ is a continuous linear functional on Y, then there is a continuous linear functional $f : X \to \mathbb{K}$ such that $f(y) = g(y)$ for all $y \in Y$, and $\|f\| = \|g\|$.*

Proof First let $\mathbb{K} := \mathbb{R}$. Let $\alpha := \|g\|$, and define $p : X \to \mathbb{R}$ by $p(x) := \alpha\|x\|$ for $x \in X$. Then p is a sublinear functional on X. By Lemma 4.1, there is a linear functional $f : X \to \mathbb{R}$ such that $f(y) = g(y)$ for all $y \in Y$, and $f(x) \le p(x)$ for all $x \in X$. Further, since

$$-f(x) = f(-x) \le p(-x) = \alpha\|-x\| = \alpha\|x\| = p(x) \quad \text{for all } x \in X,$$

we see that $|f(x)| \le \alpha\|x\|$ for all $x \in X$. Hence f is continuous on X and $\|f\| \le \alpha = \|g\|$. But since $f(y) = g(y)$ for all $y \in Y$, it follows that $\|f\| \ge \|g\|$. Thus $\|f\| = \|g\|$.

Next, let $\mathbb{K} := \mathbb{C}$. Let us regard X as a linear space over $\mathbb{R}$. Then Y is a real-linear subspace of X. Define $v : Y \to \mathbb{R}$ by $v(y) := \mathrm{Re}\, g(y)$ for $y \in Y$. By Lemma 4.3, v is a continuous real-linear functional on Y, and $\|v\| = \|g\|$. Also, by the case $\mathbb{K} = \mathbb{R}$ considered above, there is a continuous real-linear functional $u : X \to \mathbb{R}$ such that $u(y) = v(y)$ for all $y \in Y$, and $\|u\| = \|v\|$. Define $f : X \to \mathbb{C}$ by $f(x) := u(x) - iu(ix)$ for $x \in X$. By Lemma 4.3, f is a continuous complex-linear functional on X, $f(y) = u(y) - iu(iy) = v(y) - iv(iy) = g(y)$ for all $y \in Y$, and $\|f\| = \|u\| = \|v\| = \|g\|$. □

We remark that the complex case of the Hahn–Banach theorem was proved independently by Bohnenblust and by Sobczyk in 1938, about ten years after Hahn (1927) and Banach (1929) had proved the real case.

The above theorem not only gives an extension of a continuous linear functional which is simultaneously linear as well as continuous, but also guarantees a linear extension which preserves the norm of the given functional. Consider a normed space X over $\mathbb{K}$, and let Y be a subspace of X. Let g be a continuous linear functional on Y. A **Hahn–Banach extension** of g to X is a continuous linear functional f on X such that $f(y) = g(y)$ for all $y \in Y$, and $\|f\| = \|g\|$. Theorem 4.4 tells us that any continuous linear functional defined on any subspace of a normed space has at least one Hahn–Banach extension. It may or may not be unique as the following examples show.

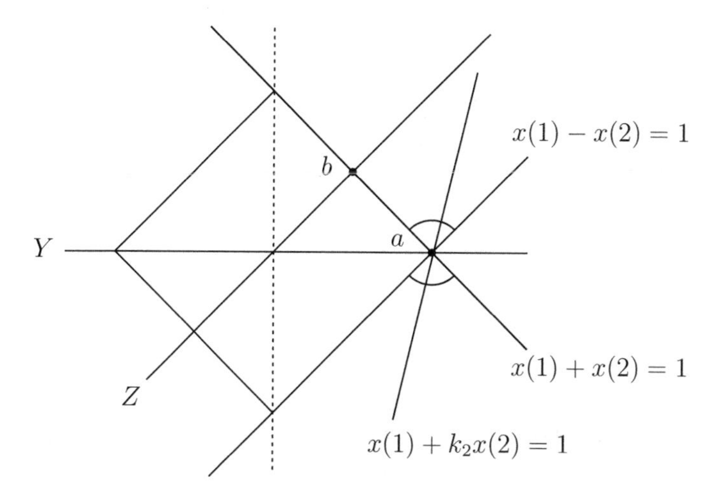

Fig. 4.1 Illustrations of $\{x \in \mathbb{R}^2 : f(x) = 1\}$, where f is a Hahn–Banach extension in Example 4.5(i) and (ii)

Examples 4.5 Let $X := \mathbb{K}^2$ with the norm $\|\cdot\|_1$. (See Fig. 4.1.)

(i) Let $Y := \{(x(1), x(2)) \in X : x(2) = 0\}$. Define $g : Y \to \mathbb{K}$ by $g(y) := y(1)$ for $y := (y(1), y(2)) \in Y$. It is clear that g is linear, g is continuous, and $\|g\| = 1 = g(a)$, where $a := (1, 0) \in Y$. Since $Y = \text{span}\,\{a\}$, a function f defined on $\mathbb{K}^2$ is a Hahn-Banach extension of g to $\mathbb{K}^2$ if and only if f is linear on $\mathbb{K}^2$ and $\|f\| = 1 = f(a)$. Now, if f is linear on $\mathbb{K}^2$, then there are $k_1, k_2 \in \mathbb{K}$ such that $f(x) = k_1x(1) + k_2x(2)$ for all $x := (x(1), x(2)) \in \mathbb{K}^2$, and $\|f\| = \max\{|k_1|, |k_2|\}$ by considering the column sums of the 1×2 matrix $\left[|k_1|\ |k_2|\right]$. Consequently, $\|f\| = 1 = f(a)$ if and only if $k_1 = 1$ and $|k_2| \le 1$. Hence $f : \mathbb{K}^2 \to \mathbb{K}$ is a Hahn–Banach extension of g to $\mathbb{K}^2$ if and only if $f(x) := x(1) + k_2x(2)$ for all $x := (x(1), x(2)) \in \mathbb{K}^2$, where $k_2 \in \mathbb{K}$ and $|k_2| \le 1$. Thus there are infinitely many Hahn–Banach extensions of g to $\mathbb{K}^2$.

(ii) Let $Z := \{(x(1), x(2)) \in \mathbb{K}^2 : x(1) = x(2)\}$. Define $h : Z \to \mathbb{K}$ by $h(z) := 2z(1)$ for $z := (z(1), z(2)) \in Z$. It is clear that h is linear, h is continuous, and $\|h\| = 1 = h(b)$, where $b := (1/2, 1/2) \in Y$. Since $Z = \text{span}\,\{b\}$, a function f defined on $\mathbb{K}^2$ is a Hahn–Banach extension of h to $\mathbb{K}^2$ if and only if f is linear on $\mathbb{K}^2$ and $\|f\| = 1 = f(b)$. As before, this is the case if and only if there are $k_1, k_2 \in \mathbb{K}$ such that $f(x) = k_1x(1) + k_2x(2)$ for all $x := (x(1), x(2)) \in \mathbb{K}^2$ with $\max\{|k_1|, |k_2|\} = 1$ and $k_1 + k_2 = 2$, that is, $k_1 = 1 = k_2$. Thus the unique Hahn–Banach extension of h to $\mathbb{K}^2$ is given by $f(x) := x(1) + x(2)$ for all $x := (x(1), x(2)) \in \mathbb{K}^2$. ◊

In Exercise 4.4, we give a necessary and sufficient condition for a normed space to have unique Hahn–Banach extensions. Also, in Theorem 4.17, we show that Hahn–Banach extensions to a Hilbert space are unique.

The Hahn–Banach extension theorem is one of the most fundamental results in linear functional analysis. We shall now consider some important consequences of this theorem.

Proposition 4.6 *Let X be a normed space over $\mathbb{K}$.*

(i) *Let a be a nonzero element of X. Then there is a continuous linear functional f on X such that $f(a) = \|a\|$ and $\|f\| = 1$.*

(ii) *Let Y be a subspace of X, and let $\overline{Y}$ denote the closure of Y in X. Let $a \in X$ but $a \notin \overline{Y}$. Then there is a continuous linear functional f on X such that $f(y) = 0$ for all $y \in Y$, $f(a) = |||a + \overline{Y}|||$ and $\|f\| = 1$.*

(iii) *Let $\{a_1, \ldots, a_m\}$ be a linearly independent subset of X, and let $Y_j :=$ span $\{a_i : i = 1, \ldots, m,\ i \neq j\}$ for $j = 1, \ldots, m$. Then there are continuous linear functionals $f_1, \ldots, f_m$ on X such that $f_j(a_i) = \delta_{i,j}$ for $i, j = 1, \ldots, m$, and $\|f_j\| = 1/|||a_j + Y_j|||$ for $j = 1, \ldots, m$.*

Proof (i) Let $Y := \{ka : k \in \mathbb{K}\}$, and define $g : Y \to \mathbb{K}$ by $g(ka) := k\|a\|$ for $k \in \mathbb{K}$. Then Y is a subspace of X, g is a continuous linear functional on Y and $\|g\| = 1$. By Theorem 4.4, there is a continuous linear functional f on X such that $f(y) = g(y)$ for all $y \in Y$, and $\|f\| = \|g\|$, and so $f(a) = g(a) = \|a\|$ and $\|f\| = \|g\| = 1$.

(ii) Consider the quotient space $X/\overline{Y}$ with the quotient norm $|||\cdot|||$. Since $a \notin \overline{Y}$, we see that $a + \overline{Y} \neq 0 + \overline{Y}$. By (i) above, there is a continuous linear functional $\tilde{f} : X/\overline{Y} \to \mathbb{K}$ such that $\tilde{f}(a + \overline{Y}) = |||a + \overline{Y}|||$ and $\|\tilde{f}\| = 1$. Let $Q : X \to X/\overline{Y}$ denote the quotient map given by $Q(x) := x + \overline{Y}$, $x \in X$. Thus

$$X \xrightarrow{Q} X/\overline{Y} \xrightarrow{\tilde{f}} \mathbb{K}.$$

Define $f := \tilde{f} \circ Q$. Then f is a continuous linear functional on X such that $f(y) = \tilde{f}(y + \overline{Y}) = \tilde{f}(0 + \overline{Y}) = 0$ for $y \in Y$, and $f(a) = \tilde{f}(a + \overline{Y}) = |||a + \overline{Y}|||$. Also, as in the proof of Corollary 3.4, for $\alpha > 0$, $|f(x)| \leq \alpha\|x\|$ for all $x \in X$ if and only if $|\tilde{f}(x + \overline{Y})| \leq \alpha|||x + \overline{Y}|||$ for all $x + \overline{Y} \in X/\overline{Y}$. Hence $\|f\| = \|\tilde{f}\| = 1$.

(iii) Let $j \in \{1, \ldots, m\}$. The subspace Y_j of X is closed in X since it is finite dimensional (Remark 2.9), and so $a_j \notin \overline{Y_j}$. By (ii) above, there is a continuous linear functional h_j on X such that $h_j(y) = 0$ for all $y \in Y_j$, $h_j(a_j) = |||a_j + Y_j|||$ and $\|h_j\| = 1$. Let $f_j := h_j/|||a_j + Y_j|||$. □

Remarks 4.7 (i) Let X be a nonzero normed space over $\mathbb{K}$. Part (i) of the above proposition shows that $BL(X, \mathbb{K})$ is nonzero. In fact, if Y is a nonzero normed space over $\mathbb{K}$, then $BL(X, Y)$ is nonzero. To see this, let f be a nonzero element in $BL(X, \mathbb{K})$, and let b be a nonzero element in Y. If we let $F(x) := f(x)\,b$ for $x \in X$, then F is a nonzero element in $BL(X, Y)$. Since F is of finite rank, in fact F is a nonzero element in $CL(X, Y)$.

(ii) Part (ii) of the above proposition is often used in approximation theory in the following way. Suppose we wish to show that an element a of a normed space X can be approximated, as closely as we please, by elements of a subspace Y of X, that is, we wish to show that $a \in \overline{Y}$. This is the case if and only if $f(a) = 0$ for every continuous linear functional f on X that vanishes on Y. In Exercise 4.8, we consider an arbitrarily close approximation of $a \in X$ by elements of a convex subset of X. Of course, this procedure entails a knowledge of how all continuous linear functionals on X behave.

(iii) The following result which is ‘dual’ to part (iii) of the above proposition can be proved by using mathematical induction. Let X be a normed space, and let $\{f_1, \ldots, f_m\}$ be a linearly independent subset of $BL(X, \mathbb{K})$. Then there are $a_1, \ldots, a_m$ in X such that $f_j(a_i) = \delta_{i,j}$ for $i, j = 1, \ldots, m$. $\diamond$

Extension of a Continuous Linear Map

Let Y be a closed subspace of a normed space X. Suppose Z is a normed space, and let $G \in BL(Y, Z)$. Let us ask whether there is $F \in BL(X, Z)$ such that $F(y) = G(y)$ for all $y \in Y$, and whether, in addition, $\|F\| = \|G\|$, that is, whether there is a norm-preserving extension of G to X. The Hahn–Banach extension theorem gives an affirmative answer if $Z := \mathbb{K}$. The same holds if $Z := \ell^\infty$. (See Exercise 4.6) Suppose there is a projection operator P in $BL(X)$ with $R(P) = Y$, and define $F := G \circ P$. Then $F \in BL(X, Z)$, $F(y) = G(y)$ for all $y \in Y$, and $\|G\| \le \|F\| \le \|G\|\|P\|$. If in fact $\|P\| = 1$, then $\|F\| = \|G\|$. In particular, if X is a Hilbert space, then the projection theorem (Theorem 2.35) shows that every $G \in BL(Y, Z)$ has a norm-preserving linear extension to X. (See Exercise 3.6(i).) On the other hand, let $Z := Y$ and $G := I_Y$. If $F : X \to Y$ is an extension of G to X, then it is easy to see that $F^2 = F$ and $R(F) = Y$. Thus if $X := \ell^\infty$ and $Y := c_0$, then $G := I_Y \in BL(Y)$ has no extension belonging to $BL(X, Y)$ since there is no continuous projection operator on ℓ^∞ whose range is c_0 (Remark 3.34).

Now suppose the subspace Y of X is finite dimensional, and let $\{a_1, \ldots, a_m\}$ be a basis for Y. Let $f_1, \ldots, f_m$ be continuous linear functionals on X such that $f_j(a_i) = \delta_{i,j}$ for $i, j = 1, \ldots, m$. (See Proposition 4.6(iii).) Define $P(x) := f_1(x)a_1 + \cdots + f_m(x)a_m$ for $x \in X$. Clearly, P is a projection operator in $BL(X)$ with $R(P) = Y$. It follows that every G in $BL(Y, Z)$ has an extension to $F \in BL(X, Z)$. In particular, if $\dim Y = 1$, then every $G \in BL(Y, Z)$ is of rank at most 1, and so we may treat it as a continuous linear functional, and use the Hahn–Banach extension theorem to obtain a norm-preserving linear extension of G to X. Thus if $\dim X = 2$, then for every subspace Y of X and for every $G \in BL(Y, Z)$, there is a norm-preserving linear extension to X. This does not hold if $\dim X = 3$. Exercise 4.7 gives a two-dimensional subspace Y of $(\mathbb{K}^3, \|\cdot\|_1)$ such that $I_Y \in BL(Y)$ has no norm-preserving linear extension belonging to $BL(\mathbb{K}^3, Y)$.

Separation Theorem

Let X be a linear space over $\mathbb{K}$. We say that a subset E of X is **convex** if

$$(1-t)x_1 + tx_2 \in E \text{ whenever } x_1, x_2 \in E \text{ and } t \in (0, 1).$$

A subspace of X is clearly a convex subset of X. Also, if $x_1, \ldots, x_n \in X$, then

$$E := \{t_1 x_1 + \cdots + t_n x_n : t_1, \ldots, t_n \ge 0 \text{ and } t_1 + \cdots + t_n = 1\}$$

is a convex subset of X. We intend to prove a result which says, roughly speaking, that two disjoint convex subsets of a linear space can be ‘separated’ by a hyperplane. Such a result is of much use in Mathematical Economics. We shall use the Hahn–Banach extension theorem to obtain this result. First, we need some preparation.

A subset E of X is called **absorbing** if for every $x \in X$, there is $r > 0$ such that $r^{-1}x \in E$. Let E be an absorbing subset of X. Clearly, $0 \in E$. Since the set $\{r > 0 : r^{-1}x \in E\}$ is nonempty for every $x \in X$, we define

$$p_E(x) := \inf\{r > 0 : r^{-1}x \in E\} \quad \text{for } x \in X.$$

The function $p_E : X \to \mathbb{R}$ is called the **Minkowski gauge** of E.

Lemma 4.8 *Let E be a convex and absorbing subset of a linear space X. Then the Minkowski gauge p_E of E is a nonnegative sublinear functional on X, and*

$$\{x \in X : p_E(x) < 1\} \subset E \subset \{x \in X : p_E(x) \leq 1\}.$$

If, in addition, $c\,x \in E$ whenever $x \in E$ and $c \in \mathbb{K}$ with $|c| = 1$, then p_E is a seminorm on X.

Proof It is clear that $p_E(x) \geq 0$ for all $x \in X$, and $p_E(0) = 0$.

Let $x, \ y \in X$, and let $r, \ s > 0$ be such that $r^{-1}x, \ s^{-1}y \in E$. Then

$$(r+s)^{-1}(x+y) = \frac{r}{r+s}\,r^{-1}x + \frac{s}{r+s}\,s^{-1}y \in E,$$

since E is convex. Hence $p_E(x+y) \leq r+s$. Thus $p_E(x+y) - s$ is a lower bound for the set $\{r > 0 : r^{-1}x \in E\}$, and so $p_E(x+y) - s \leq p_E(x)$, that is, $p_E(x+y) \leq p_E(x) + s$. Further, $p_E(x+y) - p_E(x)$ is a lower bound for the set $\{s > 0 : s^{-1}y \in E\}$, and so $p_E(x+y) - p_E(x) \leq p_E(y)$, that is, $p_E(x+y) \leq p_E(x) + p_E(y)$.

Next, let $x \in X$ and $t > 0$. Consider $r > 0$ such that $r^{-1}x \in E$. Since $(t\,r)^{-1}t\,x = r^{-1}x \in E$, we obtain $p_E(t\,x) \leq t\,r$. Hence $p_E(t\,x) \leq t\,p_E(x)$. Replacing x by $t\,x$, and t by t^{-1}, we obtain $p_E(x) \leq t^{-1}p_E(t\,x)$, that is, $t\,p_E(x) \leq p_E(t\,x)$. Thus $p_E(t\,x) = t\,p_E(x)$. Hence p_E is a sublinear functional.

Let $x \in X$ with $p_E(x) < 1$. Then there is $r \in (0,1)$ such that $r^{-1}x \ \in E$. Since $0 \in E$, and since E is convex, $x = r(r^{-1}x) + (1-r)0 \in E$. Thus $\{x \in X : p_E(x) < 1\} \subset E$. Next, let $x \in E$. Since $1 \in \{r > 0 : r^{-1}x \in E\}$, we get $p_E(x) \leq 1$. Thus $E \subset \{x \in X : p_E(x) \leq 1\}$.

Finally, suppose $c\,x \in E$ whenever $x \in E$ and $c \in \mathbb{K}$ with $|c| = 1$. Let $x \in X, \ k \in \mathbb{K}$, and $k \neq 0$. By considering $c := k/|k|$ and $c := |k|/k$, we obtain $p_E(kx) = p_E(|k|x) = |k|p_E(x)$. Hence p_E is a seminorm on X. □

Proposition 4.9 *Let X be a normed space over $\mathbb{K}$, and let E be a nonempty convex open subset of X such that $0 \notin E$. Then there is a continuous linear functional f on X such that* $\operatorname{Re} f(x) > 0$ *for all $x \in E$.*

Proof First let $\mathbb{K} := \mathbb{R}$. Let $y_0 \in E$, and $E_0 := y_0 - E$, where $y_0 - E := \{y_0 - y : y \in E\}$. Then $0 \in E_0$, and E_0 is convex. Also, E_0 is open, and so there is $r > 0$ such that $U(0, r) \subset E_0$. It follows that E_0 is absorbing. By Lemma 4.8, the Minkowski gauge p_{E_0} of E_0 is a nonnegative sublinear functional on X. Also, since $0 \notin E$, we see that $y_0 \notin E_0$, and hence $p_{E_0}(y_0) \geq 1$. Let $Y := \text{span}\,\{y_0\}$, and define $g : Y \to \mathbb{R}$ by $g(k\, y_0) := k$ for $k \in \mathbb{R}$. If $k > 0$, then $g(k\, y_0) = k \leq k\, p_{E_0}(y_0) = p_{E_0}(k\, y_0)$, and if $k \leq 0$, then $g(k\, y_0) = k \leq 0 \leq p_{E_0}(k\, y_0)$. Thus $g(y) \leq p_{E_0}(y)$ for all $y \in Y$. By Lemma 4.1, there is a linear functional f on X such that $f(y_0) = g(y_0) = 1$ and $f(x) \leq p_{E_0}(x)$ for all $x \in X$. If $x \in E_0$, then $f(x) \leq p_{E_0}(x) \leq 1$, and if $-x \in E_0$, then $-f(x) = f(-x) \leq p_{E_0}(-x) \leq 1$, that is, $f(x) \geq -1$ by Lemma 4.8. Thus $|f(x)| \leq 1$ for all x in the open subset $E_0 \cap -E_0$, which contains 0. Hence f is bounded on the closed unit ball of X, and so f is continuous on X by Proposition 3.2. Let $x \in E$. Then $y_0 - x \in E_0$, and so $1 \geq f(y_0 - x) = f(y_0) - f(x) = 1 - f(x)$, that is, $f(x) \geq 0$. Thus $f(E) \subset [0, \infty)$. But f is a continuous linear map from the normed space X onto the finite dimensional normed space $\mathbb{K}$, and so f is an open map, as pointed out in Remark 3.40. Since E is an open subset of X, $f(E)$ is an open subset of $\mathbb{R}$. This shows that $f(E) \subset (0, \infty)$, that is, $f(x) > 0$ for all $x \in E$.

Next, let $\mathbb{K} := \mathbb{C}$. Treating X as a linear space over $\mathbb{R}$, we may find, as above, a real-linear continuous functional u from X to $\mathbb{R}$ such that $u(x) > 0$ for all $x \in E$. Define $f(x) := u(x) - iu(ix)$ for $x \in X$. As we have seen in Lemma 4.3, f is a continuous complex-linear functional from X to $\mathbb{C}$ such that $\text{Re}\, f(x) = u(x)$ for all $x \in X$, and so $\text{Re}\, f(x) > 0$ for all $x \in E$. □

The following result is often considered as a companion result of the Hahn–Banach extension theorem.

Theorem 4.10 (Hahn–Banach separation theorem) *Let E_1 and E_2 be nonempty disjoint convex subsets of a normed space X over $\mathbb{K}$, and let E_1 be open. Then there is a continuous linear functional f on X and there is $t \in \mathbb{R}$ such that*

$$\text{Re}\, f(x_1) < t \leq \text{Re}\, f(x_2) \quad \textit{for all } x_1 \in E_1 \textit{ and } x_2 \in E_2.$$

Proof Let $E := E_2 - E_1 = \{x_2 - x_1 : x_1 \in E_1 \text{ and } x_2 \in E_2\}$. Now $0 \notin E$ since $E_1 \cap E_2 = \emptyset$. Also, it is easy to see that E is a convex subset of X. Further, since the set $x_2 - E_1$ is open for each $x_2 \in E_2$, and E is their union, it follows that E is an open subset of X. By Proposition 4.9, there is a continuous linear functional f on X such that $\text{Re}\, f(x) > 0$ for all $x \in E$, that is, $\text{Re}\, f(x_1) < \text{Re}\, f(x_2)$ for all $x_1 \in E_1$ and $x_2 \in E_2$. Since E_1 and E_2 are convex subsets of X, we see that $\text{Re}\, f(E_1)$ and $\text{Re}\, f(E_2)$ are convex subsets of $\mathbb{R}$, that is, they are intervals in $\mathbb{R}$. Also, since E_1 is open in X and $f \neq 0$, $f(E_1)$ is open in $\mathbb{K}$. Thus $\text{Re}\, f(E_1)$ is an open interval in $\mathbb{R}$. Hence there is $t \in \mathbb{R}$ such that $\text{Re}\, f(x_1) < t \leq \text{Re}\, f(x_2)$ for all $x_1 \in E_1$ and $x_2 \in E_2$. □

Let f be a linear functional on a linear space X over $\mathbb{K}$, and let $t \in \mathbb{R}$. The subset $\{x \in X : \text{Re}\, f(x) = t\}$ of X is called a **real hyperplane** in X. If X is a normed space, and if f is continuous, then this real hyperplane is closed in X. By the Hahn–Banach separation theorem, disjoint convex subsets E_1 and E_2 of X, of which E_1 is open, are 'separated' by a closed real hyperplane because $E_1 \subset \{x \in X : \text{Re}\, f(x) < t\}$

and $E_2 \subset \{x \in X : \text{Re}\, f(x) \geq t\}$. This result is useful in determining the closure of a convex subset of X in terms of the elements of X'. (See Exercise 4.8, and compare Remark 4.7(ii).)

4.2 Dual Spaces and Their Representations

Let X be a normed space over $\mathbb{K}$. The space $BL(X, \mathbb{K})$ of all bounded linear functionals on X is called the **dual space** of X, and it is denoted by X'.

Let X be finite dimensional, and let $x_1, \ldots, x_n$ constitute a basis for X. There are linear functionals $f_1, \ldots, f_n$ on X which satisfy $f_j(x_i) = \delta_{i,j}$, $i, j = 1, \ldots, n$. Clearly, $f_1, \ldots, f_n$ constitute a linearly independent subset of X'. In fact, they constitute a basis for X' since $f = f(x_1)f_1 + \cdots + f(x_n)f_n$ for every $f \in X'$. It is called the **dual basis** for X' relative to the basis $x_1, \ldots, x_n$ of X. In particular, the dimension of X' is equal to the dimension of X.

Let $x_1', x_2' \in X'$. By the very definition of a function, $x_1' = x_2'$ if and only if $x_1'(x) = x_2'(x)$ for all $x \in X$. Next, let $x_1, x_2 \in X$. If $x_1 \neq x_2$, that is, $x_1 - x_2 \neq 0$, then by Proposition 4.6(i), there is $x' \in X'$ such that $x'(x_1 - x_2) \neq 0$, that is, $x'(x_1) \neq x'(x_2)$. Thus $x_1 = x_2$ if and only if $x'(x_1) = x'(x_2)$ for all $x' \in X'$. This interchangeability between $x \in X$ and $x' \in X'$ explains the nomenclature 'dual space' of X'. Let us now establish the duality between the norms of elements of X and of X'. Recall that the operator norm of $x' \in X'$ is defined by $\|x'\| := \sup\{|x'(x)| : x \in X \text{ and } \|x\| \leq 1\}$.

Proposition 4.11 *Let X be a normed space. Then*

$$\|x\| = \sup\{|x'(x)| : x' \in X' \text{ and } \|x'\| \leq 1\} \quad \text{for all } x \in X.$$

Further, if Y is a normed space, and $F \in BL(X, Y)$, then

$$\|F\| = \sup\{|y'(F(x))| : x \in X,\ \|x\| \leq 1 \text{ and } y' \in Y',\ \|y'\| \leq 1\}.$$

Proof Let $x \in X$. By the basic inequality for the operator norm (Theorem 3.10(i)), $|x'(x)| \leq \|x'\|\,\|x\|$ for all $x' \in X'$. Also, by a consequence of the Hahn–Banach extension theorem (Proposition 4.6(i)), there is $x' \in X'$ such that $x'(x) = \|x\|$ and $\|x'\| \leq 1$. Hence the desired expression for $\|x\|$ follows.

Next, let $F \in BL(X, Y)$. Then for $x \in X$, it follows that $\|F(x)\| = \sup\{|y'(F(x))| : y' \in Y' \text{ and } \|y'\| \leq 1\}$. Hence

$$\begin{aligned}\|F\| &= \sup\{\|F(x)\| : x \in X \text{ and } \|x\| \leq 1\} \\ &= \sup\{|y'(F(x))| : x \in X,\ \|x\| \leq 1 \text{ and } y' \in Y',\ \|y'\| \leq 1\},\end{aligned}$$

as desired. □

We shall find the dual spaces of some important normed spaces and indicate how the dual spaces of some others can be found. This task is important because if we know the dual space of a normed space, we can rephrase questions about the given space in a way that can throw light from a different angle, and possibly lead us to answers. For example, in Remark 4.7(ii), we have seen that a criterion for the approximation of an element of a normed space by elements belonging to a fixed subspace of that normed space can be given in terms of the elements of the dual space that vanish on that subspace.

The 'duality' between X and X' does not mean that results about X and X' are always symmetric in nature. For instance, since $\mathbb{K}$ is a Banach space, Proposition 3.17(i) shows that $X' = BL(X, \mathbb{K})$ is a Banach space, even if X is not a Banach space. The consideration of a dual space gives us a neat way of constructing a completion of a normed space, and of an inner product space, as we now show.

Let X be a normed space, and let $x \in X$. Define $j_x : X' \to \mathbb{K}$ by

$$j_x(x') := x'(x) \quad \text{for } x' \in X'.$$

Clearly, j_x is linear, and $\|j_x\| := \sup\{|x'(x)| : x' \in X' \text{ and } \|x'\| \leq 1\} = \|x\|$ by Proposition 4.11. Hence j_x belongs to the dual space $(X')'$ of X'. We shall denote the **second dual space** $(X')'$ of the normed space X by X''. Define $J : X \to X''$ by $J(x) := j_x$ for $x \in X$. Then J is linear and $\|J(x)\| = \|x\|$ for all $x \in X$, that is, J is a linear isometry from X into X''. It is called the **canonical embedding** of X into X''. If $J(X) = X''$, then we say that the normed space X is **reflexive**. If X is finite dimensional, then $\dim X'' = \dim X' = \dim X$, and so X is reflexive. Also, we shall show that every Hilbert space is reflexive.

Let X_c denote the closure of $J(X)$ in X''. (If X is a Banach space, then $J(X)$ is a closed subspace of X'', and so $X_c = J(X)$.) Since $X'' := BL(X', \mathbb{K})$ is a Banach space, and since X_c is closed in X'', we see that X_c is a Banach space, and $J : X \to X_c$ is a linear isometry such that $J(X)$ is dense in X_c. The Banach space X_c is called the **completion** of the normed space X. It is unique in the following sense. If X_1 is a Banach space, and $J_1 : X \to X_1$ is a linear isometry such that $J_1(X)$ is dense in X_1, then there is a linear isometry Φ from X_c onto X_1. This can be seen as follows.

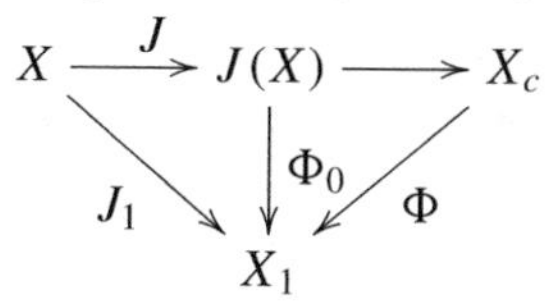

Define $\Phi_0 : J(X) \to X_1$ by

$$\Phi_0(J(x)) := J_1(x) \quad \text{for } x \in X.$$

Then $\Phi_0 \in BL(J(X), X_1)$, where $J(X)$ is dense in X_c, and X_1 is a Banach space. By Proposition 3.17(ii), there is $\Phi \in BL(X_c, X_1)$ such that $\Phi(J(x)) = \Phi_0(J(x))$ for all $x \in X$. Since $\|\Phi_0(J(x))\| = \|J_1(x)\| = \|x\| = \|J(x)\|$ for all $x \in X$, it follows that Φ is also a linear isometry. Also, since $\Phi(X_c)$ is both closed as well as dense in X_1, we obtain $\Phi(X_c) = X_1$.

For example, let $p \in \{1, 2\}$, and let $X := C([a, b])$ with the norm given by

$$\|x\|_p := \left(\int_a^b |x(t)|^p dt \right)^{1/p} \quad \text{for } x \in X.$$

By Proposition 1.26, X is dense in $L^p([a, b])$, and $L^p([a, b])$ is complete as we have seen in Example 2.24(iv). Hence the completion of X can be identified with $L^p([a, b])$.

Now let $(X, \langle \cdot \, , \cdot \rangle)$ be an inner product space. We introduce an inner product on the completion X_c of X as follows. Let $x'', \ y''$ in X_c, and let $J(x_n) \to x''$ and $J(y_n) \to y''$ in X''. Then $(J(x_n))$ and $(J(y_n))$ are Cauchy sequences in X_c, and so (x_n) and (y_n) are Cauchy sequences in X. It follows that $(\langle x_n, y_n \rangle)$ is a Cauchy sequence in $\mathbb{K}$. Hence it converges in $\mathbb{K}$, and its limit is equal to the limit of any sequence $(\langle \tilde{x}_n, \tilde{y}_n \rangle)$ such that $J(\tilde{x}_n) \to x''$ and $J(\tilde{y}_n) \to y''$ in X''. Define $\langle x'', y'' \rangle_c := \lim_{n\to\infty} \langle x_n, y_n \rangle$. It is easy to see that $\langle \cdot \, , \cdot \rangle_c$ an inner product on X_c, and $\langle J(x), J(y) \rangle_c = \langle x, y \rangle$ for all $x, y \in X$. Thus the completion of an inner product space is a Hilbert space.

In Exercise 4.19, we give three results which are initially proved only for a Hilbert space, and are then extended to an incomplete inner product space by considering its completion.

The canonical embedding $J : X \to X''$ introduced earlier is useful in a variety of ways. We give an instance below.

Theorem 4.12 (i) **(Resonance theorem)** *Let X be a normed space and $E \subset X$. Then E is a bounded subset of X if and only if $x'(E)$ is a bounded subset of $\mathbb{K}$ for every $x' \in X'$.*

(ii) *Let X and Y be normed spaces, and let $F : X \to Y$ be a linear map. Then F is continuous if and only if $y' \circ F$ is continuous for every $y' \in Y'$.*

Proof (i) Suppose E is a bounded subset of X. Then there is $\alpha > 0$ such that $\|x\| \le \alpha$ for all $x \in E$. If $x' \in X'$, then $|x'(x)| \le \|x'\| \, \|x\| \le \|x'\| \, \alpha$ for all $x \in E$, and so $x'(E)$ is a bounded subset of $\mathbb{K}$.

Conversely, suppose $x'(E)$ is a bounded subset of $\mathbb{K}$ for every x' in X'. Let $J : X \to X''$ be the canonical embedding of X into X''. Then $\{J(x) : x \in E\}$ is a subset of $X'' := BL(X', \mathbb{K})$. Now X' is a Banach space, and for each fixed $x' \in X'$, the set $\{J(x)(x') : x \in E\} = x'(E)$ is bounded. By the uniform boundedness principle (Theorem 3.23), $J(E)$ is a bounded subset of X''. But $\|J(x)\| = \|x\|$ for each $x \in X$. Hence E is a bounded subset of X.

(ii) Let the linear map $F : X \to Y$ be continuous. Since every $y' \in Y'$ is continuous, we see that $y' \circ F$ is continuous. Conversely, let $y' \circ F$ be continuous for every $y' \in Y'$. Let $E := \{F(x) : x \in X \text{ and } \|x\| \le 1\} \subset Y$. Then $y'(E)$ is bounded for every $y' \in Y'$ by Proposition 3.2. By (i) above, E is a bounded subset of Y. Thus the linear map F is continuous by Proposition 3.2. □

The resonance theorem proved above says that the boundedness of a subset of a normed space X is the same thing as the boundedness of its image under every continuous linear functional on X. The same goes for the continuity of a linear map from X to a normed space. However, a similar result does not hold for the convergence of a sequence in X. Let (x_n) be a sequence in X, and let $x \in X$. Even if $x'(x_n) \to x'(x)$ for every $x' \in X'$, we may not be able to conclude that $x_n \to x$ in X. For example, Example 4.18(i) shows that $(x'(e_n))$ converges to 0 for every $x' \in (\ell^2)'$, but (e_n) does not converge in ℓ^2, since $\|e_n - e_m\|_2 = \sqrt{2}$ for all $n \neq m \in \mathbb{N}$. (See Exercise 4.17(ii).) On the other hand, a result of **Schur** [3, p. 137] says that if $x'(x_n) \to x'(x)$ for every $x' \in (\ell^1)'$, then $x_n \to x$ in ℓ^1. Compare Exercises 4.16 and 4.17.

Before we find dual spaces of some well-known normed spaces, we prove two additional facts about X and X'.

Proposition 4.13 *Let X be a normed space.*

(i) *Let Y be a subspace of X. For $x' \in X'$, let $\Psi(x')$ denote the restriction of x' to Y. Then Ψ is a linear map from X' onto Y', and $\|\Psi(x')\| \leq \|x'\|$ for all $x' \in X'$. If Y is dense in X, then Ψ is an isometry.*
(ii) *If X' is separable, then so is X.*

Proof (i) It is clear that $\Psi(x') \in Y'$ and $\|\Psi(x')\| \leq \|x'\|$ for all $x' \in X'$. Also, the map $\Psi : X' \to Y'$ is linear. Let $y' \in Y'$. If x' is a Hahn–Banach extension of y' to X, then $x' \in X'$ and $\Psi(x') = y'$. Hence the map Ψ is onto.

Suppose Y is dense in X. Let $x \in X$, and let (y_n) be a sequence in Y such that $y_n \to x$ in X. If $x' \in X'$, then

$$|x'(x)| = \big|\lim_{n\to\infty} x'(y_n)\big| \leq \|\Psi(x')\| \lim_{n\to\infty} \|y_n\| = \|\Psi(x')\|\, \|x\|,$$

so that $\|x'\| \leq \|\Psi(x')\|$. Thus Φ is an isometry.

(ii) Let X' be separable. If $X = \{0\}$, then there is nothing to prove. Suppose $X \neq \{0\}$. As mentioned in Sect. 1.3, a nonempty subset of a separable metric space is separable. Hence the unit sphere S of X' is separable. Let $\{x_1', x_2', \ldots\}$ be a countable dense subset of S. Since $\|x_j'\| = 1$, there is $y_j \in X$ such that $\|y_j\| = 1$ and $|x_j'(y_j)| \geq 1/2$ for each $j \in \mathbb{N}$. Let

$$D := \{k_1 y_1 + \cdots + k_n y_n : n \in \mathbb{N} \text{ and } \operatorname{Re} k_j, \operatorname{Im} k_j \in \mathbb{Q} \text{ for } j = 1, \ldots, n\}.$$

Then D is a countable subset of X since $\mathbb{Q}$ is countable, and D is dense in the subspace $Y := \operatorname{span}\{y_1, y_2, \ldots\}$ of X since $\mathbb{Q}$ is dense in $\mathbb{R}$. To show that Y is dense in X, we consider $x' \in X'$ such that $x'(y) = 0$ for all $y \in Y$. Assume for a moment that $x' \neq 0$. Without loss of generality, we may assume that $\|x'\| = 1$, that is, $x' \in S$. Then there is $j \in \mathbb{N}$ such that $\|x_j' - x'\| < 1/2$, and so $|x_j'(y_j) - x'(y_j)| < 1/2$. But $x'(y_j) = 0$ since $y_j \in Y$. Thus $1/2 \leq |x_j'(y_j)| < 1/2$, which is impossible. Hence $x' = 0$. By Proposition 4.6(ii), $\overline{Y} = X$, that is, Y is dense in X. Thus D is a countable dense subset of X, and so X is separable. □

The converse of part (ii) of the above proposition does not hold, as we shall see in Example 4.18(ii).

We shall now find the dual of a Hilbert space by using the projection theorem (Theorem 2.35).

Theorem 4.14 (Riesz representation theorem, 1907) *Let H be a Hilbert space, and let $f \in H'$. Then there is a unique $y_f \in H$, called the* **representer** *of f, such that $f(x) = \langle x, y_f \rangle$ for all $x \in H$. Further, the map $T : H' \to H$ defined by $T(f) := y_f$, $f \in H'$, is a conjugate-linear isometry from H' onto H.*

Proof If $f = 0$, then let $y_f := 0$, so that $f(x) = 0 = \langle x, 0 \rangle$ for all $x \in H$.

Next, let $f \neq 0$. Since $Z(f)$ is a closed subspace of H, the projection theorem (Theorem 2.35) shows that $H = Z(f) \oplus Z(f)^\perp$. Now $Z(f) \neq H$, that is, $Z(f)^\perp \neq \{0\}$. Let $z \in Z(f)^\perp$ be such that $\|z\| = 1$. Then $f(z) \neq 0$. For $x \in H$, consider

$$w := x - \frac{f(x)}{f(z)} z.$$

Clearly, $w \in Z(f)$, and so $\langle w, z \rangle = 0$, that is, $\langle x, z \rangle = f(x)/f(z)$. Hence

$$f(x) = f(z)\langle x, z \rangle = \langle x, y_f \rangle \quad \text{for all } x \in X,$$

where $y_f := \overline{f(z)}z$, as desired. Further, if $y \in X$ and $f(x) = \langle x, y \rangle$ for all $x \in H$ as well, then $\langle x, y \rangle = f(x) = \langle x, y_f \rangle$ for all $x \in H$, that is, $y = y_f$.

For $f, g \in H'$,

$$(f + g)(x) = f(x) + g(x) = \langle x, y_f \rangle + \langle x, y_g \rangle = \langle x, y_f + y_g \rangle \quad \text{for } x \in H.$$

Thus $y_f + y_g$ is the representer of $f + g \in H'$, that is, $T(f + g) = T(f) + T(g)$. Similarly, for $f \in H'$ and $k \in \mathbb{K}$, $(k\, f)(x) = k\langle x, y_f \rangle = \langle x, \overline{k}\, y_f \rangle$ for $x \in H$. Thus $\overline{k}\, y_f$ is the representer of $kf \in H'$, that is, $T(k\, f) = \overline{k}\, T(f)$. Also,

$$\|f\| = \sup\{|\langle x, y_f \rangle| : x \in H \text{ and } \|x\| \leq 1\} = \|y_f\|$$

since $|\langle x, y_f \rangle| \leq \|x\|\, \|y_f\|$ for all $x \in X$ by the Schwarz inequality (Proposition 2.13(i)), and $\langle y_f, y_f \rangle = \|y_f\|^2$. Thus $\|T(f)\| = \|f\|$. Finally, given $y \in H$, let $f_y(x) := \langle x, y \rangle$ for $x \in X$. Then $f_y \in H'$, and y is the representer of f_y, that is, $T(f_y) = y$. Thus T is a conjugate-linear isometry from H' onto H. □

Remark 4.15 The Riesz representation theorem does not hold in any incomplete inner product space. (See Exercise 4.18.) As a specific example, let $X := c_{00}$ with the usual inner product. For $x \in X$, let $f(x) := \sum_{j=1}^{\infty} x(j)/j$. Then f is a linear functional on X. By letting $n \to \infty$ in the Schwarz inequality for numbers (Lemma 1.4(i)),

$$|f(x)|^2 \leq \left(\sum_{j=1}^{\infty} \frac{1}{j^2} \right) \left(\sum_{j=1}^{\infty} |x(j)|^2 \right) = \frac{\pi^2}{6} \|x\|_2^2.$$

Thus $f \in \overline{(c_{00})'}$ and $\|f\| \le \pi/\sqrt{6}$. But f has no representer in c_{00}: If $y \in c_{00}$, then $\langle e_j, y\rangle = \overline{y(j)} = 0$ for all large $j \in \mathbb{N}$, but $f(e_j) = 1/j \neq 0$ for $j \in \mathbb{N}$. ◊

Corollary 4.16 *Let H be a Hilbert space.*

(i) *Let y_f denote the representer of $f \in H'$. Define $\langle f, g\rangle' = \langle y_g, y_f\rangle$ for $f, g \in H'$. Then $\langle \cdot\,, \cdot\rangle'$ is an inner product on H', $\langle f, f\rangle' = \|f\|^2$ for all $f \in H'$, and H' is a Hilbert space.*

(ii) *H is a reflexive normed space.*

Proof (i) Let f, g, h in H', and let $k \in \mathbb{K}$. Then

$$\begin{aligned}\langle f+g, h\rangle' &= \langle y_h, y_{(f+g)}\rangle = \langle y_h, y_f + y_g\rangle = \langle f, h\rangle' + \langle f, g\rangle',\\ \langle kf, g\rangle' &= \langle y_g, y_{kf}\rangle = \langle y_g, \overline{k} y_f\rangle = k\langle f, g\rangle'.\end{aligned}$$

Also, $\langle f, g\rangle' = \langle y_g, y_f\rangle = \overline{\langle y_f, y_g\rangle} = \overline{\langle g, f\rangle'}$. Further, $\langle f, f\rangle' = \langle y_f, y_f\rangle \ge 0$, and $\langle f, f\rangle' = \langle y_f, y_f\rangle = 0$ if and only if $y_f = 0$, that is, $f = 0$. Thus $\langle \cdot\,, \cdot\rangle'$ is an inner product on H', and $\langle f, f\rangle' = \|y_f\|^2 = \|f\|^2$ for all $f \in H'$.

Since H is complete, and the map $f \longmapsto y_f$ is an isometry from H' onto H, H' is also complete. Thus H' is a Hilbert space.

(ii) Consider the canonical embedding $J : H \to H''$. Let $\phi \in H'' = (H')'$. By (i) above, H' is a Hilbert space, and by the Riesz representation theorem (Theorem 4.14), there is a representer $g \in H'$ of ϕ. But g itself has a representer $y_g \in H$. Then $\phi = J(y_g)$ since

$$\phi(f) = \langle f, g\rangle' = \langle y_g, y_f\rangle = f(y_g) = J(y_g)(f) \quad \text{for all } f \in H'.$$

Thus the map J is onto, that is, H is a reflexive normed space. □

As a consequence of the Riesz representation theorem, we give a simple proof of the Hahn–Banach extension theorem for a Hilbert space. Moreover, we show that a Hahn–Banach extension is unique in this case.

Theorem 4.17 (Unique Hahn–Banach extension) *Let H be a Hilbert space, G be a subspace of H, and let $g \in G'$. Then there is a unique $f \in H'$ such that $f(y) = g(y)$ for all $y \in G$, and $\|f\| = \|g\|$.*

Proof By Proposition 3.17(ii), we can assume, without loss of generality, that G is a closed subspace of H. Then G is a Hilbert space, and there is a representer $y_g \in G$ of g by the Riesz representation theorem (Theorem 4.14). Define $f : H \to \mathbb{K}$ by $f(x) := \langle x, y_g\rangle$ for $x \in H$. Then f is linear, $f(y) = \langle y, y_g\rangle = g(y)$ for $y \in G$, and $\|f\| = \|y_g\| = \|g\|$. Hence f is a Hahn–Banach extension of g.

To prove the uniqueness of f, let $\tilde{f} \in H'$ be such that $\tilde{f}(y) = g(y)$ for all $y \in G$, and $\|\tilde{f}\| = \|g\|$. Let $\tilde{y} \in H$ denote the representer of $\tilde{f}$. Then

$$\|y_g - \tilde{y}\|^2 = \|y_g\|^2 - 2\,\mathrm{Re}\,\langle y_g, \tilde{y}\rangle + \|\tilde{y}\|^2.$$

But $\|\tilde{y}\| = \|\tilde{f}\| = \|g\| = \|y_g\|$ and $\langle y_g, \tilde{y}\rangle = \tilde{f}(y_g) = g(y_g) = \langle y_g, y_g\rangle = \|y_g\|^2$. Hence $\|y_g - \tilde{y}\| = 0$, that is, $\tilde{y} = y_g$, and
$\tilde{f}(x) = \langle x, \tilde{y}\rangle = \langle x, y_g\rangle = f(x)$ for $x \in H$, that is, $\tilde{f} = f$. ☐

Examples 4.18 (i) Consider $y := (y(1), y(2), \ldots) \in \ell^2$, and define

$$f_y(x) := \langle x, \overline{y}\rangle = \sum_{j=1}^{\infty} x(j)y(j) \quad \text{for } x := (x(1), x(2), \ldots) \in \ell^2.$$

As in the proof of Theorem 4.14, $f_y \in (\ell^2)'$ and $\|f_y\| = \|\overline{y}\|_2 = \|y\|_2$. This also follows from the case $p = 2$ of Example 3.12. The map $\Phi(y) := f_y,\ y \in \ell^2$, gives a linear isometry from ℓ^2 to $(\ell^2)'$. Now let $f \in (\ell^2)'$. By Theorem 4.14, $f = f_y$, where $y := \overline{y_f}$. Therefore Φ is a linear isometry from ℓ^2 onto $(\ell^2)'$.

Next, let L^2 denote $L^2([a, b])$. Consider $y \in L^2$, and define

$$f_y(x) := \langle x, \overline{y}\rangle = \int_a^b x(t)y(t)dm(t) \quad \text{for } x \in L^2.$$

As in the proof of Theorem 4.14, $f_y \in (L^2)'$ and $\|f_y\| = \|\overline{y}\|_2 = \|y\|_2$. This also follows from the case $p = 2$ of Example 3.13. The map $\Phi(y) := f_y,\ y \in L^2$, gives a linear isometry from L^2 to $(L^2)'$. Now let $f \in (L^2)'$. By Theorem 4.14, $f = f_y$, where $y := \overline{y_f}$. Therefore Φ is a linear isometry from L^2 onto $(L^2)'$.

(ii) Let us find $(\ell^1)'$. Taking a cue from the way every $f \in (\ell^2)'$ is represented in (i) above, let us consider $y := (y(1), y(2), \ldots) \in \ell^\infty$, and define

$$f_y(x) := \sum_{j=1}^{\infty} x(j)y(j) \quad \text{for } x := (x(1), x(2), \ldots) \in \ell^1.$$

The case $p = 1$ of Example 3.12 shows that $f_y \in (\ell^1)'$ and $\|f_y\| = \|y\|_\infty$. The map $\Phi(y) := f_y,\ y \in \ell^\infty$, gives a linear isometry from ℓ^∞ to $(\ell^1)'$. Consider $f \in (\ell^1)'$. Then $f(x) = f\big(\sum_{j=1}^{\infty} x(j)e_j\big) = \sum_{j=1}^{\infty} x(j)f(e_j)$ for $x \in \ell^1$. Define $y := (f(e_1), f(e_2), \ldots)$. Since $|y(j)| = |f(e_j)| \le \|f\|$ for all $j \in \mathbb{N}$, we see that $y \in \ell^\infty$, and $f = f_y$. Hence Φ is a linear isometry from ℓ^∞ onto $(\ell^1)'$.

Next, let L^1 denote $L^1([a, b])$, and L^∞ denote $L^\infty([a, b])$. To find $(L^1)'$, consider $y \in L^\infty$, and define

$$f_y(x) := \int_a^b x(t)y(t)dm(t) \quad \text{for } x \in L^1.$$

The case $p = 1$ of Example 3.13 shows that $f_y \in (L^1)'$ and $\|f_y\| = \|y\|_\infty$. The map $\Phi(y) := f_y,\ y \in L^\infty$, gives a linear isometry from L^∞ to $(L^1)'$. Let $f \in (L^1)'$. For $t \in (a, b]$, let c_t denote the characteristic function of the subinterval $(a, t]$ of $[a, b]$, so that $c_t \in L^1$. Define $z : [a, b] \to \mathbb{K}$ by $z(a) := 0$, and $z(t) := f(c_t)$ for $t \in (a, b]$. It follows that the function z is absolutely continuous on $[a, b]$. By the fundamental theorem of calculus for Lebesgue integration (Theorem 1.23), $z'(t)$ exists for almost all $t \in [a, b]$, $z' \in L^1$, and $z(t) = \int_a^t z' dm$ for all $t \in [a, b]$. Define $y := z'$. Then $f(c_t) = \int_a^t y\,dm = \int_a^b c_t y\,dm$, $t \in [a, b]$. In fact, it can be seen that $y \in L^\infty$ and $f(x) = \int_a^b xy\,dm$ for $x \in L^1$, that is, $f = f_y$. Hence Φ is a linear isometry from L^∞ onto $(L^1)'$.

(iii) In an attempt to find $(\ell^\infty)'$, consider $y := (y(1), y(2), \ldots) \in \ell^1$, and define

$$f_y(x) := \sum_{j=1}^{\infty} x(j)y(j) \quad \text{for } x := (x(1), x(2), \ldots) \in \ell^\infty.$$

The case $p = \infty$ of Example 3.12 shows that $f_y \in (\ell^\infty)'$ and $\|f_y\| = \|y\|_1$. The map $\Phi(y) := f_y$, $y \in \ell^1$, gives a linear isometry from ℓ^1 to $(\ell^\infty)'$. However, this isometry is not onto. To see this, consider the closed subspace c_0 of ℓ^∞, and let $e_0 := (1, 1, \ldots) \notin c_0$. By Proposition 4.6(ii), there is $f \in (\ell^\infty)'$ such that $f(x) = 0$ for all $x \in c_0$ and $f(e_0) = 1$.[1] If $y \in \ell^1$ and $f = f_y$, then $y(j) = f_y(e_j) = f(e_j) = 0$ for all $j \in \mathbb{N}$, that is, $y = 0$, and so $f_y = 0$. Thus $f \neq f_y$ for any $y \in \ell^1$. In fact, Proposition 4.13(ii)) shows that $(\ell^\infty)'$ cannot be linearly homeomorphic to ℓ^1, since ℓ^1 is separable, but ℓ^∞ is not (Example 1.7(ii)). For an identification of $(\ell^\infty)'$, see [7, Chap. VII, Theorem 7].

In an attempt to find $(L^\infty)'$, consider $y \in L^1$, and define

$$f_y(x) := \int_a^b x(t)y(t)dm(t) \quad \text{for } x \in L^\infty.$$

The case $p = \infty$ of Example 3.13 shows that $f_y \in (\ell^\infty)'$ and $\|f_y\| = \|y\|_1$. The map $\Phi(y) := f_y$, $y \in L^1$, gives a linear isometry from L^1 to $(L^\infty)'$. However, this isometry is not onto. To see this, define $g : C([a, b]) \to \mathbb{K}$ by $g(x) := x(b)$ for $x \in C([a, b])$, and let f be a Hahn–Banach extension of g to L^∞. Let (b_n) be a sequence in $[a, b)$ such that $b_n \to b$. For $n \in \mathbb{N}$, let $x_n(t) := 0$ if $t \in [a, b_n]$ and $x_n(t) := (t - b_n)/(b - b_n)$ if $t \in (b_n, b]$. Then $x_n \in C([a, b])$ and $\|x_n\|_\infty \leq 1$ for all $n \in \mathbb{N}$. Also, $x_n(t) \to 0$ for each $t \in [a, b)$ and $x_n(b) = 1$. If $y \in L^1$ and $f = f_y$, then $f_y(x_n) = f(x_n) = g(x_n) = x_n(b) = 1$ for all $n \in \mathbb{N}$, whereas $f_y(x_n) = \int_a^b x_n(t)y(t)dm(t) \to 0$ by the dominated convergence theorem (Theorem 1.18(ii)). Thus $f \neq f_y$ for any $y \in L^1$. In fact, Proposition 4.13(ii)) shows that $(L^\infty)'$ cannot be linearly homeomorphic to L^1, since L^1 is separable, but L^∞ is not (Proposition 1.26(iv)). For an identification of $(L^\infty)'$, see [16, (20.35) Theorem].

(iv) Let $X := C([a, b])$ with the sup norm $\|\cdot\|_\infty$. To find the dual space of X, let $Y := BV([a, b])$, the linear space of all $\mathbb{K}$-valued functions of **bounded variation** defined on $[a, b]$. For $y \in Y$, consider the **total variation** $V(y)$ of y introduced just after Theorem 1.23 in Sect. 1.4. It is easy to see that $y \longmapsto V(y)$ is a seminorm

[1] Let $r \in \{1, 2, \infty\}$, and for $x \in \ell^\infty$, let $F(x) := (f(x), 0, 0, \ldots) \in \ell^r$. Then F is a nonzero continuous linear map from ℓ^∞ to ℓ^r, but it is not a matrix transformation, since $F(e_j)(i) = 0$ for all $i, j \in \mathbb{N}$.

on Y, and that $V(y) = 0$ if and only if the function y is constant on $[a, b]$. Let $\|y\|_{BV} := |y(a)| + V(y)$ for $y \in Y$, and observe that $\|\cdot\|_{BV}$ is a norm on Y. Let $y \in Y$.

$$|y(t)| \leq |y(a)| + |y(t) - y(a)| \leq |y(a)| + V(y) = \|y\|_{BV} \quad \text{for all } t \in [a, b].$$

Hence $\|y\|_\infty \leq \|y\|_{BV}$. Thus the norm $\|\cdot\|_{BV}$ on Y is stronger than the norm $\|\cdot\|_\infty$ on the subspace Y of $B([a, b])$. Fix $y \in Y$, and define

$$f_y(x) := \int_a^b x(t)dy(t) \quad \text{for } x \in X,$$

where dy denotes the Riemann–Stieltjes integration with respect to y. Then f_y is a linear functional on X. Also, $|f_y(x)| \leq \|x\|_\infty \|y\|_{BV}$ for all $x \in X$. Hence $f \in X'$ and $\|f_y\| \leq \|y\|_{BV}$, but equality may not hold here.

We, therefore, consider the subspace $NBV([a, b])$ of $BV([a, b])$ consisting of those functions of bounded variation which vanish at a, and which are right continuous on (a, b). These are known as the **normalized functions of bounded variation**. It can be shown that $\|f_y\| = \|y\|_{BV}$ for $y \in NBV([a, b])$, and so the map $\Phi(y) := f_y$, $y \in NBV([a, b])$, gives a linear isometry from $NBV([a, b])$ to $(C([a, b]))'$. Further, if $f \in (C([a, b]))'$, then there is a (unique) $y \in NBV([a, b])$ such that $f = f_y$. Therefore Φ is a linear isometry from $NBV([a, b])$ onto $(C([a, b]))'$. We refer the reader to [20, Theorem 6.4.4] for details. ◊

Let $p \in \{1, 2, \infty\}$, and let q be the conjugate exponent. The results in Example 4.18(i)–(iii) can be summarized as follows. If $p \in \{1, 2\}$, then the dual of ℓ^p is linearly isometric to ℓ^q, and the dual of L^p is linearly isometric to L^q, but this is not the case if $p = \infty$.

The results in Example 4.18(ii) and (iv) are also known as **Riesz representation theorems** for ℓ^1, $L^1([a, b])$ and $C([a, b])$.

4.3 Transposes and Adjoints

Let X and Y be normed spaces, and let $F \in BL(X, Y)$. Then $y' \circ F \in X'$ for every $y' \in Y'$. The map $F' : Y' \to X'$ defined by $F'(y') := y' \circ F$ for $y' \in Y'$ is called the **transpose** of F. Thus the following diagram is commutative for every $y' \in Y'$:

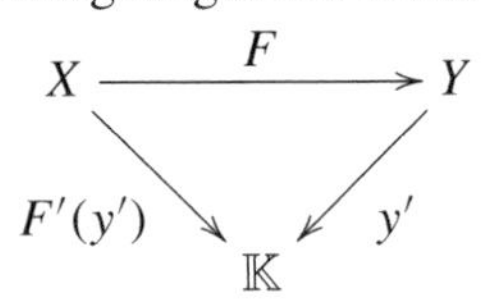

We first explain the nomenclature 'transpose'. Let X be a normed space of dimension n, and let $x_1, \ldots, x_n$ constitute a basis for X. Further, let $x_1', \ldots, x_n'$ constitute the corresponding dual basis for X'. Let Y be a normed space of dimension m, and let

$y_1, \ldots, y_m$ constitute a basis for Y. Further, let $y'_1, \ldots, y'_m$ constitute the corresponding dual basis for Y'. Then

$$F(x_j) = y'_1(F(x_j))y_1 + \cdots + y'_m(F(x_j))y_m \quad \text{for } j = 1, \ldots, n.$$

Thus the $m \times n$ matrix M having $y'_i(F(x_j))$ as the entry in the jth column and the ith row defines the linear map $F : X \to Y$ with respect to the basis $x_1, \ldots, x_n$ for X and the basis $y_1, \ldots, y_m$ for Y, as seen in Sect. 1.2. Now

$$F'(y'_i) = F'(y'_i)(x_1)x'_1 + \cdots + F'(y'_i)(x_n)x'_n \quad \text{for } i = 1, \ldots, m.$$

Hence the $n \times m$ matrix M' having $F'(y'_i)(x_j)$ as the entry in the ith column and the jth row defines the linear map $F' : Y' \to X'$ with respect to the basis $y'_1, \ldots, y'_m$ for Y' and the basis $x'_1, \ldots, x'_n$ for X'. Since $F'(y'_i)(x_j) = y'_i(F(x_j))$ for all $i = 1, \ldots, m$, $j = 1, \ldots, n$, we see that $M' = M^t$, the transpose of the matrix M. For this reason, the linear map F' is called the transpose of F.

Examples 4.19 (i) Let $p, r \in \{1, 2\}$, and let $F \in BL(\ell^p, \ell^r)$. Let q satisfy $(1/p) + (1/q) = 1$, and let s satisfy $(1/r) + (1/s) = 1$. Then $q, s \in \{2, \infty\}$. The transpose $F' : (\ell^r)' \to (\ell^p)'$ of F can be identified with a map F^t from ℓ^s to ℓ^q as follows. For $y \in \ell^s$ and $z \in \ell^q$, define

$$\Psi(y)(x) := \sum_{j=1}^{\infty} x(j)y(j), \quad x \in \ell^r, \quad \text{and} \quad \Phi(z)(x) := \sum_{j=1}^{\infty} x(j)z(j), \quad x \in \ell^p.$$

As in Example 4.18(i) and (ii), $\Psi : \ell^s \to (\ell^r)'$ and $\Phi : \ell^q \to (\ell^p)'$ are linear isometries, and they are onto. Define $F^t : \ell^s \to \ell^q$ by $F^t := (\Phi)^{-1}F'\Psi$. Then $\Phi F^t = F'\Psi$, that is, the following diagram is commutative:

$$\begin{array}{ccc} (\ell^r)' & \xrightarrow{F'} & (\ell^p)' \\ \Psi \uparrow & & \uparrow \Phi \\ \ell^s & \xrightarrow[F^t]{} & \ell^q \end{array}$$

As we have seen in Proposition 3.30, the infinite matrix $M := [k_{i,j}]$, where $k_{i,j} := F(e_j)(i)$ for $i, j \in \mathbb{N}$, defines the continuous linear map F.

Fix $y \in \ell^s$. Then $z := F^t(y) \in \ell^q$, and $z(i) = \Phi(z)(e_i)$ for $i \in \mathbb{N}$, and so

$$\begin{aligned} F^t(y)(i) &= \Phi(F^t(y))(e_i) = F'(\Psi(y))(e_i) = \Psi(y)(F(e_i)) \\ &= \sum_{j=1}^{\infty} F(e_i)(j)y(j) = \sum_{j=1}^{\infty} k_{j,i}\, y(j). \end{aligned}$$

Hence the matrix $M^t := [k_{j,i}]$ defines the map F^t. We identify F' with F^t.

For instance, let $F(x) := (0, x(1), x(2), \ldots)$ for $x := (x(1), x(2), \ldots) \in \ell^p$. Then $F \in BL(\ell^p)$, and its transpose F' is identified with $F^t \in BL(\ell^q)$ defined by $F^t(x) := (x(2), x(3), \ldots)$ for $x := (x(1), x(2), \ldots)$ in ℓ^q.

We also mention that an infinite matrix M defines a (continuous linear) map from ℓ^∞ to ℓ^∞ if and only if the set of all row sums of the matrix $|M|$ is bounded, that is, if and only if the set of all column sums of the matrix $|M^t|$ is bounded, and this is the case exactly when the infinite matrix M^t defines a (continuous linear) map from ℓ^1 to ℓ^1. (See Remark 3.32.)

(ii) Let us denote $L^1([a, b])$ by L^1, and $L^\infty([a, b])$ by L^∞. Let $k(\cdot, \cdot)$ be a measurable function on $[a, b] \times [a, b]$. Further, let us suppose the function $t \longmapsto \int_a^b |k(s, t)|dm(s)$ is essentially bounded on $[a, b]$, and let

$$\alpha_1 := \text{ess sup}_{[a,b]} \int_a^b |k(s, \cdot)|dm(s),$$

as defined in Example 3.16(i). Let $F \in BL(L^1)$ denote the Fredholm integral operator with kernel $k(\cdot, \cdot)$. The transpose $F' : (L^1)' \to (L^1)'$ of F can be identified with a map from L^∞ to L^∞ as follows. For $y \in L^1$, define

$$\Phi(y)(x) := \int_a^b x(t)y(t)dm(t), \quad x \in L^1.$$

As in Example 4.18(ii), $\Phi : L^\infty \to (L^1)'$ is a linear isometry, and it is onto. Define $F^t : L^\infty \to L^\infty$ by $F^t := (\Phi)^{-1} F'\Phi$. Then $\Phi F^t = F'\Phi$, that is, the following diagram is commutative:

$$\begin{array}{ccc} (L^1)' & \xrightarrow{F'} & (L^1)' \\ \Phi \uparrow & & \uparrow \Phi \\ L^\infty & \xrightarrow[F^t]{} & L^\infty \end{array}$$

Fix $y \in L^\infty$. Then $F^t(y) \in L^\infty$. Let $x \in L^1$. Now

$$\Phi(F^t(y))(x) = F'(\Phi(y))(x) = \Phi(y)(F(x)) = \int_a^b F(x)(t)y(t)dm(t).$$

Now Theorem 1.20 shows that

$$\begin{aligned} \int_a^b F(x)(t)y(t)dm(t) &= \int_a^b \Big(\int_a^b k(t, u)x(u)dm(u) \Big) y(t)dm(t) \\ &= \int_a^b x(u) \Big(\int_a^b k(t, u)y(t)dm(t) \Big) dm(u), \end{aligned}$$

since

$$\iint_{[a,b]\times[a,b]} |k(t,u)x(u)y(t)|d(m\times m)(t,u) \le \alpha_1 \|x\|_1 \|y\|_\infty.$$

Let us define $k'(s,t) := k(t,s)$ for $s, t \in [a,b]$. Note that

$$\beta_1' := \text{ess sup}_{[a,b]} \int_a^b |k'(\cdot,t)|dm(t) = \alpha_1 < \infty,$$

and let $G \in BL(L^\infty)$ be the Fredholm integral operator having kernel $k'(\cdot,\cdot)$, that is, define

$$G(y)(u) := \int_a^b k'(u,t)y(t)dm(t) \quad \text{for } y \in L^\infty \text{ and } u \in [a,b].$$

It follows that $\Phi(F^t(y))(x) = \Phi(G(y))(x)$ for all $y \in L^\infty$ and all $x \in L^1$, that is, $\Phi(F^t(y)) = \Phi(G(y))$ for all $y \in L^\infty$. Since Φ is one-one, we see that $F^t(y) = G(y)$ for all $y \in L^\infty$, that is, F^t is the Fredholm integral operator on L^∞ having kernel $k'(\cdot,\cdot)$. We identify F' with F^t.

For instance, let $a := 0,\ b := 1$, and let

$$F(x)(s) := \int_0^s x(t)dm(t) \quad \text{for } x \in L^1 \text{ and } s \in [0,1].$$

Then $F \in BL(L^1)$ is the Fredholm integral operator having kernel $k(\cdot,\cdot)$ defined by $k(s,t) := 1$ if $0 \le t \le s \le 1$ and $k(s,t) := 0$ if $0 \le s < t \le 1$. The transpose F' of F can be identified with $F^t \in BL(L^\infty)$ defined by

$$F^t(y)(s) := \int_s^1 y(t)dm(t) \quad \text{for } y \in L^\infty \text{ and } s \in [0,1].$$

We shall consider the case of a Fredholm integral operator on $L^2([a,b])$ in Example 4.24(ii). ◊

Proposition 4.20 *Let X and Y be normed spaces, and let $F \in BL(X,Y)$. Then $F' \in BL(Y',X')$; in fact, $\|F'\| = \|F\|$. The map $F \longmapsto F'$ is a linear isometry from $BL(X,Y)$ to $BL(Y',X')$.*

Also, if Z is a normed space, and $G \in BL(Y,Z)$, then $(G\circ F)' = F'\circ G'$.

Proof Clearly, F' is a linear map. Also, we see that F' is continuous since $\|F'(y')\| = \|y'\circ F\| \le \|F\|\|y'\|$ for all $y' \in Y'$. In fact,

$$\begin{aligned}\|F'\| &= \sup\{\|F'(y')\| : y' \in Y' \text{ and } \|y'\| \leq 1\} \\ &= \sup\{|F'(y')(x)| : y' \in Y',\ \|y'\| \leq 1 \text{ and } x \in X,\ \|x\| \leq 1\} \\ &= \sup\{|y'(F(x))| : y' \in Y',\ \|y'\| \leq 1 \text{ and } x \in X,\ \|x\| \leq 1\} \\ &= \|F\|\end{aligned}$$

by Proposition 4.11.

Next, let $F_1, F_2 \in BL(X, Y)$. Then $(F_1 + F_2)'(y') = y' \circ (F_1 + F_2) = y' \circ F_1 + y' \circ F_2 = F_1'(y') + F_2'(y') = (F_1' + F_2')(y')$ for $y' \in Y'$. Hence $(F_1 + F_2)' = F_1' + F_2'$. Similarly, $(kF_1)' = kF_1'$ for all $k \in \mathbb{K}$. It follows that $F \longmapsto F'$ is a linear isometry.

Also, let Z be a normed space, let $G \in BL(Y, Z)$ and let $z' \in Z'$. Then

$$(G \circ F)'(z') = z' \circ (G \circ F) = (z' \circ G) \circ F = F'(z' \circ G) = F'(G'(z')) = (F' \circ G')(z').$$

Hence $(G \circ F)' = F' \circ G'$. □

Now we consider the transpose of a compact linear map.

Theorem 4.21 (Schauder, 1930) *Let X and Y be normed spaces, and let $F \in CL(X, Y)$. Then $F' \in CL(Y', X')$.*

Proof Let $\overline{U}$ denote the closed unit ball of X, and let $T := F(\overline{U})$. First we show that T is a totally bounded subset of Y. Let (y_n) be a sequence in T. For each $n \in \mathbb{N}$, there is $x_n \in \overline{U}$ such that $y_n = F(x_n)$. Since F is a compact map, there is a subsequence (x_{n_k}) of the bounded sequence (x_n) such that $(F(x_{n_k}))$ is convergent in Y. Hence (y_{n_k}) is a Cauchy subsequence of the sequence (y_n). By Proposition 1.8, the metric space T is totally bounded.

Now consider a bounded sequence (y_n') in Y'. We show that the sequence $(F'(y_n'))$ in X' has a Cauchy subsequence. Let $\alpha > 0$ be such that $\|y_n'\| \leq \alpha$ for all $n \in \mathbb{N}$. Then for all $n \in \mathbb{N}$ and $x_1, x_2 \in X$,

$$|y_n'(F(x_1)) - y_n'(F(x_2))| \leq \alpha \|F(x_1) - F(x_2)\|.$$

For $n \in \mathbb{N}$, let z_n' denote the restriction of y_n' to the subset T of Y. It follows that $E := \{z_n' : n \in \mathbb{N}\}$ is a uniformly bounded, equicontinuous set of functions defined on the totally bounded metric space T. By the Arzelà theorem (Theorem 1.17(ii)), the sequence (z_n') has a uniformly convergent subsequence (z_{n_k}'), and so it is uniformly Cauchy on T. But for $j, k \in \mathbb{N}$,

$$\begin{aligned}\|F'(y_{n_j}') - F'(y_{n_k}')\| &= \sup\{|F'(y_{n_j}')(x) - F'(y_{n_k}')(x)| : x \in \overline{U}\} \\ &= \sup\{|y_{n_j}'(F(x)) - y_{n_k}'(F(x))| : x \in \overline{U}\} \\ &= \sup\{|z_{n_j}'(y) - z_{n_k}'(y)| : y \in T\}.\end{aligned}$$

Thus $(F'(y_{n_k}'))$ is a Cauchy sequence in X', and since X' is a Banach space, it converges in X'. Thus $F' \in CL(Y', X')$. □

The converse of the above result of Schauder holds if Y is a Banach space. (See Exercise 4.26.)

Just as the study of the dual space of a normed space X allows us to view X from a different angle, the study of the transpose of a bounded linear map F on X allows us to view F from a different perspective. We give a simple instance of this kind.

Proposition 4.22 *Let X and Y be normed spaces, and let $F \in BL(X, Y)$. Then $Z(F') = \{y' \in Y' : y'(y) = 0$ for all $y \in R(F)\}$. Consequently, $R(F)$ is dense in Y if and only if F' is one-one.*

If X and Y are Banach spaces and $R(F) = Y$, then F' is bounded below.

Proof Let $y' \in Y'$. Then $y' \in Z(F')$, that is, $F'(y') = 0$ if and only if $y'(F(x)) = F'(y')(x) = 0$ for all $x \in X$, that is, $y'(y) = 0$ for all $y \in R(F)$, as desired. In particular, F' is one-one, that is, $Z(F') = \{0\}$ if and only if $y' = 0$ whenever $y'(y) = 0$ for all $y \in R(F)$. By Proposition 4.6(ii), this happens if and only if the closure of $R(F)$ is Y, that is, $R(F)$ is dense in Y.

Suppose X and Y are Banach spaces and $R(F) = Y$. By the open mapping theorem (Theorem 3.39) and Proposition 3.41, there is $\gamma > 0$ satisfying the following condition: For every $y \in Y$, there is $x \in X$ such that $F(x) = y$ and $\|x\| \le \gamma\|y\|$. Let $y' \in Y'$. For $y \in Y$, there is $x \in X$ such that

$$|y'(y)| = |y'(F(x))| = |F'(y')(x)| \le \|F'(y')\|\|x\| \le \gamma\|F'(y')\|\|y\|,$$

and so, $\|y'\| \le \gamma\|F'(y')\|$. Thus F' is bounded below. □

The converse of the last statement of Proposition 4.22 also holds. (See [28, Chap. IV, Theorem 9.4].) We shall prove this converse when X and Y are Hilbert spaces (Theorem 4.27(ii)). We thus obtain a powerful method of establishing the existence of a solution of the operator equation $F(x) = y$ for each 'free term' $y \in Y$. It consists of finding $\beta > 0$ such that $\beta\|y'\| \le \|F'(y')\|$ for all $y' \in Y'$.

Adjoint of a Bounded Linear Map on a Hilbert Space

Let H and G be Hilbert spaces, and let $A \in BL(H, G)$. Since there is a conjugate-linear isometry from the dual space of a Hilbert space onto that Hilbert space (Theorem 4.14), we can consider a bounded linear map from G to H which plays the role of the transpose $A' \in BL(G', H')$ of A.

Theorem 4.23 *Let H and G be Hilbert spaces, and let $A \in BL(H, G)$. Then there is a unique $A^* \in BL(G, H)$, called the **adjoint** of A, such that*

$$\langle A(x), y\rangle_G = \langle x, A^*(y)\rangle_H \quad \text{for all } x \in H \text{ and } y \in G.$$

The map $A \longmapsto A^*$ is a conjugate-linear isometry from $BL(H, G)$ onto $BL(G, H)$. In fact, $(A^*)^* = A$. Further, $\|A^*A\| = \|A\|^2 = \|AA^*\|$.

Proof Fix $y \in G$, and let $g_y(x) := \langle A(x), y \rangle_G$ for $x \in H$. Then g_y is linear, and $|g_y(x)| \le \|A(x)\| \, \|y\| \le \|A\| \, \|y\| \, \|x\|$ for all $x \in H$. Hence $g_y \in H'$ and $\|g_y\| \le \|A\| \, \|y\|$. By the Riesz representation theorem (Theorem 4.14), there is a unique representer of g_y in H. Let us denote it by $A^*(y)$. Thus $A^*(y) \in H$, $\|A^*(y)\| = \|g_y\| \le \|A\| \, \|y\|$, and

$$\langle A(x), y \rangle_G = g_y(x) = \langle x, A^*(y) \rangle_H \quad \text{for all } x \in H.$$

Let $k \in \mathbb{K}$. Since $g_{ky} = \overline{k} g_y$, we obtain $A^*(k\,y) = \overline{\overline{k}} A^*(y) = k A^*(y)$. Let $y_1, y_2 \in G$. Since $g_{y_1+y_2} = g_{y_1} + g_{y_2}$, we obtain $A^*(y_1 + y_2) = A^*(y_1) + A^*(y_2)$. Hence the map $A^* : G \to H$ is linear. Also, since $\|A^*(y)\| \le \|A\| \, \|y\|$ for all $y \in Y$, we see that A^* is in $BL(G, H)$, and $\|A^*\| \le \|A\|$.

To prove the uniqueness of A^*, let $\widetilde{A} \in BL(G, H)$ satisfy $\langle A(x), y \rangle_G = \langle x, \widetilde{A}(y) \rangle_H$ for all $x \in H$ and $y \in G$. Then $\langle x, \widetilde{A}(y) \rangle_H = \langle A(x), y \rangle_G = \langle x, A^*(y) \rangle_H$, that is, $\langle x, (\widetilde{A} - A^*)(y) \rangle_H = 0$ for all $x \in H$ and all $y \in G$. Thus $(\widetilde{A} - A^*)(y) = 0$ for all $y \in G$, that is, $\widetilde{A} = A^*$.

Let $A, B \in BL(H, G)$. Then for $x \in H$ and $y \in G$, $\langle (A + B)(x), y \rangle_G = \langle A(x), y \rangle_G + \langle B(x), y \rangle_G = \langle x, A^*(y) \rangle_H + \langle x, B^*(y) \rangle_H = \langle x, (A^* + B^*)(y) \rangle_H$. By the uniqueness of the adjoint of $A + B$, $(A + B)^* = A^* + B^*$. Similarly, $(kA)^* = \overline{k} A^*$. Thus the map $A \longmapsto A^*$ is conjugate-linear. Also,

$$\langle A^*(y), x \rangle_H = \overline{\langle x, A^*(y) \rangle_H} = \overline{\langle A(x), y \rangle_G} = \langle y, A(x) \rangle_G \quad \text{for all } x \in H, \ y \in G.$$

By the uniqueness of the adjoint of A^*, we obtain $(A^*)^* = A$. In particular, the map $A \longmapsto A^*$ is onto.

We have already noted that $\|A^*\| \le \|A\|$. Hence $\|A\| = \|(A^*)^*\| \le \|A^*\|$, and so the map $A \longmapsto A^*$ is an isometry from $BL(H, G)$ to $BL(G, H)$. Further, by Theorem 3.10(ii), $\|A^* A\| \le \|A^*\| \, \|A\| = \|A\|^2$. Also,

$$\|A(x)\|^2 = \langle A(x), A(x) \rangle_G = \langle A^* A(x), x \rangle_H \le \|A^* A(x)\| \, \|x\| \le \|A^* A\| \, \|x\|^2$$

for all $x \in H$. Hence $\|A\|^2 \le \|A^* A\|$. Thus $\|A^* A\| = \|A\|^2$. Replacing A by A^*, and noting that $(A^*)^* = A$, we obtain $\|A A^*\| = \|A\|^2$. □

Examples 4.24 (i) Let $A \in BL(\ell^2)$. Then A is defined by the infinite matrix $M := [k_{i,j}]$, where $k_{i,j} := A(e_j)(i) = \langle A(e_j), e_i \rangle$ for $i, j \in \mathbb{N}$. (See Example 4.19(i).) Since $\langle A^*(e_j), e_i \rangle = \langle e_j, A(e_i) \rangle = \overline{k_{j,i}}$ for $i, j \in \mathbb{N}$, we see that the adjoint A^* of A is defined by the **conjugate-transpose** $\overline{M^t}$ of M.

Suppose A is the **diagonal operator** on ℓ^2 given by $A(x(1), x(2), \ldots) := (k_1 x(1), k_2 x(2), \ldots)$. Then A is defined by the infinite diagonal matrix $M := \operatorname{diag}(k_1, k_2, \ldots)$. It is easy to see that the adjoint A^* of A is the diagonal operator on ℓ^2 given by $A(x(1), x(2), \ldots) := (\overline{k_1} x(1), \overline{k_2} x(2), \ldots)$, and it is defined by the infinite diagonal matrix $\overline{M^t} := \operatorname{diag}(\overline{k_1}, \overline{k_2}, \ldots)$.

Next, let A be the **right shift operator** on ℓ^2 given by $A(x(1), x(2), \ldots) := (0, x(1), x(2), \ldots)$. Then A is defined by the infinite matrix having 1's on the subdiagonal, and 0's elsewhere. It is easy to see that the adjoint A^* of A is the **left shift operator** on ℓ^2 given by $A(x(1), x(2), \ldots) := (x(2), x(3), \ldots)$, and it is defined by the infinite matrix having 1's on the superdiagonal, and 0's elsewhere.

(ii) Let us denote $L^2([a, b])$ by L^2. Let $k(\cdot, \cdot)$ be a measurable function on $[a, b] \times [a, b]$ such that

$$\gamma_{2,2} := \left(\int_a^b \int_a^b |k(s, t)|^2 dm(t) dm(s) \right)^{1/2} < \infty,$$

as defined in Example 3.16(iii). Let $A \in BL(L^2)$ be the Fredholm integral operator having kernel $k(\cdot, \cdot)$, and let $x, y \in L^2$. Then

$$\begin{aligned}
\langle A^*(x), y \rangle = \langle x, A(y) \rangle &= \int_a^b x(t) \overline{A(y)(t)} dm(t) \\
&= \int_a^b x(t) \left(\int_a^b \overline{k(t, u)}\, \overline{y(u)} dm(u) \right) dm(t) \\
&= \int_a^b \left(\int_a^b \overline{k(t, u)} x(t) dm(t) \right) \overline{y(u)} dm(u).
\end{aligned}$$

The interchange of the order of integration is justified (Theorem 1.20), since

$$\int\int_{[a,b] \times [a,b]} |\overline{k(t, u)} x(t) \overline{y(u)}| d(m \times m)(t, u) \le \gamma_{2,2} \|x\|_2 \|y\|_2.$$

Let us define $k^*(s, t) := \overline{k(t, s)}$ for $s, t \in [a, b]$. Then $k^*(\cdot, \cdot)$ is a measurable function on $[a, b] \times [a, b]$, and

$$\gamma_{2,2}^* := \left(\int_a^b \int_a^b |k^*(s, t)|^2 dm(t) dm(s) \right)^{1/2} = \gamma_{2,2} < \infty.$$

Let $B \in BL(L^2)$ be the Fredholm integral operator having kernel $k^*(\cdot, \cdot)$:

$$B(x)(u) = \int_a^b k^*(u, t) x(t) dm(t) \quad \text{for } x \in L^2 \text{ and } u \in [a, b].$$

Hence $\langle A^*(x), y \rangle = \langle B(x), y \rangle$ for all $x, y \in L^2$, that is, $A^* = B$. ◊

Remarks 4.25 (i) Let H and G be Hilbert spaces, and $A \in BL(H, G)$. Let us relate the adjoint A^* of A to the transpose A' of A. Let $y' \in G'$, and let $y \in G$ be the representer of y' (Theorem 4.14). Then

$$A'(y')(x) = y'(A(x)) = \langle A(x), y \rangle_G = \langle x, A^*(y) \rangle_H \quad \text{for all } x \in H.$$

Hence $A^*(y) \in H$ is the representer of $A'(y') \in H'$. For $x' \in H'$, let $T(x') \in H$ denote the representer of x', and for $y' \in G'$, let $S(y') \in G$ denote the representer of y'. Then the following diagram is commutative:

$$\begin{array}{ccc} G & \xrightarrow{A^*} & H \\ S\uparrow & & \uparrow T \\ G' & \xrightarrow[A']{} & H' \end{array}$$

Since $T\circ A' = A^*\circ S$, and S is a conjugate-linear isometry from G' onto G, we obtain $A^* = T\circ A'\circ S^{-1}$. As a consequence, in view of the Schauder theorem (Theorem 4.21), we see that A^* is a compact linear map from G to H if A is a compact linear map from H to G. This result can also be proved without appealing to Theorem 4.21 of Schauder. (See Exercise 4.31(ii).)

(ii) We note that if X is an incomplete inner product space, and if A is in $BL(X)$, then there may not exist B in $BL(X)$ such that $\langle A(x), y\rangle = \langle x, B(y)\rangle$ for all x, y in H. For example, let $X := c_{00}$ with the usual inner product. For $x \in X$, let $A(x) := y$, where $y(1) := \sum_{j=1}^{\infty} x(j)/j$ and $y(i) := 0$ for all $i \geq 2$. Then $A \in BL(X)$. In fact, $\|A\|^2 \leq \left(\sum_{j=1}^{\infty} 1/j^2\right) = \pi^2/6$. If $B \in BL(X)$ and $\langle A(x), y\rangle = \langle x, B(y)\rangle$ for all $x, y \in H$, then

$$\langle A(e_j), e_1\rangle = \langle e_j, B(e_1)\rangle = \overline{B(e_1)(j)},$$

and $B(e_1)(j) = 1/j \neq 0$ for $j \in \mathbb{N}$, which is impossible since $B(e_1) \in c_{00}$. ◊

For simplicity of notation, we shall write BA instead of $B\circ A$ from now on.

Corollary 4.26 *Let H and G be Hilbert spaces, and suppose $A \in BL(H, G)$ and $B \in BL(G, H)$. Then*

(i) $(BA)^* = A^*B^*$.

(ii) *Let A^* be one-one, onto, and let its inverse be continuous. Then A is one-one, onto, and its inverse is continuous.*

Proof (i) For all $x,\ y \in H$,

$$\langle (BA)(x), y\rangle_H = \langle A(x), B^*(y)\rangle_G = \langle x, A^*(B^*(y))\rangle_H = \langle x, (A^*B^*)(y)\rangle_H.$$

By the uniqueness of the adjoint of BA, we obtain $(BA)^* = A^*B^*$.

(ii) Let C be the inverse of A^*. Then $CA^* = I_G$ and $A^*C = I_H$. Since $(A^*)^* = A$, we obtain $AC^* = I_G^* = I_G$ and $C^*A = I_H^* = I_H$ by (i) above. Hence A is one-one, onto, and its inverse $A^{-1} = C^*$ is continuous. □

Finally, we prove some results which indicate the utility of the adjoint operation. Consider an operator equation $A(x) = y$, where A is a bounded linear map from a

Hilbert space H to a Hilbert space G. Let us ask whether for every $y \in G$, there is $x \in H$ such that $A(x) = y$, that is, whether the range space of the operator A is equal to G. In general, this can be a very difficult question. Even to find whether the range space of A is dense in G can be difficult. The following theorem shows that these questions are equivalent to some questions about the adjoint A^* which are admittedly more easy to answer. (Compare Proposition 4.22 and the comment after its proof.)

Theorem 4.27 *Let H and G be Hilbert spaces, and let $A \in BL(H, G)$. Then*

(i) $Z(A^*) = R(A)^\perp$, *and* $Z(A^*)^\perp$ *is the closure of* $R(A)$ *in* G. *In particular,* $R(A)$ *is dense in* G *if and only if* A^* *is one-one.*
(ii) $R(A) = G$ *if and only if* A^* *is bounded below.*

Proof (i) Let $y \in G$. Then $y \in Z(A^*)$, that is, $A^*(y) = 0$ if and only if $\langle A(x), y\rangle_G = \langle x, A^*(y)\rangle_H = 0$ for all $x \in H$, that is, $y \in R(A)^\perp$. Thus $Z(A^*) = R(A)^\perp$, and so $Z(A^*)^\perp = R(A)^{\perp\perp}$, where $R(A)^{\perp\perp}$ is the closure of $R(A)$ in G. (Theorem 2.35).

In particular, the closure of $R(A)$ in G equals G if and only if $Z(A^*)^\perp = G$, that is, $Z(A^*) = \{0\}$. Thus $R(A)$ is dense in G if and only if A^* is one-one.

(ii) Suppose $R(A) = G$. Since H and G are Banach spaces, and A is continuous, A is an open map by Theorem 3.39. By Proposition 3.41, there is $\gamma > 0$ satisfying the following condition: For every $y \in G$, there is $x \in H$ such that $F(x) = y$ and $\|x\| \le \gamma\|y\|$. For $y \in G$, there is $x \in H$ such that

$$\|y\|^2 = \langle y, y\rangle_G = \langle A(x), y\rangle_G = \langle x, A^*(y)\rangle_H \le \|x\|\|A^*(y)\| \le \gamma\|y\|\|A^*(y)\|.$$

Let $\beta := 1/\gamma$. Then $\beta\|y\| \le \|A^*(y)\|$ for all $y \in G$. Thus A^* is bounded below.

Conversely, suppose A^* is bounded below. Then there is $\beta > 0$ such that $\beta\|y\| \le \|A^*(y)\|$ for all $y \in G$. We first prove that $R(A^*)$ is closed in H. Let (y_n) be a sequence in G, and let $A^*(y_n) \to z$ in H. Since

$$\beta\|y_n - y_m\| \le \|A^*(y_n - y_m)\| = \|A^*(y_n) - A^*(y_m)\| \quad \text{for all } m, n \in \mathbb{N},$$

we see that (y_n) is a Cauchy sequence in G. There is $y \in G$ such that $y_n \to y$ in G. Then $A^*(y_n) \to A^*(y)$ in H, and so $z = A^*(y) \in R(A^*)$. Thus $R(A^*)$ is closed in H. As a result, $R(A^*)$ is a Hilbert space.

Fix $y \in G$, and let $g(A^*(z)) := \langle z, y\rangle_G$ for $z \in G$. Then the functional g is well defined on $R(A^*)$ since the operator A^* is one-one. Clearly, g is linear. Also, if $w \in R(A^*)$ and $w := A^*(z)$, then $\beta\|z\| \le \|A^*(z)\| = \|w\|$, and so

$$|g(w)| = |g(A^*(z))| = |\langle z, y\rangle_G| \le \|z\|\|y\| \le \frac{1}{\beta}\|w\|\|y\|.$$

Hence $g \in (R(A^*))'$. By Theorem 4.14 for the Hilbert space $R(A^*)$, there is x in $R(A^*)$ such that $g(w) = \langle w, x\rangle$ for all $w \in R(A^*)$. Now

$$\langle z, y\rangle_G = g(A^*(z)) = \langle A^*(z), x\rangle = \langle z, A(x)\rangle \quad \text{for all } z \in G.$$

This shows that $y = A(x) \in R(A)$. Thus $R(A) = H$. □

4.4 Self-Adjoint, Normal and Unitary Operators

Let us study bounded operators on a Hilbert space that are 'well behaved' with respect to the adjoint operation. Let H be a Hilbert space over $\mathbb{K}$, and let $A \in BL(H)$. If $A^* = A$, then A is called **self-adjoint**. If A is invertible and $A^* = A^{-1}$, that is, if $A^*A = I = AA^*$, then A is called **unitary**. More generally, if $A^*A = AA^*$, then A is called **normal**.

Let us also consider an important subclass of self-adjoint operators. A self-adjoint operator A is called **positive** if $\langle A(x), x\rangle \geq 0$ for all $x \in H$. If A is positive, then we write $A \geq 0$. If A and B are self-adjoint operators, and if $A - B \geq 0$, then we write $A \geq B$. The relation $\geq$ on the set of all self-adjoint operators on H is clearly reflexive and transitive. Also, it is antisymmetric, since the conditions '$A \geq B$' and '$B \geq A$' imply that $\langle A(x), x\rangle = \langle B(x), x\rangle$ for all $x \in H$, and so $A = B$ as we shall see in Theorem 4.30(i). Thus $\geq$ is a partial order. We note that A^*A is a positive operator for every $A \in BL(H)$ since $(A^*A)^* = A^*A$ and $\langle A^*A(x), x\rangle = \langle A(x), A(x)\rangle \geq 0$ for all $x \in H$.

We observe that the set of all normal operators is closed under the adjoint operation and under multiplication by a scalar. The set of all self-adjoint operators is a real-linear subspace of $BL(H)$. The set of all positive operators on H is closed under addition and under multiplication by a nonnegative scalar. The set of all unitary operators is closed under the operations of composition, inversion and multiplication by a scalar of absolute value 1.

Let (A_n) be a sequence in $BL(H)$. If each A_n is normal and $A_n \to A$ in $BL(H)$, then A is normal; if in fact each A_n is unitary, then A is unitary. This follows by noting that $A_n^* \to A^*$, $A_n^*A_n \to A^*A$ and $A_nA_n^* \to AA^*$ in $BL(H)$. Next, if each A_n is self-adjoint, and if there is $A \in BL(H)$ such that $A_n(x) \to A(x)$ in H for every $x \in H$, then A is self-adjoint; if in fact each A_n is positive, then A is positive. This follows by noting that $\langle A_n(x), y\rangle \to \langle A(x), y\rangle$ and $\langle x, A_n(y)\rangle \to \langle x, A(y)\rangle$ for all $x, y \in H$.

Since $\langle A(x), y\rangle = \langle x, A^*(y)\rangle$ for all $x, y \in H$, we see that A is self-adjoint if and only if $\langle A(x), y\rangle = \langle x, A(y)\rangle$ for all $x, y \in H$. Also, A is normal if and only if $\langle A(x), A(y)\rangle = \langle A^*(x), A^*(y)\rangle$ for all $x, y \in H$. Further, A is unitary if and only if $\langle A(x), A(y)\rangle = \langle x, y\rangle = \langle A^*(x), A^*(y)\rangle$ for all $x, y \in H$. It is clear that $A \in BL(H)$ is unitary if and only if it is a Hilbert space isomorphism from H onto H.

Examples 4.28 (i) Let $H := \ell^2$, and let $A \in BL(H)$. We have seen in Example 4.24(i) that A is defined by the infinite matrix $M := [k_{i,j}]$, where $k_{i,j} := A(e_j)(i)$ for $i, j \in \mathbb{N}$, and that A^* is defined by the conjugate-transpose $\overline{M^t} := [\,\overline{k_{j,i}}\,]$ of M.

Let $\widetilde{A} \in BL(H)$ be defined by a matrix $\widetilde{M} := [\widetilde{k}_{i,j}]$. Then $\widetilde{A} = A$ if and only if $\widetilde{k}_{i,j} = \widetilde{A}(e_j)(i) = A(e_j)(i) = k_{i,j}$ for all $i, j \in \mathbb{N}$, that is, $\widetilde{M} = M$. Hence A is self-adjoint if and only if $\overline{M^t} = M$. Also, A is positive if and only if $\overline{M^t} = M$, and

$$\langle A(x), x\rangle = \sum_{i=1}^{\infty} A(x)(i)\overline{x(i)} = \sum_{i=1}^{\infty}\Big(\sum_{j=1}^{\infty} k_{i,j}x(j)\Big)\overline{x(i)} \geq 0 \quad \text{for all } x \in H.$$

For $n \in \mathbb{N}$, let $A_n(x) := \sum_{i=1}^{n}\sum_{j=1}^{n} k_{i,j}x(j)$, $x \in H$, and consider the quadratic form $q_n(x(1), \ldots, x(n)) := \sum_{i=1}^{n}\sum_{j=1}^{n} k_{i,j}x(j)\overline{x(i)}$ in the n variables $x(1), \ldots,$ $x(n)$. It can be seen that $\langle A(x), x\rangle \geq 0$ for all $x \in H$ if and only if the quadratic form q_n is nonnegative for each $n \in \mathbb{N}$. This happens exactly when every principal minor of the $n \times n$ matrix $[k_{i,j}]$ is nonnegative for each $n \in \mathbb{N}$. (See [18, Chap. 7, p. 405].)

Next, for all $i, j \in \mathbb{N}$,

$$\langle A^*A(e_j), e_i\rangle = \langle A(e_j), A(e_i)\rangle = \Big\langle \sum_{\ell=1}^{\infty} k_{\ell,j}e_\ell, \sum_{\ell=1}^{\infty} k_{\ell,i}e_\ell\Big\rangle = \sum_{\ell=1}^{\infty} k_{\ell,j}\overline{k}_{\ell,i}.$$

Thus A^*A is defined by the infinite matrix $\overline{M^t}M$. Similarly, AA^* is defined by the infinite matrix $M\overline{M^t}$. Hence A is normal if and only if $\overline{M^t}M = M\overline{M^t}$. Also, A is unitary if and only if both $\overline{M^t}M$ and $M\overline{M^t}$ equal the identity matrix, that is, the set of all columns of M is an orthonormal subset of H, and the set of all rows of M is an orthonormal subset of H. The right shift operator A on H mentioned in Example 4.24(i) is not unitary since $\overline{M^t}M$ is the identity matrix, but $M\overline{M^t}$ is not.

In particular, let (k_n) be a bounded sequence in $\mathbb{K}$, and define $A(x) := (k_1x(1),$ $k_2x(2), \ldots)$ for $x := (x(1), x(2), \ldots) \in H$. Then $A \in BL(H)$ and $\|A\| \leq \sup\{|k_n| :$ $n \in \mathbb{N}\}$. (Compare Exercise 3.22(iii).) Since A is defined by the matrix $M :=$ diag $(k_1, k_2, \ldots)$, it follows that A is a normal operator. (Conversely, if a normal operator A is defined by a triangular matrix M, then M is in fact a diagonal matrix. See Exercise 5.29.) Further, A is self-adjoint if and only if $k_n \in \mathbb{R}$ for all $n \in \mathbb{N}$, A is positive if and only if $k_n \geq 0$ for all $n \in \mathbb{N}$, and A is unitary if and only if $|k_n| = 1$ for all $n \in \mathbb{N}$.

(ii) Let $H := L^2([a, b])$. Let $x_0 \in L^\infty([a, b])$, and define $A(x) := x_0x$ for $x \in H$. Then $A \in BL(H)$ and $\|A\| \leq \|x_0\|_\infty$ since for all $x \in H$,

$$\|A(x)\|_2^2 = \int_a^b |x_0(t)x(t)|^2 dm(t) \leq \|x_0\|_\infty^2 \int_a^b |x(t)|^2 dm(t) = \|x_0\|_\infty^2 \|x\|_2^2.$$

(Compare Exercise 3.23(iii).) Now for all $x, y \in H$,

$$\langle A^*(x), y\rangle = \langle x, A(y)\rangle = \langle x, x_0y\rangle = \int_a^b x(t)\overline{x_0}(t)\,\overline{y}(t)dm(t) = \langle \overline{x_0}x, y\rangle.$$

Hence $A^*(x) = \overline{x_0}x$ for all $x \in H$. Since $A^*A(x) = |x_0|^2 x = AA^*(x)$ for all $x \in H$, we see that A is a normal operator. Further, A is self-adjoint if and only if $x_0(t) \in \mathbb{R}$ for almost all $t \in [a, b]$, A is positive if and only if $x_0(t) \geq 0$ for almost all $t \in [a, b]$ and A is unitary if and only if $|x_0(t)| = 1$ for almost all $t \in [a, b]$. (Note that the function $x(t) := 1$, $t \in [a, b]$, is in H.)

(iii) Let $H := L^2([a, b])$. Let $k(\cdot, \cdot) \in L^2([a, b] \times [a, b])$, so that

$$\gamma_{2,2} := \left(\int_a^b \int_a^b |k(s, t)|^2 dm(t) dm(s) \right)^{1/2} < \infty,$$

and let A denote the Fredholm integral operator with kernel $k(\cdot, \cdot)$. We have seen in Example 4.24(ii) that A^* is the Fredholm integral operator with kernel $k^*(s, t) := \overline{k(t, s)}$ for $s, t \in [a, b]$. Hence A is self-adjoint if $\overline{k(t, s)} = k(s, t)$ for $s, t \in [a, b]$. Also, A is a positive operator if $\overline{k(t, s)} = k(s, t)$ for $s, t \in [a, b]$, and

$$\begin{aligned} \langle A(x), x \rangle &= \int_a^b \left(\int_a^b k(s, t) x(t) dm(t) \right) \overline{x(s)} dm(s) \\ &= \int_a^b \int_a^b k(s, t) x(t) \overline{x(s)} d(m \times m)(s, t) \geq 0 \quad \text{for all } x \in H. \end{aligned}$$

Next, we consider the normality of A. Let $x \in H$ and $s \in [a, b]$. Then

$$\begin{aligned} (A^*A)(x)(s) &= \int_a^b k^*(s, u) A(x)(u) dm(u) \\ &= \int_a^b k^*(s, u) \left(\int_a^b k(u, t) x(t) dm(t) \right) dm(u) \\ &= \int_a^b \left(\int_a^b k^*(s, u) k(u, t) dm(u) \right) x(t) dm(t). \end{aligned}$$

We note that by Theorem 1.20, the change of the order of integration above is justified for almost all $s \in [a, b]$ since

$$\begin{aligned} &\left(\int_a^b \int_a^b |k^*(s, u)|\, |k(u, t)|\, |x(t)| d(m \times m)(t, u) \right)^2 \\ &\leq \left(\int_a^b \int_a^b |k^*(s, u)|^2 |x(t)|^2 d(m \times m)(t, u) \right) \left(\int_a^b \int_a^b |k(u, t)|^2 d(m \times m)(t, u) \right) \\ &= \left(\int_a^b |k^*(s, u)|^2 dm(u) \right) \|x\|_2^2 \, \gamma_{2,2}^2, \end{aligned}$$

where $\int_a^b \int_a^b |k^*(s, u)|^2 dm(u) dm(s) = \gamma_{2,2}^2 < \infty$. Similarly, we obtain

$$(AA^*)(x)(s) = \int_a^b \left(\int_a^b k(s,u)k^*(u,t)dm(u) \right) x(t)dm(t)$$

for almost all $s \in [a, b]$. This shows that A is normal if

$$\int_a^b \overline{k(u,s)} k(u,t)dm(u) = \int_a^b k(s,u)\overline{k(t,u)}dm(u) \quad \text{for } s, t \in [a, b].$$

We have seen in Example 3.46 that A is a compact operator. Since H is infinite dimensional, A cannot be unitary. ◊

It is evident from the above examples that there is an analogy between the complex numbers and normal operators on a complex Hilbert space H, in which the adjoint operation plays the role of complex conjugation. Self-adjoint operators correspond to real numbers, positive operators correspond to nonnegative real numbers and unitary operators correspond to complex numbers of absolute value 1. This is made precise in the following result.

Proposition 4.29 *Let $\mathbb{K} := \mathbb{C}$, and let $A \in BL(H)$. There are unique self-adjoint operators B and C in $BL(H)$ such that $A = B + i\,C$.*

Further, A is normal if and only if $BC = CB$, A is unitary if and only if $BC = CB$ and $B^2 + C^2 = I$, A is self-adjoint if and only if $C = 0$, and A is positive if and only if $C = 0$ and B is positive.

Proof Let

$$B := \frac{A + A^*}{2} \quad \text{and} \quad C = \frac{A - A^*}{2i}.$$

Then B and C are self-adjoint operators, and $A = B + i\,C$. If we also have $A = B_1 + i\,C_1$, where B_1 and C_1 are self-adjoint, then $A^* = B_1 - i\,C_1$, and so $B_1 = (A + A^*)/2 = B$ and $C_1 = (A - A^*)/2i = C$.

Now A is normal if and only if $(B - i\,C)(B + i\,C) = A^*A = AA^* = (B + i\,C)(B - i\,C)$, that is, $BC = CB$. Further, A is unitary if and only if A is normal and $A^*A = I$, that is, $BC = CB$ and $(B^2 + C^2) = (B - i\,C)(B + i\,C) = I$. Next, A is self-adjoint if and only if $B - i\,C = A^* = A = B + i\,C$, that is, $C = 0$. Clearly, A is positive if and only if $C = 0$ and B is positive. □

To facilitate the study of these operators, we associate a subset of scalars to each of them. Let X be a nonzero inner product space over $\mathbb{K}$, and let $A \in BL(X)$. The set $\omega(A) := \{\langle A(x), x\rangle : x \text{ in } X \text{ and } \|x\| = 1\}$ is called the **numerical range** of A. It is the range of the scalar-valued continuous function $x \longmapsto \langle A(x), x\rangle$ defined on the unit sphere $\{x \in X : \|x\| = 1\}$ of X.

If $\lambda \in \omega(A)$, and $\lambda = \langle A(x), x\rangle$, where $x \in X$ with $\|x\| = 1$, then by the Schwarz inequality, $|\lambda| \leq \|A\|\|x\|^2 = \|A\|$. Thus $\omega(A) \subset \{k \in \mathbb{K} : |k| \leq \|A\|\}$, and so $\omega(A)$ is a bounded subset of $\mathbb{K}$. If X is finite dimensional, then $\omega(A)$ is a closed subset of $\mathbb{K}$ since the unit sphere of X is compact. In general, $\omega(A)$ may not be a closed subset of $\mathbb{K}$. For example, if $X := c_{00}$, and $A(x) := (x(1), x(2)/2, x(3)/3, \ldots)$ for

$x \in c_{00}$, then $\langle A(e_n), e_n \rangle = 1/n$ belongs to $\omega(A)$ for each $n \in \mathbb{N}$, but $0 \notin \omega(A)$. The **Toeplitz–Hausdorff theorem** says that $\omega(A)$ is a convex subset of $\mathbb{K}$. We refer to [14] for a short geometric proof of this result. See Exercise 4.39 for a specific example.

Let H be a nonzero Hilbert space over $\mathbb{K}$, $A \in BL(H)$ and $k \in \mathbb{K}$. Then $k \in \omega(A)$ if and only if $\overline{k} \in \omega(A^*)$ since $\langle A^*(x), x \rangle = \langle x, A(x) \rangle = \overline{\langle A(x), x \rangle}$ for all $x \in H$. Suppose A is self-adjoint. It follows that $\langle A(x), x \rangle \in \mathbb{R}$ for all $x \in H$, and so $\omega(A)$ is a subset of $\mathbb{R}$. Clearly, $\omega(A) \subset [-\|A\|, \|A\|]$. We define $m_A := \inf \omega(A)$ and $M_A := \sup \omega(A)$. Since $\omega(A)$ is a convex subset of $\mathbb{R}$,

$$(m_A, M_A) \subset \omega(A) \subset [m_A, M_A].$$

Conversely, if $\mathbb{K} := \mathbb{C}$ and $\omega(A) \subset \mathbb{R}$, then A is self-adjoint. (See Exercise 4.40.) As a result, $A \in BL(H)$ is positive if and only if $\omega(A) \subset [0, \infty)$.

We show that the norm of a self-adjoint operator is determined by its numerical range and deduce characterizations of normal and unitary operators.

Theorem 4.30 *Let H be a Hilbert space and $A \in BL(H)$.*

(i) *Let A be self-adjoint. Then*

$$\|A\| = \sup\{|\langle A(x), x \rangle| : x \in H \text{ and } \|x\| \leq 1\}.$$

If H is nonzero, then $\|A\| = \sup\{|k| : k \in \omega(A)\}$. In particular, $A = 0$ if and only if $\langle A(x), x \rangle = 0$ for all $x \in H$.

(ii) *A is normal if and only if $\|A^*(x)\| = \|A(x)\|$ for all $x \in H$. In this case,*

$$\|A^2\| = \|A^* A\| = \|A\|^2.$$

(iii) *A is unitary if and only if $\|A(x)\| = \|x\|$ for all $x \in H$ and A is onto. In this case,*

$$\|A\| = 1 = \|A^{-1}\|.$$

Proof (i) By Theorem 4.23,

$$\|A\| = \sup\{|\langle A(x), y \rangle| : x, y \in H \text{ and } \|x\| \leq 1, \|y\| \leq 1\}.$$

Let $\alpha := \sup\{|\langle A(x), x \rangle| : x \in H \text{ and } \|x\| \leq 1\}$. Clearly, $\alpha \leq \|A\|$. Next, let $x, y \in H$. Then $\langle A(x+y), x+y \rangle - \langle A(x-y), x-y \rangle = 2\langle A(x), y \rangle + 2\langle A(y), x \rangle = 4$ Re $\langle A(x), y \rangle$, since A is self-adjoint. Hence

$$4|\text{Re} \langle A(x), y \rangle| \leq \alpha(\|x+y\|^2 + \|x-y\|^2) = 2\alpha(\|x\|^2 + \|y\|^2)$$

by the parallelogram law. (See Remark 2.15.) Suppose $\|x\| \leq 1$ and $\|y\| \leq 1$. Let $k := \text{sgn}\,(\langle A(x), y \rangle)$, and $x_0 := kx$. Then $\|x_0\| \leq \|x\| \leq 1$, and

$$|\langle A(x), y\rangle| = k\langle A(x), y\rangle = \langle A(x_0), y\rangle = \text{Re}\,\langle A(x_0), y\rangle \leq \frac{2\alpha}{4}(1+1) = \alpha.$$

Taking supremum over $x, y \in H$ with $\|x\| \leq 1$, $\|y\| \leq 1$, we obtain $\|A\| \leq \alpha$. Thus $\|A\| = \alpha$. If $H \neq \{0\}$, then $\alpha = \sup\{|\langle A(x), x\rangle| : x \in H \text{ and } \|x\| = 1\}$, which is equal to $\sup\{|k| : k \in \omega(A)\}$.

In particular, if $\langle A(x), x\rangle = 0$ for all $x \in H$, then $\|A\| = \alpha = 0$, that is, $A = 0$. Conversely, if $A = 0$, then clearly $\langle A(x), x\rangle = 0$ for all $x \in H$.

(ii) For $x \in H$, $\|A(x)\|^2 = \langle A(x), A(x)\rangle = \langle A^*A(x), x\rangle$ and $\|A^*(x)\|^2 = \langle A^*(x), A^*(x)\rangle = \langle AA^*(x), x\rangle$. If A is normal, then it follows that $\|A^*(x)\| = \|A(x)\|$ for all $x \in H$. Conversely, suppose $\|A^*(x)\| = \|A(x)\|$ for all $x \in H$. Define $B := A^*A - AA^*$. Then B is self-adjoint and $\langle B(x), x\rangle = 0$ for all $x \in H$. By (i) above, $B = 0$, that is, A is normal.

In this case, $\|A^2(x)\| = \|A(A(x))\| = \|A^*(A(x))\| = \|(A^*A)(x)\|$ for all $x \in H$, and so $\|A^2\| = \|A^*A\|$. Also, $\|A^*A\| = \|A\|^2$ by Theorem 4.23.

(iii) For $x \in H$, $\|A(x)\|^2 = \langle A(x), A(x)\rangle = \langle A^*A(x), x\rangle$. If A is unitary, then $A^*A = I$, and so $\|A(x)\| = \|x\|$ for all $x \in H$, and $AA^* = I$, and so A is onto. Conversely, suppose $\|A(x)\| = \|x\|$ for all $x \in H$, and A is onto. Define $C := A^*A - I$. Then C is self-adjoint and $\langle C(x), x\rangle = 0$ for all $x \in H$. By (i) above, $C = 0$, that is, $A^*A = I$. Let $y \in H$. Since A is onto, there is $x \in H$ such that $A(x) = y$, and so $AA^*(y) = AA^*(A(x)) = A(A^*A)(x) = A(x) = y$. Hence $AA^* = I$. Thus A is unitary. In this case, $\|A(x)\| = \|x\|$ for all $x \in H$, and so $\|A\| = 1$. Also, $\|A^{-1}\| = 1$ since A^{-1} is unitary. □

The following result generalizes the Schwarz inequality given in Proposition 2.13(i). It will be used in determining the bounds of the spectrum of a self-adjoint operator in Proposition 5.31.

Proposition 4.31 **(Generalized Schwarz inequality)** *Let A be a positive operator on a Hilbert space H, and let $x, y \in H$. Then*

$$|\langle A(x), y\rangle| \leq \langle A(x), x\rangle^{1/2}\langle A(y), y\rangle^{1/2}.$$

In particular, $\|A(x)\| \leq \langle A(x), x\rangle^{1/4}\langle A^2(x), A(x)\rangle^{1/4}$.

Proof Let $\epsilon > 0$, and define $\langle x, y\rangle_\epsilon := \langle A(x), y\rangle + \epsilon\langle x, y\rangle$ for $x, y \in H$. It is easy to see that $\langle\cdot\,,\cdot\rangle_\epsilon$ is an inner product on H. Let $x, y \in H$. By the Schwarz inequality (Proposition 2.3(i)), $|\langle x, y\rangle_\epsilon|^2 \leq \langle x, x\rangle_\epsilon\langle y, y\rangle_\epsilon$, that is,

$$|\langle A(x), y\rangle + \epsilon\langle x, y\rangle|^2 \leq \big(\langle A(x), x\rangle + \epsilon\langle x, x\rangle\big)\big(\langle A(y), y\rangle + \epsilon\langle y, y\rangle\big).$$

Let $\epsilon \to 0$. Then $|\langle A(x), y\rangle|^2 \leq \big(\langle A(x), x\rangle\big)\big(\langle A(y), y\rangle\big)$, as desired.

Put $y := A(x)$, and obtain $\|A(x)\| \leq \langle A(x), x\rangle^{1/4}\langle A^2(x), A(x)\rangle^{1/4}$. □

For another proof of the generalized Schwarz inequality, see Exercise 4.37.

Before concluding this section, we give two important examples, one of a positive operator and another of a unitary operator.

Examples 4.32 (i) Let H be a Hilbert space, and let P be an orthogonal projection operator on H, that is, $P \in BL(H)$, $P^2 = P$ and $R(P) \perp Z(P)$. We show that P is a positive operator. Let $x_1, x_2 \in H$. Then $x_j = y_j + z_j$, where $P(x_j) = y_j$, $P(z_j) = 0$ for $j = 1, 2$, and $\{y_1, y_2\} \perp \{z_1, z_2\}$. Hence

$$\langle P(x_1), x_2\rangle = \langle y_1, y_2 + z_2\rangle = \langle y_1, y_2\rangle = \langle y_1 + z_1, y_2\rangle = \langle x_1, P(x_2)\rangle.$$

Thus P is self-adjoint. Further, $P = P^2 = P^*P$, and so P is positive. In fact, $\langle P(x), x\rangle = \langle P^2(x), x\rangle = \langle P(x), P(x)\rangle = \|P(x)\|^2$ for all $x \in H$.

Conversely, suppose a projection operator P on H is normal. Consider $y \in R(P)$ and $z \in Z(P)$. Since P is normal, Theorem 4.30(ii) shows that $\|P^*(z)\| = \|P(z)\| = \|0\| = 0$, that is, $P^*(z) = 0$. Hence $\langle y, z\rangle = \langle P(y), z\rangle = \langle y, P^*(z)\rangle = \langle y, 0\rangle = 0$. Thus $R(P) \perp Z(P)$, that is, P is orthogonal.

We note that if a projection operator P on H is unitary, then $R(P) = H$, so that $Z(P) = \{0\}$, and $P = I$.

(ii) Since a unitary operator U on a Hilbert space H is a Hilbert space isomorphism, it allows us to view the Hilbert space from a different perspective. We illustrate this comment by considering a well-known unitary operator on the Hilbert space $L^2(\mathbb{R})$ over $\mathbb{K} := \mathbb{C}$.

In analogy with the Fourier series for a function in $L^1([-\pi, \pi])$ considered in Sect. 1.4, we may treat a function $x \in L^1(\mathbb{R})$ as follows. Define

$$\hat{x}(u) := \frac{1}{\sqrt{2\pi}} \int_{\mathbb{R}} x(t)e^{-iut}\, dm(t) \quad \text{for } u \in \mathbb{R}.$$

The (formal) integral

$$\int_{-\infty}^{\infty} \hat{x}(u)e^{iut}\, dm(t)$$

is called the **Fourier integral** of x. It can be shown that $\hat{x} \in C_0(\mathbb{R})$, and if $\hat{x}(u) = 0$ for all $u \in \mathbb{R}$, then $x(t) = 0$ for almost all $t \in \mathbb{R}$. Further, the **inversion theorem** says that if $\hat{x} \in L^1(\mathbb{R})$, then

$$x(t) = \frac{1}{\sqrt{2\pi}} \int_{\mathbb{R}} \hat{x}(u)e^{iut}\, dm(u) \quad \text{for almost all } t \in \mathbb{R}.$$

(See [26, 9.6, 9.11 and 9.12].) The linear map $T : L^1(\mathbb{R}) \to C_0(\mathbb{R})$ given by $T(x) := \hat{x}$, $x \in L^1(\mathbb{R})$, is called the **Fourier transform**.

These results no longer hold if $x \in L^2(\mathbb{R})$. Instead of the kernel e^{-iut} used in the definition of $\hat{x}(u)$ above, let us consider the kernel

$$k(u, t) := \int_0^u e^{-ist}\, ds = \frac{1 - e^{-iut}}{it}, \quad (u, t) \in \mathbb{R} \times (\mathbb{R} \setminus \{0\}),$$

and then differentiate the resulting integral with respect to u. Define

$$U(x)(u) := \frac{1}{\sqrt{2\pi}} \frac{d}{du} \int_{-\infty}^{\infty} x(t)k(u,t)\,dm(t), \quad x \in L^2(\mathbb{R}),\ u \in \mathbb{R}.$$

Then $U(x)(u)$ is well defined for almost all $u \in \mathbb{R}$, and $U(x) \in L^2(\mathbb{R})$. The map U is a unitary operator on $L^2(\mathbb{R})$, and its inverse V is given by

$$V(y)(u) := \frac{1}{\sqrt{2\pi}} \frac{d}{du} \int_{-\infty}^{\infty} y(t)k(u,-t)\,dm(t), \quad y \in L^2(\mathbb{R}),\ u \in \mathbb{R}.$$

Further, if $x \in L^1(\mathbb{R}) \cap L^2(\mathbb{R})$, then $U(x) = \hat{x}$ and $V(\hat{x}) = x$. For a proof, we refer to [23, p. 294]. T his result is due to Plancherel. It is of significance in **signal analysis**. A signal is represented in the 'time domain' by a square-integrable function x on $\mathbb{R}$: the signal at a time $t \in \mathbb{R}$ is determined by the scalar $x(t)$. The **total energy** of the signal is given by

$$\int_{-\infty}^{\infty} |x(t)|^2\,dm(t).$$

The **Fourier–Plancherel transform** U represents signals in the 'frequency domain': a signal x at a frequency u is determined by the scalar $U(x)(u)$. Consider signals $x, y \in L^2(\mathbb{R})$. Then $\langle x, y\rangle = \langle U(x), U(y)\rangle$ since the operator U is unitary. Thus

$$\int_{-\infty}^{\infty} x(t)\overline{y(t)}\,dm(t) = \int_{-\infty}^{\infty} U(x)(u)\overline{U(y)(u)}\,dm(u).$$

This is known as the **time-frequency equivalence**. If the signals x and y are also in $L^1(\mathbb{R})$, then the above equality can be written as

$$\int_{-\infty}^{\infty} x(t)\overline{y(t)}\,dm(t) = \int_{-\infty}^{\infty} \hat{x}(u)\overline{\hat{y}(u)}\,dm(u).$$

Letting $y := x$, we obtain the so-called **energy principle**

$$\int_{-\infty}^{\infty} |x(t)|^2\,dm(t) = \int_{-\infty}^{\infty} |U(x)(u)|^2\,dm(u).$$

These results can be compared with the Parseval identity given in Remark 2.34(i) and the Parseval formula given in Theorem 2.31(v). ◊

Exercises

4.1. Let $X := \mathbb{K}^2$ with the norm $\|\cdot\|_\infty$.

(i) Let $Y := \{(x(1), x(2)) \in X : x(2) = 0\}$, and define $g(y) := y(1)$ for $y := (y(1), y(2)) \in Y$. The only Hahn–Banach extension of g to X is given by $f(x) := x(1)$ for all $x := (x(1), x(2)) \in X$.

(ii) Let $Z := \{(x(1), x(2)) \in X : x(1) = x(2)\}$, and define $h(z) := z(1)$ for $z := (z(1), z(2)) \in Z$. Then f is a Hahn–Banach extension of h to X if and only if there is $t \in [0, 1]$ such that $f(x) := t\,x(1) + (1-t)x(2)$ for all $x := (x(1), x(2)) \in X$.

4.2. **(Helly, 1912)** Let X be a normed space. For each s in an index set S, let $x_s \in X$ and $k_s \in \mathbb{K}$. There exists $f \in X'$ such that $f(x_s) = k_s$ for each $s \in S$ if and only if there is $\alpha > 0$ such that $\big|\sum_s c_s k_s\big| \le \alpha \big\|\sum_s c_s x_s\big\|$, where $\sum_s$ is an arbitrary finite sum and $c_s \in \mathbb{K}$ with $s \in S$.

4.3. Let Y be a subspace of a normed space X, and let $g \in Y'$. Let E denote the set of all Hahn–Banach extensions of g to X. Then E is a nonempty, convex, closed and bounded subset of X', and E does not contain any open ball of X'. However, E may not be a compact subset of X'.

4.4. **(Taylor–Foguel, 1958)** Let X be a normed space over $\mathbb{K}$. There is a unique Hahn–Banach extension to X of every continuous linear functional on every subspace of X if and only if the normed space X' is **strictly convex**. (Compare Exercise 2.10 and Theorem 4.17.)

4.5. Let X and Y be normed spaces, and let $X \neq \{0\}$. Then the following conditions are equivalent: (i) Y is a Banach space. (ii) $BL(X, Y)$ is a Banach space. (iii) $CL(X, Y)$ is a Banach space.

4.6. Let X be a normed space, let Y be a subspace of X and let $m \in \mathbb{N}$. Then every $G \in BL(Y, \mathbb{K}^m)$ has an extension $F \in BL(X, \mathbb{K}^m)$. If, in particular, the norm on $\mathbb{K}^m$ is the norm $\|\cdot\|_\infty$, then every $G \in BL(Y, \mathbb{K}^m)$ has a norm-preserving extension $F \in BL(X, \mathbb{K}^m)$. Also, every $G \in BL(Y, \ell^\infty)$ has a norm-preserving extension $F \in BL(X, \ell^\infty)$.

4.7. Let $X := \mathbb{K}^3$ with $\|\cdot\|_1$, and let $Y := \{x \in \mathbb{K}^3 : x(1) + x(2) + x(3) = 0\}$. If $G := I_Y$, the identity operator on Y, and if $F : X \to Y$ is a linear extension of G, then $\|F\| > \|G\|$. (This example is due to S.H. Kulkarni.)

4.8. Let E be a convex subset of a normed space X, and let $a \in X$. Then $a \in \overline{E}$ if and only if (i) $\mathrm{Re}\, f(a) \ge 1$ whenever $f \in X'$ and $\mathrm{Re}\, f(x) \ge 1$ for all $x \in E$ and (ii) $\mathrm{Re}\, f(a) \le 1$ whenever $f \in X'$ and $\mathrm{Re}\, f(x) \le 1$ for all $x \in E$. (Compare Remark 4.7(ii).)

4.9. Let X be a normed space over $\mathbb{K}$, and let Y be a subspace of X. Suppose E is a nonempty convex open subset of X such that $Y \cap E = \emptyset$. Then there is a continuous linear functional f on X such that $f(y) = 0$ for all $y \in Y$ and $\mathrm{Re}\, f(x) > 0$ for all $x \in E$.

4.10. Let E_1 and E_2 be nonempty disjoint convex subsets of a normed space X. Let E_1 be compact, and let E_2 be closed. Then there are $f \in X'$, and t_1, t_2 in $\mathbb{R}$ such that $\mathrm{Re}\, f(x_1) \le t_1 < t_2 \le \mathrm{Re}\, f(x_2)$ for all $x_1 \in E_1$ and $x_2 \in E_2$.

4.11. (i) For $p \in \{1, 2, \infty\}$, the dual space of $(\mathbb{K}^n, \|\cdot\|_p)$ is linearly isometric to $(\mathbb{K}^n, \|\cdot\|_q)$, where $(1/p) + (1/q) = 1$.

(ii) The dual space of $(c_0, \|\cdot\|_\infty)$ is linearly isometric to ℓ^1.

(iii) For $p \in \{1, 2, \infty\}$, the dual space of $(c_{00}, \|\cdot\|_p)$ is linearly isometric to ℓ^q, where $(1/p) + (1/q) = 1$.

4.12. (i) For $x' \in c'$, let $\Psi(x')$ be the restriction of x' to c_0. Then $\Psi : c' \to (c_0)'$ is linear, and $\|\Psi(x')\| \leq \|x'\|$ for all $x' \in c'$, but Ψ is not an isometry.

(ii) Let $y := (y(1), y(2), \ldots) \in \ell^1$. For $x \in c$, define $f_y(x) := \ell_x\, y(1) + \sum_{j=1}^{\infty} x(j)y(j+1)$, where $\ell_x := \lim_{j\to\infty} x(j)$. Then $f_y \in c'$. If we let $\Phi(y) := f_y$ for $y \in \ell^1$, then Φ is a linear isometry from ℓ^1 onto c'.

4.13. Let $W^{1,2} := W^{1,2}([a, b])$, as in Example 2.28(iv). For $y \in W^{1,2}$, define

$$f_y(x) = \int_a^b x(t)y(t)dt + \int_a^b x'(t)y'(t)dm(t), \quad x \in W^{1,2}.$$

Then $f_y \in (W^{1,2})'$. If we let $\Phi(y) := f_y$ for $y \in W^{1,2}$, then Φ is a linear isometry from $W^{1,2}$ onto $(W^{1,2})'$.

4.14. For $p \in \{1, 2\}$, the dual space of $(C([a, b]), \|\cdot\|_p)$ is linearly isometric to $L^q([a, b])$, where $(1/p) + (1/q) = 1$.

4.15. Let X be a reflexive normed space. Then X is a Banach space. If, in addition, X is separable, then so is X'. Not every Banach space is reflexive.

4.16. Let X be a normed space, and let (x_n) be a sequence in X. We say that (x_n) **converges weakly** in X if there is $x \in X$ such that $x'(x_n) \to x'(x)$ for every $x' \in X'$. In this case, we write $x_n \stackrel{w}{\to} x$ in X. If $x_n \stackrel{w}{\to} x$ in X, and $x_n \stackrel{w}{\to} \tilde{x}$ in X, then $\tilde{x} = x$, which is called the **weak limit** of (x_n).

(i) Let $x \in X$. Then $x_n \stackrel{w}{\to} x$ in X if and only if (x_n) bounded in X and $x'(x_n) \to x'(x)$ for all x' in a subset of X' whose span is dense in X'.

(ii) If $x_n \to x$ in X, then $x_n \stackrel{w}{\to} x$ in X. In case X is an inner product space, $x_n \to x$ in X if and only if $x_n \stackrel{w}{\to} x$ in X and $\|x_n\| \to \|x\|$.

(iii) Let X be a Hilbert space. If (x_n) is bounded, then a subsequence of (x_n) converges weakly in X. If $(\langle x_n, \tilde{x}\rangle)$ converges in $\mathbb{K}$ for every $\tilde{x} \in X$, then (x_n) is bounded and there is $x \in X$ such that $x_n \stackrel{w}{\to} x$ in X

(iv) Let $x_n \stackrel{w}{\to} 0$ in X, and let Y be a normed space. If $F \in BL(X, Y)$, then $F(x_n) \stackrel{w}{\to} 0$ in Y. Further, if $F \in CL(X, Y)$, then $F(x_n) \to 0$ in Y.

(v) If X is a Hilbert space, Y is a normed space, $F \in CL(X, Y)$, and (u_n) is an orthonormal sequence in X, then $F(u_n) \to 0$. In particular, if $M := [k_{i,j}]$ defines a map $F \in CL(\ell^2)$, then $\alpha_2(j) \to 0$ and $\beta_2(i) \to 0$. (Compare Exercises 3.34 and 4.21.)

4.17. Let X be a normed space, and let (x_n) be a sequence in X.

(i) Let $X := \ell^1$. Define $x'(x) := \sum_{j=1}^{\infty} x(j)$ for $x \in X$. Then $x' \in X'$, and $x'(e_n) = 1$ for all $n \in \mathbb{N}$. In particular, $e_n \stackrel{w}{\not\to} 0$ in X. with the projection

(ii) Let $X := \ell^2$. Then $x_n \stackrel{w}{\to} x$ in X if and only if (x_n) is bounded in X, and $x_n(j) \to x(j)$ in $\mathbb{K}$ for every $j \in \mathbb{N}$. In particular, $e_n \stackrel{w}{\to} 0$ in ℓ^2.

(iii) Let $X := \ell^\infty$. Suppose there are $x \in X$ and $\alpha > 0$ such that for all $j \in \mathbb{N}$, $\sum_{n=1}^{\infty} |x_n(j) - x(j)| \leq \alpha$. Then $x_n \xrightarrow{w} x$ in X. In particular, $e_n \xrightarrow{w} 0$ in ℓ^∞.

(iv) Let $X := c_0$, and $x_n := e_1 + \cdots + e_n$ for $n \in \mathbb{N}$. Then $(x'(x_n))$ is convergent in $\mathbb{K}$ for every $x' \in X'$, but (x_n) has no weak limit in X.

(v) Let $X := C([a, b])$. Then $x_n \xrightarrow{w} x$ in X if and only if (x_n) is uniformly bounded on $[a, b]$, and $x_n(t) \to x(t)$ for every $t \in [a, b]$.

4.18. Let X be an inner product space. The following conditions are equivalent:

(i) X is complete.
(ii) $X = Y \oplus Y^\perp$ for every closed subspace Y of X.
(iii) $Y = Y^{\perp\perp}$ for every closed subspace Y of X.
(iv) $Y^\perp \neq \{0\}$ for every proper closed subspace Y of X.
(v) For every $f \in X'$, there is $y \in X$ such that $f(x) = \langle x, y \rangle$ for all $x \in X$.

4.19. Let X be an inner product space.

(i) Let $\{u_\alpha\}$ be an orthonormal subset of X. Then span $\{u_\alpha\}$ is dense in X if and only if $x = \sum_n \langle x, u_n \rangle u_n$ for every $x \in X$, where $\{u_1, u_2, \ldots\} := \{u_\alpha : \langle x, u_\alpha \rangle \neq 0\}$. (Compare Theorem 2.31.)

(ii) Let Y be a closed subspace of X. Then there is an inner product $\langle\langle \cdot, \cdot \rangle\rangle$ on X/Y such that $\langle\langle x + Y, x + Y \rangle\rangle = |||x + Y|||^2$ for all $x \in X$. (Compare Exercise 2.40.)

(iii) There is an inner product $\langle \cdot, \cdot \rangle'$ on X' such that $\langle f, f \rangle' = \|f\|^2$ for all $f \in X'$. (Compare Corollary 4.16(i).)

4.20. Let $F \in BL(X, Y)$. Then F is of finite rank if and only if there are $x_1', \ldots, x_m'$ in X' and $y_1, \ldots, y_m$ in Y such that $F(x) = \sum_{i=1}^{m} x_i'(x) y_i$ for all $x \in X$. Then $F' \in BL(Y', X')$ is of finite rank, and $F'(y') = \sum_{i=1}^{m} y'(y_i) x_i'$ for all $y' \in Y'$.

4.21. Let $p, r \in \{1, 2\}$. If an infinite matrix M defines a map in $CL(\ell^p, \ell^r)$, then the sequence of the rows of M tends to 0 in ℓ^q, where $(1/p) + (1/q) = 1$. This does not hold if $p \in \{1, 2, \infty\}$ and $r := \infty$. (Compare Exercise 3.34.)

4.22. **(Left shift operator** on L^p**)** Let $p \in \{1, 2, \infty\}$, and let L^p denote $L^p([0, \infty))$. For $x \in L^p$, define $A(x)(t) := x(t + 1)$, $t \in [0, \infty)$. Then $A \in BL(L^p)$. If $p \in \{1, 2\}$, then $A' \in BL((L^p)')$ can be identified with the **right shift operator** $A^t \in BL(L^q)$, where $(1/p) + (1/q) = 1$, and for $y \in L^q$, $A^t(y)(s) := 0$ if $s \in [0, 1)$, while $A^t(y)(s) := y(s - 1)$ if $s \in [1, \infty)$.

4.23. Let $p \in \{1, 2\}$, and $n \in \mathbb{N}$. For $x := (x(1), x(2), \ldots) \in \ell^p$, define $P_n(x) := (x(1), \ldots, x(n), 0, 0, \ldots)$. Then $P_n \in BL(\ell^p)$ is a projection, and P_n' can be identified with the projection $P_n^t \in BL(\ell^q)$, where $(1/p) + (1/q) = 1$, and $P_n^t(y) := (y(1), \ldots, y(n), 0, 0, \ldots)$ for $y := (y(1), y(2), \ldots) \in \ell^q$.

4.24. Let X be a normed space, and let $X = Y \oplus Z$, where Y and Z are closed subspaces of X. If $P \in BL(X)$ is the projection operator onto Y along Z, then $P' \in BL(X')$ is the projection operator onto $Z^0 := \{x' \in X' : x'(z) = 0 \text{ for all } z \in Z\}$ along $Y^0 := \{x' \in X' : x'(y) = 0 \text{ for all } y \in Y\}$.

4.25. Let X and Y be normed spaces, let J_X and J_Y denote their canonical embeddings into X'' and Y'' respectively, and let $F \in BL(X, Y)$. Define $F'' := (F')' \in BL(X'', Y'')$. Then $F''J_X = J_Y F$ and $\|F''\| = \|F\|$, and so F'' yields a norm-preserving linear extension of F to X''. Further, if X_c and Y_c denote the respective completions of X and Y, then there is a unique $F_c \in BL(X_c, Y_c)$ such that $F_c J_X = J_Y F$, and it satisfies $\|F_c\| = \|F\|$.

4.26. Let X be a normed space, Y be a Banach space, and let $F \in BL(X, Y)$. If $F' \in CL(Y', X')$, then $F \in CL(X, Y)$. (Compare Theorem 4.21.)

4.27. Let H and G be Hilbert spaces, and let $A : H \to G$ and $B : G \to H$ be linear maps such that $\langle A(x), y \rangle_G = \langle x, B(y) \rangle_H$ for all $x \in H$ and all $y \in G$. Then $A \in BL(H, G)$, $B \in BL(G, H)$, and $B = A^*$.

4.28. Let H and G be Hilbert spaces, and let $A \in BL(H, G)$, $B \in BL(G, H)$. Then $AB = 0$ if and only if $R(A^*) \perp R(B)$.

4.29. Let H be a Hilbert space, G be a closed subspace of H, and let $A \in BL(H)$. Then $A(G) \subset G$ if and only if $A^*(G^\perp) \subset G^\perp$.

4.30. **(Bounded inverse theorem)** Let H and G be Hilbert spaces, and suppose $A \in BL(H, G)$. If A is one-one and onto, then $A^{-1} \in BL(G, H)$. (Note: This result can be proved without using the Baire theorem, the Zabreiko theorem and the closed graph theorem.) The open mapping theorem and the closed graph theorem can be deduced. (Compare Exercise 3.27.)

4.31. Let H and G be Hilbert spaces.

(i) Let $A \in BL(H, G)$. Then $A \in CL(H, G)$ if and only if $A^*A \in CL(H)$.
(ii) If $A \in CL(H, G)$, then $A^* \in CL(G, H)$. (Do not use Theorem 4.21.)
(iii) Let H and G be separable. If $A \in BL(H, G)$ is a Hilbert-Schmidt map, then A^* is a Hilbert–Schmidt map. In fact, if $\{u_1, u_2, \ldots\}$ is a countable orthonormal basis for H, and $\{v_1, v_2, \ldots\}$ is a countable orthonormal basis for G, then $\sum_n \|A(u_n)\|^2 = \sum_m \|A^*(v_m)\|^2$.

4.32. A self-adjoint operator on a Hilbert space H is also known as a **hermitian operator**. Also, $A \in BL(H)$ is called a **skew-hermitian operator** if $A^* = -A$. Let $A \in BL(H)$. There are unique B and C in $BL(H)$ such that B is hermitian, C is skew-hermitian, and $A = B + C$. Further, A is normal if and only if $BC = CB$, A is hermitian if and only if $C = 0$, A is skew-hermitian if and only if $B = 0$, and A is unitary if and only if $BC = CB$ as well as $B^2 - C^2 = I$. (Compare Proposition 4.29.)

4.33. Let H be a Hilbert space, and let $A \in BL(H)$. If $A^*A - AA^* \geq 0$, then A is called **hyponormal**. A is hyponormal if and only if $\|A(x)\| \geq \|A^*(x)\|$ for all $x \in H$, and A is normal if and only if A and A^* are hyponormal. The right shift operator on ℓ^2 is hyponormal, but not normal.

4.34. Let H denote the Hilbert space of all 'doubly infinite' square-summable scalar sequences $x := (\ldots, x(-2), x(-1), x(0), x(1), x(2), \ldots)$ with the inner product $\langle x, y \rangle := \sum_{j=-\infty}^{\infty} x(j)\overline{y(j)}$, $x, y \in H$. For x in H, let $A(x)(j) := x(j-1)$ for all $j \in \mathbb{Z}$. Then the **right shift operator** A is a unitary operator on H. (Compare the right shift operator on ℓ^2.)

4.35. **(Fourier matrix)** Let $n \in \mathbb{N}$, and $\omega_n := e^{2\pi i/n}$. Let $A \in BL(\mathbb{C}^n)$ be defined by the matrix $M_n := [k_{p,j}]$, where $k_{p,j} := \omega_n^{(p-1)(j-1)}/\sqrt{n}$ for $p, j = 1, \ldots, n$. Then A is unitary. Also, M_n is a symmetric matrix. (Note: This matrix is crucial in the development of the **fast Fourier transform**.)

4.36. Let (A_n) be a sequence of self-adjoint operators on a Hilbert space H.

(i) Suppose there is $\alpha \in \mathbb{R}$ such that $A_n \leq A_{n+1} \leq \alpha I$ for all $n \in \mathbb{N}$. Then there is a unique self-adjoint operator A on H such that $A_n \leq A$ for all $n \in \mathbb{N}$, and $A \leq \widetilde{A}$ whenever $\widetilde{A}$ is a self-adjoint operator on H satisfying $A_n \leq \widetilde{A}$ for all $n \in \mathbb{N}$. In fact, $A_n(x) \to A(x)$ for every $x \in H$.

(ii) Suppose there is $\beta \in \mathbb{R}$ such that $\beta I \leq A_{n+1} \leq A_n$ for all $n \in \mathbb{N}$. Then there is a unique self-adjoint operator A on H such that $A \leq A_n$ for all $n \in \mathbb{N}$, and $\widetilde{A} \leq A$ whenever $\widetilde{A}$ is a self-adjoint operator on H satisfying $\widetilde{A} \leq A_n$ for all $n \in \mathbb{N}$. In fact, $A_n(x) \to A(x)$ for every $x \in H$.

4.37. Let A be a positive operator on a Hilbert space H, and let $G := \{x \in H : \langle A(x), x\rangle = 0\}$. Then G is a closed subspace of H. For $x + G,\ y + G \in H/G$, let $\langle\langle x + G, y + G\rangle\rangle := \langle A(x), y\rangle$. Then $\langle\langle \cdot, \cdot\rangle\rangle$ is an inner product on H/G. In particular, $|\langle A(x), y\rangle|^2 \leq \langle A(x), x\rangle\langle A(y), y\rangle$ for all $x, y \in H$. (Compare Exercise 2.13 and Proposition 4.31.)

4.38. Let H be a Hilbert space, and let (P_n) be a sequence of orthogonal projection operators on H such that $P_n P_m = 0$ if $n \neq m$. Then $\sum_n P_n(x)$ is summable in H for every $x \in H$. For $x \in H$, let $P(x)$ denote this sum. Then P is an orthogonal projection operator on H, and $R(P)$ is the closure of the linear span of $\bigcup_{n=1}^{\infty} R(P_n)$, and $Z(P) = \bigcap_{n=1}^{\infty} Z(P_n)$.

4.39. Let $X := c_{00}$, or $X := \ell^2$ with the usual inner product. For $x := (x(1), x(2), \ldots)$ in X, let $A(x) := (0, x(1), x(2), \ldots)$. Then $A \in BL(X)$, and $\omega(A) = \{k \in \mathbb{K} : |k| < 1\}$.

4.40. **(Generalized polarization identity)** Let $(X, \langle\cdot\, , \cdot\rangle)$ be an inner product space over $\mathbb{C}$, and let $A : X \to X$ be a linear map. Then for $x, y \in X$,

$$\begin{aligned} 4\langle A(x), y\rangle &= \langle A(x+y), x+y\rangle - \langle A(x-y), x-y\rangle \\ &\quad + i\langle A(x+iy), x+iy\rangle - i\langle A(x-iy), x-iy\rangle. \end{aligned}$$

Consequently, if H is a nonzero Hilbert space over $\mathbb{C}$, and $A \in BL(H)$ satisfies $\omega(A) \subset \mathbb{R}$, then A is self-adjoint. (Compare Exercise 2.14.)

Chapter 5
Spectral Theory

To a bounded operator A on normed space X over $\mathbb{K}$, we associate a subset of $\mathbb{K}$, known as the spectrum of A. It is intimately related to the invertibility of a specific linear combination of the operator A and the identity operator. Eigenvalues and approximate eigenvalues of A form a part of the spectrum of A. Determining the spectrum of a bounded operator is one of the central problems in functional analysis. In case X is a Banach space, we show that the spectrum of a bounded operator A on X is a closed and bounded subset of $\mathbb{K}$. We explore special properties of the spectrum of a compact operator on a normed space. We find relationships between the spectrum of a bounded operator A and the spectra of the transpose A' and the adjoint A^*. They yield particularly interesting results when the operator A is 'well behaved' with respect to the adjoint operation. In the last section of this chapter, we show how a compact self-adjoint operator can be represented in terms of its eigenvalues and eigenvectors. This is used in obtaining explicit solutions of operator equations.

5.1 Spectrum of a Bounded Operator

Let X be a normed space, $A \in BL(X)$, and let I denote the identity operator on X. We shall associate with A certain subsets of $\mathbb{K}$ that naturally arise while solving the operator equation $A(x) - k\,x = y$, where $y \in X$ and $k \in \mathbb{K}$ are given, and $x \in X$ is to be found. Let us fix $k \in \mathbb{K}$. We want to investigate whether the above equation has a unique solution for every 'free term' $y \in X$, and whether this unique solution depends continuously on y. It is easy to see that there exists a solution of this equation for every $y \in X$ if and only if $R(A - kI) = X$, that is, the operator $A - kI$ is onto. Also, when there is a solution for a given $y \in X$, it is unique if and only if the only solution of the 'homogeneous' equation $A(x) - k\,x = 0$ is the zero solution $x = 0$, that is, the operator $A - kI$ is one-one. Further, when there is a unique solution for every $y \in X$, it depends continuously on y if and only if the linear map $(A - kI)^{-1} : X \to X$ is continuous. Keeping these facts in mind, we now make the following definition.

© Springer Science+Business Media Singapore 2016, corrected publication 2023

B.V. Limaye, *Linear Functional Analysis for Scientists and Engineers*,
https://doi.org/10.1007/978-981-10-0972-3_5

A bounded operator A on a normed space X is called **invertible** if A is one-one and onto, and if the inverse map A^{-1} is in $BL(X)$, that is, there is $B \in BL(X)$ such that $AB = I = BA$. We give a necessary and sufficient condition for the invertibility of A. Let us recall from Sect. 3.1 that A is bounded below if there is $\beta > 0$ such that $\beta\|x\| \leq \|A(x)\|$ for all $x \in X$.

Proposition 5.1 *Let X be a normed space, and let $A \in BL(X)$. Then A is invertible if and only if A is bounded below and onto.*

Proof Suppose A is invertible. Then $\|x\| = \|A^{-1}(A(x))\| \leq \|A^{-1}\|\|A(x)\|$ for all $x \in X$ (Theorem 3.10(i)). Hence $\beta\|x\| \leq \|A(x)\|$ for all $x \in X$, where $\beta := 1/\|A^{-1}\|$. Clearly, A maps X onto X. Conversely, suppose A is bounded below and onto. Then A is one-one. Also, the inverse map $A^{-1} : X \to X$ is linear. Further, there is $\beta > 0$ such that $\beta\|A^{-1}(y)\| \leq \|A(A^{-1}(y))\| = \|y\|$ for all $y \in X$. Thus $A^{-1} \in BL(X)$ and $\|A^{-1}\| \leq 1/\beta$. □

We may compare the above result with Proposition 3.5 which says that a linear map $A : X \to X$ is a homeomorphism if and only if A is bounded as well as bounded below. Also, see our comment after Proposition 3.41.

Let $A \in BL(X)$. The set

$$\sigma(A) := \{\lambda \in \mathbb{K} : A - \lambda I \text{ is not invertible}\}$$

is called the **spectrum** of A, and the set

$$\sigma_e(A) := \{\lambda \in \mathbb{K} : A - \lambda I \text{ is not one-one}\}$$

is called the **eigenspectrum** of A. Further, the set

$$\sigma_a(A) := \{\lambda \in \mathbb{K} : A - \lambda I \text{ is not bounded below}\}$$

is called the **approximate eigenspectrum** of A. If $X \neq \{0\}$, then $\lambda \in \sigma_e(A)$ if and only if there is $x \in X$ such that $A(x) = \lambda x$ and $\|x\| = 1$, and $\lambda \in \sigma_a(A)$ if and only if there is a sequence (x_n) in X such that $A(x_n) - \lambda x_n \to 0$ and $\|x_n\| = 1$ for all $n \in \mathbb{N}$. Clearly, $\sigma_e(A) \subset \sigma_a(A)$ and $\sigma_a(A) \subset \sigma(A)$. We shall study the sets $\sigma_e(A)$, $\sigma_a(A)$, $\sigma(A)$ in that order.

Eigenspectrum

A scalar $\lambda \in \sigma_e(A)$ is called an **eigenvalue** of A. A nonzero $x \in X$ satisfying $A(x) = \lambda x$ is called an **eigenvector** of A corresponding to the eigenvalue λ of A, and the subspace $E_\lambda := \{x \in X : A(x) = \lambda x\}$ is called the corresponding **eigenspace** of A. Let $m \in \mathbb{N}$. Suppose $\lambda_1, \ldots, \lambda_m$ are distinct eigenvalues of A, and $x_1, \ldots, x_m$ are corresponding eigenvectors. By mathematical induction, it follows that $\{x_1, \ldots, x_m\}$ is a linearly independent subset of X.

Although the eigenspectrum of A seems to be more tractable than the rest of the spectrum of A, it is by no means easy to find eigenvalues of A. This is partly because both sides of the **eigenequation** $A(x) = \lambda x$, $x \neq 0$, involve unknown elements, unlike the operator equation $A(x) = y$, where $y \in X$ is known and $x \in X$ is to be found. If, however, $\lambda \in \mathbb{K}$ is known to be an eigenvalue of A, then finding the corresponding eigenvectors is reduced to solving the 'homogeneous' operator equation $A(x) - \lambda x = 0$. Similarly, if a nonzero element x of X is known to be an eigenvector of A, then we can find the corresponding eigenvalue λ by considering a linear functional f on X such that $f(x) \neq 0$, and so $\lambda = f(A(x))/f(x)$. In particular, if X is an inner product space, then $\lambda = \langle A(x), x\rangle / \langle x, x\rangle$.

Let λ be an eigenvalue of A. If x is a corresponding eigenvector of A, then $|\lambda|\|x\| = \|\lambda x\| = \|A(x)\| \leq \|A\|\|x\|$. Hence $|\lambda| \leq \|A\|$.

Proposition 5.2 *Let X be a normed space, and $A \in BL(X)$.*

(i) *If X is finite dimensional, then $\sigma(A) = \sigma_e(A)$.*
(ii) *If X is infinite dimensional and A is of finite rank, then $0 \in \sigma_e(A)$ and $\sigma(A) = \sigma_e(A)$, and the eigenspace corresponding to a nonzero eigenvalue of A is finite dimensional.*

Proof Since $\sigma_e(A) \subset \sigma(A)$, we need only show that $\sigma(A) \subset \sigma_e(A)$.

(i) Suppose $\dim X := n < \infty$. Let $k \in \mathbb{K} \setminus \sigma_e(A)$. By the rank-nullity theorem stated in Sect. 1.2, $\dim Z(A - kI) + \dim R(A - kI) = n$. Since $A - kI$ is one-one, we see that $\dim Z(A - kI) = 0$. Hence $R(A - kI)$ is an n dimensional subspace of X, that is, $R(A - kI) = X$. Thus $A - kI$ is one-one and onto. By Proposition 3.6(i), $(A - kI)^{-1}$ is continuous. Hence $k \notin \sigma(A)$.

(ii) Suppose X is infinite dimensional, and A is of finite rank. Let $\{x_1, x_2, \ldots\}$ be a denumerable linearly independent subset of X. If A is one-one, then $\{A(x_1), A(x_2), \ldots\}$ would be an infinite linearly independent subset of the finite dimensional subspace $R(A)$, which is not possible. Hence A is not one-one, that is, $0 \in \sigma_e(A)$.

Let $k \in \mathbb{K} \setminus \sigma_e(A)$. Then $k \neq 0$. We show that $A - kI$ is bounded below and onto. Assume for a moment that $A - kI$ is not bounded below. Then there is a sequence (x_n) in X such that $A(x_n) - kx_n \to 0$ and $\|x_n\| = 1$ for all $n \in \mathbb{N}$. Now $(A(x_n))$ is a bounded sequence in the finite dimensional normed space $R(A)$. By Theorem 2.10, there is a subsequence $(A(x_{n_j}))$ and there is $y \in X$ such that $A(x_{n_j}) \to y$. Then $kx_{n_j} \to y$ as well. Now $\|y\| = \lim_{j\to\infty} \|kx_{n_j}\| = |k| \neq 0$. On the other hand,

$$A(y) = A\Big(\lim_{j\to\infty} kx_{n_j}\Big) = k \lim_{j\to\infty} A(x_{n_j}) = ky,$$

and so $y \in Z(A - kI)$, that is, $y = 0$. This contradiction shows that $A - kI$ is bounded below. Next, let $B : R(A) \to R(A)$ denote the restriction of $A - kI$ to $R(A)$. Since $A - kI$ is one-one, so is B. Also, since $R(A)$ is finite dimensional, B maps $R(A)$ onto $R(A)$ by (i) above. Consider $y \in X$. Then $A(y) \in R(A)$, and there is $u \in R(A)$ with $B(u) = A(y)$, that is, $A(u) - ku = A(y)$. Noting that

$A(u-y)=ku$, we let $x:=(u-y)/k$, and obtain $A(x)=u=kx+y$, that is, $(A-kI)(x)=y$. Thus $A-kI$ is onto. By Proposition 5.1, $A-kI$ is invertible, that is, $k\notin\sigma(A)$.

Let $\lambda\in\sigma_e(A)$ and $\lambda\neq 0$. If x is an eigenvector of A corresponding to λ, then $x=A(x)/\lambda$, which is in $R(A)$. Hence the eigenspace of A corresponding to λ is a subspace of $R(A)$, and so it is finite dimensional. □

Remarks 5.3 (i) Let $\dim X:=n<\infty$, and let $x_1,\ldots,x_n$ be a basis for X. As we have seen in Example 1.3(i), there are functionals $f_1,\ldots,f_n$ on X such that $f_i(x_j)=\delta_{i,j}$ for all $i,j=1,\ldots,n$ and $x=f_1(x)x_1+\cdots+f_n(x)x_n$ for every $x\in X$. For $x\in X$, let $u:=(f_1(x),\ldots,f_n(x))\in\mathbb{K}^n$, and for $u=(u(1),\ldots,u(n))\in\mathbb{K}^n$, let $x:=u(1)x_1+\cdots+u(n)x_n$. Consider an operator $A:X\to X$. Let $k_{i,j}:=f_i(A(x_j))$ for $i,j=1,\ldots,n$, and let $M:=[k_{i,j}]$. Then the $n\times n$ matrix M defines the operator A. For $(k_1,\ldots,k_n)\in\mathbb{K}^n$, let $[k_1,\ldots,k_n]^t$ denote the $n\times 1$ matrix with entries $k_1,\ldots,k_n$. Then for $x\in X$, $u\in\mathbb{K}^n$ and $\lambda\in\mathbb{K}$, $A(x)=\lambda x$ if and only if $Mu^t=\lambda u^t$. Hence the problem of finding the eigenvalues and the corresponding eigenvectors of A is reduced to a matrix eigenvalue problem.

(ii) Let X be infinite dimensional, and let $A\in BL(X)$ be of finite rank. Then there are $x_1,\ldots,x_n$ in X and linear functionals $g_1,\ldots,g_n$ on X such that $A(x)=g_1(x)x_1+\cdots+g_n(x)x_n$ for all $x\in X$. (Compare Exercise 4.20.) The problem of finding the nonzero eigenvalues and the corresponding eigenvectors of A can be reduced to the eigenvalue problem for the $n\times n$ matrix $M:=[g_i(x_j)]$. (See Exercise 5.5 and [21, Corollary 4.2].) ◊

Examples 5.4 (i) Let $n\in\mathbb{N}$, $X:=\mathbb{K}^n$ with a given norm, and let M be an $n\times n$ matrix with entries in $\mathbb{K}$. For $x:=(x(1),\ldots,x(n))\in\mathbb{K}^n$, define $A(x)$ to be the matrix multiplication of M and the column vector $[x(1),\ldots,x(n)]^t$. Then $A:\mathbb{K}^n\to\mathbb{K}^n$ is linear, and it is continuous by Proposition 3.6. For $k\in\mathbb{K}$, the operator $A-kI$ is invertible if and only if the matrix $M-kI$ is nonsingular, that is, $\det(M-kI)\neq 0$. Thus $\lambda\in\sigma(A)$ if and only if λ is a root of the characteristic polynomial $p(t):=\det(M-tI)$. Since the characteristic polynomial of M is of degree n, there are at most n distinct eigenvalues of A. If $\mathbb{K}:=\mathbb{C}$, then the fundamental theorem of algebra shows that there is at least one eigenvalue of A. On the other hand, if $\mathbb{K}:=\mathbb{R}$, then A may not have an eigenvalue. A simple example is provided by the 2×2 matrix $M:=[k_{i,j}]$, where $k_{1,1}=0=k_{2,2}$ and $k_{1,2}=1=-k_{2,1}$. The characteristic polynomial of this matrix is t^2+1.

If M is a triangular matrix, and $\lambda_1,\ldots,\lambda_n$ are its diagonal entries, then

$$\det(M-tI)=(\lambda_1-t)\cdots(\lambda_n-t),$$

where det denotes the determinant. Hence $\sigma(A)=\sigma_e(A)=\{\lambda_1,\ldots,\lambda_n\}$. If M is not triangular, then the problem of finding the eigenvalues of A is difficult. Algorithms are developed to reduce M to an 'approximately triangular' matrix. The most notable among these is the Basic QR algorithm. The main idea is to write $M=QR$, where Q is a unitary matrix and R is an upper triangular matrix, reverse the order of

Q and R to obtain $M_1 := RQ$, and repeat this process. (Compare Exercise 2.19. See [29, pp. 356–358].)

If λ is an eigenvalue of A, then $|\lambda| \le \|A\|$, where $\|\cdot\|$ is the operator norm on $BL(\mathbb{K}^n)$ induced by the given norm on $\mathbb{K}^n$. Various choices of the norms on $\mathbb{K}^n$ yield upper bounds for the eigenspectrum of A. Let $M := [k_{i,j}]$, $i, j = 1, \ldots, n$. Let α_1 denote the maximum of the column sums of the matrix $|M|$, β_1 denote the maximum of the row sums of the matrix $|M|$, and let $\gamma_{2,2} := \left(\sum_{i=1}^{n}\sum_{j=1}^{n}|k_{i,j}|^2\right)^{1/2}$. Then $|\lambda| \le \min\{\alpha_1, \beta_1, \sqrt{\alpha_1\beta_1}, \gamma_{2,2}\}$ for every eigenvalue λ of A. Exercise 5.4 gives a well-known 'localization' result for eigenvalues of A due to Gershgorin.

(ii) Let M denote the $n \times n$ matrix with all entries equal to 1, and let A denote the operator on $\mathbb{K}^n$ defined by M. We observe that $e_0 := (1, \ldots, 1)$ is an eigenvector of A since $A(e_0) = ne_0$, and so the corresponding eigenvalue of A is n. Also, A has rank 1, and hence the nullity of A is $n - 1$. The eigenspace of A corresponding to the eigenvalue 0 is the $n - 1$ dimensional subspace $\{(x(1), \ldots, x(n)) \in \mathbb{K}^n : x(1) + \cdots + x(n) = 0\}$ of $\mathbb{K}^n$. It follows that $\sigma(A) = \{0, n\}$. We observe that the Helmert basis described in Exercise 2.18 constitutes an orthonormal basis for $\mathbb{K}^n$ consisting of eigenvectors of A.

(iii) Let $X := L^2([0, 1])$, and for $x \in X$, define

$$A(x)(s) := \int_0^1 \min\{s, t\}x(t)dm(t), \quad s \in [0, 1].$$

Then $A \in BL(X)$. Let $x_1(t) := \sin \pi t/2$ for $t \in [0, 1]$. Clearly, $x_1 \in X$. Suppose we somehow know that x_1 is an eigenvector of A. Let us find the corresponding eigenvalue of A. Consider $f \in X'$ defined by

$$f(x) := \int_0^1 x(s)dm(s), \quad x \in X.$$

It is easy to see that $f(x_1) = 2/\pi \neq 0$. Also, since

$$A(x_1)(s) = \int_0^s t \sin\frac{\pi t}{2}dt + s\int_s^1 \sin\frac{\pi t}{2}dt, \quad s \in [0, 1],$$

we find that $f(A(x_1)) = 8/\pi^3$. Hence $\lambda_1 := f(A(x_1))/f(x_1) = 4/\pi^2$ is the eigenvalue of A corresponding to the eigenvector x_1. In general, let $n \in \mathbb{N}$, and $x_n(t) := \sin(2n-1)\pi t/2$, $t \in [0, 1]$. If x_n is known to be an eigenvector of A, then we can similarly show that $\lambda_n := 4/(2n-1)^2\pi^2$ is the eigenvalue of A corresponding to the eigenvector x_n. (Compare Exercise 5.20.)

(iv) Let $X := C([a, b])$. Consider $x_1, \ldots, x_n, y_1, \ldots, y_n$ in X. Let $k(s, t) := x_1(s)y_1(t) + \cdots + x_n(s)y_n(t)$ for $s, t \in [a, b]$, and let $A \in BL(X)$ denote the Fredholm integral operator having kernel $k(\cdot, \cdot)$. Then for $x \in X$,

$$A(x)(s) = \int_a^b k(s,t)x(t)dt = \sum_{i=1}^n \left(\int_a^b y_i(t)x(t)dt \right) x_i(s), \quad s \in [a,b].$$

Thus A is a bounded operator of finite rank, $A(x) = g_1(x)x_1 + \cdots + g_n(x)x_n$, where $g_i(x) := \int_a^b y_i(t)x(t)dt$, $i = 1, \ldots, n$, for $x \in X$. As we have mentioned in Remark 5.3(ii), the problem of finding nonzero eigenvalues and corresponding eigenvectors of A is reduced to the eigenvalue problem for the $n \times n$ matrix $[g_i(x_j)]$, where $g_i(x_j) = \int_a^b y_i(t)x_j(t)dt$ for $i, j = 1, \ldots, n$. ♢

Approximate Eigenspectrum

A scalar $\lambda \in \sigma_a(A)$ is called an **approximate eigenvalue** of A. Suppose (λ_n) is a sequence of eigenvalues of A and $\lambda_n \to \lambda$ in $\mathbb{K}$. Then λ may not be an eigenvalue of A. (See Example 5.6(i).) However, λ is certainly an approximate eigenvalue of A. This follows by considering $x_n \in X$ such that $A(x_n) = \lambda_n x_n$ and $\|x_n\| = 1$ for $n \in \mathbb{N}$, and noting that

$$A(x_n) - \lambda x_n = A(x_n) - \lambda_n x_n + (\lambda_n - \lambda)x_n = (\lambda_n - \lambda)x_n \to 0.$$

On the other hand, not every approximate eigenvalue of A is a limit of a sequence of eigenvalues of A. (See Example 5.12(i).)

Proposition 5.5 *Let X be a normed space and let $A \in BL(X)$. Then the approximate eigenspectrum $\sigma_a(A)$ of A is a bounded and closed subset of $\mathbb{K}$.*

Proof Let $\lambda \in \sigma_a(A)$. Then there is a sequence (x_n) in X such that $\|x_n\| = 1$ for all $n \in \mathbb{N}$ and $A(x_n) - \lambda x_n \to 0$. Now for all $n \in \mathbb{N}$,

$$|\lambda| = \|\lambda x_n\| \leq \|\lambda x_n - A(x_n)\| + \|A(x_n)\| \leq \|A(x_n) - \lambda x_n\| + \|A\|.$$

Let $n \to \infty$ to obtain $|\lambda| \leq \|A\|$. Hence $\sigma_a(A)$ is a bounded subset of $\mathbb{K}$. (See Exercise 5.8 for a possibly sharper bound.)

To show that $\sigma_a(A)$ is a closed subset of $\mathbb{K}$, consider a sequence (λ_n) in $\sigma_a(A)$ such that $\lambda_n \to \lambda$ in $\mathbb{K}$. Assume for a moment that $\lambda \notin \sigma_a(A)$. Then there is $\beta > 0$ such that $\|A(x) - \lambda x\| \geq \beta\|x\|$ for all $x \in X$. Since $\lambda_n \to \lambda$, there is $n_0 \in \mathbb{N}$ such that $|\lambda - \lambda_{n_0}| < \beta$. Then for all $x \in X$,

$$\|A(x) - \lambda_{n_0} x\| \geq \|A(x) - \lambda x\| - |\lambda - \lambda_{n_0}|\,\|x\| \geq (\beta - |\lambda - \lambda_{n_0}|)\|x\|,$$

and so, $\lambda_{n_0} \notin \sigma_a(A)$. This contradiction shows that $\sigma_a(A)$ is closed. □

Examples 5.6 (i) Let $X := \ell^p$ with $p \in \{1, 2, \infty\}$. Consider $A \in BL(\ell^p)$ defined by

$$A(x) := \left(x(1), \frac{x(2)}{2}, \frac{x(3)}{3}, \cdots \right) \quad \text{for } x := (x(1), x(2), \ldots) \in X.$$

Then $A \in BL(X)$ and $\|A\| = 1$. Since $A(e_j) = e_j/j$, we see that $1/j$ is an eigenvalue of A, and e_j is a corresponding eigenvector for each $j \in \mathbb{N}$. Since $A(x) = 0$ if and only if $x = 0$, we see that 0 is not an eigenvalue of A. However, since $\|A(e_j)\|_p \to 0$, and $\|e_j\|_p = 1$ for all $j \in \mathbb{N}$, we see that A is not bounded below, that is, 0 is an approximate eigenvalue of A.

Let $E := \{0, 1, 1/2, 1/3, \ldots\}$, and let $k \in \mathbb{K} \setminus E$. For $y := (y(1), y(2), \ldots)$ in X, define $B(y) := \big(y(1)/(1-k), y(2)/(\frac{1}{2}-k), \ldots\big)$. Since E is a closed set and $k \notin E$, there is $\delta > 0$ such that $|(1/j) - k| \geq \delta$ for all $j \in \mathbb{N}$. Then

$$|B(y)(j)| = \frac{|y(j)|}{|(1/j) - k|} \leq \frac{1}{\delta}|y(j)| \quad \text{for all } j \in \mathbb{N}.$$

It follows that $B(y) \in X$ for all $y \in X$. Also, $\|B\| \leq (1/\delta)$, and $(A - kI)B = I = B(A - kI)$. Hence $A - kI$ is invertible. Thus

$$\sigma_e(A) = \{1, 1/2, 1/3, \ldots\} \quad \text{and} \quad \sigma_a(A) = \{0, 1, 1/2, 1/3, \ldots\} = \sigma(A).$$

(ii) Let $X := C([a, b])$ with the sup norm. Fix $x_0 \in X$, and consider the **multiplication operator** defined by $A(x) := x_0 x$ for $x \in X$. Then $A \in BL(X)$ and $\|A\| = \|x_0\|_\infty$. Let $E := \{x_0(t) \in \mathbb{K} : t \in [a, b]\}$. We show that $\sigma_a(A) = E = \sigma(A)$.

Let $t_0 \in [a, b]$, and $\lambda := x_0(t_0)$. For $n \in \mathbb{N}$, define $x_n(t) := 1 - n|t - t_0|$ if $t \in [a, b]$ with $|t - t_0| \leq 1/n$, and $x_n(t) := 0$ otherwise. Then $x_n \in X$ and $\|x_n\|_\infty = 1$ for all $n \in \mathbb{N}$. We show that $\|A(x_n) - \lambda x_n\|_\infty \to 0$. Let $\epsilon > 0$. Since x_0 is continuous at t_0, there is $\delta > 0$ such that $|x_0(t) - x_0(t_0)| < \epsilon$ for all $t \in [a, b]$ with $|t - t_0| < \delta$. Choose n_0 such that $n_0 > 1/\delta$. Then $x_n(t) = 0$ for all $n \geq n_0$ and $t \in [a, b]$ with $|t - t_0| \geq \delta$. Hence for all $n \geq n_0$ and $t \in [a, b]$,

$$|A(x_n)(t) - \lambda x_n(t)| = |x_0(t) - x_0(t_0)|\, |x_n(t)| < \epsilon,$$

and so $\|A(x_n) - \lambda x_n\|_\infty \leq \epsilon$ for all $n \geq n_0$. This shows that $E \subset \sigma_a(A)$.

Next, suppose $k \notin E$, that is, $k \neq x_0(t)$ for any $t \in [a, b]$. Then the function $1/(x_0 - k)$ belongs to X. For $y \in X$, define $B(y) := y/(x_0 - k)$. Then $B \in BL(X)$ and $(A - kI)B = I = B(A - kI)$. This shows that $\sigma(A) \subset E$. Since $\sigma_a(A) \subset \sigma(A)$, we obtain $\sigma_a(A) = E = \sigma(A)$.

In Exercise 5.6, we describe $\sigma_e(A)$, and in Exercise 5.7, we treat the multiplication operator defined on $L^2([a, b])$.

Spectrum of a Bounded Operator on a Banach Space

Let X be a normed space, and let $A \in BL(X)$. A scalar in $\sigma(A)$ is called a **spectral value** of A. In general, it is difficult to find all spectral values of A. If X is a Banach space, some properties of $\sigma(A)$ turn out to be helpful in determining it. First we improve Proposition 5.1 as follows.

Proposition 5.7 *Let X be a Banach space, and let $A \in BL(X)$. Then the following conditions are equivalent.*

(i) *A is invertible.*
(ii) *A is bounded below and the range $R(A)$ of A is dense in X.*
(iii) *A is one-one and onto.*

Proof (i) $\Longrightarrow$ (ii) by Proposition 5.1.

(ii) $\Longrightarrow$ (iii): Since A is bounded below, there is $\beta > 0$ such that $\beta\|x\| \le \|A(x)\|$ for all $x \in X$. It is clear that A is one-one. To show that A is onto, consider $y \in X$. Since $R(A)$ is dense in X, there is a sequence (x_n) in X such that $A(x_n) \to y$ in X. Then $(A(x_n))$ is a Cauchy sequence in X, and

$$\beta\|x_n - x_m\| \le \|A(x_n - x_m)\| = \|A(x_n) - A(x_m)\|$$

for all $n, m \in \mathbb{N}$. Hence (x_n) is also a Cauchy sequence in X. Since X is a Banach space, there is $x \in X$ such that $x_n \to x$ in X. Then $A(x_n) \to A(x)$ by the continuity of A. Thus $y = A(x) \in R(A)$. Hence A is onto.

(iii) $\Longrightarrow$ (i): Since X is a Banach space, the inverse operator $A^{-1} : X \to X$ is continuous by the bounded inverse theorem (Theorem 3.35). □

Let $k \in \mathbb{K}$. If $|k| < 1$, then the following geometric series expansion holds:

$$(1-k)^{-1} = \sum_{n=0}^{\infty} k^n, \quad \text{and} \quad |(1-k)^{-1}| \le \frac{1}{1-|k|}.$$

We shall now obtain an analogue of this for a bounded operator on a Banach space and derive some important properties of its spectrum.

Lemma 5.8 *Let X be a Banach space, and let $A \in BL(X)$. If $\|A\| < 1$, then $I - A$ is invertible,*

$$(I - A)^{-1} = \sum_{n=0}^{\infty} A^n \quad \textit{and} \quad \|(I-A)^{-1}\| \le \frac{1}{1-\|A\|}.$$

Proof Suppose $\|A\| < 1$. Consider the series $\sum_{n=0}^{\infty} A^n$ with terms in the Banach space $BL(X)$ (Proposition 3.17(i)). It is absolutely summable, and

$$\sum_{n=0}^{\infty} \|A^n\| \le \sum_{n=0}^{\infty} \|A\|^n = \frac{1}{1-\|A\|}.$$

By Theorem 2.23, the series is summable in $BL(X)$. Let $B \in BL(X)$ denote its sum. For $m \in \mathbb{N}$, define $B_m := \sum_{n=0}^{m} A^n$. Then $B_m \to B$ in $BL(X)$, and

$$B_m(I - A) = I - A^{m+1} = (I - A)B_m \quad \text{for each } m \in \mathbb{N}.$$

Since $\|A^{m+1}\| \le \|A\|^{m+1} \to 0$, we obtain $B(I - A) = I = (I - A)B$. Hence $I - A$ is invertible, and B is its inverse:

$$(I - A)^{-1} = B = \sum_{n=0}^{\infty} A^n.$$

As a result, $\|(I - A)^{-1}\| \le \sum_{n=0}^{\infty} \|A^n\| \le 1/(1 - \|A\|)$. □

Corollary 5.9 *Let X be a Banach space, and $A \in BL(X)$. If $\lambda \in \sigma(A)$, then $|\lambda| \le \|A\|$. In fact, $|\lambda| \le \inf\{\|A^n\|^{1/n} : n \in \mathbb{N}\}$ for every $\lambda \in \sigma(A)$. Consequently, $\sigma(A)$ is a bounded subset of $\mathbb{K}$ for every $A \in BL(X)$.*

Proof Suppose $k \in \mathbb{K}$ with $|k| > \|A\|$. Let $\widetilde{A} := A/k \in BL(X)$. Then $\|\widetilde{A}\| < 1$, and so $I - \widetilde{A}$ is invertible by Lemma 5.8. Hence $A - kI = -k(I - \widetilde{A})$ is invertible, that is, $k \notin \sigma(A)$.

Let $\lambda \in \sigma(A)$. Then $|\lambda| \le \|A\|$. Further, let $n \ge 2$. We claim that $\lambda^n \in \sigma(A^n)$. Assume for a moment that $A^n - \lambda^n I$ is invertible, and let $B \in BL(X)$ denote its inverse. Then $(A^n - \lambda^n I)B = I = B(A^n - \lambda^n I)$. Now $A^n - \lambda^n I = (A - \lambda I)C$, where $C := A^{n-1} + \lambda A^{n-2} + \cdots + \lambda^{n-2}A + \lambda^{n-1}I$. Thus $(A - \lambda I)CB = I = BC(A - \lambda I)$. But $(A^n - \lambda^n I)A = A(A^n - \lambda^n I)$, and so $AB = BA$. As a result, $CB = BC$, and so $(A - \lambda I)BC = I = BC(A - \lambda I)$, that is, BC is the inverse of $A - \lambda I$. This contradiction shows that $\lambda^n \in \sigma(A^n)$. Hence $|\lambda^n| \le \|A^n\|$, and so $|\lambda| \le \|A^n\|^{1/n}$. □

Theorem 5.10 *Let X be a Banach space, and let $A, B \in BL(X)$. Suppose A is invertible and $\|(A - B)A^{-1}\| < 1$. Then B is invertible, and*

$$B^{-1} = A^{-1} \sum_{n=0}^{\infty} \big((A - B)A^{-1}\big)^n.$$

Further, if $\epsilon := \|(A - B)A^{-1}\|$, then

$$\|B^{-1}\| \le \frac{1}{1 - \epsilon}\|A^{-1}\| \quad \textit{and} \quad \|B^{-1} - A^{-1}\| \le \frac{\epsilon}{1 - \epsilon}\|A^{-1}\|.$$

Proof Lemma 5.8 shows that $I - (A - B)A^{-1} = BA^{-1}$ is invertible, and

$$(BA^{-1})^{-1} = \sum_{n=0}^{\infty} \big((A - B)A^{-1}\big)^n.$$

Since A is invertible, it follows that $B = (BA^{-1})A$ is invertible, and

$$B^{-1} = A^{-1}(BA^{-1})^{-1} = A^{-1} \sum_{n=0}^{\infty} \big((A - B)A^{-1}\big)^n.$$

Again, since $\epsilon = \|(A - B)A^{-1}\| < 1$, we obtain

$$\|B^{-1}\| \le \|A^{-1}\| \sum_{n=0}^{\infty} \|(A - B)A^{-1}\|^n = \frac{1}{1-\epsilon}\|A^{-1}\|.$$

Also, $B^{-1} - A^{-1} = A^{-1}\sum_{n=1}^{\infty}\left((A - B)A^{-1}\right)^n$, and so

$$\|B^{-1} - A^{-1}\| \le \|A^{-1}\| \sum_{n=1}^{\infty} \|(A - B)A^{-1}\|^n = \frac{\epsilon}{1-\epsilon}\|A^{-1}\|.$$

This completes the proof. □

A similar result can be proved if X is a Banach space, $A, B \in BL(X)$, A is invertible and $\|A^{-1}(A - B)\| < 1$.

Corollary 5.11 *Let X be a Banach space. The set of all invertible operators on X is open in $BL(X)$, and the inversion map $A \longmapsto A^{-1}$ is continuous on the set $\mathcal{I}$ of all invertible operators on X. Consequently, $\sigma(A)$ is a closed subset of $\mathbb{K}$ for every A in $BL(X)$.*

Proof Let $A \in BL(X)$ be invertible. If $B \in BL(X)$ and $\|A - B\| < 1/\|A^{-1}\|$, then

$$\|(A - B)A^{-1}\| \le \|A - B\|\,\|A^{-1}\| < 1,$$

and so B is invertible by Theorem 5.10. Hence the open ball in $BL(X)$ about A of radius $1/\|A^{-1}\|$ is contained in the set $\mathcal{I}$ of all invertible operators. The set $\mathcal{I}$ is, therefore, open in $BL(X)$. (For example, if $A := I$ and $\|I - B\| < 1$, then B is invertible. We have already seen this in Lemma 5.8.)

Next, let $A \in \mathcal{I}$, and consider a sequence (A_n) in $\mathcal{I}$ such that $A_n \to A$ in $BL(X)$. Then

$$\epsilon_n := \|(A - A_n)A^{-1}\| \le \|A - A_n\|\,\|A^{-1}\| \to 0.$$

Hence there is $n_0 \in \mathbb{N}$ such that $\epsilon_n < 1$ for all $n \ge n_0$. As a consequence,

$$\|A_n^{-1} - A^{-1}\| \le \frac{\epsilon_n}{1-\epsilon_n}\|A^{-1}\| \quad \text{for all } n \ge n_0$$

by Theorem 5.10, and so $A_n^{-1} \to A^{-1}$ in $BL(X)$. Thus the inversion map is continuous on the set $\mathcal{I}$.

Suppose $A \in BL(X)$, and let (λ_n) be a sequence in $\sigma(A)$ such that $\lambda_n \to \lambda$ in $\mathbb{K}$. Then $A - \lambda_n I \to A - \lambda I$ in $BL(X)$. Since the set of all noninvertible operators is closed in $BL(X)$, we see that $A - \lambda I$ is not invertible, that is, $\lambda \in \sigma(A)$. Thus $\sigma(A)$ is a closed subset of $\mathbb{K}$. □

Corollaries 5.9 and 5.11 show that the spectrum of a bounded operator on a Banach space is a bounded and closed subset of $\mathbb{K}$. Conversely, if E is a bounded and closed subset of $\mathbb{K}$, then there is $A \in BL(\ell^2)$ such that $\sigma(A) = E$. (See Exercise 5.12.)

Examples 5.12 (i) We give an example to illustrate how $\sigma(A)$ can be determined by making use of the fact that it is bounded and closed. Let X denote one of the sequence spaces ℓ^1, ℓ^2, ℓ^∞, c, c_0 with the usual norm. Then X is a Banach space. Consider the **right shift operator** A on X defined by $A(x) := (0, x(1), x(2), \ldots)$ for $x := (x(1), x(2), \ldots) \in X$. Clearly, $\|A\| = 1$. By Corollary 5.9, $|\lambda| \le \|A\| = 1$ for every $\lambda \in \sigma(A)$. Hence $\sigma(A) \subset \{\lambda \in \mathbb{K} : |\lambda| \le 1\}$.

Let $\lambda \in \mathbb{K}$ and $|\lambda| < 1$. Then $A - \lambda I$ is not onto. In fact, there is no $x \in \ell^p$ such that $(A - \lambda I)(x) = e_1$. This is obvious if $\lambda = 0$. If $\lambda \ne 0$ and $(A - \lambda I)(x) = e_1$ for $x \in X$, then $-\lambda x(1) = 1$, $x(1) - \lambda x(2) = 0$, $x(2) - \lambda x(3) = 0, \ldots,$ and so $x(j) = -1/\lambda^j$ for all $j \in \mathbb{N}$. In particular, $x(j) \to \infty$, which is impossible. Hence $\lambda \in \sigma(A)$. By Corollary 5.11, $\sigma(A)$ is a closed subset of $\mathbb{K}$, and so $\{\lambda \in \mathbb{K} : |\lambda| \le 1\} \subset \sigma(A)$. Thus $\sigma(A) = \{\lambda \in \mathbb{K} : |\lambda| \le 1\}$.

Let us look for eigenvalues of A. Since A is one-one, $0 \notin \sigma_e(A)$. Now consider $k \in \mathbb{K} \setminus \{0\}$. If $A(x) - kx = 0$, then $-kx(1) = 0$, $x(1) - kx(2) = 0, x(2) - kx(3) = 0, \ldots,$ and so $x(j) = 0$ for all $j \in \mathbb{N}$, that is, $x = 0$. Hence $A - kI$ is one-one. Thus $\sigma_e(A) = \emptyset$.

Next, let us find approximate eigenvalues of A. If $k \in \mathbb{K}$, then

$$\|A(x) - kx\|_p \ge \big|\|A(x)\|_p - |k|\|x\|_p\big| = |1 - |k||\,\|x\|_p \quad \text{for all } x \in X.$$

Hence if $|k| \ne 1$, then $A - kI$ is bounded below, that is, $k \notin \sigma_a(A)$. Conversely, let $\lambda \in \mathbb{K}$ and $|\lambda| = 1$. First let $p \in \{1, 2\}$, and for $n \in \mathbb{N}$, define

$$x_n := \frac{1}{n^{1/p}}(1, \overline{\lambda}, (\overline{\lambda})^2, \ldots, (\overline{\lambda})^{n-1}, 0, 0, \ldots) \in X, \quad \text{so that } \|x_n\|_p = 1.$$

Then $\|A(x_n) - \lambda x_n\|_p^p = \|(-\lambda, 0, \ldots, 0, (\overline{\lambda})^{n-1}, 0, 0, \ldots)\|_p^p/n = 2/n \to 0$. Hence $\lambda \in \sigma_a(A)$. Next, let $p := \infty$, and for $n \in \mathbb{N}$, define

$$x_n := \frac{1}{n}(1, 2\overline{\lambda}, \ldots, n(\overline{\lambda})^{n-1}, (n-1)(\overline{\lambda})^n, \ldots, 2(\overline{\lambda})^{2n-3}, (\overline{\lambda})^{2n-2}, 0, 0, \ldots) \in X.$$

Then $A(x_n) - \lambda x_n = (-\lambda, -1, -\overline{\lambda}, \ldots, -(\overline{\lambda})^{n-2}, (\overline{\lambda})^{n-1}, \ldots, (\overline{\lambda})^{2n-2}, 0, 0, \ldots)/n$, and so $\|x_n\|_\infty = 1$ for each $n \in \mathbb{N}$, but $\|A(x_n) - \lambda x_n\|_\infty \to 0$. Hence $\lambda \in \sigma_a(A)$. Thus $\sigma_a(A) = \{\lambda \in \mathbb{K} : |\lambda| = 1\}$ in all cases.

The **left shift operator** B on X is defined by $B(x) := (x(2), x(3), \ldots)$ for $x := (x(1), x(2), \ldots) \in X$. Exercise 5.14 describes $\sigma(B)$, $\sigma_e(B)$ and $\sigma_a(B)$.

(ii) We give an example to show that the spectrum of a bounded operator on an incomplete normed space may be neither bounded nor closed.

Let X denote the linear space of all doubly infinite scalar sequences $x:=(\ldots, x(-2), x(-1), x(0), x(1), x(2), \ldots)$ such that there is $j_x \in \mathbb{N}$ with $x(j) = 0$ for all $|j| \geq j_x$. Consider the norm $\|\cdot\|_p$ on X, where $p \in \{1, 2, \infty\}$. Let A denote the **right shift operator** on X defined by $A(x)(j) := x(j-1),\ j \in \mathbb{Z}$, for $x \in X$. Clearly, $A \in BL(X)$, and $\|A\| = 1$. If B denotes the **left shift operator** on X defined by $B(x)(j) := x(j+1),\ j \in \mathbb{Z}$, for $x \in X$, then $B \in BL(X)$ and $AB = I = BA$. Hence $0 \notin \sigma(A)$.

On the other hand, $A - \lambda I$ is not onto for any nonzero $\lambda \in \mathbb{K}$. In fact, if $\lambda \in \mathbb{K} \setminus \{0\}$, then there is no $x \in X$ such that $(A - \lambda I)(x) = e_0$, where $e_0 := (\ldots, 0, 0, 1, 0, 0, \ldots)$, where 1 occurs only in the 0th entry. If $(A - \lambda I)(x) = e_0$ for $x \in X$, then $x(-1) - \lambda x(0) = 1$ and $x(j-1) - \lambda x(j) = 0$ for all nonzero $j \in \mathbb{Z}$. If $x(0) = 0$, then $x(-j) = \lambda^{j-1} \neq 0$ for all $j \in \mathbb{N}$, and if $x(0) \neq 0$, then $x(j) = \lambda^{-j} x(0) \neq 0$ for all $j \in \mathbb{N}$, neither of which is possible. Hence $\sigma(A) = \mathbb{K} \setminus \{0\}$, a subset of $\mathbb{K}$ which is neither bounded nor closed. ◊

Remark 5.13 Let X be a Banach space over $\mathbb{K}$, and let $A \in BL(X)$. In Corollaries 5.9 and 5.11, we have obtained some important properties of $\sigma(A)$ without first checking whether $\sigma(A)$ is nonempty. Example 5.4(i) shows that if $\mathbb{K} := \mathbb{R}$, then $\sigma(A)$ can very well be empty. On the other hand, if $\mathbb{K} := \mathbb{C}$ and $X \neq \{0\}$, then the **Gelfand–Mazur theorem** says that $\sigma(A)$ is nonempty for every $A \in BL(X)$. Further, the **spectral radius formula** of Gelfand, says that

$$\max\{|\lambda| : \lambda \in \sigma(A)\} = \inf\{\|A^n\|^{1/n} : n \in \mathbb{N}\} = \lim_{n\to\infty} \|A^n\|^{1/n}.$$

(Compare Corollary 5.9.) These results can be proved by employing Liouville's theorem and Laurent's theorem in complex analysis along with the Hahn–Banach extension theorem (Theorem 4.4) and the uniform boundedness principle (Theorem 3.23). We refer the interested reader to [28, Theorems V.3.2 and V.3.5]. ◊

The numerical range of a bounded operator on an inner product space introduced in Sect. 4.4 is intimately related to its spectrum.

Proposition 5.14 *Let X be a nonzero inner product space. Let $A \in BL(X)$. Then $\sigma_e(A) \subset \omega(A)$ and $\sigma_a(A) \subset \overline{\omega(A)}$. Further, if X is in fact a Hilbert space, then $\sigma(A) \subset \overline{\omega(A)}$.*

Proof Let $\lambda \in \sigma_e(A)$. There is $x \in X$ with $\|x\| = 1$ such that $A(x) = \lambda x$, and so $\lambda = \langle \lambda x, x\rangle = \langle A(x), x\rangle \in \omega(A)$. Next, let $\lambda \in \sigma_a(A)$. There is a sequence (x_n) in X with $\|x_n\| = 1$ for each $n \in \mathbb{N}$ such that $A(x_n) - \lambda x_n \to 0$. But

$$|\langle A(x_n), x_n\rangle - \lambda| = |\langle A(x_n) - \lambda x_n, x_n\rangle| \leq \|A(x_n) - \lambda x_n\|$$

for all $n \in \mathbb{N}$. Thus $\langle A(x_n), x_n\rangle \to \lambda$, and hence $\lambda \in \overline{\omega(A)}$.

Finally, suppose X is in fact a Hilbert space, and let $\lambda \in \sigma(A)$. Then, by Proposition 5.7, either $A - \lambda I$ is not bounded below or $R(A - \lambda I)$ is not dense in X. If $A - \lambda I$ is not bounded below, that is, if $\lambda \in \sigma_a(A)$, then λ belongs to the closure of $\omega(A)$ as we have seen above. Next, suppose $R(A - \lambda I)$ is not dense in X. Then, by Proposition 4.22, $A' - \lambda I$ is not one-one, that is, there is $x' \in X'$ such that $A'(x') = \lambda x'$ and $\|x'\| = 1$. By the Riesz representation theorem (Theorem 4.14), there is a unique $y \in X$ such that $x'(x) = \langle x, y \rangle$ for all $x \in X$. Then $\|y\| = \|x'\| = 1$ and

$$\lambda = \lambda\langle y, y \rangle = \lambda x'(y) = A'(x')(y) = x'(A(y)) = \langle A(y), y \rangle \in \omega(A).$$

Thus we see that in both cases $\lambda \in \overline{\omega(A)}$. □

Example 5.12(ii) shows that if an inner product space is not complete, then $\sigma(A)$ may not be contained in the closure of $\omega(A)$.

An element of the numerical range of an operator is also known as a **Rayleigh quotient**. Rayleigh quotients possess the so-called minimum residual property stated in Exercise 5.16. In view of this property, they can serve as approximations of eigenvalues of an operator.

5.2 Spectrum of a Compact Operator

In Sect. 3.4, we have introduced a compact linear map as a natural and useful generalization of a bounded linear map of finite rank. In this section, we study the spectrum of a compact linear operator on an infinite dimensional normed space. We shall see that it resembles the spectrum of a bounded operator of finite rank described in Proposition 5.2(ii): 0 is a spectral value of A, and every nonzero spectral value of A is in fact an eigenvalue of A, the corresponding eigenspace being finite dimensional. We shall make use of the Riesz lemma (Lemma 2.7) several times in this section.

Let X be a normed space over $\mathbb{K}$, and let $A \in CL(X)$, the set of all compact operators on X. Recall that a linear operator $A : X \to X$ is called compact if for every bounded sequence (x_n) in X, the sequence $(A(x_n))$ has a subsequence which converges in X. We begin by exploring the eigenspectrum $\sigma_e(A)$ of a compact operator A.

Lemma 5.15 *Let X be a normed space, and let $A \in CL(X)$. Suppose (λ_n) is a sequence of eigenvalues of A, and suppose there is an eigenvector x_n corresponding to λ_n for each $n \in \mathbb{N}$ such that $\{x_1, x_2, \ldots\}$ is an infinite linearly independent subset of X. Then $\lambda_n \to 0$.*

Proof For $n \in \mathbb{N}$, let $Y_n := \text{span}\,\{x_1, \ldots, x_n\}$. Fix $n \geq 2$. Since Y_{n-1} is finite dimensional, it is a closed subspace of X (Lemma 2.8). Also, Y_{n-1} is a proper subspace of Y_n, since the set $\{x_1, \ldots, x_n\}$ is linearly independent. By the Riesz lemma (Lemma

2.7), there is $y_n \in Y_n$ such that $\|y_n\| = 1$ and $d(y_n, Y_{n-1}) \geq 1/2$. Since $A(x_1) = \lambda_1 x_1, \ldots, A(x_n) = \lambda_n x_n$, we see that $(A - \lambda_n I)(Y_n) \subset Y_{n-1}$ and $A(Y_{n-1}) \subset Y_{n-1}$. Considering the cases $\lambda_n = 0$ and $\lambda_n \neq 0$ separately, we obtain

$$\|A(y_n) - A(y)\| = \|\lambda_n y_n + (A - \lambda_n I)(y_n) - A(y)\| \geq |\lambda_n|/2 \quad \text{for all } y \in Y_{n-1}.$$

In particular, $\|A(y_n) - A(y_m)\| \geq |\lambda_n|/2$ for all $m = 1, \ldots, n-1$.

Assume for a moment that $\lambda_n \not\to 0$. Then there is $\delta > 0$ such that $|\lambda_{n_k}| \geq \delta$, where $2 \leq n_1 < n_2 < \cdots$ are in $\mathbb{N}$. Hence $\|A(y_{n_k}) - A(y_m)\| \geq |\lambda_{n_k}|/2 \geq \delta/2$ for all $k \in \mathbb{N}$ and $m = 1, \ldots, n_k - 1$. It follows that $\|A(y_{n_k}) - A(y_{n_j})\| \geq \delta/2$ for all $k, j \in \mathbb{N}$ with $k \neq j$. Now (y_{n_k}) is a bounded sequence in X, but the sequence $(A(y_{n_k}))$ has no convergent subsequence. This contradicts the compactness of the operator A. Hence $\lambda_n \to 0$. □

Proposition 5.16 *Let X be a normed space, and let $A \in CL(X)$. Then the eigenspectrum $\sigma_e(A)$ of A is countable. Also, if $k \in \mathbb{K}$ and (λ_n) is any sequence in $\sigma_e(A) \setminus \{k\}$ such that $\lambda_n \to k$, then $k = 0$. Further, for every nonzero eigenvalue λ of A, the corresponding eigenspace E_λ is finite dimensional.*

Proof If $\lambda_1, \lambda_2, \ldots$ are distinct eigenvalues of A, and if x_n is an eigenvector corresponding to λ_n for $n \in \mathbb{N}$, then $\{x_1, x_2, \ldots\}$ is an infinite linearly independent subset of X, and so $\lambda_n \to 0$ by Lemma 5.15. Hence for every $\epsilon > 0$, the set $S_\epsilon := \{\lambda \in \sigma_e(A) : |\lambda| \geq \epsilon\}$ is finite. It follows that the set $\sigma_e(A) \setminus \{0\} = \bigcup_{n=1}^{\infty} S_{1/n}$ is countable, and so is the set $\sigma_e(A)$. Let $k \in \mathbb{K}$ be nonzero. Letting $\epsilon := |k|/2$, we see that only finitely many eigenvalues of A have their absolute values greater than or equal to $|k|/2$. Hence no sequence in $\sigma_e(A) \setminus \{k\}$ can converge to k. (This means that 0 is the only possible 'limit point' of the set $\sigma_e(A)$.)

Next, consider a nonzero eigenvalue λ of A. Assume for a moment that the corresponding eigenspace E_λ is infinite dimensional. Let $\{x_1, x_2, \ldots\}$ be an infinite linearly independent subset of E_λ. Letting $\lambda_n := \lambda$ for each $n \in \mathbb{N}$ in Lemma 5.15, we obtain $\lambda_n \to 0$, that is, $\lambda = 0$, contrary to our assumption. Hence E_λ is finite dimensional. □

Now let us turn to the approximate eigenspectrum $\sigma_a(A)$ of a compact operator A on a normed space X. First we prove a preliminary result which is of independent interest.

Lemma 5.17 *Let X be a normed space, and let $A \in CL(X)$. Suppose (x_n) is a bounded sequence in X, and let $\lambda \in \mathbb{K}$ with $\lambda \neq 0$. If $(A(x_n) - \lambda x_n)$ is a convergent sequence in X, then (x_n) has a convergent subsequence. In fact, if $A(x_n) - \lambda x_n \to y$ and $x_{n_k} \to x$ in X, then $A(x) - \lambda x = y$.*

Proof Define $y_n := A(x_n) - \lambda x_n$ for $n \in \mathbb{N}$, and let $y_n \to y$ in X. Since A is compact and (x_n) is a bounded sequence, there is a subsequence (x_{n_k}) of (x_n) such that $(A(x_{n_k}))$ converges in X. Since

$$x_{n_k} = \frac{1}{\lambda}\big(A(x_{n_k}) - y_{n_k}\big) \quad \text{for all } k \in \mathbb{N},$$

we see that the subsequence (x_{n_k}) of the sequence (x_n) is convergent. Let $x_{n_k} \to x$ in X. Then $A(x_{n_k}) - \lambda x_{n_k} \to A(x) - \lambda x$. But $A(x_{n_k}) - \lambda x_{n_k} = y_{n_k} \to y$ as well. Hence $A(x) - \lambda x = y$. □

Let $A \in CL(X)$. Consider $y \in X$ and nonzero $\lambda \in \mathbb{K}$. In an attempt to find a solution $x \in X$ of the operator equation $A(x) - \lambda x = y$, suppose we are able to find a bounded sequence (x_n) of approximate solutions of this equation in the sense that $A(x_n) - \lambda x_n \to y$. Then the above lemma says that there is a convergent subsequence (x_{n_k}) of the sequence (x_n) of approximate solutions, and if $x_{n_k} \to x$ in X, then $A(x) - \lambda x = y$. Thus the limit x of a subsequence of the bounded sequence (x_n) of approximate solutions is indeed an exact solution of the given operator equation.

Proposition 5.18 *Let X be a normed space, and let $A \in CL(X)$. Every nonzero approximate eigenvalue of A is in fact an eigenvalue of A.*

Proof Let $\lambda \in \sigma_a(A)$ and $\lambda \neq 0$. Then there is a sequence (x_n) in X such that $\|x_n\| = 1$ for all $n \in \mathbb{N}$ and $A(x_n) - \lambda x_n \to 0$. By Lemma 5.17, there is a convergent subsequence (x_{n_k}), and if $x_{n_k} \to x$ in X, then $A(x) - \lambda x = 0$. Thus $A(x) = \lambda x$. Also, $x \neq 0$ since $\|x\| = 1$. Hence $\lambda \in \sigma_e(A)$. □

Next, we study the spectrum $\sigma(A)$ of a compact operator A on a normed space X. We first consider the range space $R(A - kI)$, where $k \neq 0$.

Lemma 5.19 *Let X be a normed space, and let $A \in CL(X)$. Suppose $k \in \mathbb{K}$ is nonzero, and suppose $A - kI$ is bounded below. Then $R((A - kI)^n)$ is a closed subspace of X for every $n \in \mathbb{N}$.*

Proof Let $(A(x_n) - kx_n)$ be a sequence in $R(A - kI)$ which converges in X to, say, y. Then there is $\alpha > 0$ such that $\|A(x_n) - kx_n\| \leq \alpha$ for all $n \in \mathbb{N}$. Also, since $A - kI$ is bounded below, there is $\beta > 0$ such that $\beta\|x\| \leq \|(A - kI)(x)\|$ for all $x \in X$. It follows that $\|x_n\| \leq \alpha/\beta$ for all $n \in \mathbb{N}$. Thus (x_n) is a bounded sequence in X. By Lemma 5.17, (x_n) has a convergent subsequence, and if it converges to x in X, then $A(x) - kx = y$. Thus $y \in R(A - kI)$. Hence $R(A - kI)$ is a closed subspace of X.

Now let $n \geq 2$. By the binomial expansion, $(A - kI)^n = A_n + k_n I$, where

$$A_n := A^n - k\binom{n}{1}A^{n-1} + \cdots + (-k)^{n-1}\binom{n}{n-1}A \quad \text{and} \quad k_n := (-k)^n.$$

By Proposition 3.43 and Remark 3.44(i), $A_n \in CL(X)$. Clearly $k_n \neq 0$, and

$$\|(A - kI)^n(x)\| \geq \|(A - kI)^{n-1}(x)\|\beta \geq \cdots \geq \|(A - kI)(x)\|\beta^{n-1} \geq \|x\|\beta^n$$

for all $x \in X$. It follows that $A_n + k_n I = (A - kI)^n$ is bounded below. Replacing A by A_n, and k by $-k_n$ in the argument given in the first paragraph, we see that $R((A - kI)^n) = R(A_n + k_n I)$ is a closed subspace of X. □

Let $A \in CL(X)$. If X is finite dimensional, then the spectrum of A is described in Proposition 5.2(i)). We now consider the spectrum of A when X is infinite dimensional.

Proposition 5.20 *Let X be an infinite dimensional normed space, and suppose A is a compact operator on X. Then $0 \in \sigma_a(A)$ and $\sigma(A) = \sigma_a(A)$.*

Proof Let $\{x_1, x_2, \ldots\}$ be an infinite linearly independent subset of X. For $n \in \mathbb{N}$, let $Y_n := \text{span}\{x_1, \ldots, x_n\}$. As in the proof of Lemma 5.15, for each $n \geq 2$, there is $y_n \in Y_n$ such that $\|y_n\| = 1$ and $d(y_n, Y_{n-1}) \geq 1/2$. In particular, $\|y_n - y_m\| \geq 1/2$ for all $n, m \geq 2$ with $n \neq m$. Assume for a moment that A is bounded below. Then there is $\beta > 0$ such that $\|A(x)\| \geq \beta\|x\|$ for all $x \in X$, and so

$$\|A(y_n) - A(y_m)\| = \|A(y_n - y_m)\| \geq \beta\|y_n - y_m\| \geq \frac{\beta}{2}.$$

This shows that the sequence $(A(y_n))$ does not have a convergent subsequence, although (y_n) is a bounded sequence in X. This contradicts the compactness of A. Hence A is not bounded below, that is, $0 \in \sigma_a(A)$.

Since $\sigma_a(A) \subset \sigma(A)$ always, we need only show $\sigma(A) \subset \sigma_a(A)$. Let us consider $k \in \mathbb{K}$ such that $A - kI$ is bounded below, and prove that $A - kI$ is invertible. By Proposition 5.1, it is enough to show that $R(A - kI) = X$.

For $n \in \mathbb{N}$, let $\widetilde{Y}_n := R((A - kI)^n)$, and note that $\widetilde{Y}_{n+1} \subset \widetilde{Y}_n$. Since $k \neq 0$ and $A - kI$ is bounded below, Lemma 5.19 shows that each $\widetilde{Y}_n$ is a closed subspace of X. Assume for a moment that $\widetilde{Y}_{n+1}$ is a proper subspace of $\widetilde{Y}_n$ for each $n \in \mathbb{N}$. Fix $n \in \mathbb{N}$. By the Riesz lemma (Lemma 2.7), there is $y_n \in \widetilde{Y}_n$ such that $\|y_n\| = 1$ and $d(y_n, \widetilde{Y}_{n+1}) \geq 1/2$. Since $(A - kI)(\widetilde{Y}_n) \subset \widetilde{Y}_{n+1}$ and $A(\widetilde{Y}_{n+1}) \subset \widetilde{Y}_{n+1}$, we see that

$$\|A(y_n) - A(y)\| = \|ky_n + (A - kI)(y_n) - A(y)\| \geq |k|/2 > 0 \quad \text{for all } y \in \widetilde{Y}_{n+1}.$$

In particular, $\|A(y_n) - A(y_m)\| \geq |k|/2$ for all $n, m \in \mathbb{N}$ with $n \neq m$. Now (y_n) is a bounded sequence in X, but the sequence $(A(y_n))$ has no convergent subsequence. This contradicts the compactness of A. Hence there is $m \in \mathbb{N}$ such that $\widetilde{Y}_{m+1} = \widetilde{Y}_m$.

Let $\widetilde{Y}_0 := R((A - kI)^0) = X$. We claim that $\widetilde{Y}_m = \widetilde{Y}_{m-1}$. Let $y \in \widetilde{Y}_{m-1}$. Then there is $x \in X$ such that $y = (A - kI)^{m-1}(x)$, and so $(A - kI)(y) = (A - kI)^m(x)$ is in $\widetilde{Y}_m = \widetilde{Y}_{m+1}$. Hence there is $z \in X$ with $(A - kI)(y) = (A - kI)^{m+1}(z)$. Since $(A - kI)(y - (A - kI)^m(z)) = 0$ and since $A - kI$ is one-one, it follows that $y - (A - kI)^m(z) = 0$, that is, $y = (A - kI)^m(z) \in \widetilde{Y}_m$. Hence $\widetilde{Y}_{m-1} \subset \widetilde{Y}_m$. Since $\widetilde{Y}_m \subset \widetilde{Y}_{m-1}$ for every $m \in \mathbb{N}$, our claim is justified. Similarly, we obtain $\widetilde{Y}_{m+1} = \widetilde{Y}_m = \widetilde{Y}_{m-1} = \widetilde{Y}_{m-2} = \cdots = \widetilde{Y}_1 = \widetilde{Y}_0$. Thus $R(A - kI) = \widetilde{Y}_1 = \widetilde{Y}_0 = X$, as desired. □

Finally, we state a comprehensive result describing the eigenspectrum, the approximate eigenspectrum and the spectrum of a compact operator on an infinite dimensional normed space.

Theorem 5.21 *Let X be an infinite dimensional normed space, and let A be a compact operator on X. Then*

(i) *Every nonzero spectral value of A is an eigenvalue of A.*
(ii) *The set of all eigenvalues of A is countable, and the eigenspace corresponding to each nonzero eigenvalue is finite dimensional. If $\sigma_e(A)$ is denumerable, and $\lambda_1, \lambda_2, \ldots$ is an enumeration of (distinct) eigenvalues of A, then $\lambda_n \to 0$.*
(iii) *0 is an approximate eigenvalue of A.*

Proof (i) Let $\lambda \in \sigma(A)$ and $\lambda \neq 0$. By Proposition 5.20, $\lambda \in \sigma_a(A)$, and by Proposition 5.18, $\lambda \in \sigma_e(A)$.

(ii) By Proposition 5.16, $\sigma_e(A)$ is countable, and the eigenspace E_λ is finite dimensional for every nonzero $\lambda \in \sigma_e(A)$. Suppose $\sigma_e(A)$ is denumerable, and $\lambda_1, \lambda_2, \ldots$ is an enumeration of (distinct) eigenvalues of A. Then $\lambda_n \to 0$, as we have seen in the proof of Proposition 5.16.

(iii) By Proposition 5.20, $0 \in \sigma_a(A)$. □

Remark 5.22 Suppose X is an infinite dimensional inner product space, and let A be a compact operator on X. We give simpler proofs of a couple of results stated in Theorem 5.21. Suppose λ is a nonzero eigenvalue of A. Assume for a moment that the eigenspace $Z(A - \lambda I)$ corresponding to λ is infinite dimensional. By the Gram–Schmidt process (Theorem 2.17), there is an infinite orthonormal subset $\{u_n : n \in \mathbb{N}\}$ of $Z(A - \lambda I)$. Then

$$\|A(u_n) - A(u_m)\|^2 = |\lambda|^2 \|u_n - u_m\|^2 = 2|\lambda|^2 > 0 \quad \text{for all } n \neq m.$$

Hence the sequence $(A(u_n))$ does not have a convergent subsequence, which contradicts the compactness of A. Thus $Z(A - \lambda I)$ is finite dimensional.

Next, assume for a moment that A is bounded below. Then there is $\beta > 0$ such that $\|A(x)\| \geq \beta \|x\|$ for all $x \in X$. If $\{v_n : n \in \mathbb{N}\}$ is an infinite orthonormal subset of X, then

$$\|A(v_n) - A(v_m)\|^2 \geq \beta^2 \|v_n - v_m\|^2 = 2\beta^2 \quad \text{for all } n \neq m.$$

Hence the sequence $(A(v_n))$ does not have a convergent subsequence, which contradicts the compactness of A. Hence $0 \in \sigma_a(A)$. ◊

Examples 5.23 (i) As far as the spectrum of a compact operator on an infinite dimensional normed space is concerned, the scalar 0 has a special status. Although 0 is an approximate eigenvalue of such an operator, it need not be an eigenvalue, and if 0 is in fact an eigenvalue, then the corresponding eigenspace E_0 may be either finite dimensional or infinite dimensional. The following examples illustrate these phenomena. Let $X := \ell^p$ with $p \in \{1, 2, \infty\}$. Define

$$A(x) := \left(x(1), \frac{x(2)}{2}, \frac{x(3)}{3}, \cdots\right) \quad \text{for } x := (x(1), x(2), \ldots) \in X.$$

For $n \in \mathbb{N}$, let $A_n(x) := (x(1), x(2)/2, \ldots, x(n)/n, 0, 0 \ldots)$ for $x \in X$. Then $A_n \in BL(X)$ is of finite rank for each $n \in \mathbb{N}$, and so it is a compact operator (Theorem 3.42).

Also, X is a Banach space, and $\|A - A_n\| = 1/(n+1) \to 0$. Hence A is a compact operator (Proposition 3.43). Clearly, $\sigma_e(A) = \{1, 1/2, \ldots\}$, and so $0 \notin \sigma_e(A)$. Next, let

$$B(x) := \left(0, \frac{x(2)}{2}, \frac{x(3)}{3}, \cdots\right) \quad \text{for } x := (x(1), x(2), \ldots) \in X.$$

Then $B \in CL(X)$. Now $B(x) = 0$ for $x \in X$ if and only if $x(j) = 0$ for all $j \geq 2$, that is, x is a scalar multiple of e_1. Hence 0 is an eigenvalue of B, and the corresponding eigenspace is one dimensional. Also, if we let

$$C(x) := \left(0, \frac{x(2)}{2}, 0, \frac{x(4)}{4}, \cdots\right) \quad \text{for } x := (x(1), x(2), \ldots) \in X,$$

then $C \in CL(X)$, and $C(x) = 0$ for $x \in X$ if and only if $x(2j) = 0$ for all $j \in \mathbb{N}$. Hence 0 is an eigenvalue of C, and the corresponding eigenspace equals span $\{e_1, e_3, \ldots\}$, which is infinite dimensional.

(ii) Let $X := C([0, 1])$ with the sup norm. For $s, t \in [0, 1]$, let $k(s, t) := (1 - s)t$ if $t \leq s$ and $k(s, t) := s(1 - t)$ if $s < t$, and let A denote the Fredholm integral operator on X with the continuous kernel $k(\cdot, \cdot)$. Then $A \in CL(X)$. (See Example 3.46.) Let $x \in X$, and define

$$y(s) := A(x)(s) = (1 - s)\int_0^s t\, x(t)\, dt + s\int_s^1 (1 - t)x(t)\, dt, \quad s \in [0, 1].$$

Then it is clear that $y(0) = 0 = y(1)$. Since the functions $t \longmapsto t\, x(t)$ and $t \longmapsto (1 - t)x(t)$ are continuous on $[0, 1]$, the fundamental theorem of calculus for Riemann integration (Theorem 1.22) shows that $y \in C^1([0, 1])$, and

$$\begin{aligned} y'(s) &= (1 - s)s\, x(s) - \int_0^s t\, x(t)\, dt - s(1 - s)x(s) + \int_s^1 (1 - t)x(t)\, dt \\ &= -\int_0^1 t\, x(t)dt + \int_s^1 x(t)dt \quad \text{for all } s \in [0, 1]. \end{aligned}$$

Since the function $t \longmapsto x(t)$ is continuous on $[0, 1]$, $y' \in C^1([0, 1])$, and $y''(s) = -x(s)$ for all $s \in [0, 1]$. Thus if $x \in X$ and $y := A(x)$, then $y \in C^2([0, 1])$, $y'' = -x$ and $y(0) = 0 = y(1)$.

Conversely, suppose $x \in X$, and let $y \in C^2([0, 1])$ satisfy $y'' = -x$ and $y(0) = 0 = y(1)$. Integrating by parts, we obtain

$$
\begin{aligned}
A(y'')(s) &= (1-s)\int_0^s t\,y''(t)dt + s\int_s^1 (1-t)y''(t)dt \\
&= (1-s)\Big(s\,y'(s) - \int_0^s y'(t)dt\Big) + s\Big(-(1-s)y'(s) + \int_s^1 y'(t)dt\Big) \\
&= (1-s)\big(s\,y'(s) - y(s)\big) + s\big(-(1-s)y'(s) - y(s)\big) \\
&= -y(s)
\end{aligned}
$$

for all $s \in [0, 1]$. Hence $A(y'') = -y$, that is, $A(x) = y$.

Let $x, y \in X$. We have proved above that x is a solution of the integral equation $A(x) = y$ if and only if y a solution of the **Sturm–Liouville boundary value problem** $y \in C^2([0, 1])$, $y'' = -x$ and $y(0) = 0 = y(1)$. The kernel $k(\cdot\,, \cdot)$ of the Fredholm integral operator A is known as **Green's function** of the associated Sturm–Liouville problem.

Let $x \in X$ be such that $A(x) = 0$. Then $0'' = -x$, that is, $x = 0$. Hence 0 is not an eigenvalue of A. Next, suppose $\lambda \in \mathbb{K}$ and $\lambda \neq 0$. Let $x \in X$ be such that $A(x) = \lambda x$. Then it follows that $\lambda x'' = -x$ and $\lambda x(0) = 0 = \lambda x(1)$, that is, $\lambda x'' + x = 0$ and $x(0) = 0 = x(1)$. Now the differential equation $\lambda x'' + x = 0$ has a nonzero solution satisfying $x(0) = 0 = x(1)$ if and only if $\lambda = 1/n^2\pi^2$, $n \in \mathbb{N}$. In this case, the general solution is given by $x(s) := c_n \sin n\pi s$, $s \in [0, 1]$, where $c_n \in \mathbb{K}$.[1] Fix $n \in \mathbb{N}$, let $\lambda_n := 1/n^2\pi^2$, $x_n(s) := \sin n\pi s$, $s \in [0, 1]$, and let $y_n := \lambda_n x_n$. Then $y_n'' = \lambda_n x_n'' = -x_n$ and $y_n(0) = 0 = y_n(1)$. Hence $A(x_n) = y_n = \lambda_n x_n$. It follows that λ_n is an eigenvalue of A, and the corresponding eigenspace of A is spanned by the function x_n. There are no other eigenvalues of A.

Since A is compact and X is infinite dimensional, we obtain

$$
\sigma_e(A) = \{1/n^2\pi^2 : n \in \mathbb{N}\} \quad \text{and} \quad \sigma_a(A) = \sigma(A) = \sigma_e(A) \cup \{0\}
$$

by Theorem 5.21.

Next, let $Y := L^p([0, 1])$, where $p \in \{1, 2, \infty\}$, and let B denote the Fredholm integral operator on Y with the continuous kernel $k(\cdot\,, \cdot)$ given above. Then B is in $CL(Y)$, and $B(x) \in C([0, 1])$ for every $x \in Y$, as we have noted in Example 3.46. Hence if λ is a nonzero eigenvalue of B and $x \in Y$ is a corresponding eigenvector, then in fact $x = B(x)/\lambda$ is continuous on $[0, 1]$, and so λ is an eigenvalue of the compact operator A on $X := C([0, 1])$ treated above. Conversely, it is obvious that every nonzero eigenvalue of A is also an eigenvalue of B.

---

[1] If $\mathbb{K} := \mathbb{C}$, then we must first show that $\lambda \in \mathbb{R}$. Let $\mu := 1/\lambda$. Since $\mu x = -x''$ and $\overline{\mu}\,\overline{x} = -\overline{x}''$, we obtain $(\mu - \overline{\mu})|x|^2 = (\mu - \overline{\mu})x\,\overline{x} = \overline{x}''x - x''\overline{x}$. Hence

$$
(\mu - \overline{\mu})\int_0^1 |x(t)|^2 = \int_0^1 \big(\overline{x}''(t)x(t) - x''(t)\overline{x}(t)\big)dt = \big(\overline{x}'(t)x(t) - x'(t)\overline{x}(t)\big)\Big|_{t=0}^{t=1} = 0,
$$

since $x(0) = 0 = x(1)$. If $x \neq 0$, then $\mu = \overline{\mu}$, that is, $\lambda \in \mathbb{R}$.

Let us show $0 \notin \sigma_e(B)$. Let $x \in Y$ and $B(x) = 0$. Define

$$y(s) := (1-s)\int_0^s t\,x(t)dm(t) + s\int_s^1 (1-t)x(t)dm(t) \quad \text{for } s \in [0,1].$$

Then $y(s) = B(x)(s) = 0$ for almost all $s \in [0, 1]$. But since $B(x) \in C([0, 1])$, we see that $y(s) = 0$ for all $s \in [0, 1]$. Since the functions $t \longmapsto t\,x(t)$ and $t \longmapsto (1-t)x(t)$ belong to $L^1([0, 1])$, the fundamental theorem of calculus for Lebesgue integration (Theorem 1.23) shows that y is absolutely continuous on $[0, 1]$, and $y'(s) = -\int_0^1 t\,x(t)dt + \int_s^1 x(t)dt$ for almost all $s \in [0, 1]$. Again, since the function $t \longmapsto x(t)$ is continuous on $[0, 1]$, we see that y' is absolutely continuous on $[0, 1]$, and $(y')'(s) = -x(s)$ for almost all $s \in [0, 1]$. Since $y = 0$, we obtain $y' = 0$ and $x = -y'' = 0$. Thus we see that if $x \in Y$ and $B(x) = 0$, then $x = 0$. Hence $0 \notin \sigma_e(B)$. Thus

$$\sigma_e(B) = \{1/n^2\pi^2 : n \in \mathbb{N}\} \quad \text{and} \quad \sigma_a(B) = \sigma(B) = \sigma_e(B) \cup \{0\}.$$

It is very rare that one is able to solve the eigenvalue problem for a compact integral operator by reducing it to the eigenvalue problem for a differential operator, as we have done above. As a matter of fact, eigenvalue problems for differential operators are often converted into eigenvalue problems for compact integral operators, and then the latter are solved approximately. This is done by approximating a compact integral operator by a sequence of bounded operators of finite rank. We refer the reader to [1, Sects. 4.1 and 4.2] and to [21, Examples 2.4] for a variety of such approximations. ◊

5.3 Spectra of Transposes and Adjoints

Let us first study the invertibility of the transpose of a bounded operator.

Proposition 5.24 *Let X be a normed space, and let $A \in BL(X)$. If A is invertible, then A' is invertible. Conversely, if A' is invertible, then A is bounded below and $R(A)$ is dense in X.*

In case X is a Banach space, A' is invertible if and only if A is invertible.

Proof Suppose A is invertible. If $B \in BL(X)$ is the inverse of A, then $AB = I = BA$, where I denotes the identity operator on X. By Proposition 4.20, we see that $B'A' = I' = A'B'$, where I' denotes the identity operator on X'. Thus $B' \in BL(X')$ is the inverse of A', and so $A' \in BL(X')$ is invertible.

Conversely, suppose A' is invertible, and let $\beta := \|(A')^{-1}\|$. We show that $\|x\| \le \beta\|A(x)\|$ for all $x \in X$. Clearly, this holds if $x = 0$. Let $x \in X$ be nonzero. By Proposition 4.6(i), there is $x' \in X'$ such that $x'(x) = \|x\|$ and $\|x'\| = 1$. Then

$$\begin{aligned}\|x\| &= |x'(x)| = |\big((A')^{-1}A'\big)(x')(x)| = |(A')^{-1}(x')(A(x))| \\ &\le \beta\|x'\|\|A(x)\| = \beta\|A(x)\|\end{aligned}$$

by the basic inequality for the operator norm (Theorem 3.10(i)). Hence A is bounded below. Also, since A' is one-one, we see that $R(A)$ is dense in X by Proposition 4.22.

Suppose X is a Banach space. Then A is invertible if and only if A is bounded below and $R(A)$ is dense in X (Proposition 5.7). Thus A' is invertible if and only if A is invertible by what we have shown above. □

Let X be a normed space, and let $A \in BL(X)$. The set $\sigma_c(A) := \{k \in \mathbb{K} : R(A - kI)$ is not dense in $X\}$ is called the **compression spectrum** of A.

Theorem 5.25 *Let X be a normed space, and let $A \in BL(X)$. Then*

$$\sigma_c(A) = \sigma_e(A') \quad \textit{and} \quad \sigma_a(A) \cup \sigma_e(A') \subset \sigma(A') \subset \sigma(A).$$

If X is a Banach space, then $\sigma(A') = \sigma_a(A) \cup \sigma_e(A') = \sigma_a(A) \cup \sigma_c(A) = \sigma(A)$.

Proof Let $k \in \mathbb{K}$. By Proposition 4.22, $R(A - kI)$ is dense in X if and only if $A' - kI$ is one-one. Hence $\sigma_c(A) = \sigma_e(A')$.

Replacing A by $A - kI$ in Proposition 5.24, we obtain

$$\sigma_a(A) \cup \sigma_e(A') = \sigma_a(A) \cup \sigma_c(A) \subset \sigma(A') \subset \sigma(A).$$

Also, if X is a Banach space, then $\sigma(A') = \sigma(A)$ by Proposition 5.24, and $\sigma(A) = \sigma_a(A) \cup \sigma_c(A) = \sigma_a(A) \cup \sigma_e(A')$ by Proposition 5.7. □

Let X be a Banach space, and $A \in BL(X)$. Then the equality $\sigma(A) = \sigma_a(A) \cup \sigma_e(A')$ proved above says that a scalar is a spectral value of A if and only if either it is an approximate eigenvalue of A or it is an eigenvalue of A'. Also, $A - kI$ is invertible if and only if $A - kI$ is one-one and onto (Proposition 5.7). But $A - kI$ is onto if and only if $A' - kI$ is bounded below. (See our comment after Proposition 4.22.) Hence $\sigma(A) = \sigma_e(A) \cup \sigma_a(A')$. Thus a scalar is a spectral value of A if and only if either it is an eigenvalue of A or it is an approximate eigenvalue of A'.

Examples 5.26 (i) Let X be a finite dimensional normed space, and let $A \in BL(X)$. Suppose $x_1, \ldots, x_n$ constitute a basis for X, and let $x_1', \ldots, x_n'$ constitute the corresponding dual basis for X'. Right after defining the transpose of a bounded linear map on a linear space in Sect. 5.2, we have seen that the $n \times n$ matrix $M := [x_i'(A(x_j))]$ defines the operator A on X, and the transpose $M^t := [x_j'(A(x_i)]$ of M defines the operator A' on X'. By Proposition 5.2, $\sigma(A) = \sigma_e(A)$ and $\sigma(A') = \sigma_e(A')$. Now $\lambda \in \mathbb{K}$ is an eigenvalue of A if and only if $\det(M - \lambda I) = 0$, as we have seen in Example 5.4(i). Since $\det(M^t - \lambda I) = \det(M - \lambda I)$, we see that $\sigma(A') = \sigma_e(A') = \sigma_e(A) = \sigma(A)$.

(ii) Let X denote the linear space of all doubly infinite scalar sequences $x := (\ldots, x(-2), x(-1), x(0), x(1), x(2), \ldots)$ such that there is $j_x \in \mathbb{N}$ with $x(j) = 0$ for all $|j| \geq j_x$. Consider the norm $\|\cdot\|_p$ on X, where $p \in \{1, 2, \infty\}$. Let A denote the **right shift operator** on X defined by $A(x)(j) := x(j-1)$, $j \in \mathbb{Z}$, for $x \in X$. Then $A \in BL(X)$ and $\|A\| = 1$. We have seen in Example 5.12

(ii) that $\sigma(A) = \mathbb{K} \setminus \{0\}$. But since X' is a Banach space, and $\|A'\| = \|A\| = 1$, it follows that $\sigma(A') \subset \{k \in \mathbb{K} : |k| \le 1\}$ (Corollary 5.9). Also, $0 \notin \sigma(A')$ by Proposition 5.24. Thus $\sigma(A')$ is strictly contained in $\sigma(A)$.

(iii) Let $p \in \{1, 2\}$, and let $A \in BL(\ell^p)$. Let $k_{i,j} := A(e_j)(i)$ for $i, j \in \mathbb{N}$. We saw in Example 4.19(i) that the infinite matrix $M := [k_{i,j}]$ defines A, and if q satisfies $(1/p) + (1/q) = 1$, then the infinite matrix $M^t := [k_{j,i}]$ defines a map $A^t \in BL(\ell^q)$ which can be identified with the transpose $A' \in BL((\ell^p)')$. Clearly, $\sigma_e(A') = \sigma_e(A^t)$, $\sigma_a(A') = \sigma_a(A^t)$ and $\sigma(A') = \sigma(A^t)$.

For instance, let $A(x) := (0, x(1), x(2), \ldots)$ for $x := (x(1), x(2), \ldots) \in \ell^p$. As in Example 5.12(i), $\sigma_e(A) = \emptyset$, $\sigma_a(A) = \{\lambda \in \mathbb{K} : |\lambda| = 1\}$ and $\sigma(A) = \{\lambda \in \mathbb{K} : |\lambda| \le 1\}$. Now A^t is given by $A^t(x) := (x(2), x(3), \ldots)$ for $x := (x(1), x(2), \ldots)$ in ℓ^q. By Exercise 5.14, $\sigma_a(A^t) = \{\lambda \in \mathbb{K} : |\lambda| \le 1\} = \sigma(A^t)$, while $\sigma_e(A^t) = \{\lambda \in \mathbb{K} : |\lambda| < 1\}$ if $q = 2$ and $\sigma_e(A^t) = \{\lambda \in \mathbb{K} : |\lambda| \le 1\}$ if $q = \infty$. Hence $\sigma(A') = \sigma(A)$, but $\sigma_e(A)$ is strictly contained in $\sigma_e(A')$, and $\sigma_a(A)$ is strictly contained in $\sigma_a(A')$. ◊

Let us now study the spectrum of the transpose of a compact operator on a normed space X.

Theorem 5.27 *Let X be a normed space, and let $A \in CL(X)$. For every nonzero $\lambda \in \mathbb{K}$, $\lambda \in \sigma_e(A')$ if and only if $\lambda \in \sigma_e(A)$. Also, $\sigma(A') = \sigma_a(A') = \sigma_a(A) = \sigma(A)$.*

Proof If X is finite dimensional, then $\sigma(A') = \sigma_a(A') = \sigma_e(A') = \sigma_e(A) = \sigma_a(A) = \sigma(A)$, as shown in Example 5.26(i).

Now suppose X is infinite dimensional. By Theorem 4.21, $A' \in CL(X')$. Let $\lambda \in \mathbb{K}$ and $\lambda \ne 0$. If $\lambda \in \sigma_e(A')$, then $\lambda \in \sigma(A') \subset \sigma(A)$ by Theorem 5.25, and so $\lambda \in \sigma_e(A)$ by Theorem 5.21(i). Conversely, let $\lambda \in \sigma_e(A)$. To show that $\lambda \in \sigma_e(A')$, it is enough to show that $\lambda \in \sigma_e(A'')$, by what we have just proved. Let $x \in X$ be an eigenvector of A corresponding to λ. Consider the canonical embedding J of X into X''. Then $J(x) \ne 0$, and for every $x' \in X'$,

$$A''(J(x))(x') = J(x)(A'(x')) = A'(x')(x) = x'(A(x)) = x'(\lambda x) = \lambda J(x)(x'),$$

that is, $A''J(x) = \lambda J(x)$. (Compare the equality $F''J_X = J_Y F$ stated in Exercise 4.25.) This shows that λ is an eigenvalue of A'', as desired.

Next, let $\lambda := 0$. Since X and X' are infinite dimensional, $0 \in \sigma_a(A)$ and $\sigma(A) = \sigma_a(A)$, and $0 \in \sigma_a(A')$ and $\sigma(A') = \sigma_a(A')$ by Proposition 5.20.

Thus $\sigma(A') = \sigma_a(A') = \sigma_a(A) = \sigma(A)$. □

Remarks 5.28 (i) The scalar 0 can be eigenvalue of the transpose of a compact operator A without being an eigenvalue of A itself. It is easy to see that 0 is not an eigenvalue of the compact operator A on ℓ^2 defined by

$$A(x) := \left(0, x(1), \frac{x(2)}{2}, \frac{x(3)}{3}, \cdots\right) \quad \text{for } x := (x(1), x(2), \ldots) \in \ell^2.$$

As we have seen in Example 4.19(i), the transpose A' of A can be identified with the operator A^t on ℓ^2 defined by

$$A^t(x) := \Big(\frac{x(2)}{2}, \frac{x(3)}{3}, \cdots\Big) \quad \text{for } x := (x(1), x(2), \ldots) \in \ell^2.$$

Clearly, 0 is an eigenvalue of A^t, and e_1 is a corresponding eigenvector of A^t. Further, since the transpose A'' of A' can be identified with A itself, we see that 0 is an eigenvalue of $B := A'$, but not of $B' = A''$.

(ii) By Theorem 5.27, the nonzero eigenvalues of a compact operator A on a normed space X are the same as the nonzero eigenvalues of its transpose A' on the Banach space X'. Also, it can be shown that for each such nonzero eigenvalue, the corresponding eigenspaces of A and of A' have the same finite dimension. The proof is rather involved, and we refer the reader to [28, Theorem V.7.14 (a)]. ◊

Spectrum of the Adjoint of an Operator

Let H be a Hilbert space. For $x' \in H'$, let $T(x') \in H$ denote the representer of x' given by the Riesz representation theorem. Then T is a conjugate-linear isometry from H' onto H (Theorem 4.14). Let $A \in BL(H)$. We have seen in Remark 4.25(i) that $A^* = TA'T^{-1}$, where A^* is the adjoint of A and A' is the transpose of A. Hence for $k \in \mathbb{K}$, $A' - kI$ is one-one if and only if $A^* - \overline{k}I$ is one-one, $A' - kI$ is bounded below if and only if $A^* - \overline{k}I$ is bounded below, and $A' - kI$ is invertible if and only if $A^* - \overline{k}I$ is invertible. As a result, $\sigma_e(A^*) = \{\overline{\lambda} \in \mathbb{K} : \lambda \in \sigma_e(A')\}$, $\sigma_a(A^*) = \{\overline{\lambda} \in \mathbb{K} : \lambda \in \sigma_a(A')\}$ and $\sigma(A^*) = \{\overline{\lambda} \in \mathbb{K} : \lambda \in \sigma(A')\}$.

Theorem 5.29 *Let H be a Hilbert space, and let $A \in BL(H)$. Then A is invertible if and only if A^* is invertible, and then $(A^*)^{-1} = (A^{-1})^*$. In particular, $\sigma(A^*) = \{\overline{\lambda} : \lambda \in \sigma(A)\}$. Also,*

$$\sigma(A) = \sigma_a(A) \cup \{\lambda \in \mathbb{K} : \overline{\lambda} \in \sigma_e(A^*)\} = \sigma_e(A) \cup \{\lambda \in \mathbb{K} : \overline{\lambda} \in \sigma_a(A^*)\}.$$

Suppose $A \in CL(H)$. For every nonzero $\lambda \in \mathbb{K}$, $\lambda \in \sigma_e(A)$ if and only if $\overline{\lambda} \in \sigma_e(A^*)$. Also, $\sigma(A^*) = \sigma_a(A^*) = \{\overline{\lambda} : \lambda \in \sigma_a(A)\} = \{\overline{\lambda} : \lambda \in \sigma(A)\}$.

Proof Let A be invertible. Then there is $B \in BL(H)$ such that $AB = I = BA$, and so $B^*A^* = I = A^*B^*$ by Corollary 4.26. Hence A^* is invertible, and $B^* = (A^{-1})^*$ is its inverse. Conversely, let A^* be invertible. By Theorem 4.23, $A = (A^*)^*$. Hence A is invertible, and if B^* is the inverse of A^*, then $B = (B^*)^*$ is the inverse of A. Now for $k \in \mathbb{K}$, $A - kI$ is invertible if and only if $(A - kI)^* = A^* - \overline{k}I$ is invertible, that is, $\sigma(A^*) = \{\overline{\lambda} : \lambda \in \sigma(A)\}$. (This result can also be deduced from Theorem 5.25.)

By Theorem 5.25, $\sigma(A) = \sigma_a(A) \cup \sigma_e(A') = \sigma_a(A) \cup \{\lambda \in \mathbb{K} : \overline{\lambda} \in \sigma_e(A^*)\}$. Also, if $k \in \mathbb{K}$, then $A - kI$ is invertible if and only if it is one-one and onto by Proposition 5.7. But $A - kI$ is onto if and only if $A^* - kI$ is bounded below by Theorem 4.27(ii). Hence $\sigma(A) = \sigma_e(A) \cup \{\lambda \in \mathbb{K} : \overline{\lambda} \in \sigma_a(A^*)\}$.

If $A \in CL(H)$, then the desired results follow from Theorem 5.27. □

Normal operators introduced in Sect. 4.4 have special spectral properties as stated below. Subspaces G_1 and G_2 of H are called **mutually orthogonal** if $x_1 \perp x_2$ for all $x_1 \in G_1$ and $x_2 \in G_2$.

Proposition 5.30 *Let H be a Hilbert space, and let $A \in BL(H)$ be normal.*

(i) *Let $\lambda \in \sigma_e(A)$. Then $\overline{\lambda} \in \sigma_e(A^*)$, and the eigenspace of A^* corresponding to $\overline{\lambda}$ is the same as the eigenspace of A corresponding to λ.*
(ii) *The eigenspaces of A corresponding to distinct eigenvalues are mutually orthogonal.*
(iii) $\sigma(A) = \sigma_a(A)$.

Proof (i) Let $x \in H$. Since $A - \lambda I$ is normal, Theorem 4.30(ii) shows that

$$\|A^*(x) - \overline{\lambda}x\| = \|(A - \lambda I)^*(x)\| = \|(A - \lambda I)(x)\|.$$

Thus $A(x) = \lambda x$ if and only if $A^*(x) = \overline{\lambda}x$.

(ii) Let λ and μ be distinct eigenvalues of A, and let $x, y \in H$ be such that $A(x) = \lambda x$ and $A(y) = \mu\, y$. Then $A^*(y) = \overline{\mu}\, y$ by (i) above and

$$\lambda\langle x, y\rangle = \langle \lambda x, y\rangle = \langle A(x), y\rangle = \langle x, A^*(y)\rangle = \langle x, \overline{\mu}\, y\rangle = \mu\langle x, y\rangle.$$

Since $\lambda \neq \mu$, we obtain $\langle x, y\rangle = 0$, that is, $x \perp y$.

(iii) Let $\lambda \in \sigma(A)$. Then by Theorem 5.29, $\lambda \in \sigma_a(A)$ or $\overline{\lambda} \in \sigma_e(A^*)$. But if $\overline{\lambda} \in \sigma_e(A^*)$, then $\lambda \in \sigma_e(A)$ by (i) above. Since $\sigma_e(A) \subset \sigma_a(A) \subset \sigma(A)$, we see that $\sigma(A) = \sigma_a(A)$. □

Consider now a self-adjoint operator A on a nonzero Hilbert space H. Just before Theorem 4.30, we have seen that $(m_A, M_A) \subset \omega(A) \subset [m_A, M_A]$, where m_A is the infimum and M_A is the supremum of the numerical range $\omega(A)$ of A.

Proposition 5.31 *Let H be a nonzero Hilbert space over $\mathbb{K}$, and let A be a self-adjoint operator on H. Then*

$$\{m_A, M_A\} \subset \sigma_a(A) = \sigma(A) \subset [m_A, M_A].$$

Proof By Proposition 5.14, $\sigma(A) \subset \overline{\omega(A)} \subset [m_A, M_A]$. Also, since A is normal, $\sigma_a(A) = \sigma(A)$ by Proposition 5.30(iii).

Next, we show that $m_A \in \sigma_a(A)$. By the definition of m_A, there is a sequence (x_n) in H such that $\|x_n\| = 1$ for all $n \in \mathbb{N}$, and $\langle A(x_n), x_n\rangle \to m_A$. We claim that $A(x_n) - m_A x_n \to 0$. Let $B := A - m_A I$. Then B is a positive operator. By the generalized Schwarz inequality (Proposition 4.31),

$$\|B(x_n)\| \leq \langle B(x_n), x_n\rangle^{1/4}\langle B^2(x_n), B(x_n)\rangle^{1/4} \leq \langle B(x_n), x_n\rangle^{1/4}\|B\|^{3/4}.$$

Since $\langle B(x_n), x_n\rangle = \langle A(x_n), x_n\rangle - m_A \to 0$, we see that $\|A(x_n) - m_A x_n\| = \|B(x_n)\| \to 0$. Hence $m_A \in \sigma_a(A)$. Similarly, by considering $C := M_A I - A$, and noting that C is a positive operator, we see that $M_A \in \sigma_a(A)$. □

The above proposition shows that the spectrum of every self-adjoint operator on a nonzero Hilbert space is nonempty. However, the eigenspectrum of a self-adjoint operator can very well be empty. For example, let $H := L^2([a, b])$, and $A(x)(t) := t\,x(t)$ for $x \in H$ and $t \in [a, b]$. As we have seen in Example 4.28(iii), A is self-adjoint. Also, $\sigma_e(A) = \emptyset$. To see this, let $\lambda \in \mathbb{R}$, and suppose there is $x \in H$ such that $t\,x(t) = A(x)(t) = \lambda x(t)$ for almost all $t \in [a, b]$. Then it follows that $x(t) = 0$ for almost all $t \in [a, b]$. (Compare Exercise 5.7.)

Corollary 5.32 *Let H be a nonzero Hilbert space, and let $A \in BL(H)$.*

(i) *If A is self-adjoint, then $\|A\| = \max\{|m_A|, |M_A|\} = \sup\{|\lambda| : \lambda \in \sigma(A)\}$.*
(ii) *If $\lambda \in \sigma(A^*A)$, then $\lambda \geq 0$, and $\|A\| = \sup\left\{\sqrt{\lambda} : \lambda \in \sigma(A^*A)\right\}$.*

Proof (i) By Theorem 4.30(i), $\|A\| = \sup\{|k| : k \in \omega(A)\}$. Since $\omega(A)$ is a subset of $\mathbb{R}$, and since $m_A = \inf \omega(A)$ and $M_A = \sup \omega(A)$, it follows that $\|A\| = \max\{|m_A|, |M_A|\}$. Also, by Proposition 5.31, $m_A,\ M_A \in \sigma(A)$ and $\sigma(A)$ is a subset of $[m_A, M_A]$. Hence $\max\{|m_A|, |M_A|\} = \sup\{|\lambda| : \lambda \in \sigma(A)\}$.

(ii) Since A^*A is a positive operator, we see that $\omega(A^*A) \subset [0, \infty)$, and so $\sigma(A^*A) \subset \overline{\omega(A^*A)} \subset [0, \infty)$ by Proposition 5.14. By (i) above, $\|A^*A\| = \sup\{\lambda : \lambda \in \sigma(A^*A)\}$. But $\|A^*A\| = \|A\|^2$ as we have seen in Theorem 4.23. Hence $\|A\| = \|A^*A\|^{1/2} = \sup\left\{\sqrt{\lambda} : \lambda \in \sigma(A^*A)\right\}$. □

Part (ii) of the above corollary gives a formula for the norm of a bounded operator on $\mathbb{K}^n$, or on ℓ^2, or on $L^2([a, b])$ with the norm $\|\cdot\|_2$. Any such formula had eluded us so far. (See Examples 3.14(iii), 3.15(iii) and 3.16(iii).) For instance, let $A(x(1), x(2)) := (x(1) + x(2), x(1))$ for $(x(1), x(2)) \in \mathbb{K}^2$. Then it follows that $A^*A(x(1), x(2)) = (2x(1) + x(2), x(1) + x(2))$ for all $(x(1), x(2))$ in $\mathbb{K}^2$, and $\sigma(A^*A) = \{(3 - \sqrt{5})/2, (3 + \sqrt{5})/2\}$. Hence $\|A\| = (3 + \sqrt{5})/2$. (Compare Example 3.15(iii).) Of course, in general, it is by no means easy to calculate the norm of a bounded operator using this formula.

Example 5.33 Let $H := \ell^2$, and define $A(x) := (\lambda_1 x(1), \lambda_2 x(2), \ldots)$ for $x := (x(1), x(2), \ldots) \in H$, where $\lambda_n := (-1)^n(n-1)/n$ for $n \in \mathbb{N}$. Then it is easy to see that A is self-adjoint, and $A(e_n) = \lambda_n e_n$, so that $\lambda_n \in \sigma_e(A)$ for each $n \in \mathbb{N}$. Also, if $\lambda \in \sigma_e(A)$ and x is a corresponding eigenvector of A, then there is $n \in \mathbb{N}$ such that $x(n) \neq 0$, and so the equation $\lambda_n x(n) = \lambda x(n)$ shows that $\lambda = \lambda_n$. Thus $\sigma_e(A) = \{\lambda_n : n \in \mathbb{N}\}$. Further, since $\sigma_e(A) \subset \sigma_a(A)$, and $\sigma_a(A)$ is a closed subset of $\mathbb{K}$, the closure $E := \{\lambda_n : n \in \mathbb{N}\} \cup \{-1, 1\}$ of $\{\lambda_n : n \in \mathbb{N}\}$ is contained in $\sigma_a(A)$. Also, if $k \notin E$, and $\beta := d(k, E)$, then $\beta > 0$ and $\|A(x) - kx\|_2 = \|((\lambda_1 - k)x(1), (\lambda_2 - k)x(2), \ldots)\|_2 \geq \beta\|x\|_2$ for all $x \in H$, and so $k \notin \sigma_a(A)$. Thus $\sigma(A) = \sigma_a(A) = E$.

Let $x \in H$ with $\|x\|_2 = 1$. Then $\langle A(x), x\rangle = \lambda_1|x(1)|^2 + \lambda_2|x(2)|^2 + \cdots$, and since $-1 < \lambda_n < 1$ for all $n \in \mathbb{N}$, we see that

$$-1 = -\|x\|_2^2 < \langle A(x), x\rangle < \|x\|_2^2 = 1.$$

Also, $\langle A(e_{2n-1}), e_{2n-1}\rangle = -(2n-2)/(2n-1) \to -1$, while $\langle A(e_{2n}), e_{2n}\rangle = (2n-1)/2n \to 1$. Hence $m_A = -1$ and $M_A = 1$. Also, since $\omega(A)$ is a convex subset of $\mathbb{R}$, we obtain $\omega(A) = (-1, 1)$, ◊

We shall conclude this section by giving a procedure for approximating m_A and M_A for a self-adjoint operator A on a nonzero Hilbert space.

Proposition 5.34 (Ritz method) *Let H be a nonzero Hilbert space, and let A be a self-adjoint operator on H. For $x_1, x_2, \ldots$ in H, and for $n \in \mathbb{N}$, let $G_n :=$ span $\{x_1, \ldots, x_n\}$, $\alpha_n := \inf\{\langle A(x), x\rangle : x \in G_n$ and $\|x\| = 1\}$ and $\beta_n := \sup\{\langle A(x), x\rangle : x \in G_n$ and $\|x\| = 1\}$. Then*

$$m_A \le \alpha_{n+1} \le \alpha_n \le \cdots \le \alpha_1 \le \beta_1 \le \beta_2 \le \cdots \le \beta_n \le \beta_{n+1} \le M_A.$$

Further, if span $\{x_1, x_2, \ldots\}$ *is dense in H, then*

$$m_A = \lim_{n\to\infty} \alpha_n \quad \text{and} \quad \lim_{n\to\infty} \beta_n = M_A.$$

Proof It is easy to see that (α_n) is a nonincreasing sequence in $\mathbb{R}$ which is bounded below by m_A. Hence (α_n) converges in $\mathbb{R}$ to, say, m_0. Clearly, $m_A \le m_0$.

Suppose span $\{x_1, x_2, \ldots\}$ is dense in H. Let, if possible, $m_A < m_0$. By the definition of m_A, there is $x \in H$ with $\|x\| = 1$ and $\langle A(x), x\rangle < m_0$. Find a sequence (y_n) in span $\{x_1, x_2, \ldots\}$ such that $y_n \to x$ as $n \to \infty$. Note that $\|y_n\| \to \|x\|$. Letting $z_n := y_n/\|y_n\|$ for all large $n \in \mathbb{N}$, we see that $\|z_n\| = 1$ and $z_n \to x$. Also, for each large $n \in \mathbb{N}$, there is an integer j_n such that $z_n \in G_{j_n} =$ span $\{x_1, \ldots, x_{j_n}\}$. Since $m_0 \le \alpha_{j_n} \le \langle A(z_n), z_n\rangle$ for all $n \in \mathbb{N}$,

$$m_0 \le \lim_{n\to\infty} \langle A(z_n), z_n\rangle = \langle A(x), x\rangle < m_0.$$

This contradiction shows that $m_A = m_0 = \lim_{n\to\infty} \alpha_n$.

Similarly, we can show that (β_n) is a nondecreasing sequence in $\mathbb{R}$, and $\lim_{n\to\infty} \beta_n = M_A$. Since $\alpha_1 \le \beta_1$, the proof is complete. □

Example 5.35 We illustrate the Ritz method by considering the self-adjoint Fredholm integral operator B on $L^2([0, 1])$ with kernel $k(s, t) := (1 - s)t$ if $0 \le t \le s \le 1$ and $k(s, t) := s(1 - t)$ if $0 \le s < t \le 1$, treated in Example 5.23(ii). We have found that $\sigma(B) = \{0\} \cup \{1/n^2\pi^2 : n \in \mathbb{N}\}$. Hence $m_B = 0$ and $M_B = 1/\pi^2$. Let $x_j(t) := t^{j-1}$ for $j \in \mathbb{N}$. We let $\mathbb{K} := \mathbb{R}$, $G_2 :=$ span $\{x_1, x_2\}$ and $E_2 := \{\langle B(x), x\rangle : x \in G_2$ and $\|x\|_2 = 1\}$. Let us calculate $\alpha_2 := \inf E_2$ and $\beta_2 := \sup E_2$.

Let $x \in G_2$. Then there are $a, b \in \mathbb{R}$ such that $x := ax_1 + bx_2$, and so

$$
\begin{aligned}
B(x)(s) &= (1-s)\int_0^s t\,x(t)\,dt + s\int_s^1 (1-t)x(t)\,dt \\
&= (1-s)\left(a\frac{s^2}{2}+b\frac{s^3}{3}\right) + s\left(a\frac{(1-s)^2}{2}+b\left(\frac{1}{6}-\frac{s^2}{2}+\frac{s^3}{3}\right)\right)
\end{aligned}
$$

for $s \in [0, 1]$. Hence

$$
\langle B(x), x\rangle = \int_0^1 B(x)(s)\big(ax_1(s)+bx_2(s)\big)ds = \frac{a^2}{12}+\frac{ab}{12}+\frac{b^2}{45},
$$

while

$$
\langle x, x\rangle = \int_0^1 (a+bs)^2\,ds = a^2+ab+\frac{b^2}{3}.
$$

Thus

$$
E_2 = \left\{\frac{a^2}{12}+\frac{ab}{12}+\frac{b^2}{45} : a, b \in \mathbb{R} \text{ and } a^2+ab+\frac{b^2}{3}=1\right\}.
$$

For $b \in \mathbb{R}$, there is $a \in \mathbb{R}$ such that $a^2+ab+b^2/3 = 1$ if and only if $b^2 \le 12$, and then

$$
\frac{a^2}{12}+\frac{ab}{12}+\frac{b^2}{45} = \frac{1}{12}\left(1-\frac{b^2}{3}\right)+\frac{b^2}{45} = \frac{1}{12}-\frac{b^2}{180}.
$$

Hence

$$
E_2 = \left\{\frac{1}{12}-\frac{b^2}{180} : b \in \mathbb{R} \text{ and } b^2 \le 12\right\},
$$

and so $\alpha_2 = \inf E_2 = 1/60$, while $\beta_2 = \sup E_2 = 1/12$. ◊

5.4 Spectral Theorem

The concepts of eigenspectrum, approximate eigenspectrum and spectrum of a bounded operator A on a normed space X are introduced in Sect. 5.1 by considering the operator equation $A(x) - k\,x = y$, where the free term $y \in X$ and the scalar $k \in \mathbb{K}$ are given, and the solution $x \in X$ is to be found. If k is a spectral value of A, then we cannot find a unique solution of this equation which depends continuously on the free term. From this point of view, eigenvalues, approximate eigenvalues and spectral values are 'undesirable' scalars. In the present section, we shall show how these very scalars can be used to represent the operator A itself. We shall limit ourselves to the representation of a bounded operator on a Hilbert space by means of its eigenvalues and eigenvectors. This remarkable result is known as the spectral theorem, of which we shall give several versions. The main theme is to investigate situations where eigenvectors of an operator form an orthonormal basis for the Hilbert space. The following result shows that if an operator is 'diagonal' with respect to an orthonormal basis, then it is normal. Subsequently, we shall address the question: When can a normal operator be so 'diagonalized'?

Proposition 5.36 *Let H be a Hilbert space over $\mathbb{K}$, and let $A \in BL(H)$. Suppose H has an orthonormal basis $\{u_\alpha\}$ consisting of eigenvectors of A. Let $A(u_\alpha) = \lambda_\alpha u_\alpha$, where $\lambda_\alpha \in \mathbb{K}$ for each α. Then*

(i) *$A^*(u_\alpha) = \overline{\lambda_\alpha} u_\alpha$ for each α. Also, A is normal, $\sigma_e(A) = \{\lambda_\alpha\}$, and $\sigma(A) = \sigma_a(A)$ is the closure of $\{\lambda_\alpha\}$.*
(ii) *A is self-adjoint if and only if $\lambda_\alpha \in \mathbb{R}$ for all α, A is positive if and only if $\lambda_\alpha \geq 0$ for all α, and A is unitary if and only if $|\lambda_\alpha| = 1$ for all α.*
(iii) *A is compact if and only if $S := \{u_\alpha : \lambda_\alpha \neq 0\}$ is a countable subset of H. In this case, let $S := \{u_1, u_2, \ldots\}$, and $A(u_n) = \lambda_n u_n$ for each n. Then*

$$A(x) = \sum_n \lambda_n \langle x, u_n \rangle u_n \quad \text{for all } x \in H.$$

If in fact the set S is denumerable, then $\lambda_n \to 0$.

Proof Let $x, y \in H$. Since $\{u_\alpha\}$ be an orthonormal basis for H, $x = y$ if and only if $\langle x, u_\alpha \rangle = \langle y, u_\alpha \rangle$ for all α by condition (ii) of Theorem 2.31.

(i) The orthogonality of the set $\{u_\alpha\}$ shows that for all α and β,

$$\langle A^*(u_\alpha), u_\beta \rangle = \langle u_\alpha, A(u_\beta) \rangle = \langle u_\alpha, \lambda_\beta u_\beta \rangle = \overline{\lambda_\beta} \langle u_\alpha, u_\beta \rangle = \langle \overline{\lambda_\alpha} u_\alpha, u_\beta \rangle.$$

Hence $A^*(u_\alpha) = \overline{\lambda_\alpha} u_\alpha$ for each α. Let $x \in H$. Then

$$\begin{aligned} \langle A^*A(x), u_\alpha \rangle &= \langle A(x), A(u_\alpha) \rangle = \overline{\lambda_\alpha} \langle A(x), u_\alpha \rangle \\ &= \overline{\lambda_\alpha} \langle x, A^*(u_\alpha) \rangle = \overline{\lambda_\alpha} \lambda_\alpha \langle x, u_\alpha \rangle = |\lambda_\alpha|^2 \langle x, u_\alpha \rangle. \end{aligned}$$

Similarly, $\langle AA^*(x), u_\alpha \rangle$ is equal to $\lambda_\alpha \overline{\lambda_\alpha} \langle x, u_\alpha \rangle = |\lambda_\alpha|^2 \langle x, u_\alpha \rangle$ for all α. Hence $A^*A(x) = AA^*(x)$ for all $x \in H$. Thus A is normal.

It is clear that $\lambda_\alpha \in \sigma_e(A)$ for each α. Conversely, suppose there is $\lambda \in \mathbb{K}$ such that $\lambda \neq \lambda_\alpha$ for any α, and suppose $x \in H$ satisfies $A(x) = \lambda x$. Since A is normal, Proposition 5.30(ii) shows that $x \perp u_\alpha$ for all α. Since $\{u_\alpha\}$ is an orthonormal basis for H, it follows that $x = 0$ by condition (iii) of Theorem 2.31, and so $\lambda \notin \sigma_e(A)$. Thus $\sigma_e(A) = \{\lambda_\alpha\}$. Let Λ denote the closure of $\{\lambda_\alpha\}$ in $\mathbb{K}$. Since $\sigma_e(A) \subset \sigma_a(A)$, and $\sigma_a(A)$ is a closed subset of $\mathbb{K}$ (Proposition 5.5), we see that $\Lambda \subset \sigma_a(A)$. Conversely, let $k \in \mathbb{K} \setminus \Lambda$. We show that $A - kI$ is bounded below. Let $\delta := d(k, \Lambda)$. For $x \in H$, let $\{v_1, v_2, \ldots\} := \{u_\alpha : \langle x, u_\alpha \rangle \neq 0\}$, and let $A(v_j) := \mu_j v_j$, where $\mu_j \in \{\lambda_\alpha\}$. By Theorem 2.31, $x = \sum_j \langle x, v_j \rangle v_j$, and so $A(x) = \sum_j \langle x, v_j \rangle A(v_j) = \sum_j \mu_j \langle x, v_j \rangle v_j$. Hence

$$\|A(x) - kx\|^2 = \Big\| \sum_j (\mu_j - k)\langle x, v_j\rangle v_j \Big\|^2 = \sum_j |\mu_j - k|^2 |\langle x, v_j\rangle|^2$$
$$\geq \delta^2 \sum_j |\langle x, v_j\rangle|^2 = \delta^2 \|x\|^2$$

by the Parseval formula given in condition (v) of Theorem 2.31. Since $\delta > 0$, we see that $k \notin \sigma_a(A)$. Thus $\sigma_a(A) = \Lambda$. Since A is normal, $\sigma(A) = \sigma_a(A) = \Lambda$ by Proposition 5.30(iii).

(ii) A is self-adjoint if and only if $A^*(x) = A(x)$ for all $x \in H$, that is, $\langle A^*(x), u_\alpha\rangle = \langle A(x), u_\alpha\rangle$ for all $x \in H$ and all α. But $\langle A^*(x), u_\alpha\rangle = \langle x, A(u_\alpha)\rangle = \overline{\lambda_\alpha}\langle x, u_\alpha\rangle$ and $\langle A(x), u_\alpha\rangle = \langle x, A^*(u_\alpha)\rangle = \lambda_\alpha\langle x, u_\alpha\rangle$. It follows that A is self-adjoint if and only if $\overline{\lambda_\alpha} = \lambda_\alpha$, that is, $\lambda_\alpha \in \mathbb{R}$ for all α. Further, if A is positive, then clearly $\lambda_\alpha = \langle A(u_\alpha), u_\alpha\rangle \geq 0$ for all α. Conversely, if $\lambda_\alpha = \langle A(u_\alpha), u_\alpha\rangle \geq 0$ for all α, then by considering the Fourier expansion of each $x \in H$ given in Theorem 2.31, we see that $\langle A(x), x\rangle \geq 0$ for all $x \in H$.

Since A is normal, it is unitary if and only if $A^*A(x) = x$ for all $x \in H$, that is, $\langle A^*A(x), u_\alpha\rangle = \langle x, u_\alpha\rangle$ for all $x \in H$ and all α. But $\langle A^*A(x), u_\alpha\rangle = |\lambda_\alpha|^2\langle x, u_\alpha\rangle$ as in (i) above. Thus A is unitary if and only if $|\lambda_\alpha| = 1$ for all α.

(iii) Let A be a compact operator. The eigenspace of A corresponding to each fixed nonzero eigenvalue λ_α is finite dimensional, and the set of all such eigenvalues of A is countable by Theorem 5.21(ii). Since a countable union of finite sets is countable, it follows that the set $S := \{u_\alpha : \lambda_\alpha \neq 0\}$ is countable.

Conversely, suppose S is countable. Let $S := \{u_1, u_2, \ldots\}$, and let $A(u_n) = \lambda_n u_n$ for each n. Consider $x \in H$. By Bessel's inequality (Proposition 2.19),

$$\sum_{n=1}^{m} |\lambda_n|^2 |\langle x, u_n\rangle|^2 \leq \|A\|^2 \|x\|^2 \quad \text{for all } m \in \mathbb{N}.$$

Hence $\sum_n \lambda_n\langle x, u_n\rangle u_n$ converges in H by the Riesz–Fischer theorem (Theorem 2.29). Define $B(x) := \sum_n \lambda_n\langle x, u_n\rangle u_n$. If $u_\alpha \notin S$, then $\lambda_\alpha = 0$ and $\langle u_n, u_\alpha\rangle = 0$ for all $n \in \mathbb{N}$, and so $\langle A(x), u_\alpha\rangle = \langle x, A^*(u_\alpha)\rangle = \lambda_\alpha\langle x, u_\alpha\rangle = 0$ and $\langle B(x), u_\alpha\rangle = \sum_n \lambda_n\langle x, u_n\rangle\langle u_n, u_\alpha\rangle = 0$. Also, if $u_\alpha \in S$ and $u_\alpha = u_n$ for some $n \in \mathbb{N}$, then $\langle A(x), u_\alpha\rangle = \langle x, A^*(u_\alpha)\rangle = \lambda_n\langle x, u_n\rangle$ and $\langle B(x), u_\alpha\rangle = \sum_n \lambda_n\langle x, u_n\rangle\langle u_n, u_n\rangle = \lambda_n\langle x, u_n\rangle$. Thus we obtain $\langle A(x), u_\alpha\rangle = \langle B(x), u_\alpha\rangle$ for all α. Hence

$$A(x) = B(x) = \sum_n \lambda_n\langle x, u_n\rangle u_n \quad \text{for all } x \in H.$$

First, let S be finite. Then there is $m \in \mathbb{N}$ such that $S := \{u_1, \ldots, u_m\}$, and $A(x) = \sum_{n=1}^{m} \lambda_n\langle x, u_n\rangle u_n$ for all $x \in H$, and so $R(A) = \text{span}\,\{u_1, \ldots, u_m\}$. Thus A is a bounded operator of finite rank. Therefore, it is compact (Theorem 3.42). Next, let S

be denumerable. Then by Lemma 5.15, $\lambda_n \to 0$.[2] Let $\epsilon > 0$. There is $m_0 \in \mathbb{N}$ such that $|\lambda_n| < \epsilon$ for $n > m_0$. For $m \in \mathbb{N}$, let $A_m(x) := \sum_{n=1}^{m} \lambda_n \langle x, u_n \rangle u_n,\ x \in H$. Then for all $m \geq m_0$,

$$\|A(x) - A_m(x)\|^2 = \Big\| \sum_{n=m+1}^{\infty} \lambda_n \langle x, u_n \rangle u_n \Big\|^2 = \sum_{n=m+1}^{\infty} |\lambda_n|^2 |\langle x, u_n \rangle|^2$$

$$\leq \epsilon^2 \sum_{n=m+1}^{\infty} |\langle x, u_n \rangle|^2 \leq \epsilon^2 \|x\|^2, \quad x \in H,$$

by Bessel's inequality (Proposition 2.19). Hence $\|A - A_m\| \leq \epsilon$ for all $m \geq m_0$. Thus $A_m \to A$ in $BL(H)$. Since each A_m is a bounded operator of finite rank, it is a compact operator on H, and so A is compact by Proposition 3.43. □

Examples 5.37 (Diagonal operator) Let $H := \ell^2$, and let (λ_n) be a bounded sequence in $\mathbb{K}$. Then $\|(\lambda_1 x(1), \lambda_2 x(2), \ldots)\|_2 \leq \sup\{|\lambda_n| : n \in \mathbb{N}\}\ \|x\|_2$ for all $x := (x(1), x(2), \ldots) \in H$. Let $A(x) := (\lambda_1 x(1), \lambda_2 x(2), \ldots)$. Then $A(e_n) = \lambda_n e_n$ for $n \in \mathbb{N}$, and $\{e_1, e_2, \ldots\}$ is an orthonormal basis for H. By Proposition 5.36(i), $A^*(e_n) = \overline{\lambda_n} e_n$ for each $n \in \mathbb{N}$, and A is a normal operator on H. Also, $\sigma_e(A) = \{\lambda_n : n \in \mathbb{N}\}$, and $\sigma(A)$ is the closure of $\{\lambda_n : n \in \mathbb{N}\}$. Let $S := \{e_n : \lambda_n \neq 0\}$. Now S is a countable subset of H, and

$$A(x) = \sum_{\lambda_n \neq 0} \lambda_n \langle x, e_n \rangle e_n \quad \text{for all } x \in H.$$

Also, A is self-adjoint if and only if $\lambda_n \in \mathbb{R}$ for all $n \in \mathbb{N}$, A is positive if and only if $\lambda_n \geq 0$ for all $n \in \mathbb{N}$, and A is unitary if and only if $|\lambda_n| = 1$ for all $n \in \mathbb{N}$. Also, A is compact if and only if either S is a finite set, or the set S is denumerable and $\lambda_n \to 0$. ◊

Proposition 5.36(i) says that if a Hilbert space H has an orthonormal basis consisting of eigenvectors of $A \in BL(H)$, then A is normal. The converse does not hold in general. For example, let $H := L^2([0, 1])$, and for $x \in H$, define $A(x)(t) := t\,x(t),\ t \in [0, 1]$. It is easy to see that A is self-adjoint and has no eigenvalues. We shall now find conditions under which a normal operator on a Hilbert space will have eigenvectors that form an orthonormal basis.

We begin our investigation by proving the following useful result.

---

[2] Using the orthonormality of the eigenvectors $u_1, u_2, \ldots$ of A, we can give a shorter proof of '$\lambda_n \to 0$' as follows. Assume for a moment that $\lambda_n \not\to 0$. Then there is $\delta > 0$, and there are $n_1 < n_2 < \cdots$ in $\mathbb{N}$ such that $|\lambda_{n_k}| \geq \delta$ for all $k \in \mathbb{N}$, and so

$$\|A(u_{n_k}) - A(u_{n_j})\|^2 = \|\lambda_{n_k} u_{n_k} - \lambda_{n_j} u_{n_j}\|^2 = |\lambda_{n_k}|^2 + |\lambda_{n_j}|^2 \geq 2\delta^2$$

for all $k, j \in \mathbb{N}$ with $k \neq j$ by the Pythagoras theorem (Proposition 2.16(i)). Now (u_{n_k}) is a bounded sequence in H, and the sequence $(A(u_{n_k}))$ has no convergent subsequence, which contradicts the compactness of the operator A.

Lemma 5.38 *Let H be a Hilbert space, and let A be a normal operator on H. If G is a subspace of H spanned by eigenvectors of A, then $A(G^{\perp}) \subset G^{\perp}$.*

Proof Let $x \in G^{\perp}$ and $y \in G$. There are eigenvectors $y_1, \ldots, y_m$ of A in G, and $k_1, \ldots, k_m$ in $\mathbb{K}$ such that $y = k_1 y_1 + \cdots + k_m y_m$. If $A(y_j) = \lambda_j y_j$ for $j = 1, \ldots, m$, then by Proposition 5.30(i), $A^*(y_j) = \overline{\lambda_j} y_j$ for $j = 1, \ldots, m$. Since $\langle x, y_j \rangle = 0$ for $j = 1, \ldots, m$, we see that

$$\langle A(x), y \rangle = \langle x, A^*(y) \rangle = \langle x, k_1 \overline{\lambda_1} y_1 + \cdots + k_m \overline{\lambda_m} y_m \rangle = \sum_{j=1}^{m} \overline{k_j} \lambda_j \langle x, y_j \rangle = 0,$$

that is, $A(x) \perp y$. Thus $A(G^{\perp}) \subset G^{\perp}$. □

We first consider the case where the Hilbert space is finite dimensional and the scalars are complex numbers.

Theorem 5.39 (Finite dimensional spectral theorem over $\mathbb{C}$) *Let H be a finite dimensional nonzero Hilbert space over $\mathbb{C}$, and let $A \in BL(H)$. Then A is normal if and only if H has an orthonormal basis consisting of eigenvectors of A.*

In this case, let $\{u_1, \ldots, u_m\}$ be an orthonormal basis for H, and let $\lambda_1, \ldots, \lambda_m$ be complex numbers such that $A(u_n) = \lambda_n u_n$ for $n = 1, \ldots, m$. Then $\sigma(A) = \sigma_e(A) = \{\lambda_1, \ldots, \lambda_m\}$, and

$$A(x) = \sum_{n=1}^{m} \lambda_n \langle x, u_n \rangle u_n \quad \textit{for all } x \in H.$$

Further, A is self-adjoint if and only if $\lambda_n \in \mathbb{R}$ for $n = 1, \ldots, m$, A is positive if and only if $\lambda_n \geq 0$ for $n = 1, \ldots, m$, and A is unitary if and only if $|\lambda_n| = 1$ for $n = 1, \ldots, m$.

Proof Let $m := \dim H$. Now $m \in \mathbb{N}$, as $H \neq \{0\}$ and H is finite dimensional.

Suppose A is normal. We have seen in Remark 5.3(i) that there is an $m \times m$ matrix $M := [k_{i,j}]$ with $k_{i,j} \in \mathbb{C}$ for $i, j = 1, \ldots, m$ such that the eigenvalues of A and the eigenvalues of M are the same. Further, in Example 5.4(i), we have seen that the eigenvalues of M are the roots of the characteristic polynomial $\det(M - tI)$ of degree m. Since the coefficients of this polynomial are complex numbers, it has at least one complex root, and so A has at least one eigenvalue. Let $\mu_1, \ldots, \mu_k$ be the distinct eigenvalues of A. For $j = 1, \ldots, k$, let $E_j := Z(A - \mu_j I)$ denote the eigenspace of A corresponding to μ_j, and define $G := E_1 + \cdots + E_k$.

We claim that $G^{\perp} = \{0\}$. Since the subspace G of H is spanned by eigenvectors of A, Lemma 5.38 shows that $A(G^{\perp}) \subset G^{\perp}$. Assume for a moment that $G^{\perp} \neq \{0\}$. Now $G^{\perp}$ is a finite dimensional nonzero Hilbert space over $\mathbb{C}$. Hence the restriction of A to $G^{\perp}$ has an eigenvalue $\mu \in \mathbb{C}$ along with a corresponding eigenvector $x \in G^{\perp}$. Since μ is an eigenvalue of A as well, there is $j \in \{1, \ldots, k\}$ such that $\mu =$

μ_j, and so $x \in E_j \subset G$. But this is impossible since $x \neq 0$, and $x \in G \cap G^\perp = \{0\}$. Hence $G^\perp = \{0\}$, and $H = G + G^\perp = G + \{0\} = G$ by the projection theorem (Theorem 2.35).

Let $\{u_{j,1}, \ldots, u_{j,m_j}\}$ be an orthonormal basis for E_j for $j = 1, \ldots, k$. Since A is normal, $\{u_{1,1} \ldots, u_{1,m_1}, \ldots, u_{k,1}, \ldots, u_{k,m_k}\}$ is an orthonormal subset of H by Proposition 5.30(ii), and clearly it spans $E_1 + \cdots + E_m = G = H$. By condition (i) of Theorem 2.31, it is in fact an orthonormal basis for H. Since $m = m_1 + \cdots + m_k$, we rename the elements $u_{1,1} \ldots, u_{1,m_1}, \ldots, u_{k,1}, \ldots, u_{k,m_k}$ of H as $u_1, \ldots, u_m$, and let λ_n denote the eigenvalue of A corresponding to the eigenvector u_n for $n = 1, \ldots, m$. Then $\sigma(A) = \sigma_e(A) = \{\mu_1, \ldots, \mu_k\} = \{\lambda_1, \ldots, \lambda_m\}$. Also,

$$x = \sum_{n=1}^{m} \langle x, u_n \rangle u_n, \quad \text{and} \quad A(x) = \sum_{n=1}^{m} \lambda_n \langle x, u_n \rangle u_n \quad \text{for all } x \in H.$$

The converse part and the last statement of the theorem follow from Proposition 5.36(i) and (ii) for the case $\mathbb{K} := \mathbb{C}$. □

For a version of the above result in terms of orthogonal projection operators, see Exercise 5.36.

Theorem 5.40 (Spectral theorem for a compact self-adjoint operator) *Let H be a Hilbert space over $\mathbb{K}$, and let $A \in BL(H)$ be nonzero. Then A is compact and self-adjoint if and only if H has an orthonormal basis $\{u_\alpha\}$ consisting of eigenvectors of A such that (i) the eigenvalue λ_α of A corresponding to u_α is real for each α, (ii) the subset $S := \{u_\alpha : \lambda_\alpha \neq 0\}$ of H is countable, say $S = \{u_1, u_2, \ldots\}$ with $A(u_n) = \lambda_n u_n$, and (iii) $\lambda_n \to 0$ if S is denumerable.*

In this case, $\sigma(A) \setminus \{0\} = \sigma_e(A) \setminus \{0\} = \{\lambda_1, \lambda_2, \ldots\}$, and

$$A(x) = \sum_{n} \lambda_n \langle x, u_n \rangle u_n \quad \textit{for all } x \in H,$$

where any λ_n appears only a finite number of times in the summation.

Proof H is nonzero since $A \neq 0$. Suppose A is compact and self-adjoint. Since A is self-adjoint, $\sigma(A) \subset \mathbb{R}$, and both m_A and M_A are approximate eigenvalues of A (Proposition 5.31). Also, $\|A\| = \max\{|m_A|, |M_A|\}$ by Corollary 5.32(ii). Hence at least one of m_A and M_A is nonzero and its absolute value is equal to $\|A\|$. Further, since A is compact, every nonzero spectral value of A is in fact an eigenvalue of A (Theorem 5.21(i)). Hence there is a nonzero real eigenvalue of A. Also, the set of all nonzero eigenvalues of A is countable, and the eigenspace corresponding to each nonzero eigenvalue is finite dimensional by Theorem 5.21(ii).

Let $\mu_1, \mu_2, \ldots$ be the distinct nonzero eigenvalues of A. Then each $\mu_j \in \mathbb{R}$. For each j, let $E_j := Z(A - \mu_j I)$ denote the eigenspace of A corresponding to μ_j, and let G denote the closure of the linear span of $\cup_j E_j$.

Since the subspace G of H is the closure of the linear span of eigenvectors of A, it follows from Lemma 5.38 that $A(G^\perp) \subset G^\perp$. Let A_0 denote the restriction of A to $G^\perp$. We claim that $A_0 = 0$. Assume for a moment that $A_0 \neq 0$. Now $G^\perp$ is a Hilbert space over $\mathbb{K}$, and it is easy to see that A_0 is a nonzero compact self-adjoint operator on $G^\perp$. Hence A_0 has a real eigenvalue μ and a corresponding eigenvector $x \in G^\perp$. Since μ is an eigenvalue of A as well, there is $j \in \mathbb{N}$ such that $\mu = \mu_j$, and so $x \in E_j \subset G$. But this is impossible since $x \neq 0$, and $G \cap G^\perp = \{0\}$. Hence $A_0 = 0$, and so $G^\perp \subset Z(A)$. Conversely, if $x \in Z(A)$, then $x \in E_j^\perp$ for every j by Proposition 5.30(ii), and so $x \in G^\perp$. Thus $G^\perp = Z(A)$, and $H = G \oplus G^\perp = G + Z(A)$ by the projection theorem (Theorem 2.35).

Let $\{u_{j,1}, \ldots, u_{j,m_j}\}$ be an orthonormal basis for E_j for each j. Since A is normal, $\{u_{1,1}, \ldots, u_{1,m_1}, u_{2,1}, \ldots, u_{2,m_2}, \ldots\}$ is an orthonormal subset of H by Proposition 5.30(ii), and clearly its span is dense in G. Let us rename the elements $u_{1,1}, \ldots, u_{1,m_1}, \ldots, u_{2,1}, \ldots, u_{2,m_2}, \ldots$ of H as $u_1, u_2 \ldots$. Let λ_n denote the eigenvalue of A corresponding to the eigenvector u_n for $n \in \mathbb{N}$.

If $Z(A) = \{0\}$, then $H = G$, and so $\{u_1, u_2, \ldots\}$ is an orthonormal basis for H consisting of eigenvectors of A by condition (i) of Theorem 2.31. Next, if $Z(A) \neq \{0\}$, and $\{v_\alpha\}$ is an orthonormal basis for the nonzero Hilbert space $Z(A)$, then $\{u_\alpha\} := \{u_1, u_2, \ldots\} \cup \{v_\alpha\}$ is an orthonormal basis for $G \oplus Z(A) = H$ consisting of eigenvectors of A. Also, if $A(u_\alpha) = \lambda_\alpha u_\alpha$, then $\lambda_\alpha \in \{\mu_1, \mu_2, \ldots\}$ or $\lambda_\alpha = 0$, and so $\lambda_\alpha \in \mathbb{R}$. We note that the set $S := \{u_\alpha : \lambda_\alpha \neq 0\} = \{u_1, u_2, \ldots\}$ is countable. Finally, suppose the set S is denumerable. Since A is compact, $\mu_n \to 0$ by Theorem 5.21(ii). Let $\epsilon > 0$. Then there is $n_0 \in \mathbb{N}$ such that $|\mu_n| < \epsilon$ for all $n > n_0$. It follows that $|\lambda_n| < \epsilon$ for all $n > m_1 + \cdots + m_{n_0}$. Hence $\lambda_n \to 0$.

Conversely, suppose H has an orthonormal basis $\{u_\alpha\}$ consisting of eigenvectors of A such that the eigenvalue λ_α of A corresponding to u_α is real for each α, the subset $S := \{u_\alpha : \lambda_\alpha \neq 0\}$ of H is countable, say $S = \{u_1, u_2, \ldots\}$ with $A(u_n) = \lambda_n u_n$, and $\lambda_n \to 0$ if S is denumerable. Then Proposition 5.36(ii) shows that A is self-adjoint, and Proposition 5.36(iii) shows that A is compact. Further, by Proposition 5.36(i), $\sigma_e(A) = \{\lambda_\alpha\}$, and so $\sigma_e(A) \setminus \{0\} = \{\lambda_1, \lambda_2, \ldots\}$. Also, $A(x) = \sum_n \lambda_n \langle x, u_n \rangle u_n$ for all $x \in H$, and no λ_n can appear infinitely many times in this summation since $\lambda_n \neq 0$, and $\lambda_n \to 0$. Finally, by Theorem 5.21(i), $\sigma(A) \setminus \{0\} = \sigma_e(A) \setminus \{0\}$. □

For a version of the above result in terms of orthogonal projection operators, see Exercise 5.37.

We shall now use Theorem 5.40 to treat the case where the Hilbert space is finite dimensional and the scalars are real numbers.

Corollary 5.41 (Finite dimensional spectral theorem over $\mathbb{R}$) *Let H be a finite dimensional nonzero Hilbert space over $\mathbb{R}$, and let $A \in BL(H)$. Then A is self-adjoint if and only if H has an orthonormal basis consisting of eigenvectors of A.*

In this case, let $\{u_1, \ldots, u_m\}$ *is an orthonormal basis for* H, *and let* $\lambda_1, \ldots, \lambda_m$ *be real numbers such that* $A(u_n) = \lambda_n u_n$ *for* $n = 1, \ldots, m$. *Then* $\sigma(A) = \sigma_e(A) = \{\lambda_1, \ldots, \lambda_m\}$, *and*

$$A(x) = \sum_{n=1}^{m} \lambda_n \langle x, u_n \rangle u_n \textit{ for all } x \in H.$$

Further, A is positive if and only if $\lambda_n \geq 0$ *for* $n = 1, \ldots, m$.

Proof Let $m := \dim H$. Now $m \in \mathbb{N}$, as $H \neq \{0\}$ and H is finite dimensional.

Suppose A is self-adjoint. If $A = 0$, then let $\{u_1, \ldots, u_m\}$ be an orthonormal basis for H, and let $\lambda_n := 0$ for $n = 1, \ldots, m$. Next, let $A \neq 0$. Now A is a compact operator on H by Theorem 3.42. Let $\mathbb{K} := \mathbb{R}$ in Theorem 5.40. Since $\dim H = m$, we conclude that there is $n \in \mathbb{N}$ with $n \leq m$ such that

$$A(x) = \lambda_1 \langle x, u_1 \rangle u_1 + \cdots + \lambda_n \langle x, u_n \rangle u_n \quad \text{for all } x \in H,$$

where $\{u_1, \ldots, u_n\}$ is an orthonormal subset of H and $\lambda_1, \ldots, \lambda_n$ are nonzero real numbers. If $n < m$, then we can find $u_{n+1}, \ldots, u_m$ in H such that $\{u_1, \ldots, u_m\}$ is an orthonormal basis for H, and we may let $\lambda_j := 0$ for $j = n + 1, \ldots, m$. Since $A(u_j) = \lambda_j u_j$ for $j = 1, \ldots, m$, the orthonormal basis $\{u_1, \ldots, u_m\}$ for H consists of eigenvectors of A. Also, it is clear that $A(x) = \lambda_1 \langle x, u_1 \rangle u_1 + \cdots + \lambda_m \langle x, u_m \rangle u_m$ for all $x \in H$, and $\sigma(A) = \sigma_e(A) = \{\lambda_1, \ldots, \lambda_m\}$.

The converse part and the last statement of the theorem follow from Proposition 5.36(ii) if we let $\mathbb{K} := \mathbb{R}$. □

Remarks 5.42 (i) If A is a nonzero compact normal operator on a Hilbert space H over $\mathbb{R}$, then H need not have an orthonormal basis consisting of eigenvectors of A. In fact, if H is finite dimensional, and the dimension of H is even, then A need not have any eigenvector at all. For example, let $n \in \mathbb{N}$ and $H := \mathbb{R}^{2n}$. Define $A(x) := (-x(2), x(1), \ldots, -x(2n), x(2n-1))$ for $x := (x(1), \ldots, x(2n))$ in H. Since H is finite dimensional, A is a compact operator. Next, $A^*(x) = (x(2), -x(1), \ldots, x(2n), -x(2n-1))$, $A^*A(x) = (x(1), x(2), \ldots, x(2n-1), x(2n)) = AA^*(x)$ for all $x \in H$. Hence A is in fact a unitary operator. It is also easy to see that if $\lambda \in \mathbb{R}$ and $x \in H$ satisfy $A(x) = \lambda x$, then $\lambda^2 = -1$ or $x = 0$. Thus $\sigma_e(A) = \emptyset$, and so no $x \in H$ is an eigenvector of A. Next, let $\mathbb{K} := \mathbb{R}$, $H := \ell^2$, and define

$$A(x) := \Big(-x(2), x(1), -\frac{x(4)}{2}, \frac{x(3)}{2}, -\frac{x(6)}{3}, \frac{x(5)}{3}, \cdots \Big)$$

for $x := (x(1), x(2), \ldots) \in H$. Then A is compact and normal, but $\sigma_e(A) = \emptyset$. On the other hand, let $n \in \mathbb{N}$, $H := \mathbb{R}^{2n+1}$, and for $x := (x(1), \ldots, x(2n+1))$ in H, define $A(x) := (-x(2), x(1), \ldots, -x(2n), x(2n-1), x(2n+1))$. Then A is compact and normal, $\sigma_e(A) = \{1\}$, and the eigenspace of A corresponding to $\lambda_1 := 1$ is spanned by $u_1 := (0, \ldots, 0, 1) \in H$. In this case also, H does not have an orthonormal basis consisting of eigenvectors of A.

(ii) It is possible to prove an analogue of Theorem 5.40 for a nonzero compact normal operator A on a Hilbert space H over $\mathbb{C}$. To prove this, we need to know that if A is a normal operator on a Hilbert space H over $\mathbb{C}$, then there is a nonzero eigenvalue $\lambda \in \mathbb{C}$ of A. This can be seen as follows. For a bounded linear operator A on a Banach space over $\mathbb{C}$, we have stated the spectral radius formula $\max\{|\lambda| : \lambda \in \sigma(A)\} = \lim_{n\to\infty} \|A^n\|^{1/n}$ in Remark 5.13. Since A is a normal operator on a Hilbert space H, $\|A^2\| = \|A\|^2$ by Theorem 4.30(ii), and so $\|A^{2^j}\| = \|A\|^{2^j}$ for all $j \in \mathbb{N}$. Hence

$$\max\{|\lambda| : \lambda \in \sigma(A)\} = \lim_{n\to\infty} \|A^n\|^{1/n} = \lim_{j\to\infty} \|A^{2^j}\|^{1/2^j} = \|A\|.$$

Thus there is $\lambda \in \sigma(A)$ such that $|\lambda| = \|A\| \neq 0$; in fact, $\lambda \in \sigma_e(A)$, since A is compact. Hence Theorem 5.40 and its proof hold for a nonzero compact normal operator on a Hilbert space over $\mathbb{C}$, and we obtain the following result known as the **spectral theorem for a compact normal operator**:

Let H be a Hilbert space over $\mathbb{C}$, and let $A \in BL(H)$ be nonzero. Then A is compact and normal if and only if H has an orthonormal basis $\{u_\alpha\}$ consisting of eigenvectors of A such that the subset $S := \{u_\alpha : \lambda_\alpha \neq 0\}$ of H is countable, say $S = \{u_1, u_2, \ldots\}$ with $A(u_n) = \lambda_n u_n$, and $\lambda_n \to 0$ if S is denumerable. In this case, $\sigma(A) \setminus \{0\} = \sigma_e(A) \setminus \{0\} = \{\lambda_1, \lambda_2, \ldots\}$, and

$$A(x) = \sum_n \lambda_n \langle x, u_n\rangle u_n \quad \text{for all } x \in H,$$

where any λ_n appears only a finite number of times in the summation.

The spectral theorem for a compact normal operator is sometimes called a 'structure theorem' since it gives a complete description of the structure of the operator in terms of its eigenvalues and eigenvectors. If a normal operator is not compact, then eigenvalues and eigenvectors are not enough to describe its structure. One can give an integral representation of such an operator in terms of orthogonal projection operators associated with the operator, and obtain a result similar to Exercises 5.36 and 5.37. See [23, pages 275 and 288].

(iii) Suppose A is a compact self-adjoint operator on a Hilbert space H over $\mathbb{K}$ (as in Theorem 5.40), or A is a compact normal operator on a Hilbert space H over $\mathbb{C}$ (as in (ii) above). Then the nonzero eigenvalues $\{\lambda_1, \lambda_2, \ldots\}$ of A can be renumbered so that $|\lambda_1| \geq |\lambda_2| \geq \cdots$, because the set $\{\lambda \in \sigma_e(A) : |\lambda| = r\}$ is finite for every $r > 0$. For $m \in \mathbb{N}$, let $A_m(x) := \sum_{n=1}^{m} \lambda_n \langle x, u_n\rangle u_n$, $x \in H$. Then $\|A - A_m\| \leq |\lambda_{m+1}|$ for $m \in \mathbb{N}$. (Compare the proof of the compactness of A given in Proposition 5.36(iii).) This inequality gives an estimate of the rate at which the sequence (A_m) of finite rank operators converges to the operator A in $BL(H)$. ◊

Theorem 5.40 is useful in finding explicit solutions of operator equations of the type $x - \mu A(x) = y$, as the following result shows.

Theorem 5.43 *Let A be a nonzero compact self-adjoint operator on a Hilbert space H over $\mathbb{K}$, or a compact normal operator on a Hilbert space H over $\mathbb{C}$. Then there is a countable orthonormal subset $\{u_1, u_2, \ldots\}$ of H and there are nonzero scalars $\mu_1, \mu_2, \ldots$ such that $A(x) = \sum_n \mu_n^{-1}\langle x, u_n\rangle u_n$ for all $x \in H$.*

(i) *Suppose μ is a scalar such that $\mu \neq \mu_n$ for any n. Let $y \in H$. Then there is a unique $x \in H$ such that $x - \mu A(x) = y$, and in fact*

$$x = y + \mu \sum_n \frac{\langle y, u_n\rangle u_n}{\mu_n - \mu}.$$

Further, $\|x\| \leq \alpha\|y\|$, where $\alpha := 1 + |\mu|/\inf\{|\mu_n - \mu| : n \in \mathbb{N}\}$.

(ii) *Suppose μ is a scalar such that $\mu = \mu_{j_1} = \cdots = \mu_{j_m}$, and $\mu \neq \mu_n$ for any other n. Let $y \in H$. Then there is $x_0 \in H$ such that $x_0 - \mu A(x_0) = y$ if and only if $y \perp u_{j_1}, \ldots, y \perp u_{j_m}$. In this case, $x - \mu A(x) = y$ if and only if there are scalars $k_{j_1}, \ldots, k_{j_m}$ such that*

$$x = y + \mu \sum_{n \notin \{j_1,\ldots,j_m\}} \frac{\langle y, u_n\rangle u_n}{\mu_n - \mu} + k_{j_1}u_{j_1} + \cdots + k_{j_m}u_{j_m}.$$

Proof By Theorem 5.40 and Remark 5.42(ii), there is a countable orthonormal subset $\{u_1, u_2, \ldots\}$ of H, and there are nonzero scalars $\lambda_1, \lambda_2, \ldots$ such that $A(x) = \sum_n \lambda_n\langle x, u_n\rangle u_n$ for all $x \in H$. Let $\mu_n := \lambda_n^{-1}$ for $n \in \mathbb{N}$. Then $\mu_n \neq 0$ for each n, and $A(x) = \sum_n \mu_n^{-1}\langle x, u_n\rangle u_n$ for all $x \in H$. As a consequence, $\langle A(x), u_n\rangle = \mu_n^{-1}\langle x, u_n\rangle$ for all $x \in H$ and each n.

(i) Suppose μ is a scalar and $\mu \neq \mu_n$ for any n. Let $y \in H$. Then for $x \in H$, $x - \mu A(x) = y$ if and only if $x = y + \mu A(x) = y + \mu\sum_n \mu_n^{-1}\langle x, u_n\rangle u_n$. Now $\langle x, u_n\rangle = \langle y, u_n\rangle + \mu\,\mu_n^{-1}\langle x, u_n\rangle$, that is, $\langle x, u_n\rangle = \mu_n(\mu_n - \mu)^{-1}\langle y, u_n\rangle$ for each n. Hence $x = y + \mu A(x)$ if and only if $x = y + \mu\sum_n(\mu_n - \mu)^{-1}\langle y, u_n\rangle u_n$.

Let $z := \mu\sum_n(\mu_n - \mu)^{-1}\langle y, u_n\rangle u_n$. Since $\mu \neq \mu_n$ for any n, and since $|\mu_n| = 1/|\lambda_n| \to \infty$ if the orthonormal set $\{u_1, u_2, \ldots\}$ is denumerable (Theorem 5.40 and Remark 5.42(ii)), $\delta := \inf\{|\mu_n - \mu| : n \in \mathbb{N}\} > 0$. By Bessel's inequality (Proposition 2.19),

$$\|z\|^2 = |\mu|^2 \sum_n \frac{|\langle y, u_n\rangle|^2}{|\mu_n - \mu|^2} \leq \frac{|\mu|^2}{\delta^2}\|y\|^2.$$

Hence $\|x\| = \|y + z\| \leq \|y\| + \|z\| \leq \alpha\|y\|$, where $\alpha := 1 + |\mu|/\delta$.

(ii) Suppose $\mu = \mu_{j_1} = \cdots = \mu_{j_m}$, and $\mu \neq \mu_n$ for any other n. Let $y \in H$. Suppose there is $x_0 \in H$ such that $x_0 - \mu A(x_0) = y$. If $n \in \{j_1, \ldots, j_m\}$, then $\mu = \mu_n$, and

$$\langle y, u_n\rangle = \langle x_0, u_n\rangle - \mu\langle A(x_0), u_n\rangle = \langle x_0, u_n\rangle - \mu\,\mu_n^{-1}\langle x_0, u_n\rangle = 0,$$

that is, $y \perp u_n$. Conversely, suppose $y \perp u_n$ for each $n \in \{j_1, \ldots, j_m\}$. As in (i) above, there is $\delta > 0$ such that $|\mu_n - \mu| > \delta$ for all $n \notin \{j_1, \ldots, j_m\}$. Hence

$$\sum_{n\notin\{j_1,\ldots,j_m\}} \frac{|\langle y, u_n\rangle|^2}{|\mu_n - \mu|^2} \le \frac{\|y\|^2}{\delta^2} < \infty,$$

and by the Riesz–Fischer theorem (Theorem 2.29), there is $z_0 \in H$ such that

$$z_0 = \mu \sum_{n\notin\{j_1,\ldots,j_m\}} \frac{\langle y, u_n\rangle u_n}{\mu_n - \mu}.$$

Let $x_0 := y + z_0$. Then $x_0 - \mu A(x_0) = y + z_0 - \mu\big(A(y) + A(z_0)\big)$. But

$$\begin{aligned} A(y) + A(z_0) &= \sum_n \frac{\langle y, u_n\rangle u_n}{\mu_n} + \mu \sum_{n\notin\{j_1,\ldots,j_m\}} \frac{\langle y, u_n\rangle A(u_n)}{\mu_n - \mu} \\ &= \sum_{n\notin\{j_1,\ldots,j_m\}} \frac{\langle y, u_n\rangle u_n}{\mu_n} + \mu \sum_{n\notin\{j_1,\ldots,j_m\}} \frac{\langle y, u_n\rangle u_n}{\mu_n(\mu_n - \mu)} \\ &= \sum_{n\notin\{j_1,\ldots,j_m\}} \frac{\langle y, u_n\rangle u_n}{\mu_n - \mu}, \end{aligned}$$

since $\langle y, u_{j_1}\rangle = \cdots = \langle y, u_{j_m}\rangle = 0$ and $A(u_n) = \lambda_n u_n = \mu_n^{-1} u_n$ for each n. Thus $\mu\big(A(y) + A(z_0)\big) = z_0$, and so $x_0 - \mu A(x_0) = y + z_0 - z_0 = y$.

Now for $x \in H$, $x - \mu A(x) = y$ if and only if $x - x_0 - \mu A(x - x_0) = y - y = 0$, that is, $x - x_0$ is in the eigenspace of A corresponding to its eigenvalue $\mu^{-1} = \lambda_{j_1} = \cdots = \lambda_{j_m}$. Since this eigenspace is spanned by $u_{j_1}, \ldots, u_{j_m}$, we obtain the desired result. □

Examples 5.44 (i) Let $H := \ell^2$, and let $k_{i,j} \in \mathbb{K}$ be such that $k_{j,i} = \overline{k_{i,j}}$ for all $i, j \in \mathbb{N}$, and $\gamma_{2,2}^2 := \sum_{i=1}^{\infty}\sum_{j=1}^{\infty} |k_{i,j}|^2 < \infty$. Let A denote the operator on ℓ^2 defined by the matrix $M := [k_{i,j}]$. In Examples 3.45(iii) and 4.28(ii), we have seen that A is compact and self-adjoint. Suppose $A \ne 0$. Using Theorem 5.43, we can obtain explicit solutions of the nontrivial system of an infinite number of linear equations in an infinite number of variables $x(1), x(2), \ldots$ given by

$$x(i) - \mu \sum_{j=1}^{\infty} k_{i,j} x(j) = y(i), \quad i \in \mathbb{N}.$$

Let us consider a simple example. Let $k_{i,j} := 1/ij$ for $i, j \in \mathbb{N}$. Then

$$\gamma_{2,2}^2 = \sum_{i=1}^{\infty} \frac{1}{i^2} \sum_{j=1}^{\infty} \frac{1}{j^2} = \left(\frac{\pi^2}{6}\right)^2 < \infty.$$

Also, $k_{i,j} \in \mathbb{R}$ and $k_{j,i} = k_{i,j}$ for all $i, j \in \mathbb{N}$. Hence M defines a compact self-adjoint operator A on ℓ^2. If $x \in \ell^2$, then $A(x)(i) = \sum_{j=1}^{\infty} k_{i,j} x(j) = (1/i)\sum_{j=1}^{\infty} x(j)/j$ for all $i \in \mathbb{N}$, that is, $A(x) = \big(\sum_{j=1}^{\infty} x(j)/j\big)(1, 1/2, 1/3, \ldots)$. Let $u_1 := \big(\sqrt{6}/\pi\big)$ $(1, 1/2, 1/3 \ldots) \in \ell^2$. Then $\|u_1\|_2 = 1$, and

$$A(x) = \frac{\pi^2}{6}\langle x, u_1\rangle u_1 \quad \text{for all } x \in \ell^2.$$

It follows that $\sigma(A) = \sigma_e(A) = \{0, \lambda_1\}$, where $\lambda_1 := \pi^2/6$, and span $\{u_1\}$ is the eigenspace of A corresponding to the eigenvalue $\pi^2/6$.

Let $y \in \ell^2$, and $\mu \in \mathbb{K}$, $\mu \neq 0$. If $\mu \neq 6/\pi^2$, then there is a unique $x \in \ell^2$ satisfying $x - \mu A(x) = y$. In fact,

$$x = y + \frac{\mu}{\mu_1 - \mu}\langle y, u_1\rangle u_1 = y + \frac{6\mu}{6 - \pi^2\mu}\Big(\sum_{j=1}^{\infty}\frac{y(j)}{j}\Big)\Big(1, \frac{1}{2}, \frac{1}{3}, \ldots\Big),$$

since $\mu_1 = \lambda_1^{-1} = 6/\pi^2$. Also, $\|x\|_2 \leq \alpha\|y\|_2$, where $\alpha := 1 + (|\mu|\pi^2/|6 - \mu\pi^2|)$.

If $\mu := 6/\pi^2$, then there is $x \in \ell^2$ satisfying $x - \mu A(x) = y$ if and only if $\langle y, u_1\rangle = 0$, that is, $\sum_{j=1}^{\infty} y(j)/j = 0$. In this case, $x \in \ell^2$ satisfies $x - \mu A(x) = y$ if and only if there is $k_1 \in \mathbb{K}$ such that $x = y + k_1 u_1$.

(ii) Let $H := L^2([a, b])$, and $k(\cdot\,, \cdot)$ be a measurable function on $[a, b] \times [a, b]$ such that $\overline{k(t, s)} = k(s, t)$ for all $s, t \in [a, b]$, and $k(\cdot\,, \cdot) \in L^2([a, b] \times [a, b])$. Let A denote the Fredholm integral operator on $L^2([a, b])$ having kernel $k(\cdot\,, \cdot)$. Then A is compact and self-adjoint (Examples 3.46 and 4.28(iii)). Suppose $A \neq 0$. Using Theorem 5.43, we can obtain explicit solutions of the following **Fredholm integral equation of the second kind**:

$$x(s) - \mu\int_a^b k(s, t)x(t)\,dm(t) = y(s), \quad a \leq s \leq b.$$

Let us consider a specific example. Let $a := 0$, $b := 1$. Let $k(s, t) := s(1 - t)$ if $0 \leq s \leq t \leq 1$, and $k(s, t) := (1 - s)t$ if $0 \leq t \leq s \leq 1$. Then $k(\cdot\,, \cdot)$ is a real-valued continuous function on $[0, 1] \times [0, 1]$, and $k(t, s) = k(s, t)$ for all $s, t \in [0, 1]$. In Example 5.23(ii), we have seen that $0 \notin \sigma_e(A)$, and the nonzero eigenvalues of the operator A are given by $\lambda_n := 1/n^2\pi^2$, $n \in \mathbb{N}$, and the eigenspace of A corresponding to the eigenvalue λ_n equals span $\{x_n\}$, where $x_n(s) := \sin n\pi s$, $s \in [0, 1]$. Let $u_n(s) := \sqrt{2}\sin n\pi s$, $s \in [0, 1]$. It can be seen that $\{u_n : n \in \mathbb{N}\}$ is an orthonormal basis for $L^2([0, 1])$ consisting of eigenvectors of A. (Compare Exercise 2.34.) Then for each $x \in L^2([0, 1])$, $A(x) = \sum_n \lambda_n\langle x, u_n\rangle u_n$, that is,

$$\begin{aligned}(1 - s)\int_0^s t\,x(t)\,dm(t) &+ s\int_s^1 (1 - t)x(t)\,dm(t)\\ &= \frac{2}{\pi^2}\sum_{n=1}^{\infty}\frac{1}{n^2}\Big(\int_0^1 x(t)\sin n\pi t\,dm(t)\Big)\sin n\pi s.\end{aligned}$$

where the series on the right side converges in $L^2([0, 1])$. Let $y \in L^2([0, 1])$ and $\mu \in \mathbb{K}$, $\mu \neq 0$. Consider the integral equation

$$x(s) - \mu\left((1-s)\int_0^s t\,x(t)\,dm(t) + s\int_s^1 (1-t)x(t)\,dm(t)\right) = y(s), \quad s \in [0, 1].$$

If $\mu \neq n^2\pi^2$ for any $n \in \mathbb{N}$, then there is a unique $x \in L^2([0, 1])$ satisfying the above integral equation. In fact, for $s \in [0, 1]$,

$$x(s) = y(s) + 2\mu \sum_{n=1}^{\infty} \frac{1}{n^2\pi^2 - \mu}\left(\int_0^1 y(t) \sin n\pi t\,dm(t)\right) \sin n\pi s.$$

Further, $\|x\|_2 \leq \alpha\|y\|_2$, where $\alpha := 1 + (|\mu|/\inf\{|n^2\pi^2 - \mu| : n \in \mathbb{N}\})$.

Suppose $\mu = n_1^2\pi^2$, where $n_1 \in \mathbb{N}$. Then there is $x \in L^2([0, 1])$ satisfying the above integral equation if and only if $y \perp u_{n_1}$, that is,

$$\langle y, u_{n_1}\rangle = \sqrt{2}\int_0^1 y(t) \sin n_1\pi t\,dm(t) = 0.$$

In this case, $x \in L^2([0, 1])$ satisfies the above integral equation if and only if there is $k_1 \in \mathbb{K}$ such that for $s \in [0, 1]$,

$$x(s) = y(s) + 2n_1^2 \sum_{n \neq n_1} \frac{1}{n^2 - n_1^2}\left(\int_0^1 y(t) \sin n\pi t\,dm(t)\right) \sin n\pi s + k_1 \sin n_1\pi s.$$

The explicit result given above can be contrasted with the existence and uniqueness result given in Exercise 5.24. Of course, we have only obtained the explicit form of a solution of the operator equation $x - \mu A(x) = y$. To be able to compute such a solution exactly, we need to know all nonzero eigenvalues of the compact self-adjoint operator A, and the corresponding eigenvectors. This is, in general, a formidable task. A major contribution to the solution

$$x = y + \mu \sum_n \frac{\langle y, u_n\rangle}{\mu_n - \mu} u_n$$

comes from the terms $\langle y, u_n\rangle/(\mu_n - \mu)$ for which $|\mu_n - \mu|$ is small, that is, μ_n is close to, but not equal to, μ. Hence one can ignore those eigenvalues of A which are far from $1/\mu$ and focus on only the ones that are near $1/\mu$. ◊

Exercises

5.1 Let X be a normed space, and let $A \in BL(X)$ be invertible. Then $\sigma(A^{-1}) = \{\lambda^{-1} : \lambda \in \sigma(A)\}$. In fact, if $k \in \mathbb{K}$, $k \neq 0$, and $A - kI$ is invertible, then $A^{-1} - k^{-1}I$ is invertible, and $-k(A - kI)^{-1}A$ is its inverse.

5.2 Let X be a normed space, $A \in BL(X)$, and let p be a polynomial. Then $\{p(\lambda) : \lambda \in \sigma(A)\} \subset \sigma(p(A))$, where equality holds if $\mathbb{K} := \mathbb{C}$.

5.3 Let X be a normed space, and let $A, B \in BL(X)$. If $I - AB$ is invertible, then $I - BA$ is invertible, and $I + B(I - AB)^{-1}A$ is its inverse. Consequently, $\sigma(AB) \setminus \{0\} = \sigma(BA) \setminus \{0\}$.

5.4 **(Gershgorin theorem)** Let an $n \times n$ matrix $M := [k_{i,j}]$ define an operator $A \in BL(\mathbb{K}^n)$. For $i = 1, \ldots, n$, let $r_i := |k_{i,1}| + \cdots + |k_{i,i-1}| + |k_{i,i+1}| + \cdots + |k_{i,n}|$, and $D_i := \{k \in \mathbb{K} : |k - k_{i,i}| \leq r_i\}$. Then $\sigma(A) \subset D_1 \cup \cdots \cup D_n$. (Note: $D_1, \ldots, D_n$ are known as the **Gershgorin disks** of A. They localize the eigenvalues of A.)

5.5 Let X be a linear space over $\mathbb{K}$, and let $A : X \to X$ be of finite rank. Suppose $A(x) := g_1(x)x_1 + \cdots + g_n(x)x_n$ for all $x \in X$, where $x_1, \ldots, x_n$ are in X and $g_1, \ldots, g_n$ are linear functionals on X. Let M be the $n \times n$ matrix $[g_i(x_j)]$. Let $x \in X, u := (u(1), \ldots, u(n)) \in \mathbb{K}^n$, and nonzero $\lambda \in \mathbb{K}$. Then $A(x) = \lambda x$ and $u = (g_1(x), \ldots, g_n(x))$ if and only if $Mu^t = \lambda u^t$ and $x = \left(\sum_{j=1}^n u(j)x_j\right)/\lambda$. In this case, x is an eigenvector of A corresponding to λ if and only if u^t is an eigenvector of M corresponding to λ.

5.6 Let $X := C([a, b])$ with the sup norm, and let A denote the multiplication operator considered in Example 5.6(ii). Then $\sigma_e(A)$ consists of all $\lambda \in \mathbb{K}$ such that $x_0(t) = \lambda$ for all t in a nontrivial subinterval of $[a, b]$.

5.7 **(Multiplication operator)** Let $X := L^2([a, b])$, $x_0 \in L^\infty([a, b])$, and $A(x) := x_0 x$ for $x \in X$. Then $\sigma_a(A)$ and $\sigma(A)$ equal the **essential range** of x_0 consisting of $\lambda \in \mathbb{K}$ such that $m(\{t \in [a, b] : |x_0(t) - \lambda| < \epsilon\}) > 0$ for every $\epsilon > 0$. Also, $\sigma_e(A)$ is the set of all $\lambda \in \mathbb{K}$ such that $m(\{t \in [a, b] : x_0(t) = \lambda\} > 0$.

5.8 Let X be a normed space, and let $A \in BL(X)$. If $\lambda \in \sigma_a(A)$, and $n \in \mathbb{N}$, then $\lambda^n \in \sigma_a(A^n)$, and so $|\lambda| \leq \inf\left\{\|A^n\|^{1/n} : n \in \mathbb{N}\right\}$.

5.9 Let X be a normed space, and let $A \in BL(X)$.

(i) If there is $\alpha > 0$ such that $\|A(x)\| \leq \alpha\|x\|$ for all $x \in X$, then $\sigma_a(A)$ is contained in $\{k \in \mathbb{K} : |k| \leq \alpha\}$.

(ii) If there is $\beta > 0$ such that $\|A(x)\| \geq \beta\|x\|$ for all $x \in X$, then $\sigma_a(A)$ is contained in $\{k \in \mathbb{K} : |k| \geq \beta\}$.

(iii) If A is an isometry, then $\sigma_a(A)$ is contained in $\{k \in \mathbb{K} : |k| = 1\}$.

5.10 Let $X := \ell^1$, and let $A(x) := (0, x(1), 2x(2), x(3), 2x(4), \ldots)$ for $x \in X$. Then $A \in BL(X)$, $\|A\| = 2$, but $|\lambda| \leq \sqrt{2}$ for every $\lambda \in \sigma(A)$.

5.11 **(Diagonal operator)** Let $X := \ell^p$ with $p \in \{1, 2, \infty\}$, and let (λ_j) be a bounded sequence in $\mathbb{K}$. For $x := (x(1), x(2), \ldots)$ in X, define $A(x) := (\lambda_1 x(1), \lambda_2 x(2), \ldots)$. Then $A \in BL(X)$, $\|A\| = \sup\{|\lambda_j| : j \in \mathbb{N}\}$, $\sigma_e(A) = \{\lambda_j : j \in \mathbb{N}\}$, while $\sigma_a(A)$ and $\sigma(A)$ equal the closure of $\{\lambda_j : j \in \mathbb{N}\}$.

5.12 Let E be a nonempty closed and bounded subset of $\mathbb{K}$. Then there is a diagonal operator $A \in BL(\ell^2)$ such that $\sigma(A) = E$. Further, if (λ_j) is a sequence in $\mathbb{K}$ such that $\lambda_j \to 0$, then there is a diagonal operator $A \in CL(\ell^2)$ such that $\sigma_e(A) = \{\lambda_j : j \in \mathbb{N}\}$ and $\sigma(A) = \{\lambda_j : j \in \mathbb{N}\} \cup \{0\}$.

5.13 **(Neumann expansion)** Let X be a Banach space, and let $A \in BL(X)$. If $k \in \mathbb{K}$ and $|k| > \|A\|$, then $A - kI$ is invertible,

$$(A - kI)^{-1} = -\sum_{n=0}^{\infty} \frac{A^n}{k^{n+1}} \quad \text{and} \quad \|(A - kI)^{-1}\| \leq \frac{1}{|k| - \|A\|}.$$

In particular, if A denotes the right shift operator on ℓ^p defined in Example 5.12(i), then $(A - kI)^{-1}(y)(j) = -y(j)/k - \cdots - y(1)/k^j$ for $y \in \ell^p$, $j \in \mathbb{N}$.

5.14 **(Left shift operator)** Let X denote one of the spaces $\ell^1, \ell^2, \ell^\infty, c_0, c$, and define $B(x) := (x(2), x(3), \ldots)$ for $x := (x(1), x(2), \ldots) \in X$.
If $X := \ell^1$, ℓ^2 or c_0, then $\sigma_e(B) = \{\lambda \in \mathbb{K} : |\lambda| < 1\}$; if $X := \ell^\infty$, then $\sigma_e(B) = \{\lambda \in \mathbb{K} : |\lambda| \leq 1\}$; if $X := c$, then $\sigma_e(B) = \{\lambda \in \mathbb{K} : |\lambda| < 1\} \cup \{1\}$. In all cases, $\sigma_a(B) = \{\lambda \in \mathbb{K} : |\lambda| \leq 1\} = \sigma(B)$.

5.15 Let $p \in \{1, 2, \infty\}$, and let A be the right shift operator on $L^p([0, \infty))$ defined in Exercise 4.22. If $p \in \{1, 2\}$, then $\sigma_e(A) = \{\lambda \in \mathbb{K} : |\lambda| < 1\}$, and if $p := \infty$, then $\sigma_e(A) = \{\lambda \in \mathbb{K} : |\lambda| \leq 1\}$. In all cases, $\sigma_a(A)$ and $\sigma(A)$ equal $\{\lambda \in \mathbb{K} : |\lambda| \leq 1\}$.

5.16 Let X be a nonzero inner product space, and let $A \in BL(X)$. For nonzero $x \in X$, let $q_A(x) := \langle A(x), x\rangle / \langle x, x\rangle$ and $r_A(x) := A(x) - q_A(x)x$. The scalar $q_A(x)$ is called the **Rayleigh quotient** of A at x, and the element $r_A(x)$ of X is called the corresponding **residual**. Then $r_A(x) \perp x$, and $\|r_A(x)\| = \min\{\|A(x) - k\,x\| : k \in \mathbb{K}\}$. (Note: This is known as the **minimum residual property** of the Rayleigh quotient.)

5.17 Let X be a normed space, and let $A \in CL(X)$. Then $\sigma(A)$ is a closed and bounded subset of $\mathbb{K}$. If $\lambda \in \sigma(A)$, then $|\lambda| \leq \inf\{\|A^n\|^{1/n} : n \in \mathbb{N}\}$.

5.18 **(Weighted-shift operators)** Let $X := \ell^p$ with $p \in \{1, 2, \infty\}$, and let (w_n) be a sequence in $\mathbb{K}$ such that $w_n \to 0$. For $x := (x(1), x(2), \ldots) \in X$, define $A(x) := (0, w_1 x(1), w_2 x(2), \ldots)$ and $B(x) := (w_2 x(2), w_3 x(3), \ldots)$. Then $A, B \in CL(X)$, and $\sigma(A) = \sigma(B) = \{0\}$. Further, $0 \in \sigma_e(A)$ if and only if there is $j \in \mathbb{N}$ such that $w_j = 0$, and then the dimension of the corresponding eigenspace of A is the number of times 0 occurs among $w_1, w_2, \ldots$. Also, $0 \in \sigma_e(B)$, and the dimension of the corresponding eigenspace of B is one plus the number of times 0 occurs among $w_2, w_3, \ldots$.

5.19 **(Volterra integration operator)** Let $X := L^2([a, b])$. For $s, t \in [a, b]$, let $k(s, t) := 1$ if $t \leq s$ and $k(s, t) := 0$ if $s < t$, and let A denote the Fredholm integral operator on X with kernel $k(\cdot\,, \cdot)$. Then $A \in CL(X)$, $\sigma_e(A) = \emptyset$, and $\sigma_a(A) = \sigma(A) = \{0\}$.

5.20 Let $X := C([0, 1])$, and $Y := L^p([0, 1])$ with $p \in \{1, 2, \infty\}$. For $s, t \in [0, 1]$, let $k(s, t) := \min\{s, t\}$, and let A denote the Fredholm integral operator on X, and on Y, with kernel $k(\cdot\,,\cdot)$. Then

$$\sigma_e(A) = \{4/(2n-1)^2\pi^2 : n \in \mathbb{N}\} \quad \text{and} \quad \sigma_a(A) = \sigma(A) = \sigma_e(A) \cup \{0\}.$$

Further, the eigenspace corresponding to the eigenvalue $4/(2n-1)^2\pi^2$ of A is span $\{x_n\}$, where $x_n(s) := \sin(2n-1)\pi s/2,\ 0 \le s \le 1$, for $n \in \mathbb{N}$.

5.21 **(Perturbation by a compact operator)** Let X be a normed space, $A \in BL(X)$, and let $B \in CL(X)$. Then $\sigma(A) \setminus \sigma_e(A) \subset \sigma(A+B)$.

5.22 Let X be a normed space, $A \in CL(X)$, and let $k \in \mathbb{K}$ be nonzero. Then $A - kI$ is one-one if and only if $A - kI$ is onto.

5.23 Let X be a normed space, $A \in CL(X)$, and let $k \in \mathbb{K}$ be nonzero. Then $R(A - kI)$ is a closed subspace of X. (Compare Lemma 5.19.)

5.24 **(Fredholm alternative)** Let X be a normed space, and let $A \in CL(X)$. Exactly one of the following alternatives holds.

(i) For every $y \in X$, there is a unique $x \in X$ such that $x - A(x) = y$.
(ii) There is nonzero $x \in X$ such that $x - A(x) = 0$.

If the alternative (i) holds, then the unique solution x of the operator equation $x - A(x) = y$ depends continuously on the 'free term' $y \in X$. If the alternative (ii) holds, then the solution space of the homogeneous equation $x - A(x) = 0$ is finite dimensional.

5.25 Let X be a normed space, and let $A \in CL(X)$.

(i) The **homogeneous equation** $x - A(x) = 0$ has a nonzero solution in X if and only if the **transposed homogeneous equation** $x' - A'(x') = 0$ has a nonzero solution in X'. (Note: The solution spaces of the two homogeneous equations have the same dimension, that is, dim $Z(I - A') =$ dim $Z(I - A)$. See [28, Theorem V.7.14 (a)].)
(ii) Let $y \in X$. There is $x \in X$ such that $x - A(x) = y$ if and only if $x_j'(y) = 0$ for $j = 1, \ldots, m$, where $\{x_1', \ldots, x_m'\}$ is a basis for the solution space of the transposed homogeneous equation $x' - A'(x') = 0$.
Further, if x_0 is a **particular solution** of the equation $x - A(x) = y$, then its **general solution** is given by $x := x_0 + k_1 x_1 + \cdots + k_m x_m$, where $k_1, \ldots, k_m$ are in $\mathbb{K}$, and $\{x_1, \ldots, x_m\}$ is a basis for the solution space of the homogeneous equation $x - A(x) = 0$.

5.26 Let X be a normed space, and let $A \in BL(X)$. Then $\sigma_e(A) \subset \sigma_e(A'')$ and $\sigma_a(A) \subset \sigma_a(A'')$, while $\sigma(A'') = \sigma(A') \subset \sigma(A)$.

5.27 Let X be a normed space, $A \in BL(X)$, and $\lambda \in \sigma_e(A)$. A nonzero $x \in X$ is called a **generalized eigenvector** of A corresponding to λ if there is $n \in \mathbb{N}$ such that $(A - \lambda I)^n(x) = 0$. If H is a Hilbert space, and $A \in BL(H)$ is normal, then every generalized eigenvector of A is an eigenvector of A.

5.28 Let A be a normal operator on a separable Hilbert space H. Then $\sigma_e(A)$ is a countable set.

5.29 Let $A \in BL(\ell^2)$ be defined by the infinite matrix $M := [k_{i,j}]$. Suppose either M is **upper triangular** (that is, $k_{i,j} = 0$ for all $i > j$) or M is **lower triangular** (that is, $k_{i,j} = 0$ for all $i < j$). Then A is normal if and only if M is diagonal (that is, $k_{i,j} = 0$ for all $i \neq j$).

5.30 Let H be a Hilbert space, and let $A \in BL(H)$ be unitary. Then $\sigma(A)$ is contained in $\{k \in \mathbb{K} : |k| = 1\}$. Further, if $k \in \mathbb{K}$ and $|k| \neq 1$, then $\|(A - kI)^{-1}\| \leq 1/|\,|k| - 1|$.

5.31 Let H be a nonzero Hilbert space, and let $A \in BL(H)$. If $k \in \mathbb{K} \setminus \overline{\omega(A)}$, and $\beta := d\big(k, \overline{\omega(A)}\big)$, then $A - kI$ is invertible, and $\|(A - kI)^{-1}\| \leq 1/\beta$. In particular, if A is self-adjoint and $k \in \mathbb{K} \setminus [m_A, M_A]$, then β is equal to $|\Im k|$ if $\Re k \in [m_A, M_A]$, to $|k - m_A|$ if $\Re k < m_A$ and to $|k - M_A|$ if $\Re k > M_A$. If in fact $\mathbb{K} := \mathbb{R}$, A is self-adjoint, and $k \in \mathbb{R} \setminus \sigma(A)$, then $\|(A - kI)^{-1}\| = 1/d$, where $d := d(k, \sigma(A))$. (Note: This result also holds if $\mathbb{K} := \mathbb{C}$ and A is normal.)

5.32 Let H be a Hilbert space over $\mathbb{C}$. If $A \in BL(H)$ is self-adjoint, then its **Cayley transform** $T(A) := (A - iI)(A + iI)^{-1}$ is unitary and $1 \notin \sigma(T(A))$. Conversely, if $B \in BL(H)$ is unitary and $1 \notin \sigma(B)$, then its **inverse Cayley transform** $S(B) := i(I + B)(I - B)^{-1}$ is self-adjoint. Further, $S(T(A)) = A$ and $T(S(B)) = B$. (Note: The function $z \longmapsto (z - i)(z + i)^{-1}$ maps $\mathbb{R}$ onto $E := \{z \in \mathbb{C} : |z| = 1 \text{ and } z \neq 1\}$, and its inverse function $w \longmapsto i(1 + w)(1 - w)^{-1}$ maps E onto $\mathbb{R}$.)

5.33 Let A be a self-adjoint operator on a nonzero Hilbert space. Then $A \geq 0$ if and only if $\sigma(A) \subset [0, \infty)$. In this case, $0 \in \omega(A)$ if and only if $0 \in \sigma_e(A)$.

5.34 Let $\mathbb{K} := \mathbb{R}$, and let $\theta \in [0, 2\pi)$. For $x := (x(1), x(2)) \in \mathbb{R}^2$, define

$$A(x) := (x(1)\cos\theta - x(2)\sin\theta, x(1)\sin\theta + x(2)\cos\theta).$$

Then $A^*(x) := (x(1)\cos\theta + x(2)\sin\theta, -x(1)\sin\theta + x(2)\cos\theta)$ for $x \in \mathbb{R}^2$, and A is a unitary operator. Also, $\sigma(A) = \{1\}$ if $\theta := 0$, $\sigma(A) = \{-1\}$ if $\theta := \pi$, and $\sigma(A) = \emptyset$ otherwise.

5.35 Let H be a separable Hilbert space over $\mathbb{K}$, and let A be a normal Hilbert–Schmidt operator on H. Let (λ_n) be the sequence of nonzero eigenvalues of A, each such eigenvalue being repeated as many times as the dimension of the corresponding eigenspace. Then $\sum_n |\lambda_n|^2 < \infty$.

5.36 Let H be a finite dimensional nonzero Hilbert space over $\mathbb{C}$, and let A be in $BL(H)$. Then A is normal if and only if there are orthogonal projection operators $P_1, \ldots, P_k$ on H, and distinct $\mu_1, \ldots, \mu_k$ in $\mathbb{C}$ such that $I = P_1 + \cdots + P_k$ and $A = \mu_1 P_1 + \cdots + \mu_k P_k$, where $P_i P_j = 0$ for all $i \neq j$.

5.37 Let H be a Hilbert space over $\mathbb{K}$, and let $A \in BL(H)$ be nonzero. Then A is compact and self-adjoint if and only if there are orthogonal projection operators $P_0, P_1, P_2, \ldots$ on H and distinct $\mu_1, \mu_2, \ldots$ in $\mathbb{R} \setminus \{0\}$ such that $x = P_0(x) + P_1(x) + P_2(x) + \cdots$ for all $x \in H$ and $A = \mu_1 P_1 + \mu_2 P_2 + \cdots$, where $P_i P_j = 0$ for all $i \neq j$, P_1, $P_2, \ldots$ are of finite rank, and either the set $\{\mu_1, \mu_2, \ldots\}$ is finite or $\mu_n \to 0$.

5.38 Let $H := L^2([0, 1])$. For $s, t \in [0, 1]$, let $k(s, t) := \min\{1 - s, 1 - t\}$, and let A denote the Fredholm integral operator on H with kernel $k(\cdot\,, \cdot)$. Then $A(x) = \sum_{n=1}^{\infty} \lambda_n \langle x, u_n \rangle u_n$ for $x \in H$, where $\lambda_n := 4/(2n - 1)^2\pi^2$ and $u_n(s) := \sqrt{2}\cos(2n - 1)\pi s/2$, $s \in [0, 1]$ and $n \in \mathbb{N}$.

5.39 Let A denote the Fredholm integral operator on $L^2([0, 1])$ with kernel $k(s, t) := \min\{s, t\}$, $0 \leq s, t \leq 1$. For $x \in L^2([0, 1])$ and $n \in \mathbb{N}$, let

$$s_n(x) := \int_0^1 x(t) \sin(2n - 1)\frac{\pi t}{2}\, dm(t).$$

Then for every $x \in L^2([0, 1])$,

$$A(x)(s) = \frac{8}{\pi^2} \sum_{n=1}^{\infty} \frac{s_n(x)}{(2n - 1)^2} \sin(2n - 1)\frac{\pi s}{2}, \quad s \in [0, 1],$$

where the series on the right side converges in $L^2([0, 1])$. Let $y \in L^2([0, 1])$.

(i) Suppose $\mu \in \mathbb{K}$, $\mu \neq 0$ and $\mu \neq (2n - 1)^2\pi^2/4$ for any $n \in \mathbb{N}$. Then there is a unique $x \in L^2([0, 1])$ satisfying $x - \mu A(x) = y$. In fact,

$$x(s) = y(s) + 8\mu \sum_{n=1}^{\infty} \frac{s_n(y)}{(2n - 1)^2\pi^2 - 4\mu} \sin(2n - 1)\frac{\pi s}{2}, \quad s \in [0, 1].$$

Further, $\|x\| \leq \alpha\|y\|$, where $\alpha := 1 + 4|\mu|/\min_{n\in\mathbb{N}}\{|(2n - 1)^2\pi^2 - 4\mu|\}$.

(ii) Suppose $\mu := (2n_1 - 1)^2\pi^2/4$, where $n_1 \in \mathbb{N}$. Then there is x in $L^2([0, 1])$ satisfying $x - \mu A(x) = y$ if and only if $s_{n_1}(y) = 0$. In this case,

$$x(s) = y(s) + \frac{(2n_1 - 1)^2}{2} \sum_{n \neq n_1} \frac{s_n(y)}{(n - n_1)(n + n_1 - 1)} \sin(2n - 1)\frac{\pi s}{2}$$
$$+ k_1 \sin(2n_1 - 1)\frac{\pi s}{2} \quad \text{for } s \in [0, 1], \text{ where } k_1 \in \mathbb{K}.$$

5.40 Let H be a Hilbert space over $\mathbb{K}$, and let $A \in BL(H)$ be nonzero. Then A is compact if and only if there are countable orthonormal subsets $\{u_1, u_2, \ldots\}$ and $\{v_1, v_2, \ldots\}$ of H, and there are positive real numbers $s_1, s_2, \ldots$ such that $A(x) = \sum_n s_n \langle x, u_n \rangle v_n$ for all $x \in H$ with $s_n \to 0$ if the set $\{v_1, v_2, \ldots\}$ is denumerable. In this case, $A^*A(x) = \sum_n s_n^2 \langle x, u_n \rangle u_n$ for $x \in H$, and $\sum_n s_n^2 = \sum_n \|A(u_n)\|^2$. In particular, if H is a separable Hilbert space, then A is a Hilbert–Schmidt operator if and only if $\sum_n s_n^2 < \infty$. (Note: The positive real numbers $s_1, s_2, \ldots$ are called the nonzero **singular values** of A.)

Correction to: Linear Functional Analysis for Scientists and Engineers

Correction to:
B. V. Limaye, *Linear Functional Analysis for Scientists and Engineers*, https://doi.org/10.1007/978-981-10-0972-3_6

In the original version of the book, the following belated corrections have been incorporated:
Note: "Line $+i$" means ith line from the top whereas "Line $-i$" means ith line from the bottom.

Page xi, Line -7: Change with $[k_{i,j}]$ to with $k_{i,j}$
Page 19, Line -8: Change Example 2.24 to Example 2.24(iii)
Page 51, Line -11: Change $\bigcup_{j=1}^{\infty}$ to $\bigcup_{j=1}^{\infty} E_j$
Page 70, Line $+14$: Change infinite to finite
Page 88, Line -4: Change $F(x_n)$ to $F_n(x)$
Page 94, Line $+16$: Change $\|f_{i,n}\|_p$ to $\|f_{i,n}\|$
Page 104, Line $+10$: Change such that to such that $y = f(x)$ and
Page 104, Line $+11$: Change $U_Y(F(x), \delta\beta)$ to $U_Y(y, \delta\beta)$
Page 106, Line $+16$: Change of maps to of the maps
Page 127, Line -19: Change disjoint to nonempty disjoint
Page 131, Line $+16$: Change is a linear to is linear
Page 134, Line -5: Change $f_y \in (l^1)'$ to $f_y \in (L^1)'$
Page 178, Line -5: Change $\beta\|x\| \leq \|A(x)\|$ to $\|x\| \leq \beta\|A(x)\|$
Page 248, Line $+6$: Change $A(v_m)$ to $A(u_m)$
Page 248, Line $+9$: Change $s\|x\|^2$ to $s^2\|x\|^2$
Page 252, Line -2: Change theorem...integrations to theorems...integration

The updated version of the book can be found at
https://doi.org/10.1007/978-981-10-0972-3

© Springer Science+Business Media Singapore 2023
B.V. Limaye, *Linear Functional Analysis for Scientists and Engineers*,
https://doi.org/10.1007/978-981-10-0972-3_6

Solutions to Exercises

Chapter 2

2.1. U is convex: $p\big((1-t)x+ty\big) < (1-t)+t = 1$ if $x, y \in U$ and $t \in (0,1)$.
U is absorbing: $p\big(x/2p(x)\big) = 1/2 < 1$ if $x \in X$ and $p(x) \neq 0$.
U is balanced: $p(kx) = |k|p(x) < 1$ if $x \in U$ and $|k| \leq 1$.

2.2. $p_j(x) + p_j(y) \leq \max\{p_1(x), \ldots, p_m(x)\} + \max\{p_1(y), \ldots, p_m(y)\}$ if $j = 1, \ldots, m$ and $x, y \in X$. Hence $p(x+y) \leq p(x) + p(y)$ for $x, y \in X$.
Let $X := \mathbb{K}^2$, and let $p_1(x) := |x(1)| + 2|x(2)|$, $p_2(x) = 2|x(1)| + |x(2)|$ for $x := (x(1), x(2)) \in X$. Then p_1 and p_2 are norms on X. But q is not a seminorm on X: If $x := (1,0)$ and $y := (0,1)$, then $q(x+y) > q(x) + q(y)$.

2.3. Let $p \in \{1, 2, \infty\}$, $x \in \ell^p$, $x_n := (x(1), \ldots, x(n), 0, 0, \ldots) \in c_{00}$, $n \in \mathbb{N}$.
If $p := 1$ or $p := 2$, then $\|x_n - x\|_p \to 0$.
If $p := \infty$ and $x \in c_0$, then $\|x_n - x\|_\infty \to 0$. Conversely, let (y_n) be a sequence in c_0 such that $y_n \to x$ in ℓ^∞, and let $\epsilon > 0$. There is $n_0 \in \mathbb{N}$ such that $\|y_{n_0} - x\|_\infty < \epsilon/2$. Since $y_{n_0} \in c_0$, there is $j_0 \in \mathbb{N}$ such that $|y_{n_0}(j)| < \epsilon/2$ for all $j \geq j_0$. Then $|x(j)| \leq |x(j) - y_{n_0}(j)| + |y_{n_0}(j)| \leq \|x - y_{n_0}\|_\infty + |y_{n_0}(j)| < \epsilon/2 + \epsilon/2 = \epsilon$ for all $j \geq j_0$. Hence $x \in c_0$.

2.4. For $t \in (0,1]$, let $x(t) := 1/\sqrt{t}$, and let $x(0) := 0$. Then $x \in L^1([0,1])$, but $x \notin L^2([0,1])$. Also, $\sqrt{x} \in L^2([0,1])$, but $\sqrt{x} \notin L^\infty([0,1])$.
Let $y(t) := x(t)$ if $t \in [0,1]$ and $y(t) := 0$ if either $t < 0$ or $t > 1$. Then $y \in L^1(\mathbb{R})$, but $y \notin L^2(\mathbb{R}) \cup L^\infty(\mathbb{R})$. Let $z(t) := \sqrt{x(t)}$ if $t \in [0,1]$, $z(t) := 0$ if $t < 0$, and $z(t) := 1/t$ if $t > 1$. Then $z \in L^2(\mathbb{R})$, but $z \notin L^1(\mathbb{R}) \cup L^\infty(\mathbb{R})$.
Let $u(t) := 1$ if $t \in \mathbb{R}$. Then $u \in L^\infty(\mathbb{R})$, but $u \notin L^1(\mathbb{R}) \cup L^2(\mathbb{R})$.

2.5. For $x \in C([0,1])$, $\|x\|_1 \leq \|x\|_2 \leq \|x\|_\infty$. For $n \in \mathbb{N}$, let $x_n(t) := 1 - nt$ if $0 \leq t \leq 1/n$, and $x_n(t) := 0$ if $(1/n) < t \leq 1$. Then $x_n \in C([0,1])$ for $n \in \mathbb{N}$, $\|x_n\|_\infty = 1$, $\|x_n\|_1 = 1/2n$ and $\|x_n\|_2 = 1/\sqrt{3n}$.

2.6. Let $x \in X$. If $\|x\|' = 0$, then x is a constant function with $x(a) = 0$, that is, $x = 0$. For $t \in (a,b]$, there is $s \in (a,t)$ such that $x(t) - x(a) = (t-a)x'(s)$ by the mean value theorem, and so $|x(t)| \leq |x(a)| + (b-a)\|x'\|_\infty$.

© Springer Science+Business Media Singapore 2016, corrected publication 2023

B.V. Limaye, *Linear Functional Analysis for Scientists and Engineers*,
https://doi.org/10.1007/978-981-10-0972-3

Hence $\|x\|_{1,\infty} = \max\{\|x\|_\infty, \|x'\|_\infty\} \le \max\{1, b-a\}\|x\|'$, $\|x\|' \le \|x\|_{1,\infty}$ and $\|x\|_\infty \le \|x\|_{1,\infty}$ for all $x \in X$.

Also, if for $n \in \mathbb{N}$, we let $x_n(t) := (t-a)^n/(b-a)^n$, $t \in [a, b]$, then $x_n \in X$, $\|x_n\|_\infty = 1$, and $\|x_n'\|_\infty = n/(b-a) \to \infty$.

2.7. For $x \in X$, let us write $|||x + Z(F)||| := \inf\{\|x+z\| : z \in Z(F)\}$. Let $y \in Y$. There is $x \in X$ such that $F(x) = y$, and then $|||x + Z(F)||| = \inf\{\|u\| : u \in X \text{ and } u - x \in Z(F)\} = \inf\{\|u\| : u \in X \text{ and } F(u) = y\} = q(y)$. If $k \in \mathbb{K}$, then $F(kx) = ky$. Also, if $y_1, y_2 \in Y$ and $F(x_1) = y_1$, $F(x_2) = y_2$, then $F(x_1 + x_2) = y_1 + y_2$. As in the proof of Proposition 2.5(i), $q(ky) = |||kx + Z(F)||| = |k|\, |||x + Z(F)||| = |k|q(y)$, and $q(y_1 + y_2) = |||x_1 + x_2 + Z(F)||| \le |||x_1 + Z(F)||| + |||x_2 + Z(F)||| = q(y_1) + q(y_2)$.

Suppose $Z(F)$ is a closed subset of X. Let $y \in Y$ be such that $q(y) = 0$. If $x \in X$ and $F(x) = y$, then $|||x + Z(F)||| = q(y) = 0$, and so $x \in Z(F)$, that is, $y = F(x) = 0$. Hence q is a norm on Y. Conversely, suppose q is a norm on Y. Let (x_n) be a sequence in $Z(F)$ such that $x_n \to x$ in X. Then $|||x + Z(F)||| = |||x - x_n + Z(F)||| \le \|x - x_n\| \to 0$, and so $|||x + Z(F)||| = 0$. Let $y := F(x)$. Then $q(y) = |||x + Z(F)||| = 0$, so that $F(x) = y = 0$, that is, $x \in Z(F)$. Thus $Z(F)$ is a closed subset of X.

2.8. Suppose there are $x_0 \in X$ and $r > 0$ such that $U(x_0, r) \subset E$. Since E is compact, it is closed in X, and so $\overline{U}(x_0, r) \subset E$. But then the closed unit ball of X is a closed subset of the compact set $s(E - x_0)$, where $s := 1/r$.

2.9. The closed unit ball E of ℓ^2 is not compact since the normed space ℓ^2 is infinite dimensional. Let (x_n) be a sequence in the Hilbert cube C. There is a subsequence $x_{n,n}$ of (x_n) such that for each $j \in \mathbb{N}$, the sequence $x_{n,n}(j)$ converges in $\mathbb{K}$ to $x(j)$, say. Let $x := (x(1), x(2), \ldots)$. Then $x \in \ell^2$ since $|x(j)| \le 1/j$ for each $j \in \mathbb{N}$. Also,

$$\|x_{n,n} - x\|_2^2 \le \sum_{j=1}^{m} |x_{n,n}(j) - x(j)|^2 + \sum_{j=m+1}^{\infty} \left(\frac{2}{j}\right)^2 \quad \text{for every } m \in \mathbb{N}.$$

Let $\epsilon > 0$. Choose $m \in \mathbb{N}$ such that the second term above is less than $\epsilon^2/2$, and then choose $n_0 \in \mathbb{N}$ such that $|x_{n,n}(j) - x(j)|^2 < \epsilon^2/2m$ for all $n \ge n_0$ and $j = 1, \ldots, m$. It follows that $\|x_{n,n} - x\|_2 < \epsilon$ for all $n \ge n_0$.

2.10. Suppose a norm $\|\cdot\|$ on a linear space X is induced by an inner product. If $x, y \in X$, $\|x\| = 1 = \|y\|$ and $x \ne y$, then

$$\|x+y\|^2 = 2\|x\|^2 + 2\|y\|^2 - \|x-y\|^2 < 2\|x\|^2 + 2\|y\|^2 = 4.$$

Thus $(X, \|\cdot\|)$ is strictly convex. The norm $\|\cdot\|_2$ on ℓ^2, and the norm $\|\cdot\|_2$ on $L^2([0, 1])$ are induced by inner products. Hence they are strictly convex. On the other hand, let $x := e_1$ and $y := e_2$. Then $\|x\|_1 = 1 = \|y\|_1$ and $\|x+y\|_1 = 2$, and $\|x+y\|_\infty = 1 = \|x-y\|_\infty$ and $\|(x+y) + (x-y)\|_\infty = 2$. Hence ℓ^1 and ℓ^∞ are not strictly convex. Similarly, if x and y denote the characteristic functions of $[0, 1/2]$ and $(1/2, 1]$, then $\|x\|_1 = 1/2 = \|y\|_1$ and $\|x+y\|_1 =$

1, and $\|x+y\|_\infty = 1 = \|x-y\|_\infty$ and $\|(x+y)+(x-y)\|_\infty = 2$. Hence $L^1([0,1)$ and $L^\infty([0,1])$ are not strictly convex. Thus if $p \in \{1,\infty\}$, then the norms on ℓ^p and $L^p([0,1])$ are not induced by an inner product.

2.11. Suppose $|\langle x, y\rangle|^2 = \langle x, x\rangle\langle y, y\rangle$, and define $z := \langle y, y\rangle x - \langle x, y\rangle y$. Proceeding as in the proof of Proposition 2.13(i), we see that $\langle z, z\rangle = 0$, and so $z = 0$, that is, $\langle y, y\rangle x = \langle x, y\rangle y$. Conversely, if $\langle y, y\rangle x = \langle x, y\rangle y$, then we can readily check that $|\langle x, y\rangle| = \|x\|\|y\|$.
Suppose $\|x+y\|^2 = (\|x\| + \|y\|)^2$. Then $\operatorname{Re}\langle x, y\rangle = \|x\|\|y\|$. If either $x = 0$ or $y = 0$, then clearly, $\|y\|x = \|x\|y$. Now let $x \neq 0$ and $y \neq 0$, and define $u := x/\|x\|$, $v := y/\|y\|$. Then $\|u\| = 1 = \|v\|$, and $\|u+v\| = 2$ since

$$\|u+v\|^2 = \|u\|^2 + \|v\|^2 + 2\operatorname{Re}\langle u, v\rangle = 2 + \frac{2\operatorname{Re}\langle x, y\rangle}{\|x\|\|y\|} = 4.$$

By Exercise 2.10, $(X, \|\cdot\|)$ is strictly convex, and so $u = v$, that is, $\|y\|x = \|x\|y$. Conversely, if $\|y\|x = \|x\|y$, then we can readily check that $\|x+y\| = \|x\| + \|y\|$.

2.12. Let $x, y, z \in X$. By the parallelogram law,

$$\begin{aligned}&\|x+y+z\|^2 + \|x+y-z\|^2 + \|x-y+z\|^2 + \|x-y-z\|^2\\ &\quad= 2\|x+y\|^2 + 2\|z\|^2 + 2\|x-y\|^2 + 2\|z\|^2\\ &\quad= 4(\|x\|^2 + \|y\|^2 + \|z\|^2).\end{aligned}$$

2.13. Let $x \in Z$. If $k \in \mathbb{K}$, then clearly $kx \in Z$. Let $y \in X$, and let $\epsilon > 0$. Then

$$0 \le \langle x+\epsilon y, x+\epsilon y\rangle = \langle x, x\rangle + \epsilon^2\langle y, y\rangle + 2\epsilon\operatorname{Re}\langle x, y\rangle = \epsilon^2\langle y, y\rangle + 2\epsilon\operatorname{Re}\langle x, y\rangle,$$

and so $0 \le \epsilon\langle y, y\rangle + 2\operatorname{Re}\langle x, y\rangle$. Let $\epsilon \to 0$ to obtain $0 \le \operatorname{Re}\langle x, y\rangle$. Replace x by $-x$ to obtain $\operatorname{Re}\langle x, y\rangle \le 0$, and so $\operatorname{Re}\langle x, y\rangle = 0$. Replace x by ix to obtain $\operatorname{Im}\langle x, y\rangle = 0$, and so $\langle x, y\rangle = 0$.
In particular, if $x, y \in Z$, then $\langle x+y, x+y\rangle = \langle x, x\rangle + \langle y, y\rangle + 2\operatorname{Re}\langle x, y\rangle = 0+0+0 = 0$, that is, $(x+y) \in Z$. Thus Z is a subspace of X.
Suppose $x + Z = x_1 + Z$ and $y + Z = y_1 + Z$. Then $x_1 - x, y_1 - y \in Z$, and so $\langle x_1, y_1\rangle = \langle (x_1 - x) + x, (y_1 - y) + y\rangle$, which is equal to

$$\langle (x_1 - x), (y_1 - y)\rangle + \langle (x_1 - x), y\rangle + \langle x, (y_1 - y)\rangle + \langle x, y\rangle = \langle x, y\rangle.$$

Let $\langle\langle x+Z, y+Z\rangle\rangle := \langle x, y\rangle$ for $x, y \in X$. If $\langle\langle x+Z, x+Z\rangle\rangle = 0$, then $x \in Z$, that is, $x + Z = 0 + Z$. It follows that $\langle\langle\cdot,\cdot\rangle\rangle$ is an inner product on X/Z. Now $|\langle x, y\rangle|^2 = |\langle\langle x+Z, y+Z\rangle\rangle|^2 \le \langle\langle x+Z, x+Z\rangle\rangle\langle\langle y+Z, y+Z\rangle\rangle = \langle x, x\rangle\langle y, y\rangle$ for all $x, y \in X$.

2.14. Let $x, y \in X$. Then $\langle x+y, x+y\rangle - \langle x-y, x-y\rangle = 4\operatorname{Re}\langle x, y\rangle$. Replacing y by iy, $\langle x+iy, x+iy\rangle - \langle x-iy, x-iy\rangle = 4\operatorname{Re}\langle x, iy\rangle = 4\operatorname{Im}\langle x, y\rangle$. Hence the right side is equal to $4\operatorname{Re}\langle x, y\rangle + i\,4\operatorname{Im}\langle x, y\rangle = 4\langle x, y\rangle$.

2.15. $\|x-y\|^2 = \|x\|^2 + \|y\|^2 - 2\,\mathrm{Re}\,\langle x, y\rangle = \|x\|^2 + \|y\|^2 - 2\,\|x\|\,\|y\|\cos\theta_{x,y}$.

2.16. In Example 2.12(i), replace n by mn. Also, $\|I_n\|_F = (1+\cdots+1)^{1/2} = \sqrt{n}$.

2.17. For $x \in \ell^2$, $\|x\|_w^2 = \sum_{j=1}^{\infty} w(j)|x(j)|^2 \le \|w\|_\infty \|x\|_2^2$. If $v \in \ell^\infty$, then $w(j) \ge 1/\|v\|_\infty$ for all $j \in \mathbb{N}$, and so $\|x\|_w^2 \ge \|x\|_2^2/\|v\|_\infty$. Conversely, suppose there is $\alpha > 0$ such that $\|x\|_2 \le \alpha\|x\|_w$ for all $x \in \ell^2$. Then $1 \le \alpha\, w(j)$ by considering $x := e_j$ for $j \in \mathbb{N}$. Hence $\|v\|_\infty \le \alpha$.

2.18. The set $\{e_1+\cdots+e_m, e_1-e_2, \ldots, e_1-e_m\}$ is linearly independent. Next, $y_1 := 0$, $z_1 := x_1 = e_1+\cdots+e_m$, $u_1 := z_1/\|z_1\|_2 = (e_1+\cdots+e_m)/\sqrt{m}$ and $y_2 := \langle x_2, u_1\rangle u_1 = 0$, $z_2 := x_2 - y_2 = e_1 - e_2$, $u_2 := z_2/\|z_2\|_2 = (e_1 - e_2)/\sqrt{2}$, as desired. Let $n \in \{2, \ldots, m-1\}$ and assume that $u_j := (e_1+\cdots+e_{j-1}-(j-1)e_j)/\sqrt{(j-1)j}$ for $j = 2, \ldots, n$. Then

$$y_{n+1} := \sum_{j=1}^{n} \langle x_{n+1}, u_j\rangle u_j = \sum_{j=2}^{n} \langle x_{n+1}, u_j\rangle u_j = \sum_{j=2}^{n} \frac{u_j}{\sqrt{(j-1)j}},$$

and $z_{n+1} := x_{n+1} - y_{n+1} = e_1 - e_{n+1} - \sum_{j=2}^{n} u_j/\sqrt{(j-1)j}$. Now

$$e_1 - \sum_{j=2}^{n} \frac{u_j}{\sqrt{(j-1)j}} = e_1 - \sum_{j=2}^{n} \frac{e_1+\cdots+e_{j-1}-(j-1)e_j}{(j-1)j} = \frac{1}{n}\sum_{j=1}^{n} e_j.$$

Hence $z_{n+1} = (e_1+\cdots+e_n)/n - e_{n+1} = (e_1+\cdots+e_n - ne_{n+1})/n$, and $u_{n+1} := z_{n+1}/\|z_{n+1}\|_2 = (e_1+\cdots+e_n - ne_{n+1})/\sqrt{n(n+1)}$ as desired.

2.19. For $j = 1, \ldots, n$, denote the jth column of A by $x_j \in \mathbb{K}^m$. As in Theorem 2.17, obtain $u_1, \ldots, u_n$ by the Gram–Schmidt orthogonalization of $x_1, \ldots, x_n$. Then $x_j = y_j + \|x_j - y_j\|u_j$, where $y_1 := 0$, and

$$y_j := \langle x_j, u_1\rangle u_1 + \cdots + \langle x_j, u_{j-1}\rangle u_{j-1} \quad \text{for } j = 2, \ldots, n.$$

Let Q denote the $m \times n$ matrix whose jth column is u_j for $j = 1, \ldots, n$, and let $R := [r_{i,j}]$ denote the $n \times n$ matrix, where $r_{i,j} := \langle x_j, u_i\rangle$ if $1 \le i \le j-1$, $r_{j,j} := \|x_j - y_j\|$ and $r_{i,j} := 0$ if $j+1 \le i \le n$ for $j = 1, \ldots, n$. Then for $j = 1, \ldots, m$, $x_j = \langle x_j, u_1\rangle u_1 + \cdots + \langle x_j, u_{j-1}\rangle u_{j-1} + \|x_j - y_j\|u_j = r_{1,j}u_1 + \cdots + r_{j-1}u_{j-1} + r_j u_j$, that is, $A = [x_1, \ldots, x_n] = [u_1, \ldots, u_n]R = QR$.

Uniqueness: Suppose $A = Q'R'$, where the columns $u_1', \ldots, u_n'$ of Q' form an orthonormal subset of $\mathbb{K}^m$, and $R' := [r_{i,j}']$ is upper triangular with positive diagonal entries. Then $x_1 = r_{1,1}'u_1'$. Hence $r_{1,1}' = \|x_1\| = r_{1,1}$, and $u_1' = x_1/r_{1,1} = u_1$. Next, let $j \in \{2, \ldots, n\}$, and suppose we have shown $u_1' = u_1, \ldots, u_{j-1}' = u_{j-1}$. Then $x_j = r_{1,j}'u_1' + \cdots + r_{j-1,j}'\,u_{j-1}' + r_{j,j}'u_j' = r_{1,j}'u_1 + \cdots + r_{j-1,j}'\,u_{j-1} + r_{j,j}'u_j'$. Hence $r_{i,j}' = \langle x_j, u_i\rangle = r_{i,j}$ for $i = 1, \ldots, j-1$ and $r_{j,j}' = \|x_j - r_{1,j}u_1 - \cdots - r_{j-1,j}\,u_{j-1}\| = r_{j,j}$, so that $u_j' = (x_j - r_{1,j}u_1 - \cdots - r_{j-1,j}\,u_{j-1})/r_{j,j} = u_j$. Thus $Q' = Q$ and $R' = R$.

If A is an infinite matrix whose columns form a linearly independent subset of ℓ^2, then the above arguments hold.

2.20. Let $m \in \mathbb{N}$. Then $\sum_{n=0}^{m} |\int_{-1}^{1} x(t)p_n(t)dt|^2 \le \int_{-1}^{1} |x(t)|^2 dt$ by the Bessel inequality. Also, equality does not hold here since $x \notin \text{span}\{p_0, p_1, \ldots, p_m\}$.

2.21. If $r := 0$, then $\int_{-T}^{T} u_r(t)dt = 2T$, and if $r \ne 0$, then $\int_{-T}^{T} u_r(t)dt = (2\sin rT)/r$ for all $T > 0$. Hence $\lim_{T\to\infty} \frac{1}{2T}\int_{-T}^{T} u_r(t)dt$ is equal to 1 if $r = 0$, and it is equal to 0 if $r \ne 0$. As a result, $\lim_{T\to\infty} \frac{1}{2T}\int_{-T}^{T} x(t)dt$ exists for every $x \in X$. Let $p, q \in X$. Then $p\overline{q} \in X$, and so $\langle p, q\rangle$ is well-defined. Clearly, the function $\langle\cdot\,,\cdot\rangle : X \times X \to \mathbb{K}$ is linear in the first variable, it is conjugate symmetric, and $\langle p, p\rangle \ge 0$ for all $p \in X$. Let $p := c_1 u_{r_1} + \cdots + c_n u_{r_n}$, where $r_1, \ldots, r_n$ are distinct real numbers, and $c_1, \ldots, c_n \in \mathbb{C}$. Then $\langle p, p\rangle = |c_1|^2 + \cdots + |c_n|^2$. It follows that $p = 0$ whenever $\langle p, p\rangle = 0$. Thus $\langle\cdot\,,\cdot\rangle$ is an inner product on X. Also, $\langle u_r, u_s\rangle = \lim_{T\to\infty} \frac{1}{2T}\int_{-T}^{T} u_{r-s}(t)dt$. Hence $\{u_r : r \in \mathbb{R}\}$ is an uncountable orthonormal subset of X. (In particular, it is a linearly independent subset of X.)

2.22. Since the map $F : X \to Y$ is linear, and since $\langle\cdot\,,\cdot\rangle_Y$ is an inner product on Y, the function $\langle\cdot\,,\cdot\rangle_X : X \times X \to \mathbb{K}$ is linear in the first variable, it is conjugate-symmetric, and $\langle x, x\rangle_X \ge 0$ for all $x \in X$.
(i) $\langle\cdot\,,\cdot\rangle_X$ is an inner product on X if and only if $x = 0$ whenever $\langle x, x\rangle_X = \langle F(x), F(x)\rangle_Y = 0$, that is, $F(x) = 0$.
(ii) Since F is one-one, if $F(u_\alpha) = F(u_\beta)$, then $u_\alpha = u_\beta$. Also, $\langle u_\alpha, u_\beta\rangle_X = \langle F(u_\alpha), F(u_\beta)\rangle_Y$ for all α, β.
(iii) Since F is one-one, $\{F(u_\alpha)\}$ is an orthonormal subset of Y. Also, since F is onto, if $\{F(u_\alpha)\}$ is a proper subset of an orthonormal subset E of Y, then $\{u_\alpha\}$ is a proper subset of the orthonormal subset $F^{-1}(E)$ of X.

2.23. There are $\alpha > 0$ and $\beta > 0$ such that $\beta\|x\| \le \|x\|' \le \alpha\|x\|$ for all $x \in X$. Hence (x_n) is a Cauchy sequence in $(X, \|\cdot\|')$ if and only if (x_n) is a Cauchy sequence in $(X, \|\cdot\|)$, and (x_n) is a convergent sequence in $(X, \|\cdot\|')$ if and only if (x_n) is a convergent sequence in $(X, \|\cdot\|)$.

2.24. c_0 is the closure of c_{00} in the Banach space ℓ^∞ (Exercise 2.3).
To show that c is a closed subspace of the Banach space ℓ^∞, let (x_n) be a sequence in c such that $x_n \to x$ in ℓ^∞, and let $\epsilon > 0$. There is $n_0 \in \mathbb{N}$ such that $\|x_{n_0} - x\|_\infty < \epsilon/3$. Since x_{n_0} is a Cauchy sequence in $\mathbb{K}$, there is $j_0 \in \mathbb{N}$ such that $|x_{n_0}(i) - x_{n_0}(j)| < \epsilon/3$ for all $i, j \ge j_0$. Then

$$|x(i) - x(j)| \le |(x - x_{n0})(i) - (x - x_{n0})(j)| + |x_{n0}(i) - x_{n0}(j)| < \frac{2\epsilon}{3} + \frac{\epsilon}{3} = \epsilon$$

for all $i, j \ge j_0$. Hence $(x(j))$ is Cauchy sequence in $\mathbb{K}$, and so $x \in c$.
To show that $C_0(T)$ is a closed subspace of the Banach space $C(T)$, let (x_n) be a sequence in $C_0(T)$ such that $x_n \to x$ in $C(T)$, and let $\epsilon > 0$. There is $n_0 \in \mathbb{N}$ such that $\|x_{n_0} - x\|_\infty < \epsilon/2$. Since $x_{n_0} \in C_0(T)$, there is a compact subset T_ϵ of T such that $|x_{n_0}(t)| < \epsilon/2$ for all $t \in T\backslash T_\epsilon$. Then $|x(t)| \le |x(t) -$

$x_{n_0}(t)| + |x_{n_0}(t)| \le \|x - x_{n_0}\|_\infty + |x_{n_0}(t)| < \epsilon/2 + \epsilon/2 = \epsilon$ for all $t \in T \backslash T_\epsilon$, and so $x \in C_0(T)$.

2.25. Since X has a denumerable basis, X is not a Banach space.

2.26. (i) Let Y be an m dimensional subspace of X. There is an orthonormal basis $u_1, \ldots, u_m$ of Y. Let (x_n) be a Cauchy sequence in Y. Then $x_n = \langle x_n, u_1\rangle u_1 + \cdots + \langle x_n, u_m\rangle u_m$ for $n \in \mathbb{N}$. Fix $j \in \{1, \ldots, m\}$. For all $n, p \in \mathbb{N}$, $|\langle x_n, u_j\rangle - \langle x_p, u_j\rangle| \le \|x_n - x_p\|$. Hence the Cauchy sequence $(\langle x_n, u_j\rangle)$ converges in $\mathbb{K}$ to k_j, say. Then $x_n \to k_1u_1 + \cdots + k_m u_m$ in Y.
(ii) Let X be infinite dimensional. By Theorem 2.17, there is a denumerable orthonormal subset $\{u_1, u_2, \ldots\}$ of X. The sequence (u_n) in the closed unit ball of X does not have a convergent subsequence since $\|u_n - u_p\| = \sqrt{2}$ for all $n \ne p$. Hence the closed unit ball of X is not compact.
Conversely, suppose X is finite dimensional. By Theorem 2.23(iii), there is an isometry from X onto an Euclidean space whose closed unit ball is compact by the classical Heine–Borel theorem.
(iii) Suppose X is complete, and it has a denumerable (Hamel) basis. By Theorem 2.17, there is an orthonormal subset $\{u_1, u_2, \ldots\}$ of X which is a (Hamel) basis for X. Then $x := \sum_{n=1}^\infty u_n/n$ belongs to X by the Riesz–Fischer theorem, but it does not belong to span $\{u_1, u_2, \ldots\}$.

2.27. Let $x := (x(1), x(2), \ldots) \in c_0$, and for $m \in \mathbb{N}$, let $s_m(x) := \sum_{n=1}^m x(n)e_n$. Then $\|s_m(x) - x\|_\infty = \sup\{|x(n)| : n = m+1, m+2, \ldots\} \to 0$ as $m \to \infty$. Thus $x = \sum_{n=1}^\infty x(n)e_n$. Also, if there are $k_1, k_2, \ldots$ in $\mathbb{K}$ such that $x = \sum_{n=1}^\infty k_n e_n$, then $k_j = \sum_{n=1}^\infty k_n e_n(j) = x(j)$ for each $j \in \mathbb{N}$.
Let $x \in c$, and $x(n) \to \ell_x$. Then $y := x - \ell_x e_0 \in c_0$, and $y = \sum_{n=1}^\infty y(n)e_n$. Hence $x = \ell_x e_0 + \sum_{n=1}^\infty (x(n) - \ell_x)e_n$. Also, if $x = \sum_{n=0}^\infty k_n e_n$, where $k_0, k_1, k_2, \ldots$ are in $\mathbb{K}$, then $y_0 := x - k_0 e_0 = \sum_{n=1}^\infty k_n e_n$ belongs to the closure of c_{00} in ℓ^∞, that is, $y_0 \in c_0$. It follows that $k_0 = \ell_x$ and $k_j = \sum_{n=1}^\infty k_n e_n(j) = x(j) - \ell_x e_0(j) = x(j) - \ell_x$ for each $j \in \mathbb{N}$.
Let $x \in \ell^1$. Define $k_n := x(2n-1) + x(2n)$ and $\ell_n := x(2n-1) - x(2n)$ for $n \in \mathbb{N}$. For $m \in \mathbb{N}$, let $s_{2m}(x) := k_1u_1 + \ell_1v_1 + \cdots + k_m u_m + \ell_m v_m$ and $s_{2m-1}(x) := k_1u_1 + \ell_1 v_1 + \cdots + k_{m-1}u_{m-1} + \ell_{m-1}v_{m-1} + k_m u_m$. Then for all m in $\mathbb{N}$, $s_{2m}(x) = x(1)e_1 + \cdots + x(2m)e_{2m}$ and $s_{2m-1}(x) = x(1)e_1 + \cdots + x(2m-2)e_{2m-2} + k_m u_m$. Since $k_m \to 0$, we obtain $x = k_1u_1 + \ell_1v_1 + k_2u_2 + \ell_2v_2 + \cdots$. Also, if $x = k_1'u_1 + \ell_1'v_1 + k_2'u_2 + \ell_2'v_2 + \cdots$, then $x(2n-1) = (k_n' + \ell_n')/2$ and $x(2n) = (k_n' - \ell_n')/2$, that is, $k_n' = x(2n-1) + x(2n) = k_n$ and $\ell_n' = x(2n-1) - x(2n) = \ell_n$ for $n \in \mathbb{N}$.
As ℓ^∞ and $L^\infty([a,b])$ are not separable, they do not have Schauder bases.

2.28. For $j \in \mathbb{N}$ and $x(j) \in X_j$, define $\|x(j)\|_j := \langle x(j), x(j)\rangle^{1/2}$. The function $\langle\cdot,\cdot\rangle : X \times X \to \mathbb{K}$ is well-defined since for $x, y \in X$,

$$\sum_{j=1}^\infty |\langle x(j), y(j)\rangle_j| \le \sum_{j=1}^\infty \|x(j)\|_j \|y(j)\|_j \le \Big(\sum_{j=1}^\infty \|x(j)\|_j^2\Big)^{1/2}\Big(\sum_{j=1}^\infty \|y(j)\|_j^2\Big)^{1/2}.$$

If $x, y \in X$, then $x + y \in X$ since $\|x(j) + y(j)\|_j \leq \|x(j)\|_j + \|y(j)\|_j$ and

$$\Big(\sum_{j=1}^{\infty}\big(\|x(j)\|_j + \|y(j)\|_j\big)^2\Big)^{1/2} \leq \Big(\sum_{j=1}^{\infty}\|x(j)\|_j^2\Big)^{1/2} + \Big(\sum_{j=1}^{\infty}\|y(j)\|_j^2\Big)^{1/2}.$$

Also, it is easy to see that if $x \in X$ and $k \in \mathbb{K}$, then $k\,x \in X$. Further, it follows that $\langle \cdot\,,, \cdot \rangle$ is an inner product on X. Suppose X is complete. Fix $j \in \mathbb{N}$. If $(x_n(j))$ is a Cauchy sequence in X_j, define $x_n := (0, \ldots, 0, x_n(j), 0, 0, \ldots)$, where $x_n(j)$ is in the jth entry, and note that (x_n) is Cauchy sequence in X. If $x_n \to x$ in X, then $x_n(j) \to x(j)$ in X_j. Thus X_j is complete. Conversely, suppose X_j is complete for each $j \in \mathbb{N}$. Then the completeness of X follows exactly as in the proof of the completeness of ℓ^2 given in Example 2.24(ii).

2.29. The case $k := 1$ is treated in Examples 2.24(iii), (v) and 2.28(iv). We consider here the case $k := 2$. The cases $k > 2$ are similar.
(i) Let (x_n) be a Cauchy sequence in $(C^2([a, b]), \|\cdot\|_{2,\infty})$. Then (x_n), (x_n') and (x_n'') are Cauchy sequences in the Banach space $(C([a, b]), \|\cdot\|_\infty)$. A well-known result in Real Analysis shows that there are x and y in $C^1([a, b])$ such that $\|x_n - x\|_\infty \to 0$, $\|x_n' - x'\|_\infty \to 0$, and $\|x_n' - y\|_\infty \to 0$, $\|x_n'' - y'\|_\infty \to 0$. Then $x' = y \in C^1([a, b])$, that is, $x \in C^2([a, b])$, and $\|x_n - x\|_{2,\infty} = \|x_n - x\|_\infty + \|x_n' - x'\|_\infty + \|x_n'' - x''\|_\infty \to 0$.
(ii) Let (x_n) be a Cauchy sequence in $(W^{2,1}([a, b]), \|\cdot\|_{2,1})$. Then (x_n), (x_n') and (x_n'') are Cauchy sequences in the Banach space $(L^1([a, b]), \|\cdot\|_1)$. As we have seen in Example 2.24 (v), there are absolutely continuous functions x and y on $[a, b]$ such that $\|x_n - x\|_1 \to 0$, $\|x_n' - x'\|_1 \to 0$, and $\|x_n' - y\|_1 \to 0$, $\|x_n'' - y'\|_1 \to 0$. Then $x' = y$ is absolutely continuous on $[a, b]$, that is, $x \in W^{2,1}([a, b])$, and $\|x_n - x\|_{2,1} = \|x_n - x\|_1 + \|x_n' - x'\|_1 + \|x_n'' - x''\|_1 \to 0$.

(iii) Let (x_n) be a Cauchy sequence in $(W^{2,2}([a, b]), \|\cdot\|_{2,2})$. Then (x_n), (x_n') and (x_n'') are Cauchy sequences in the Hilbert space $(L^2([a, b]), \|\cdot\|_2)$. As we have seen in Example 2.28(iv), there are absolutely continuous functions x and y on $[a, b]$ such that $\|x_n - x\|_2 \to 0$, $x' \in L^2([a, b])$, $\|x_n' - x'\|_2 \to 0$, and $\|x_n' - y\|_2 \to 0$, $y' \in L^2([a, b])$, $\|x_n'' - y'\|_2 \to 0$. Then $x' = y$ is absolutely continuous on $[a, b]$ and $x'' = y' \in L^2([a, b])$, that is, $x \in W^{2,2}([a, b])$, and $\|x_n - x\|_{2,2} = \|x_n - x\|_2 + \|x_n' - x'\|_2 + \|x_n'' - x''\|_2 \to 0$.

2.30. (i)$\Longrightarrow$(ii): Let $s := \sum_{n=1}^{\infty} x_n$. Then $\langle s, x_n\rangle = \langle x_n, x_n\rangle = \|x_n\|^2$ for $n \in \mathbb{N}$.
(ii)$\Longrightarrow$(iii): Let $u_n := 0$ if $x_n = 0$, and $u_n := x_n/\|x_n\|$ if $x_n \neq 0$. Then $\langle s, u_n\rangle = \|x_n\|$ for $n \in \mathbb{N}$, and $\sum_{n=1}^{\infty}\|x_n\|^2 = \sum_{n=1}^{\infty}|\langle s, u_n\rangle|^2 \leq \|s\|^2$.
(iii)$\Longrightarrow$(i): Let $s_m := \sum_{n=1}^{m} x_n$, $m \in \mathbb{N}$. Then for $m > p$, $\|s_m - s_p\|^2 = \|x_{p+1} + \cdots + x_m\|^2 = \|x_{p+1}\|^2 + \cdots + \|x_m\|^2$. Thus (s_m) is a Cauchy sequence in the Hilbert space H, and so it converges in H.

2.31. For each $n \in \mathbb{N}$, $E_n := \{u_n, v_n, w_n\}$ is an orthonormal subset of ℓ^2 and span E_n = span $\{e_{3n-2}, e_{3n-1}, e_{3n}\}$. (Let $m := 3$ in Exercise 2.18.) Hence $E := \{u_n : n \in \mathbb{N}\} \cup \{v_n : n \in \mathbb{N}\} \cup \{w_n : n \in \mathbb{N}\}$ is an orthonormal subset of ℓ^2 and span E = span $\{e_j : j \in \mathbb{N}\}$, which is dense in ℓ^2.

2.32. For each $n \in \mathbb{N}$, $E_n := \{u_n, v_n, w_n\}$ is an orthonormal subset of ℓ^2 and $\text{span}\, E_n = \text{span}\, \{e_{3n-2}, e_{3n-1}, e_{3n}\}$. (Compare Exercise 4.35.) Hence $E := \{u_n : n \in \mathbb{N}\} \cup \{v_n : n \in \mathbb{N}\} \cup \{w_n : n \in \mathbb{N}\}$ is an orthonormal subset of ℓ^2 and $\text{span}\, E = \text{span}\, \{e_j : j \in \mathbb{N}\}$, which is dense in ℓ^2.

2.33. It is easy to see that $E := \{u_0, u_1, v_1, u_2, v_2, \ldots\}$ is an orthonormal subset of $L^2([-\pi, \pi])$. Let $x \in E^\perp$. For $k \in \mathbb{Z}$,

$$\hat{x}(k) = \frac{1}{2\pi} \int_{-\pi}^{\pi} x(t) e^{-ikt} dm(t) = \frac{1}{2\pi} \int_{-\pi}^{\pi} x(t)(\cos kt - i \sin kt) dm(t) = 0$$

since $\langle x, u_0 \rangle = 0$ and $\langle x, u_n \rangle = \langle x, v_n \rangle = 0$ for all $n \in \mathbb{N}$. Hence $x = 0$ a.e. on $[-\pi, \pi]$. Thus $E^\perp = \{0\}$.
Since $\langle x, u_0 \rangle = \sqrt{2\pi}\, a_0$ and $\langle x, u_n \rangle = \sqrt{\pi}\, a_n$, $\langle x, v_n \rangle = \sqrt{\pi}\, b_n$ for all $n \in \mathbb{N}$, the Fourier expansion $x = \langle x, u_0 \rangle u_0 + \sum_{n=1}^{\infty} \big(\langle x, u_n \rangle u_n + \langle x, v_n \rangle v_n\big)$ and the Parseval formula $\|x\|_2^2 = |\langle x, u_0 \rangle|^2 + \sum_{n=1}^{\infty} \big(|\langle x, u_n \rangle|^2 + |\langle x, v_n \rangle|^2\big)$ yield the desired results.

2.34. The subsets $E := \{u_0, u_1, u_2, \ldots,\}$ and $F := \{v_1, v_2, \ldots\}$ of $L^2([0, 1])$ are orthonormal. Given $x \in L^2([0, 1])$, let $y(t) := x(t/\pi)$ if $t \in [0, \pi]$, and $y(t) := x(-t/\pi)$ if $t \in [-\pi, 0)$. Then y is an even function on $[-\pi, \pi]$. Also, $y \in L^2([-\pi, \pi])$. Further, $\int_{-\pi}^{\pi} y(t) dm(t) = 2\pi \int_0^1 x(s) dm(s) = 2\pi a_0$. For $n \in \mathbb{N}$, $\int_{-\pi}^{\pi} y(t) \cos nt\, dm(t) = 2\pi \int_0^1 x(s) \cos n\pi s\, dm(s) = \pi a_n$ and $\int_{-\pi}^{\pi} y(t) \sin nt\, dm(t) = 0$. By Exercise 2.33, $y(t) = a_0 + \sum_{n=1}^{\infty} a_n$ $\cos nt$, the series converging in the mean square on $[-\pi, \pi]$. Hence $x(s) = a_0 + \sum_{n=1}^{\infty} a_n \cos n\pi s$, the series converging in the mean square on $[0, 1]$.
Similarly, given $x \in L^2([0, 1])$, let $z(t) := x(t/\pi)$ if $t \in [0, \pi]$, and $z(t) := -x(-t/\pi)$ if $t \in [-\pi, 0)$. Then z is an odd function on $[-\pi, \pi]$. As above, we may use Exercise 2.33 to obtain $z(t) = \sum_{n=1}^{\infty} b_n \sin nt$, the series converging in the mean square on $[-\pi, \pi]$, and so $x(s) = \sum_{n=1}^{\infty} b_n \sin n\pi s$, the series converging in the mean square on $[0, 1]$.

2.35. The subspace $G := \text{span}\, \{v_0, v_1, \ldots\}$ of H consists of all polynomials in t^3 defined on $[-1, 1]$. Let x be a continuous function on $[-1, 1]$, and define $y \in X$ by $y(t) := x(t^{1/3})$, $t \in [-1, 1]$. Let $\epsilon > 0$. By the Weierstrass theorem, there is a polynomial q defined on $[-1, 1]$ such that $\|y - q\|_\infty < \epsilon$. Define $p(s) := q(s^3)$ for $s \in [-1, 1]$. Then $p \in G$ and $\|x - p\|_\infty = \sup\{|y(s^3) - q(s^3)| :$
$s \in [-1, 1]\} = \|y - q\|_\infty < \epsilon$. Further, $\|x - p\|_2 \leq \sqrt{2}\, \|x - p\|_\infty < \sqrt{2}\, \epsilon$. By Proposition 1.26(ii), G is dense in H. The calculations of v_0, v_1 v_2 are routine.

2.36. If $x \in H$, then by Bessel's inequality, $\sum_\alpha |\langle x, u_\alpha \rangle|^2 \leq \|x\|^2 < \infty$. By the Riesz–Fischer theorem, $\sum_\alpha \langle x, u_\alpha \rangle v_{\phi(\alpha)} \in G$. Thus the map $F : H \to G$ is well-defined. It is easy to see that F is linear. Let $x_j := \sum_\alpha \langle x_j, u_\alpha \rangle u_\alpha \in H$ for $j := 1, 2$. By the orthonormality of the sets $\{u_\alpha\}$ and $\{v_{\phi(\alpha)}\}$, $\langle x_1, x_2 \rangle = \sum_\alpha \langle x_1, u_\alpha \rangle \overline{\langle x_2, u_\alpha \rangle} = \langle F(x_1), F(x_2) \rangle$. In particular, F is continuous. To show that F is onto, let $y \in G$. Then $y = \sum_\beta \langle y, v_\beta \rangle v_\beta$. By Bessel's inequality,

$\sum_\beta |\langle y, v_\beta\rangle|^2 \le \|y\|^2 < \infty$. By the Riesz–Fischer theorem, $x := \sum_\beta \langle y, v_\beta\rangle u_{\phi^{-1}(\beta)} \in H$. If $\alpha := \phi^{-1}(\beta)$, then $F(u_\alpha) = v_\beta$. Hence $F(x) = \sum_\beta \langle y, v_\beta\rangle F(u_{\phi^{-1}(\beta)}) = \sum_\beta \langle y, v_\beta\rangle v_\beta = y$ by the continuity and the linearity of F.

2.37. Let G denote the closure of span E. Since $E^\perp = G^\perp$, we obtain $E^{\perp\perp} = G^{\perp\perp} = G$, as in the Projection Theorem.

2.38. Since $y \in G$, we obtain $\|x - y\| \ge d(x, G)$. Let $w \in G$. Then $(y - w) \in G$. Since $(x - y) \in G^\perp$, we obtain $(x - y) \perp (y - w)$, so that

$$\|x - w\|^2 = \|(x - y) + (y - w)\|^2 = \|x - y\|^2 + \|y - w\|^2 \ge \|x - y\|^2.$$

Thus $\|x - y\| = d(x, G)$. If $w \in G$ and $\|x - w\| = d(x, G)$, then $\|x - w\| = \|x - y\|$, and so $\|y - w\| = 0$ by the above inequality, that is, $w = y$.

2.39. (i) Let $x \in X\backslash Y$, and $Z := \text{span}\,\{x, Y\}$. Now $Y \ne Z$, and since Y is finite dimensional, Y is a closed subspace of Z. By the lemma of Riesz, there is $z_n \in Z$ such that $\|z_n\| = 1$ and $(1 - 1/n) < d(z_n, Y) \le 1$ for every $n \in \mathbb{N}$. Now (z_n) is a sequence in the closed unit ball of Z. By Theorem 2.10, there is a convergent subsequence (z_{n_k}). Let $z_{n_k} \to x_1$ in Z. Then $\|x_1\| = 1$. Also, since $(1 - 1/n_k) \le |||z_{n_k} + Y||| \le 1$, and $|||z_{n_k} + Y||| \to |||x_1 + Y|||$, we see that $d(x_1, Y) = |||x_1 + Y||| = 1$.
(ii) Let $x \in H\backslash G$, and let y be the orthogonal projection of x on G. By Exercise 2.38, $\|x - y\| = d(x, G)$. Let $x_1 := (x - y)/\|x - y\|$.

2.40. The function $\langle\langle\cdot\,,\cdot\rangle\rangle : H/G \times H/G \to \mathbb{K}$ is well-defined: Suppose $x_j + G = x'_j + G$ for $j = 1, 2$. Let $x_j = y_j + z_j$ and $x'_j = y'_j + z'_j$, where $y_j, y'_j \in G$ and $z_j, z'_j \in G^\perp$ for $j = 1, 2$. Since $z_j - z'_j = (x_j - x'_j) + (y_j - y'_j)$, where $(x_j - x'_j) \in G$ and $(y_j - y'_j) \in G$, we see that $(z_j - z'_j) \in G$ for $j = 1, 2$. Hence $(z_1 - z'_1) \perp z_2$ and $(z_2 - z'_2) \perp z'_1$, and so

$$\langle\langle x_1 + G, x_2 + G\rangle\rangle = \langle z_1, z_2\rangle = \langle z'_1, z_2\rangle = \langle z'_1, z'_2\rangle = \langle\langle x'_1 + G, x'_2 + G\rangle\rangle.$$

It is easy to see that $\langle\langle\cdot\,,\cdot\rangle\rangle$ is an inner product on H/G. Also, by Exercise 2.38, $\langle\langle x_1 + G, x_1 + G\rangle\rangle = \langle z_1, z_1\rangle = \|z_1\|^2 = \|x_1 - y_1\|^2 = \big(d(x_1, G)\big)^2 = |||x_1 + G|||^2$. Further, since H is complete, so is H/G.

Chapter 3

3.1. Suppose $L := \{x_1, x_2, \ldots\}$ is an infinite linearly independent subset of X, and let B be a (Hamel) basis for X containing L. Define $f(x_n) := n\|x_n\|$ for $n \in \mathbb{N}$, and $f(b) := 0$ for $b \in B\backslash L$. Let $f : X \to \mathbb{K}$ denote the linear extension of this function. Then f is not continuous. Similarly, define $F(x_n) := nx_n$ for $n \in \mathbb{N}$, and $F(b) := b$ for $b \in B\backslash L$. Let $F : X \to X$ denote the linear extension of this function. Then F is one-one and $R(F) = X$, but F is not continuous. Let

Y be a normed space, and $y_0 \in Y$, $y_0 \neq 0$. If $G(x) := f(x)y_0$, $x \in X$, then G is linear but not continuous.

3.2. (i) If $x := k_1x_1 + \cdots + k_nx_n$, then $\|x\| \geq |||k_jx_j + X_j||| = |k_j|\,|||x_j + X_j|||$ for $j = 1, \ldots, n$, and so $\|F(x)\| \leq |k_1|\|y_1\| + \cdots + |k_n|\|y_n\| \leq \alpha\|x\|$.
(ii) If $y := \ell_1y_1 + \cdots + \ell_my_m$, then $\|y\| \geq |||\ell_iy_i + Y_i||| = |\ell_i|\,|||y_i + Y_i|||$ for $i = 1, \ldots, m$, and if $x := \ell_1x_1 + \cdots + \ell_mx_m$, then $F(x) = y$ and $\|x\| \leq |\ell_1|\|x_1\| + \cdots + |\ell_m|\|x_m\| \leq \gamma\|y\|$.

3.3. Suppose Y is a subspace of X such that $Z(f) \subset Y$ and $Y \neq Z(f)$. Let $y_0 \in Y \backslash Z(f)$. Consider $x \in X$. Let $k := f(x)/f(y_0)$ and $y := x - k\,y_0$. Then $x = y + k\,y_0$, where $y \in Z(f) \subset Y$, and so $x \in Y$.
Next, let Y denote the closure of $Z(f)$. Suppose f is continuous. Then $Y = Z(f)$. If $Z(f)$ is dense in X, then $Z(f) = Y = X$. Conversely, suppose $Z(f)$ is not dense in X. Then $Z(f) \subset Y$ and $Y \neq X$, and so $Y = Z(f)$, that is, $Z(f)$ is a closed subspace of X. Hence f is continuous.

3.4. Consider the norm $\|\cdot\|_p$ on $X := c_{00}$, where $p \in \{1, 2, \infty\}$.
$p = 1$: If $r \geq 0$, then $|f_r(x)| \leq \sum_{j=1}^{\infty} |x(j)| = \|x\|_1$ for all $x \in X$, and $f_r(e_1) = 1$, so that $\|f\| = 1$. Conversely, if $r < 0$, then $f_r(e_n) = n^{-r} \to \infty$.
$p = 2$: If $r > 1/2$, then $\alpha := \left(\sum_{j=1}^{\infty} j^{-2r}\right)^{1/2} < \infty$, and $|f_r(x)| \leq \alpha\|x\|_2$ for all $x \in X$. Further, let $x_n := (1, 2^{-r}, \ldots, n^{-r}, 0, 0\ldots) \in X$ and $\alpha_n := \left(\sum_{j=1}^{n} j^{-2r}\right)^{1/2}$ for $n \in \mathbb{N}$. Then $\|x_n\|_2 = \alpha_n$ and $f_r(x_n) = \alpha_n^2$ for $n \in \mathbb{N}$. If $r > 1/2$, then $f_r(x_n/\alpha_n) \to \alpha$, and if $r \leq 1/2$, then $f_r(x_n/\alpha_n) \to \infty$.
$p = \infty$: If $r > 1$, then $\beta := \sum_{j=1}^{\infty} j^{-r} < \infty$, and $|f_r(x)| \leq \beta\|x\|_\infty$ for all $x \in X$. Further, let $x_n := (1, \ldots, 1, 0, 0, \ldots) \in X$ and $\beta_n := \sum_{j=1}^{n} j^{-r}$ for $n \in \mathbb{N}$. Then $\|x_n\|_\infty = 1$ and $f_r(x_n) = \beta_n$ for each $n \in \mathbb{N}$. If $r > 1$, then $f_r(x_n) \to \beta$, and if $r \leq 1$, then $f_r(x_n) \to \infty$.

3.5. For $x \in \ell^1$, $\|F(x)\|_1 \leq \sum_{i=1}^{\infty}\left(\sum_{j=i}^{\infty} |x(j)|\right)/i^2 \leq \sum_{i=1}^{\infty}\left(\sum_{j=1}^{\infty} |x(j)|\right)/i^2 = \pi^2\|x\|_1/6$. Also, $\|F(e_n)\|_1 = \|(1, 1/2^2, \ldots, 1/n^2, 0, 0, \ldots)\|_1 \to \pi^2/6$. Further, if $x \in \ell^1$ and $x(n) \neq 0$, then $\sum_{j=n+1}^{\infty} |x(j)| < \sum_{j=1}^{\infty} |x(j)|$, and so $\|F(x)\|_1 < \pi^2\|x\|_1/6$.

3.6. (i) If $P \neq 0$, then there is $y \in R(P)$ with $\|y\| = 1$, and so $\|P\| \geq \|P(y)\| = \|y\| = 1$. Suppose X is an inner product space. If P is an orthogonal projection, then $\|x\|^2 = \|P(x)\|^2 + \|x - P(x)\|^2 \geq \|P(x)\|^2$, $x \in X$, and so $\|P\| \leq 1$. Conversely, let $\|P\| = 0$ or $\|P\| = 1$. Then $\|P\| \leq 1$. Let $y \in R(P)$ and $z \in Z(P)$. If $z := 0$, then clearly $\langle y, z\rangle = 0$. Let $z \neq 0$, and assume $\|z\| = 1$ without loss of generality. Define $x := y - \langle y, z\rangle z$. Then $\|x\|^2 = \|y\|^2 - |\langle y, z\rangle|^2 = \|P(x)\|^2 - |\langle y, z\rangle|^2 \leq \|x\|^2 - |\langle y, z\rangle|^2$, and so $\langle y, z\rangle = 0$, that is $y \perp z$. Thus P is an orthogonal projection operator.
(ii) $|||Q(x)||| = |||x + Z||| \leq \|x\|$ for all $x \in X$, and so $\|Q\| \leq 1$. If $Z = X$, then $X/Z = \{0 + Z\}$, and so $\|Q\| = 0$. On the other hand, if $Z \neq X$ and $\epsilon > 0$, then by the Riesz lemma, there is $x \in X$ such that $\|x\| = 1$ and $|||Q(x)||| = d(x, Z) > 1 - \epsilon$, and so $\|Q\| > 1 - \epsilon$, which shows that $\|Q\| = 1$.

3.7. Suppose $M := [k_{i,j}]$ defines a map F from c_{00} to itself. Then the jth column $F(e_j) := (k_{1,j}, k_{2,j}, \ldots)$ is in c_{00} for each $j \in \mathbb{N}$. Conversely, sup-

pose the jth column $(k_{1,j}, k_{2,j}, \ldots)$ of M is in c_{00} for each $j \in \mathbb{N}$. Then for each $j \in \mathbb{N}$, there is $m_j \in \mathbb{N}$ such that $k_{i,j} = 0$ for all $i > m_j$. If $x := (x(1), \ldots, x(n), 0, 0, \ldots) \in c_{00}$, then $y(i) := \sum_{j=1}^{\infty} k_{i,j}x(j) = \sum_{j=1}^{n} k_{i,j}x(j) \in \mathbb{K}$ for all $i \in \mathbb{N}$, and so $y(i) = 0$ if $i > \max\{m_1, \ldots, m_n\}$. Thus $y := (y(1), y(2), \ldots) \in c_{00}$.
The result for the norm $\|\cdot\|_1$ follows as in Example 3.14(i) since $e_j \in c_{00}$ for $j \in \mathbb{N}$, and the result for the norm $\|\cdot\|_\infty$ follows as in Example 3.14(ii) by considering $x_{i,m} := (\operatorname{sgn} k_{i,1}, \ldots, \operatorname{sgn} k_{i,m}, 0, 0, \ldots) \in c_{00}$ for $i, m \in \mathbb{N}$.

3.8. (i) Let $x \in X$. For $i \in \mathbb{N}$, $\sum_{j=1}^{\infty} |k_{i,j}\langle x, u_j\rangle| \le \|x\|\beta_1(i)$, where $\beta_1(i) := \sum_{j=1}^{\infty} |k_{i,j}|$, and so let $f_i(x) := \sum_{j=1}^{\infty} k_{i,j}\langle x, u_j\rangle$. Also, writing $|k_{i,j}\langle x, u_j\rangle| = |k_{i,j}|^{1/2}\big(|k_{i,j}|^{1/2}|\langle x, u_j\rangle|\big)$ for $i, j \in \mathbb{N}$, Bessel's inequality shows that

$$\sum_{i=1}^{\infty} |f_i(x)|^2 \le \sum_{i=1}^{\infty}\Big(\sum_{j=1}^{\infty} |k_{i,j}|\Big)\Big(\sum_{j=1}^{\infty} |k_{i,j}||\langle x, u_j\rangle|^2\Big) \le \beta_1\alpha_1\|x\|^2 < \infty.$$

(ii) Let $x \in X$. For $i \in \mathbb{N}$,

$$\sum_{j=1}^{\infty} |k_{i,j}\langle x, u_j\rangle| \le \Big(\sum_{j=1}^{\infty} |k_{i,j}|^2\Big)^{1/2}\Big(\sum_{j=1}^{\infty} |\langle x, u_j\rangle|^2\Big)^{1/2} \le \gamma_{2,2}\|x\|^2,$$

and so let $f_i(x) := \sum_{j=1}^{\infty} k_{i,j}\langle x, u_j\rangle$. Also, Bessel's inequality shows that

$$\sum_{i=1}^{\infty} |f_i(x)|^2 \le \sum_{i=1}^{\infty}\Big(\sum_{j=1}^{\infty} |k_{i,j}|^2\Big)\Big(\sum_{j=1}^{\infty} |\langle x, u_j\rangle|^2\Big) \le \gamma_{2,2}^2\|x\|^2.$$

In both cases, $y := \sum_{i=1}^{\infty} f_i(x)v_i$ belongs to Y by the Riesz–Fischer theorem for the Hilbert space Y. Let $F(x) := y$. Then $\|F(x)\|^2 \le \sum_{i=1}^{\infty} |f_i(x)|^2$ again by the Bessel inequality.
Hence the matrix M defines $F \in BL(X, Y)$, and $\|F\| \le \sqrt{\alpha_1\beta_1}$ in case (i), while $\|F\| \le \gamma_{2,2}$ in case (ii).

3.9. $p = 1$: For $j \in \mathbb{N}$, $\alpha_1(j) = \sum_{i=1}^{\infty} 1/i^2 j^2 = \pi^2/6j^2$, and so $\|F\| = \alpha_1 = \pi^2/6$.
$p = \infty$: For $i \in \mathbb{N}$, $\beta_1(i) = \sum_{j=1}^{\infty} 1/i^2 j^2 = \pi^2/6i^2$, and so $\|F\| = \beta_1 = \pi^2/6$.
$p = 2$: $\gamma_{2,2}^2 = \sum_{i=1}^{\infty}\sum_{j=1}^{\infty} 1/i^4 j^4 = (\pi^4/90)^2$, and so $\|F\| \le \gamma_{2,2} = \pi^4/90$. Also, if we let $x(j) := (-1)^j/j^2$ for $j \in \mathbb{N}$, then $x \in \ell^2$, and $\|x\|_2 = \pi^2/\sqrt{90}$, whereas $\|F(x)\|_2 = \pi^6/(90)^{3/2}$. Hence $\|F\| = \pi^4/90$.

3.10. Let $p \in \{1, 2, \infty\}$, and $(1/p) + (1/q) = 1$. As in Example 3.13, f_y is a continuous linear map on $(X, \|\cdot\|_p)$ and $\|f_y\| \le \|y\|_q$.
(i) $p = 1$: Let $t_0 \in (a, b)$, and for $t \in [a, b]$, let $x_n(t) := n - n^2|t - t_0|$ if $|t - t_0| \le 1/n$ and $x_n(t) := 0$ otherwise. Clearly, $x_n \in X$, $x_n \ge 0$ and $\|x_n\|_1 \le 1$

for $n \in \mathbb{N}$. Let $A_n := \{t \in [a, b] : t_0 - (1/n) \le t \le t_0\}$ and $B_n := \{t \in [a, b] : t_0 \le t \le t_0 + (1/n)\}$ for $n \in \mathbb{N}$. Then

$$f_y(x_n) = \int_{A_n} \big(n + n^2(t - t_0)\big) y(t)dt + \int_{B_n} \big(n - n^2(t - t_0)\big) y(t)dt$$
$$= n \int_{A_n \cup B_n} y(t)dt + n^2 \int_{A_n} (t - t_0) y(t)dt - n^2 \int_{B_n} (t - t_0) y(t)dt.$$

Now $n \int_{A_n} y(t)dt \to y(t_0), n \int_{B_n} y(t)dt \to y(t_0), n^2 \int_{A_n} (t - t_0)y(t)dt \to - y(t_0)/2$, and $n^2 \int_{B_n} (t - t_0)y(t)dt \to y(t_0)/2$ by the continuity of y at t_0. Hence $f_y(x_n) \to 2y(t_0) - y(t_0)/2 - y(t_0)/2 = y(t_0)$. Thus $\|y\|_\infty \le \|f_y\|$.
(ii) $p = \infty$: Let $x_n := n\overline{y}/(1 + n|y|)$, $n \in \mathbb{N}$. Then $x_n \in X$ and $\|x_n\|_\infty \le 1$. Since $n|y|^2/(1 + n|y|) \to |y|$ pointwise and monotonically on $[a, b]$,

$$f_y(x_n) = \int_a^b \frac{n|y|^2}{1 + n|y|} dm \to \int_a^b |y| dm = \|y\|_1.$$

Thus $\|y\|_1 \le \|f_y\|$.
(iii) $p = 2$: If $x := \overline{y} \in X$, then $\|x\|_2 = \|y\|_2$ and $f_y(x) = \|y\|_2^2$. Thus $\|y\|_2 \le \|f_y\|$.

3.11. If $x \in X$, then clearly $F(x) \in X$.
(i) Let $x \in X$. Using the Fubini theorem, we obtain $\|F(x)\|_1 \le \alpha_{\|} x\|_1$. Hence $F \in BL(X)$ and $\|F\| \le \alpha_1$. On the other hand, let $t_0 \in (a, b)$, and consider the sequence (x_n) given in Exercise 3.10(i). For $s \in [a, b]$, define $y_s(t) := k(s, t)$, $t \in [a, b]$. Then $y_s \in X$, and $\int_a^b x_n(t)y_s(t)dt \to y_s(t_0)$ for each $s \in [a, b]$. By the bounded convergence theorem,

$$\|F(x_n)\|_1 = \int_a^b \left| \int_a^b k(s, t)x_n(t)dt \right| ds \to \int_a^b |k(s, t_0)| ds.$$

Thus $\int_a^b |k(s, t_0)|dt \le \|F\|$ for every $t_0 \in (a, b)$. Also, note that the function $t \longmapsto \int_a^b |k(s, t)|ds$ is continuous on $[a, b]$. Hence $\alpha_1 \le \|F\|$.
(ii) For $s \in [a, b]$, define $y_s(t) := k(s, t)$, $t \in [a, b]$. Then $y_s \in X$, and $\beta_1 = \sup\{\|y_s\|_1 : s \in [a, b]\}$. Consider the linear functional f_{y_s} on X considered in Exercise 3.10(ii). Then $\beta_1 = \sup\{\|f_{y_s}\| : s \in [a, b]\}$. Further, $F(x)(s) = f_{y_s}(x)$ for all $x \in X$ and $s \in [a, b]$. It follows that

$$\|F(x)\|_\infty = \sup\{|f_{y_s}(x)| : s \in [a, b]\} \le \beta_1 \|x\|_\infty \quad \text{for all } x \in X.$$

Hence $F \in BL(X)$ and $\|F\| \le \beta_1$. On the other hand, for each $s \in [a, b]$,

$$\|f_{y_s}\| = \sup\{|f_{y_s}(x)| : x \in X \text{ and } \|x\|_\infty \le 1\} \le \|F\|$$

since $|f_{y_s}(x)| \le \|F(x)\|_\infty$ for all $x \in X$. Hence $\beta_1 \le \|F\|$.
(iii) Let $x \in X$ and $s \in [a, b]$. By the Schwarz inequality,

$$|F(x)(s)|^2 \le \left(\int_a^b |k(s,t)|dt\right)\left(\int_a^b |k(s,t)||x(t)|^2 dt\right)$$
$$\le \beta_1\left(\int_a^b |k(s,t)||x(t)|^2 dt\right).$$

By the Fubini theorem, $\|F(x)\|_2^2 \le \beta_1 \int_a^b \big(\int_a^b |k(s,t)|ds\big)|x(t)|^2 dt \le \beta_1\alpha_1 \|x\|_2^2$. Hence $F \in BL(X)$ and $\|F\| \le (\alpha_1\beta_1)^{1/2}$.

3.12. (i) For $i, j \in \mathbb{N}$, the series $\sum_{n=1}^\infty k_1(i,n)k_2(n,j)$ converges in $\mathbb{K}$ since $\sum_{n=1}^\infty |k_1(i,n)k_2(n,j)| \le (\sum_{n=1}^\infty |k_1(i,n)|^2)^{1/2}(\sum_{n=1}^\infty |k_2(n,j)|^2)^{1/2} < \infty$. Consider $x \in \ell^2$, and let $y := F_2(x) \in \ell^2$. Then for $i \in \mathbb{N}$,

$$F(x)(i) = F_1(y)(i) = \sum_{n=1}^\infty k_1(i,n)y(n) = \sum_{n=1}^\infty k_1(i,n)\sum_{j=1}^\infty k_2(n,j)x(j)$$
$$= \sum_{j=1}^\infty\left(\sum_{n=1}^\infty k_1(i,n)k_2(n,j)\right)x(j) = \sum_{j=1}^\infty k(i,j)x(j).$$

Interchanging the order of summation is justified since

$$\left(\sum_{n=1}^\infty\sum_{j=1}^\infty |k_1(i,n)k_2(n,j)x(j)|\right)^2 \le \left(\sum_{n=1}^\infty\sum_{j=1}^\infty |k_1(i,n)x(j)|^2\right)\gamma_{2,2}^2$$
$$= \left(\sum_{n=1}^\infty |k_1(i,n)|^2\right)\|x\|_2^2\gamma_{2,2}^2,$$

where $\gamma_{2,2}^2 := \sum_{n=1}^\infty\sum_{j=1}^\infty |k_2(n,j)|^2$. (See, for example, [13, Proposition 7.21].) Hence the matrix $M := [k(i,j)]$ defines the map F. Also,

$$\|F\|^2 \le \sum_{i=1}^\infty\sum_{j=1}^\infty |k(i,j)|^2 \le \sum_{i=1}^\infty\sum_{j=1}^\infty\left(\sum_{n=1}^\infty |k_1(i,n)k_2(n,j)|\right)^2$$
$$\le \sum_{i=1}^\infty\sum_{j=1}^\infty\left(\sum_{n=1}^\infty |k_1(i,n)|^2\right)\left(\sum_{n=1}^\infty |k_2(n,j)|^2\right)$$
$$= \left(\sum_{i=1}^\infty\sum_{n=1}^\infty |k_1(i,n)|^2\right)\left(\sum_{n=1}^\infty\sum_{j=1}^\infty |k_2(n,j)|^2\right).$$

(ii) Replace $i, j, n \in \mathbb{N}$ by $s, t, u \in [a, b]$ respectively, and replace summation by Lebesgue integration in (i) above.

3.13. The seminorm p is discontinuous on X: Let $x_n(t) := t^n/n$ for $n \in \mathbb{N}$ and $t \in [0, 1]$. Then $x_n \in X$ and $\|x_n\|_{1,\infty} = 1$, but $p(x_n) = n - 1 \to \infty$.
The seminorm p is countably subadditive on X: Let $s := \sum_{k=1}^{\infty} x_k$ in X with $\sum_{k=1}^{\infty} p(x_k) = \sum_{k=1}^{\infty} \|x_k''\|_\infty < \infty$. Let $Y := C([0, 1])$ with the sup norm. Since Y is a Banach space, the absolutely summable series $\sum_{k=1}^{\infty} x_k''$ of terms in Y is summable in Y. Let $y := \sum_{k=1}^{\infty} x_k'' \in Y$. Define $s_n := \sum_{k=1}^{n} x_k$ for $n \in \mathbb{N}$. Since $s_n \to s$ in X, the sequence (s_n') in $C^1([0, 1])$ converges (uniformly) to the function $s' \in C^1([0, 1])$, and the derived sequence (s_n''), where $s_n'' = \sum_{k=1}^{n} x_k''$, $n \in \mathbb{N}$, converges uniformly to the function y. By a well-known theorem in Real Analysis, $y = (s')' = s''$. Thus $s'' = \sum_{k=1}^{\infty} x_k''$, and so $p(s) = \|s''\|_\infty \le \sum_{k=1}^{\infty} \|x_k''\|_\infty = \sum_{k=1}^{\infty} p(x_k)$.

3.14. (i) If $x_n := k_{n,1}y_1 + \cdots + k_{n,m}y_m \to x := k_1y_1 + \cdots + k_my_m$ in X, then

$$|p(x_n) - p(x)| \le p(x_n - x) \le |k_{n,1} - k_1|p(y_1) + \cdots + |k_{n,m} - k_m|p(y_m) \to 0.$$

(ii) Let p be a lower semicontinuous seminorm on a Banach space X. With notation as in Lemma 3.18, $s_n \to s$ in X, and $p(s_n) \le \sum_{k=1}^{\infty} p(x_k)$ for $n \in \mathbb{N}$. Hence $p(s) \le \lim_{n\to\infty} \inf\{p(s_m) : m \ge n\} \le \sum_{k=1}^{\infty} p(x_k)$. Thus p is countably subadditive. By the Zabreiko theorem, p is continuous.

3.15. If $F \in BL(X, \ell^r)$ and $j \in \mathbb{N}$, then $|f_j(x)| = |F(x)(j)| \le \|F(x)\|_r \le \|F\|\|x\|$ for all $x \in X$, and so $f_j \in BL(X, \mathbb{K})$.
Conversely, suppose $f_j \in BL(X, \mathbb{K})$ for all $j \in \mathbb{N}$.
$r = 1$: For $n \in \mathbb{N}$, let $p_n(x) := |f_1(x)| + \cdots + |f_n(x)|$, $x \in X$. Then each p_n is a continuous seminorm on X, and for each $x \in X$, $p_n(x) \le \|F(x)\|_1$ for all $n \in \mathbb{N}$. By Corollary 3.22, there is $\alpha > 0$ such that $p_n(x) \le \alpha\|x\|$ for all $n \in \mathbb{N}$ and $x \in X$, and so $\|F(x)\|_1 \le \alpha\|x\|$ for all $x \in X$. Hence $F \in BL(X, \ell^r)$. A similar argument holds if $r \in \{2, \infty\}$:
$r = 2$: For $n \in \mathbb{N}$, let $p_n(x) := \left(|f_1(x)|^2 + \cdots + |f_n(x)|^2\right)^{1/2}$, $x \in X$.
$r = \infty$: For $n \in \mathbb{N}$, let $p_n(x) := |f_n(x)|$, $x \in X$.
Aliter: The map $F : X \to \ell^r$ is closed: Let $x_n \to 0$ in X and $F(x_n) \to y$ in ℓ^r. Then $F(x_n)(j) \to y(j)$, and also $F(x_n)(j) = f_j(x_n) \to 0$, so that $y(j) = 0$ for each $j \in \mathbb{N}$. By the closed graph theorem, F is continuous.

3.16. Let E be a totally bounded subset of X, and let $\epsilon > 0$. Find $x_1, \ldots, x_m$ in E such that $E \subset U(x_1, \epsilon) \cup \cdots \cup U(x_m, \epsilon)$. Define $F(x) := \lim_{n\to\infty} F_n(x)$ for $x \in X$. There is n_0 such that $\|F_n(x_j) - F(x_j)\| < \epsilon$ for all $n \ge n_0$ and $j = 1, \ldots, m$. Let $x \in E$, and choose x_j such that $\|x - x_j\| < \epsilon$. By Theorem 3.24, there is $\alpha > 0$ such that $\|F_n\| \le \alpha$ for all $n \in \mathbb{N}$, and $\|F\| \le \alpha$. Hence

$$\begin{aligned}\|F_n(x) - F(x)\| &\le \|F_n(x - x_j)\| + \|F_n(x_j) - F(x_j)\| + \|F(x_j - x)\| \\ &\le \|F_n\|\|x - x_j\| + \|F_n(x_j) - F(x_j)\| + \|F\|\|x - x_j\| \\ &\le (2\alpha + 1)\epsilon\end{aligned}$$

for all $n \ge n_0$. Thus $(F_n(x))$ converges to $F(x)$ uniformly for $x \in E$.

3.17. Suppose $\|F_n\| \le \alpha$ for all $n \in \mathbb{N}$, and there is $E \subset X$ with span E dense in X and $(F_n(x))$ is Cauchy in Y for each $x \in E$. Let $X_0 :=$ span E. Since Y is a Banach space, $(F_n(x))$ converges in Y for each $x \in E$, and hence for each $x \in X_0$. Define $F_0(x) := \lim_{n\to\infty} F_n(x)$ for $x \in X_0$. Then $\|F_0(x)\| \le \lim_{n\to\infty} \|F_n(x)\| \le \alpha\|x\|$ for all $x \in X_0$. Thus $F_0 \in BL(X_0, Y)$ and $\|F_0\| \le \alpha$. By Proposition 3.17(ii), there is $F \in BL(X, Y)$ satisfying $F(x_0) = F_0(x_0)$ for all $x_0 \in X_0$ and $\|F\| = \|F_0\| \le \alpha$. Let $x \in X$ and $\epsilon > 0$. Since X_0 is dense in X, there is $x_0 \in X_0$ such that $\|x - x_0\| < \epsilon$. Also, $F_n(x_0) \to F_0(x_0) = F(x_0)$ in Y, and so there is $n_0 \in \mathbb{N}$ such that $\|F_n(x_0) - F(x_0)\| < \epsilon$ for all $n \ge n_0$. Hence

$$\begin{aligned}\|F_n(x) - F(x)\| &\le \|F_n(x - x_0)\| + \|F_n(x_0) - F(x_0)\| + \|F(x_0 - x)\| \\ &\le \|F_n\|\,\|x - x_0\| + \|F_n(x_0) - F(x_0)\| + \|F\|\|x_0 - x\| \\ &\le (2\alpha + 1)\,\epsilon\end{aligned}$$

for all $n \ge n_0$. Thus $F_n(x) \to F(x)$ in Y for all $x \in X$.

Conversely, let $F \in BL(X, Y)$ be such that $F_n(x) \to F(x)$ in Y for each $x \in X$. Then $(\|F_n\|)$ is bounded by the Banach–Steinhaus theorem.

3.18. If $n_0 \in \mathbb{N}$ and $x \in R(P_{n_0})$, then $R(P_{n_0}) \subset R(P_n)$, and so $P_n(x) = x$ for all $n \ge n_0$. Thus if $E := \bigcup_{n=1}^{\infty} R(P_n)$, then $P_n(x) \to x$ for each $x \in E$. In Exercise 3.17, let $Y := X$, $F_n := P_n$ for $n \in \mathbb{N}$, and $F = I$.

3.19. Define $x_0(t) := 1$ and $x_n(t) := t^n$ for $n \in \mathbb{N}$ and $t \in [a, b]$. In Polya's theorem, let $E := \{x_0, x_1, x_2, \ldots\}$. Then span E is the linear space of all polynomial functions on $[a, b]$, which is dense in $C([a, b])$ by the Weierstrass theorem. If all weights are nonnegative, then $\sum_{j=1}^{m_n} |w_{n,j}| = \sum_{j=1}^{m_n} w_{n,j} = Q_n(x_0) \to Q(x_0) = b - a$.

3.20. (i) $\|x\|_X \le \|x\|_F$ for all $x \in X$. Also, F is continuous if and only if there is $\alpha > 0$ with $\|F(x)\|_Y \le \alpha\|x\|_X$, that is, $\|x\|_F \le (1 + \alpha)\|x\|_X$ for all $x \in X$.
(ii) By the closed graph theorem, F is continuous, and so $\|\cdot\|_F$ is equivalent to $\|\cdot\|_X$ by (i) above. Hence $(X, \|\cdot\|_F)$ is a Banach space (Exercise 2.23).
(iii) Let (x_n) be a sequence in X such that $\|x_n\|_X \to 0$ and there is $y \in Y$ with $\|F(x_n) - y\|_Y \to 0$. Then (x_n) is a Cauchy sequence in $(X, \|\cdot\|_F)$ since (x_n) is a Cauchy sequence in $(X, \|\cdot\|_X)$ and $(F(x_n))$ is a Cauchy sequence in $(Y, \|\cdot\|_Y)$.
Since $(X, \|\cdot\|_F)$ is a Banach space, there is $x \in X$ such that $\|x_n - x\|_F \to 0$, that is, $\|x_n - x\|_X \to 0$ and $\|F(x_n) - F(x)\|_Y = \|F(x_n - x)\|_Y \to 0$. Hence $x = 0$ and $F(x) = y$, so that $y = F(0) = 0$.

(iv) The comparable norms $\|\cdot\|_X$ and $\|\cdot\|_F$ are equivalent by the two-norm theorem. Hence F is continuous by (i) above.

3.21. F is linear: Let $F(x_1) := x_2$ and $F(\tilde{x}_1) := \tilde{x}_2$. Then $F_1(x_1) = F_2(x_2)$ and $F_1(\tilde{x}_1) = F_2(\tilde{x}_2)$. Since F_1 and F_2 are linear, $F_1(x_1 + \tilde{x}_1) = F_2(x_2 + \tilde{x}_2)$, that is, $F(x_1 + \tilde{x}_1) = x_2 + \tilde{x}_2 = F(x_1) + F(\tilde{x}_1)$. Similarly, $F(k\,x_1) = kF(x_1)$.
F is a closed map: Let $x_{1,n} \to x_1$ in X_1 and $F(x_{1,n}) \to x_2$ in X_2. Define $x_{2,n} := F(x_{1,n})$ for $n \in \mathbb{N}$. Since F_1 is continuous, $F_1(x_{1,n}) \to F_1(x_1)$, and since F_2 is continuous, $F_2(x_{2,n}) \to F_2(x_2)$. But $F_1(x_{1,n}) = F_2(x_{2,n})$ since $F(x_{1,n}) = x_{2,n}$ for all $n \in \mathbb{N}$. Hence $F_1(x_1) = F_2(x_2)$, that is, $F(x_1) = x_2$.
Since X_1 and X_2 are Banach spaces, F is continuous.

3.22. (i) If $y \in \ell^q$, then $\|xy\|_1 = \sum_{j=1}^{\infty} |x(j)y(j)| \le \|x\|_p\|y\|_q$, and so $xy \in \ell^1$ for all $x \in \ell^p$. Conversely, suppose $xy \in \ell^1$ for all $x \in \ell^p$. Let $M := [k_{i,j}]$, where $k_{1,j} := y(j)$ for $j \in \mathbb{N}$, and $k_{i,j} := 0$ otherwise. Then M defines a map from ℓ^p to ℓ^1, and so its first row $y = (y(1), y(2), \ldots)$ is in ℓ^q by Corollary 3.26. Also, if we let $f_y(x) := \sum_{j=1}^{\infty} x(j)y(j)$ for $x \in \ell^p$, then $\|F\| = \|f_y\| = \|y\|_q$.
(ii) If $y \in \ell^r$, then $\|xy\|_r \le \|x\|_\infty\|y\|_r$, and so $xy \in \ell^r$ for all $x \in \ell^\infty$, and $\|F\| \le \|y\|_r$. Conversely, suppose $xy \in \ell^r$ for all $x \in \ell^\infty$. Let $x := (1, 1, \ldots)$. Since $x \in \ell^\infty$, $y = xy \in \ell^r$ and $\|y\|_r = \|xy\|_r = \|F(x)\|_r \le \|F\|$.
(iii) If $y \in \ell^\infty$ and $p \le r$, then $\|xy\|_r \le \|x\|_r\|y\|_\infty \le \|x\|_p\|y\|_\infty$, and so $xy \in \ell^r$ for all $x \in \ell^p$, and $\|F\| \le \|y\|_\infty$. Conversely, suppose $p \le r$ and $xy \in \ell^r$ for all $x \in \ell^p$. Let $M := \operatorname{diag}(y(1), y(2), \ldots)$. Then M defines the matrix transformation F from ℓ^p to ℓ^r, and so F is continuous by Proposition 3.30. Hence $|y(j)| = \|F(e_j)\|_r \le \|F\|$ for $j \in \mathbb{N}$, and so $\|y\|_\infty \le \|F\|$.

3.23. (i) If $y \in L^q$, then $\|xy\|_1 = \int_0^1 |x(t)y(t)|dm(t) \le \|x\|_p\|y\|_q$, and so $xy \in L^1$ for all $x \in L^p$. Conversely, suppose $xy \in L^1$ for all $x \in L^p$. For $n \in \mathbb{N}$, let $y_n(t) := y(t)$ if $|y(t)| \le n$ and $y_n(t) := 0$ otherwise. Then $y_n \in L^\infty \subset L^q$. For $n \in \mathbb{N}$, define $f_n(x) := \int_0^1 x(t)y_n(t)dm(t)$, $x \in L^p$, and note that $\|f_n\| = \|y_n\|_q$. If $x \in L^p$, then $f_n(x) \to \int_0^1 x(t)y(t)dm(t)$ by the dominated convergence theorem, and so there is $\alpha > 0$ such that $\|y_n\|_q = \|f_n\| \le \alpha$ for all $n \in \mathbb{N}$ by Theorem 3.24. If $p = 1$, then the set

$$\{t \in [0,1] : |y(t)| > \alpha\} = \bigcup_{n=1}^{\infty} \{t \in [0,1] : |y_n(t)| > \alpha\}$$

is of measure zero, and so $\|y\|_\infty = \operatorname{ess\,sup}|y| \le \alpha$. If $p = 2$, then $\|y_n\|_2^2 \to \int_0^1 |y(t)|^2 dm(t)$ by the monotone convergence theorem, and so $\|y\|_2 \le \alpha$. If $p = \infty$, then letting $x(t) := 1$ for $t \in [a, b]$, we see that $y = xy \in L^1$. Thus $y \in L^q$ in all cases. Also, if we let $f_y(x) := \int_0^1 x(t)y(t)dm(t)$ for $x \in L^p$, then $\|F\| = \|f_y\| = \|y\|_q$.
(ii) If $y \in L^r$, then $\|xy\|_r \le \|x\|_\infty\|y\|_r$, and so $xy \in L^r$ for all $x \in L^\infty$, and $\|F\| \le \|y\|_r$. Conversely, suppose $xy \in L^r$ for all $x \in L^\infty$. Let $x(t) := 1$ for $t \in [a, b]$. Since $x \in L^\infty$, $y = xy \in L^r$ and $\|y\|_r = \|xy\|_r = \|F(x)\|_r \le$

$\|F\|$.

(iii) If $y \in L^\infty$, then $\|xy\|_2 \le \|x\|_2\|y\|_\infty$, and so $xy \in L^2$ for all $x \in L^2$, and $\|F\| \le \|y\|_\infty$. Conversely, suppose $xy \in L^2$ for all $x \in L^2$. We show that F is a closed map. Let $x_n \to x$ in L^2 and $F(x_n) = x_n y \to z$ in L^2. Since $\|xy - z\|_1 \le \|(x - x_n)y\|_1 + \|x_n y - z\|_1 \le \|x - x_n\|_2\|y\|_2 + \|x_n y - z\|_2 \to 0$, we see that $z = xy = F(x)$. By the closed graph theorem, F is continuous.

Let $\epsilon > 0$, and let $E_\epsilon := \{t \in [0, 1] : |y(t)| > \|F\| + \epsilon\}$. Assume for a moment that $m(E_\epsilon) > 0$. If x denotes the characteristic function of E_ϵ, then $|x|\big(\|F\| + \epsilon\big) \le |xy|$ on $[0, 1]$, and $\big(\|F\| + \epsilon\big)\|x\|_2 \le \|xy\|_2 = \|F(x)\|_2 \le \|F\|\|x\|_2$, where $\|x\|_2 \ne 0$. This is impossible. Hence $|y| \le \|F\| + \epsilon$ a.e. on $[0, 1]$. Since this holds for every $\epsilon > 0$, $\|y\|_\infty = \text{ess sup}|y| \le \|F\|$.

3.24. Let $X := C([a, b])$. Then $(X, \|\cdot\|_\infty)$ is a Banach space, and $\|x_n - x\|_\infty \to 0$ if and only if (x_n) converges to x uniformly on $[a, b]$. We show that the identity map I from $(X, \|\cdot\|_\infty)$ to $(X, \|\cdot\|)$ is a closed map. Let $\|x_n\|_\infty \to 0$ and $\|x_n - y\| = \|I(x_n) - y\| \to 0$. Then $y(t) = \lim_{n\to\infty} x_n(t) = 0$ for every $t \in [a, b]$, that is, $y = 0$. By the closed graph theorem, I is continuous, and so there is $\alpha > 0$ such that $\|x\| \le \alpha\|x\|_\infty$ for all $x \in X$, that is, the norm $\|\cdot\|_\infty$ is stronger than the norm $\|\cdot\|$. The norms $\|\cdot\|_\infty$ and $\|\cdot\|$ on X are complete and comparable, and so they are equivalent.

3.25. Let $W := Y \times Z$, and $\|(y, z)\|' := \|y\| + \|z\|$ for $(y, z) \in W$. Then $(W, \|\cdot\|')$ is a Banach space. Define $F : W \to X$ by $F(y, z) := y + z$ for $(y, z) \in W$. Then F is linear, onto, and $\|F(y, z)\| = \|y + z\| \le \|y\| + \|z\| = \|(y, z)\|'$ for $(y, z) \in W$. By the open mapping theorem, F is an open map. Hence there is $\gamma > 0$ such that for every $x \in X$, there is $(y, z) \in W$ with $x = F(y, z) = y + z$ and $\|y\| + \|z\| = \|(y, z)\|' \le \gamma\|x\|$.

3.26. Since F is an open map, there is $\gamma > 0$ such that for each $n \in \mathbb{N}$, there is $z_n \in X$ with $F(z_n) = y_n - y$ and $\|z_n\| \le \gamma\|y_n - y\|$. Let $x_n := x + z_n$ for $n \in \mathbb{N}$. Then $F(x_n) = F(x) + F(z_n) = y_n$ and $\|x_n - x\| = \|z_n\| \to 0$.

3.27. Note: In this exercise, the bounded inverse theorem and the open mapping theorem are deduced directly from the Zabreiko theorem, and then closed graph theorem follows.

(i) Since F is continuous, one-one and onto, F and F^{-1} are closed maps. In Remark 3.29, replace X, Y, F and p by Y, X, F^{-1} and q respectively. If both X ad Y are Banach spaces, the continuity of q follows from the Zabreiko theorem.

(ii) Let $Z := Z(F)$. Then $\widetilde{F} \in BL(X/Z, Y)$ is one-one and onto. Also, $q(y) = \left|\left|\left|(\widetilde{F})^{-1}(y)\right|\right|\right|$ for $y \in Y$. If X is a Banach space, then q is a countably subadditive seminorm on Y, and if Y is also a Banach space, then q is continuous, as in (i) above, and so there is $\gamma > 0$ such that $q(y) < \gamma\|y\|$ for every $y \in Y$. By the definition of an infimum, for every $y \in Y$, there is $x \in X$ satisfying $F(x) = y$ and $\|x\| \le \gamma\|y\|$. Hence F is an open map by Proposition 3.41.

(iii) Φ is continuous since $\|\Phi(x, F(x))\| = \|x\| \le \|x\| + \|F(x)\|$ for $x \in X$.

If X and Y are Banach spaces, and F is a closed map, then $\text{Gr}(F)$ is a closed subspace of the Banach space $X \times Y$, and so $\Phi^{-1} \in BL(X, \text{Gr}(F))$ by (i) above. Hence there is $\alpha > 0$ such that $\|F(x)\| \le \|x\| + \|F(x)\| \le \alpha\|x\|$ for all $x \in X$.

3.28. (i) Let $p' \le p$ and $r \le r'$. Then $\|x\|_p \le \|x\|_{p'}$ and $\|F(x)\|_{r'} \le \|F(x)\|_r$ for all $x \in \ell^p$. Hence $BL(\ell^p, \ell^r) \subset BL(\ell^{p'}, \ell^{r'})$, and $\|F\|_{p',r'} = \sup\{\|F(x)\|_{r'} : x \in \ell^{p'}$ and $\|x\|_{p'} \le 1\} \le \sup\{\|F(x)\|_r : x \in \ell^p$ and $\|x\|_p \le 1\} = \|F\|_{p,r}$ for all $F \in BL(\ell^p, \ell^r)$.
Let $F \in CL(\ell^p, \ell^r)$, and let (x_n) be a bounded sequence in $\ell^{p'}$. Then (x_n) is a bounded sequence in ℓ^p, and so there is a subsequence (x_{n_k}) such that $(F(x_{n_k}))$ converges in ℓ^r, and hence in $\ell^{r'}$. Thus $F \in CL(\ell^{p'}, \ell^{r'})$.
(ii) Let $p' \ge p$ and $r \ge r'$. Replace ℓ^p, ℓ^r, $\ell^{p'}$, $\ell^{r'}$ by L^p, L^r, $L^{p'}$, $L^{r'}$ respectively in (i) above.

3.29. Let $\widetilde{F} := k_1 F + \cdots + k_n F^n$. Then $\widetilde{F} \in CL(X)$. Hence $G = k_0 I + \widetilde{F}$ belongs to $CL(X)$ if and only if $k_0 I$ belongs to $CL(X)$, that is, $k_0 = 0$.

3.30. Since $P^2 = P$ and P is a closed map, $R(P)$ is a closed subspace of X. The desired result follows from Theorem 3.42.

3.31. (i) Suppose M defines a map F from ℓ^p to ℓ^r. Then F is continuous. Since $\alpha_r(j) = \|F(e_j)\|_r$ for $j \in \mathbb{N}$, $\alpha_r = \sup\{\|F(e_j)\|_r : j \in \mathbb{N}\} \le \|F\|$. Next, let $i \in \mathbb{N}$, and define $f_i(x) := \sum_{j=1}^{\infty} k_{i,j} x(j)$, $x \in \ell^p$. Then $|f_i(x)| \le \|F(x)\|_r \le \|F\|\|x\|_p$ for all $x \in \ell^p$, and so $\|f_i\| \le \|F\|$. Since $\beta_q(i) = \|f_i\|$ for $i \in \mathbb{N}$, $\beta_q = \sup\{\|f_i\| : i \in \mathbb{N}\} \le \|F\|$. (See Corollary 3.26.)
(ii) Let $p = 1$, and $r \in \{1, 2, \infty\}$.

First suppose $\alpha_r < \infty$. We show that M defines a map from ℓ^1 to ℓ^r.
$r = 1$: This is worked out in the text (Example 3.14(i)).
$r = 2$: Let $x \in \ell^1$ and $i \in \mathbb{N}$. Then $\sum_{j=1}^{\infty} |k_{i,j} x(j)| \le \alpha_2 \|x\|_1$. Let $y(i) := \sum_{j=1}^{\infty} k_{i,j} x(j)$. Writing $|k_{i,j} x(j)| = \big(|k_{i,j}||x(j)|^{1/2}\big)|x(j)|^{1/2}$, we obtain

$$\Big(\sum_{j=1}^{\infty} |k_{i,j} x(j)|\Big)^2 \le \Big(\sum_{j=1}^{\infty} |k_{i,j}|^2 |x(j)|\Big) \sum_{j=1}^{\infty} |x(j)| = \Big(\sum_{j=1}^{\infty} |k_{i,j}|^2 |x(j)|\Big)\|x\|_1.$$

Hence $\sum_{i=1}^{\infty} |y(i)|^2 \le \|x\|_1 \sum_{j=1}^{\infty} \big(\sum_{i=1}^{\infty} |k_{i,j}|^2\big)|x(j)| \le \alpha_2^2 \|x\|_1^2$, and $y := (y(1), y(2), \ldots) \in \ell^2$. Thus M defines a map F from ℓ^1 to ℓ^2, and $\|F\| \le \alpha_2$.
$r = \infty$: Let $x \in \ell^1$ and $i \in \mathbb{N}$. Then $\sum_{j=1}^{\infty} |k_{i,j} x(j)| \le \alpha_\infty \|x\|_1$. Let $y(i) := \sum_{j=1}^{\infty} k_{i,j} x(j)$. Since $|y(i)| \le \alpha_\infty \|x\|_1$, $y := (y(1), y(2), \ldots) \in \ell^\infty$. Thus M defines a map F from ℓ^1 to ℓ^∞, and $\|F\| \le \alpha_\infty$.
Conversely, if M defines a map F from ℓ^1 to ℓ^r, then $\alpha_r \le \|F\| < \infty$.

Lastly, let $r \in \{1, 2, \infty\}$, and assume that $\alpha_r(j) \to 0$. For $n \in \mathbb{N}$, let M_n denote the infinite matrix whose first n columns are the same as those of the matrix M, and the remaining columns are zero. Then the matrices M_n

and $M - M_n$ define maps F_n and $F - F_n$ from ℓ^1 to ℓ^r respectively, and $\|F - F_n\| = \sup\{\alpha_r(j) : j = n+1, n+2, \ldots\}$ for each $n \in \mathbb{N}$. Since each F_n is of finite rank and $\|F - F_n\| \to 0$, $F \in CL(\ell^1, \ell^r)$.
(iii) Let $r = \infty$, and $p \in \{1, 2, \infty\}$.

First suppose $\beta_q < \infty$. We show that M defines a map from ℓ^p to ℓ^∞.
$p = \infty$: This is worked out in the text (Example 3.14(ii)).
$p = 2$: Let $x \in \ell^2$ and $i \in \mathbb{N}$. Then $\sum_{j=1}^\infty |k_{i,j}x(j)| \le \beta_2(i)\|x\|_2$. Let $y(i) := \sum_{j=1}^\infty k_{i,j}x(j)$. Hence $|y(i)| \le \beta_2\|x\|_2$, and $y := (y(1), y(2), \ldots) \in \ell^\infty$. Thus M defines a map F from ℓ^2 to ℓ^∞, and $\|F\| \le \beta_2$.
$p = 1$: Let $x \in \ell^1$ and $i \in \mathbb{N}$. Then $\sum_{j=1}^\infty |k_{i,j}x(j)| \le \beta_\infty(i)\|x\|_1$. Let $y(i) := \sum_{j=1}^\infty k_{i,j}x(j)$. Hence $|y(i)| \le \beta_\infty\|x\|_1$, and $y := (y(1), y(2), \ldots) \in \ell^\infty$. Thus M defines a map F from ℓ^1 to ℓ^∞, and $\|F\| \le \beta_\infty$.
Conversely, if M defines a map F from ℓ^p to ℓ^∞, then $\beta_q \le \|F\| < \infty$.

Lastly, let $p \in \{1, 2, \infty\}$, and assume that $\beta_q(i) \to 0$. For $n \in \mathbb{N}$, let M_n denote the infinite matrix whose first n rows are the same as those of the matrix M, and the remaining rows are zero. Then the matrices M_n and $M - M_n$ define maps F_n and $F - F_n$ from ℓ^p to ℓ^∞ respectively, and $\|F - F_n\| = \sup\{\beta_q(i) : i = n+1, n+2, \ldots\}$ for each $n \in \mathbb{N}$. Since each F_n is of finite rank and $\|F - F_n\| \to 0$, $F \in CL(\ell^p, \ell^\infty)$.

3.32. We use Exercise 3.31.
(i) For $j \in \mathbb{N}$, $\alpha_1(j) = j$, $\alpha_2(j) = \sqrt{j}$, and $\alpha_\infty(j) = 1$, while for $i \in \mathbb{N}$, $\beta_\infty(i) = 1$ and $\beta_2(i) = \beta_1(i) = \infty$. If M defines a map from ℓ^p to ℓ^r, then $\alpha_r < \infty$ and $\beta_q < \infty$, and so $p = 1$ and $r = \infty$. Conversely, suppose $p = 1$ and $r = \infty$. Then M defines a map $F \in BL(\ell^1, \ell^\infty)$, and $\|F\| = \alpha_\infty = \beta_\infty = 1$. But $F \notin CL(\ell^1, \ell^\infty)$, since $\|e_j\|_1 = 1$ for all $j \in \mathbb{N}$, and $\|F(e_j) - F(e_k)\|_\infty = \|e_{k+1} + \cdots + e_j\|_\infty = 1$ for all $j > k$ in $\mathbb{N}$.
(ii) For $j \in \mathbb{N}$, $\alpha_1(j) = \sum_{k=1}^j 1/k$, $\alpha_2(j) = \left(\sum_{k=1}^j 1/k^2\right)^{1/2}$, and $\alpha_\infty(j) = 1$, while for $i \in \mathbb{N}$, $\beta_\infty(i) = 1/i$, $\beta_2(i) = \beta_1(i) = \infty$. If M defines a map from ℓ^p to ℓ^r, then $\alpha_r < \infty$ and $\beta_q < \infty$, and so $p = 1$ and $r \in \{2, \infty\}$. Conversely, suppose $p = 1$ and $r \in \{2, \infty\}$. Then M defines F in $BL(\ell^1, \ell^r)$, and $\|F\| = \alpha_2 = \pi/\sqrt{6}$ if $r = 2$, and $\|F\| = \alpha_\infty = \beta_\infty = 1$ if $r = \infty$. To see that $F \in CL(\ell^1, \ell^2)$, let M_n denote the infinite matrix whose first n rows are the same as those of the matrix M, and the remaining rows are zero. Then the matrices M_n and $M - M_n$ define maps F_n and $F - F_n$ from ℓ^1 to ℓ^2 respectively, each F_n is of finite rank, and $\|F - F_n\| = \left(\sum_{k=n+1}^\infty 1/k^2\right)^{1/2} \to 0$. Also, $F \in CL(\ell^1, \ell^\infty)$ since $\beta_\infty(i) \to 0$.

3.33. (i) Let $p = 1$. By Exercise 3.31(ii), the converse of Corollary 3.31 holds.
Let $p \in \{2, \infty\}$. Define $M := [k_{i,j}]$, where $k_{1,j} := 1$ for all $j \in \mathbb{N}$, and $k_{i,j} := 0$ otherwise. Let $r \in \{1, 2, \infty\}$. Then $\alpha_r(j) = 1$ for all $j \in \mathbb{N}$, and so $\alpha_r = 1$. If $x(j) := 1/j$ for $j \in \mathbb{N}$, then $x \in \ell^p$, but the series $\sum_{j=1}^\infty k_{1,j}x(j)$ does not converge in $\mathbb{K}$. Hence M does not define a map from ℓ^p to ℓ^r.
(ii) Let $r = \infty$. By Exercise 3.31(iii), the converse of Corollary 3.26 holds.

Let $r \in \{1, 2\}$. Define $M := [k_{i,j}]$, where $k_{i,1} := 1$ for all $i \in \mathbb{N}$, and $k_{i,j} := 0$ otherwise. Let $p \in \{1, 2, \infty\}$. Then $\beta_q(i) = 1$ for all $i \in \mathbb{N}$, and so $\beta_q = 1$. If $x := e_1$, then $x \in \ell^p$, and $\sum_{j=1}^{\infty} k_{i,j} x(j) = 1$ for all $i \in \mathbb{N}$. However, $(1, 1, \ldots) \notin \ell^r$. Hence M does not define a map from ℓ^p to ℓ^r.

(iii) Let $p \in \{2, \infty\}$ and $r \in \{1, 2\}$. If $p > r$, then there is $x \in \ell^p \backslash \ell^r$, and so the identity matrix I does not define a map from ℓ^p to ℓ^r, although the r-norm of each column of I and the q-norm of each row of I is equal to 1.

If $p = 2 = r$, let M denote the matrix which has an $n \times n$ diagonal block with all entries equal to $1/\sqrt{n}$ for each $n = 1, 2, \ldots$ in that order, and whose all other entries are equal to 0. Then the 2-norm of each column as well as each row of M is equal to 1. Assume for a moment that M defines an operator F on ℓ^2. Then $F \in BL(\ell^2)$ by Proposition 3.30. For $n \in \mathbb{N}$, let $x_n := (0, \ldots, 0, 1/\sqrt{n}, \ldots, 1/\sqrt{n}, 0, 0, \ldots)$, where $1/\sqrt{n}$ occurs only in the n places numbered $\big((n-1)n/2\big) + 1, \ldots, n(n+1)/2$. Then $\|x_n\|_2 = 1$, but $\|F(x_n)\|_2 = \sqrt{n} \to \infty$, which is impossible.

3.34. Let $p \in \{2, \infty\}$, $r \in \{1, 2, \infty\}$. Suppose $M := [k_{i,j}]$ defines $F \in CL(\ell^p, \ell^r)$. Assume that $\|F(e_j)\|_r = \alpha_r(j) \not\to 0$. Then there are $j_1 < j_2 < \cdots$ in $\mathbb{N}$ and there is $\delta > 0$ such that $\alpha_r(j_k) \geq \delta$ for all $k \in \mathbb{N}$. Since $F \in CL(\ell^p, \ell^r)$, there is a subsequence (e_m) of the sequence (e_{j_k}), and there is $y \in \ell^r$ such that $F(e_m) \to y$ in ℓ^r. Fix $i \in \mathbb{N}$. Now $\beta_q(i) \leq \beta_q \leq \|F\|$ (Exercise 3.31(i)), where $q \in \{1, 2\}$. Hence $F(e_m)(i) = k_{i,m} \to 0$, and so $y(i) = \lim_{m\to\infty} F(e_m)(i) = 0$. Thus $y = 0$. But $\|y\|_r = \lim_{m\to\infty} \|F(e_m)\|_r \geq \delta$. Hence $\alpha_r(j) \to 0$.

Let $p = 1$. If $M := [k_{i,j}]$, where $k_{1,j} := 1$ for $j \in \mathbb{N}$, and $k_{i,j} := 0$ otherwise, then M defines a map in $CL(\ell^1, \ell^r)$, but $\alpha_r(j) = 1$ for all $j \in \mathbb{N}$.

3.35. (i) Define $a_0 := c_0 := 0$. For $j \in \mathbb{N}$, $\alpha_1(j) = |a_j| + |b_j| + |c_{j-1}|$ and $\alpha_2(j)^2 = |a_j|^2 + |b_j|^2 + |c_{j-1}|^2$, while for $i \in \mathbb{N}$, $\beta_1(i) = |a_{i-1}| + |b_i| + |c_i|$. Now M defines $F \in BL(\ell^1)$ if and only if $\alpha_1 := \sup\{\alpha_1(j) : j \in \mathbb{N}\} < \infty$, and M defines $F \in BL(\ell^\infty)$ if and only $\beta_1 := \sup\{\beta_1(i) : i \in \mathbb{N}\} < \infty$. Also, if $\alpha_1 < \infty$ and $\beta_1 < \infty$, then M defines $F \in BL(\ell^2)$, and conversely, if M defines $F \in BL(\ell^2)$, then $\alpha_2 < \infty$ (Exercise 3.31(i)). All these statements hold if and only if (a_j), (b_j), (c_j) are bounded sequences.

Let $a_j \to 0$, $b_j \to 0$ and $c_j \to 0$. Then $\alpha_1(j) \to 0$ and $\beta_1(i) \to 0$, so that $F \in CL(\ell^p)$ for $p \in \{1, 2, \infty\}$. To prove the converse, note that for $j \in \mathbb{N}$, $\alpha_\infty(j) = \max\{|a_j|, |b_j|, |c_{j-1}|\}$, while for $i \in \mathbb{N}$, $\beta_\infty(i) = \max\{|a_{i-1}|, |b_i|, |c_i|\}$. If $F \in CL(\ell^1)$, then $\beta_\infty(i) \to 0$ (Exercise 4.21), if $F \in CL(\ell^2)$, then $\alpha_2(j) \to 0$ (Exercise 3.34), and if $F \in CL(\ell^\infty)$, then $\alpha_\infty(j) \to 0$ (Exercise 3.34). In each case, $a_j \to 0$, $b_j \to 0$ and $c_j \to 0$.

(ii) Let $a_j := c_j := 0$ and $b_j := k_j$ for all $j \in \mathbb{N}$ in (i) above.

(iii) Let $a_j := w_j$ and $b_j := c_j := 0$ for all $j \in \mathbb{N}$ in (i) above.

3.36. Let $p \in \{2, \infty\}$ and let $r \in \{1, 2\}$. For $n \in \mathbb{N}$, let M_n denote the infinite matrix whose first n rows are the same as those of the matrix M, and the remaining rows are zero.

$p = \infty$ and $r = 1$: For $x \in \ell^\infty$, $\sum_{i=1}^\infty \sum_{j=1}^\infty |k_{i,j}x(j)| \leq \gamma_{1,1}\|x\|_\infty$. Hence M defines $F \in BL(\ell^\infty, \ell^1)$, and $\|F\| \leq \gamma_{1,1}$. Also, M_n defines $F_n \in CL(\ell^\infty, \ell^1)$, and $\|F - F_n\| \leq \sum_{i=n+1}^\infty \sum_{j=1}^\infty |k_{i,j}| \to 0$. Hence $F \in CL(\ell^\infty, \ell^1)$.

$p = \infty$ and $r = 2$: For $x \in \ell^\infty$, $\sum_{i=1}^\infty \left(\sum_{j=1}^\infty |k_{i,j}x(j)|\right)^2 \leq \gamma_{1,2}^2\|x\|_\infty^2$. Hence M defines $F \in BL(\ell^\infty, \ell^2)$, and $\|F\| \leq \gamma_{1,2}$. Also, the matrix M_n defines $F_n \in CL(\ell^\infty, \ell^2)$, and $\|F - F_n\|^2 \leq \sum_{i=n+1}^\infty \left(\sum_{j=1}^\infty |k_{i,j}|\right)^2 \to 0$. Hence $F \in CL(\ell^\infty, \ell^2)$.

$p = 2$ and $r = 1$: For $x \in \ell^2$, $\sum_{i=1}^\infty \sum_{j=1}^\infty |k_{i,j}x(j)| \leq \gamma_{2,1}\|x\|_2$. Hence M defines $F \in BL(\ell^2, \ell^1)$, and $\|F\| \leq \gamma_{2,1}$. Also, M_n defines $F_n \in CL(\ell^2, \ell^1)$, and $\|F - F_n\| \leq \sum_{i=n+1}^\infty \beta_2(i) \to 0$. Hence $F \in CL(\ell^2, \ell^1)$.

The case $p = 2$, $r = 2$ is treated in the text (Example 3.14(iii)).

Let $\gamma_{1,1} < \infty$. Then M defines $F \in CL(\ell^\infty, \ell^1)$, and $CL(\ell^\infty, \ell^1)$ is contained in $CL(\ell^p, \ell^r)$ for all $p, r \in \{1, 2, \infty\}$ (Exercise 3.28(i)).

Note that $\gamma_{2,2} \leq \gamma_{1,2}$, $\gamma_{2,1} \leq \gamma_{1,1}$.

3.37. Let $n \in \mathbb{N}$. Then $k_n(\cdot, \cdot) \in C([0, 1]\times[0, 1])$, and for $x \in X$ and $s \in [0, 1]$, $|F_n(x)(s)| \leq \|k_n(\cdot, \cdot)\|_\infty\|x\|_1$, so that $\|F_n(x)\|_\infty \leq \|k_n(\cdot, \cdot)\|_\infty\|x\|_1$. Thus $F_n \in BL(X, Y)$. For $i = 0, 1, \ldots, n$, let $y_i(s) := s^i(1-s)^{n-i}$, $s \in [0, 1]$. Then $y_i \in Y$ for each i, and

$$F_n(x)(s) = \sum_{i=0}^n c_i y_i(s), \text{ where } c_i := \binom{n}{i} \sum_{j=0}^n k\left(\frac{i}{n}, \frac{j}{n}\right)\binom{n}{j} \int_0^1 t^j(1-t)^{n-j}dt$$

for all $x \in X$ and $s \in [0, 1]$. Hence each F_n is of finite rank. Also, it follows that $\|F_n - F\| \leq \|k_n(\cdot, \cdot) - k(\cdot, \cdot)\|_\infty \to 0$. (See [6, p.10].) Since Y is a Banach space, $F \in CL(X, Y)$.

3.38. Let E denote the closure of $F(\overline{U})$.

Suppose $F \in CL(X, Y)$, and let $(\tilde{y}_n)$ be a sequence in E. Then there is $y_n \in F(\overline{U})$ such that $\|y_n - \tilde{y}_n\| < 1/n$, and there is $x_n \in \overline{U}$ such that $F(x_n) = y_n$ for each $n \in \mathbb{N}$. Let (y_{n_k}) be a subsequence of (y_n) such that $y_{n_k} \to y$ in Y. Then $\tilde{y}_{n_k} \to y$, and $y \in E$. Hence E is a compact subset of Y. In particular, every sequence in $F(\overline{U})$ has a Cauchy subsequence, that is, $F(\overline{U})$ is totally bounded.

Conversely, suppose E is a compact subset of Y. Let (x_n) be a bounded sequence in X. There is $\alpha > 0$ such that x_n/α belongs to $\overline{U}$, and let $y_n := F(x_n/\alpha)$ for $n \in \mathbb{N}$. Let (y_{n_k}) be a subsequence of (y_n) such that $y_{n_k} \to y$ in E. Then $F(x_{n_k}) \to \alpha y$ in Y. Hence we see that $F \in CL(X, Y)$. This conclusion also holds if we assume that $F(\overline{U})$ is totally bounded and Y is a Banach space, since a Cauchy subsequence in $F(\overline{U})$ converges in Y.

3.39. By Exercise 3.38, $F(\overline{U})$ is a totally bounded subset of Y, and by Exercise 3.16, $(F_n(y))$ converges uniformly to $F(y)$, $y \in F(\overline{U})$. Now $\|(F_n - F)F\| = \sup\{\|(F_n - F)(F(x))\| : x \in \overline{U}\} = \sup\{\|(F_n - F)(y)\| : y \in F(\overline{U})\} \to 0$.

3.40. By Exercise 4.31(iii), $\sum_k \|A(\tilde{u}_k)\|^2 = \sum_k \|A^*(\tilde{u}_k)\|^2 = \sum_j \|A(u_j)\|^2$.
(i) Let $A \in BL(H, G)$ be a Hilbert–Schmidt map, and let $\{u_1, u_2, \ldots\}$ be a countable orthonormal basis for H such that $\sum_j \|A(u_j)\|^2 < \infty$. Consider $x = \sum_j \langle x, u_j \rangle u_j \in H$. Then $A(x) = \sum_j \langle x, u_j \rangle A(u_j)$. For $n \in \mathbb{N}$, define $A_n(x) := \sum_{j=1}^n \langle x, u_j \rangle A(u_j)$, $x \in H$. Since $A_n \in BL(H, G)$ is of finite rank, it is a compact linear map for each $n \in \mathbb{N}$. Also, for all $x \in X$,

$$\|A(x) - A_n(x)\|^2 \leq \Big(\sum_{j>n} |\langle x, u_j \rangle| \|A(u_j)\|\Big)^2 \leq \Big(\sum_{j>n} \|A(u_j)\|^2\Big) \|x\|^2.$$

Hence $\|A - A_n\|^2 \leq \big(\sum_{j>n} \|A(u_j)\|^2\big) \to 0$, and so A is a compact map.
(ii) Let $A \in BL(\ell^2)$. Suppose A is defined by a matrix $M := [k_{i,j}]$. Since $A(e_j)(i) = k_{i,j}$ for $i, j \in \mathbb{N}$, $\sum_{j=1}^\infty \|A(e_j)\|_2^2 = \sum_{j=1}^\infty \big(\sum_{i=1}^\infty |k_{i,j}|^2\big) = \gamma_{2,2}^2$. If $\gamma_{2,2} < \infty$, then clearly A is a Hilbert–Schmidt map. Conversely, suppose A is a Hilbert–Schmidt map on ℓ^2. Then there is a denumerable orthonormal basis $\{\tilde{e}_1, \tilde{e}_2, \ldots\}$ for ℓ^2 such that $\sum_{k=1}^\infty \|A(\tilde{e}_k)\|_2^2 < \infty$. Let $k_{i,j} := A(e_j)(i)$ for $i, j \in \mathbb{N}$. Then $\gamma_{2,2}^2 = \sum_{j=1}^\infty \|A(e_j)\|_2^2 = \sum_{k=1}^\infty \|A(\tilde{e}_k)\|_2^2 < \infty$, and so A is defined by the matrix $M := [k_{i,j}]$ satisfying $\gamma_{2,2} < \infty$.
(iii) Let $A \in BL(L^2)$. Let $\{u_1, u_2, \ldots\}$ be a denumerable orthonormal basis for L^2 consisting of continuous functions on $[a, b]$. For $i, j \in \mathbb{N}$, let $w_{i,j}(s, t) := u_i(s)\overline{u_j}(t)$. Then $\{w_{i,j} : i, j \in \mathbb{N}\}$ is a denumerable orthonormal basis for $L^2([a, b] \times [a, b])$.
Suppose A is a Fredholm integral operator defined by a kernel $k(\cdot\,, \cdot)$ in $L^2([a, b] \times [a, b])$. Define

$$c_{i,j} := \int_a^b \int_a^b k(s, t)\overline{w_{i,j}}(s, t)dm(s)dm(t) \quad \text{for } i, j \in \mathbb{N}.$$

Then $\langle A(u_j), u_i \rangle = c_{i,j}$ for all $i, j \in \mathbb{N}$. By Parseval's formula,

$$\sum_{j=1}^\infty \|A(u_j)\|_2^2 = \sum_{j=1}^\infty \Big(\sum_{i=1}^\infty |\langle A(u_j), u_i \rangle|^2\Big) = \sum_{j=1}^\infty \sum_{i=1}^\infty |c_{i,j}|^2 = \|k(\cdot\,, \cdot)\|_2^2.$$

Hence A is a Hilbert–Schmidt map. Conversely, suppose A is a Hilbert–Schmidt map. Define $c_{i,j} := \langle A(u_j), u_i \rangle$ for $i, j \in \mathbb{N}$. Arguing as in (ii) above, we obtain $\sum_{i=1}^\infty \sum_{j=1}^\infty |c_{i,j}|^2 = \sum_{j=1}^\infty \|A(u_j)\|_2^2 < \infty$. The Riesz–Fischer theorem shows that the double series $\sum_{i=1}^\infty \sum_{j=1}^\infty c_{i,j} w_{i,j}$ converges in $L^2([a, b] \times [a, b])$, to say, $k(\cdot\,, \cdot)$. Let B denote the Fredholm integral operator on L^2 defined by the kernel $k(\cdot\,, \cdot)$. Then $\langle B(u_j), u_i \rangle = c_{i,j} = \langle A(u_j), u_i \rangle$ for all $i, j \in \mathbb{N}$. Hence $A = B$.

Chapter 4

4.1. (i) Let $a := (1, 0)$. Clearly, g is linear, continuous, and $\|g\| = 1 = g(a)$. A function $f : \mathbb{K}^2 \to \mathbb{K}$ is a Hahn–Banach extension of g to $\mathbb{K}^2$ if and only if f is linear on $\mathbb{K}^2$ and $\|f\| = 1 = f(a)$, that is, there are $k_1, k_2 \in \mathbb{K}$ such that $f(x) = k_1x(1) + k_2x(2)$ for all $x := (x(1), x(2)) \in \mathbb{K}^2$, $\|f\| = |k_1| + |k_2| = 1$, and $k_1 = 1$. Hence the only Hahn–Banach extension of g to $\mathbb{K}^2$ is given by $f(x) := x(1)$ for $x := (x(1), x(2)) \in \mathbb{K}^2$.
(ii) Let $b := (1, 1)$. Clearly, h is linear, continuous, and $\|h\| = 1 = h(b)$. A function $f : \mathbb{K}^2 \to \mathbb{K}$ is a Hahn–Banach extension of h to $\mathbb{K}^2$ if and only if f is linear on $\mathbb{K}^2$ and $\|f\| = 1 = f(b)$, that is, there are $k_1, k_2 \in \mathbb{K}$ such that $f(x) = k_1x(1) + k_2x(2)$ for all $x := (x(1), x(2)) \in \mathbb{K}^2$, $\|f\| = |k_1| + |k_2| = 1$, and $k_1 + k_2 = 1$. But for $k_1 \in \mathbb{K}$, $|k_1| + |1 - k_1| = 1$ if and only if $k_1 \in [0, 1]$. Hence the Hahn–Banach extensions of h to $\mathbb{K}^2$ are given by $f_t(x) := tx(1) + (1 - t)x(2)$ for $x := (x(1), x(2)) \in \mathbb{K}^2$, where $t \in [0, 1]$.

4.2. Suppose there is $\alpha > 0$ such that $\left|\sum_s c_s k_s\right| \le \alpha \left\|\sum_s c_s x_s\right\|$ as stated. Let $Y := \text{span}\,\{x_s : s \in S\}$, and for $y := \sum_s c_s x_s \in Y$, define $g(y) := \sum_s c_s k_s$. Then $g \in Y'$ and $\|g\| \le \alpha$, and so there is $f \in X'$ such that $\|f\| = \|g\| \le \alpha$ and $f(x_s) = g(x_s) = k_s$ for all $s \in S$. The converse holds with $\alpha := \|f\|$.

4.3. By the Hahn–Banach extension theorem, $E \ne \emptyset$.
E is convex: Suppose $f_1, f_2 \in E$, $t \in (0, 1)$, and $f := (1 - t)f_1 + tf_2$. Then $f(y) = (1 - t)f_1(y) + tf_2(y) = (1 - t)g(y) + tg(y) = g(y)$ for all $y \in Y$, and so $\|g\| \le \|f\| \le (1 - t)\|f_1\| + t\|f_2\| = (1 - t)\|g\| + t\|g\| = \|g\|$.
E is closed: Suppose (f_n) is in E, and $f \in X'$ such that $\|f_n - f\| \to 0$. Then $f(y) = \lim_{n\to\infty} f_n(y) = \lim_{n\to\infty} g(y) = g(y)$ for all $y \in Y$, and $\|f\| = \lim_{n\to\infty} \|f_n\| = \lim_{n\to\infty} \|g\| = \|g\|$.
E is bounded and E contains no open ball: $E \subset \{f \in X' : \|f\| = \|g\|\}$.
E may not be compact: Let $X := (C([0, 1]), \|\cdot\|_\infty)$, and let Y denote the subspace of X consisting of all constant functions. Define $g(y) := y(0)$ for $y \in Y$. Then $g \in Y'$ and $\|g\| = 1$. Given $t \in [0, 1]$, define $f_t(x) := x(t)$ for $x \in X$. Then each f_t is a Hahn–Banach extension of g, and $\|f_t - f_s\| \ge 1$ if $t \ne s$, since there is $x \in X$ such that $\|x\|_\infty = 1$, $x(t) = 0$ and $x(s) = 1$. Hence the sequence $(f_{1/n})$ in E does not have a convergent subsequence.

4.4. Suppose X' is strictly convex. Let Y be a subspace of X, $g \in Y'$ with $\|g\| = 1$, and let f_1 and f_2 be Hahn–Banach extensions of g to X. Then $\|f_1\| = \|f_2\| = \|g\| = 1$. Also, $(f_1 + f_2)/2$ is a Hahn–Banach extension of g to X, and so $\|(f_1 + f_2)/2\| = \|g\| = 1$. Hence $f_1 = f_2$.
Conversely, suppose there are $f_1 \ne f_2$ in X' such that $\|f_1\| = 1 = \|f_2\|$ and $\|f_1 + f_2\| = 2$. Let $Y := \{x \in X : f_1(x) = f_2(x)\}$, and define $g : Y \to \mathbb{K}$ by $g(y) := f_1(y)$ for $y \in Y$. Then $\|g\| \le 1$. It can also be shown that $\|g\| \ge 1$. (See [11].) Hence f_1 and f_2 are distinct Hahn–Banach extensions of g.

4.5. If Y is a Banach space, then $BL(X, Y)$ is a Banach space (Proposition 3.17(i)). If $BL(X, Y)$ is a Banach space, then its closed subspace $CL(X, Y)$ is a

Banach space. Now suppose $CL(X, Y)$ is a Banach space. Let $a \in X$ be nonzero, and let $f \in X'$ be such that $f(a) = \|a\|$ and $\|f\| = 1$. Let (y_n) be a Cauchy sequence in Y, and for $n \in \mathbb{N}$, define $F_n : X \to Y$ by $F_n(x) := f(x)y_n$, $x \in X$. Then $F_n \in CL(X, Y)$ and $\|F_n - F_m\| = \|y_n - y_m\|$ for all $n, m \in \mathbb{N}$. Hence there is $F \in CL(X, Y)$ such that $\|F_n - F\| \to 0$. In particular, $\|a\| y_n = F_n(a) \to F(a)$. Hence Y is a Banach space.

4.6. For $j \in \{1, \ldots, m\}$, let us define $g_j(y) := G(y)(j)$, $y \in Y$. Then $G(y) = (g_1(y), \ldots, g_m(y))$ for $y \in Y$. By Lemma 2.8(ii), $g_j \in Y'$, and so there is a Hahn–Banach extension $f_j \in X'$ of g_j for $j = 1, \ldots, m$. Define $F : X \to \mathbb{K}^m$ by $F(x) := (f_1(x), \ldots, f_m(x))$ for $x \in X$. Then $F(y) = G(y)$ for $y \in Y$, and by Lemma 2.8(ii), $F \in BL(X, \mathbb{K}^m)$. Note that $\|F\| \geq \|G\|$, and $|F(x)(j)| = |f_j(x)| \leq \|f_j\|\|x\| = \|g_j\|\|x\|$ for all $x \in X$ and $j = 1, \ldots, m$.
Consider the norm $\|\cdot\|_\infty$ on $\mathbb{K}^m$. We show that $\|F\| \leq \|G\|$. For $x \in X$,

$$\|F(x)\|_\infty = \max\{|F(x)(1)|, \ldots, |F(x)(m)|\} \leq \max\{\|g_1\|, \ldots, \|g_m\|\}\|x\|,$$

while for $y \in Y$, $|g_j(y)| \leq \max\{|g_1(y)|, \ldots, |g_m(y)|\} = \|G(y)\|_\infty$, and so $\|g_j\| \leq \|G\|$ for each $j = 1, \ldots, m$. Thus $\|F(x)\|_\infty \leq \|G\|\|x\|$ for all $x \in X$.
Finally, suppose $G \in BL(X, \ell^\infty)$. For $j \in \mathbb{N}$, define $f_j \in X'$ as above, and let $F(x) := (f_1(x), f_2(x), \ldots)$ for $x \in X$. Since $|F(x)(j)| = |f_j(x)| \leq \|f_j\|\|x\| = \|g_j\|\|x\| \leq \|G\|\|x\|$ for $x \in X$ and $j \in \mathbb{N}$, $F(x) \in \ell^\infty$. Clearly, $F : X \to \ell^\infty$ is linear and $F(y) = G(y)$ for all $y \in Y$. Also, on replacing 'max' by 'sup' in the earlier argument, it follows that $\|F\| = \|G\|$.

4.7. Suppose $F : \mathbb{K}^3 \to Y$ is linear, and $F(y) = y$ for all $y \in Y$. Let $F(e_1) := (k_1, k_2, k_3)$, where $k_1 + k_2 + k_3 = 0$. Then $F(e_2) = F(e_2 - e_1) + F(e_1) = e_2 - e_1 + (k_1, k_2, k_3) = (k_1 - 1, k_2 + 1, k_3)$, and $F(e_3) = F(e_3 - e_1) + F(e_1) = e_3 - e_1 + (k_1, k_2, k_3) = (k_1 - 1, k_2, k_3 + 1)$. Hence $\|F(e_1)\|_1 = |k_1| + |k_2| + |k_3|$, $\|F(e_2)\|_1 = |k_1 - 1| + |k_2 + 1| + |k_3|$ and $\|F(e_3)\|_1 = |k_1 - 1| + |k_2| + |k_3 + 1|$.
We show that at least one of $\|F(e_1)\|_1$, $\|F(e_2)\|_1$, and $\|F(e_3)\|_1$ is greater than 1. Suppose $\|F(e_3)\|_1 \leq 1$. Then $k_3 \neq 0$, for otherwise $|k_1 - 1| + |k_2| \leq 0$, that is, $k_1 = 1$ and $k_2 = 0$, and so $k_1 + k_2 + k_3 \neq 0$. If $\|F(e_1)\|_1 \leq 1$ also, then $\|F(e_2)\|_1 \geq 1 - |k_1| + 1 - |k_2| + |k_3| \geq 1 + 2|k_3| > 1$. Thus $\|F\| > 1$.

4.8. Suppose the stated condition holds. Assume for a moment that $a \notin \overline{E}$. Then there is $r > 0$ such that $U(a, r) \cap E = \emptyset$. Now $U(a, r)$ and E are disjoint convex subsets of X, and $U(a, r)$ is open. By the Hahn–Banach separation theorem, there are $f \in X'$ and $t \in \mathbb{R}$ such that $\text{Re } f(a) < t \leq \text{Re } f(x)$ for all $x \in E$. Let $t \neq 0$, and $g := f/t$. If $t > 0$, then $\text{Re } g(a) < 1$ and $\text{Re } g(x) \geq 1$ for all $x \in E$, while if $t < 0$, then $\text{Re } g(a) > 1$ and $\text{Re } g(x) \leq 1$ for all $x \in E$, contrary to the stated condition. If $t = 0$, there is $s \in \mathbb{R}$ such that $\text{Re } f(a) < s < t$, and we may consider $g := f/s$. Hence $a \in \overline{E}$. The converse follows by the continuity of f at a.

4.9. $E \cap \overline{Y} = \emptyset$ since E is open and $E \cap Y = \emptyset$. Let $Z := X/\overline{Y}$ with the quotient norm $|||\cdot|||$, and let $\widetilde{E} := Q(E)$, where $Q : X \to Z$ is the quotient map. Then

$\widetilde{E}$ is an open convex subset of Z and $0 + \overline{Y} \notin \widetilde{E}$. Hence there is $\tilde{f} \in Z'$ such that $\operatorname{Re} \tilde{f}(x + \widetilde{E}) > 0$ for all $x + \overline{Y} \in \widetilde{E}$. Let $f := \tilde{f} \circ Q$.

4.10. Let $r := \inf\{\|x_1 - x_2\| : x_1 \in E_1 \text{ and } x_2 \in E_2\}$. Assume for a moment that $r = 0$. Then there is a convergent sequence $(x_{1,n})$ in E_1 and a sequence $(x_{2,n})$ in E_2 such that $\|x_{1,n} - x_{2,n}\| \to 0$. Also, if $x_{1,n} \to x_1$, then $x_{2,n} \to x_1$ as well, and so $x_1 \in E_1 \cap E_2$, contrary to the hypothesis. Hence $r > 0$. Define $E_r := E_1 + U(0, r)$. Then $E_r = \bigcup\{x_1 + U(0, r) : x_1 \in E_1\}$ is an open convex subset of X. If $x_1 \in E_1$, $x \in U(0, r)$, and $x_2 := x_1 + x \in E_2$, then $r \leq \|x_1 - x_2\| = \|x\| < r$, a contradiction. Hence $E_r \cap E_2 = \emptyset$. By the Hahn–Banach separation theorem, there are $f \in X'$ and $t_2 \in \mathbb{R}$ such that $\operatorname{Re} f(x_1) < t_2 \leq \operatorname{Re} f(x_2)$ for $x_1 \in E_r$ and $x_2 \in E_2$. Since E_1 is a compact subset of X, $\operatorname{Re} f(E_1)$ is a compact subset of $\mathbb{R}$, and so it is closed in $\mathbb{R}$. Hence there is $t_1 < t_2$ such that $\operatorname{Re} f(x_1) \leq t_1 < t_2$ for all $x_1 \in E_1$.

4.11. (i) Replace the summation $\sum_{j=1}^{\infty}$ in Example 4.18(ii), (i) and (iii) by the summation $\sum_{j=1}^{n}$ to obtain an isometry Φ from $(\mathbb{K}^n, \|\cdot\|_q)$ into the dual of $(\mathbb{K}^n, \|\cdot\|_p)$, where $(1/p) + (1/q) = 1$, and $p = 1, 2, \infty$. This isometry is onto since both $(\mathbb{K}^n, \|\cdot\|_q)$ and the dual of $(\mathbb{K}^n, \|\cdot\|_p)$ have dimension n.
(ii) Let $y := (y(1), y(2), \ldots) \in \ell^1$, and define $f_y(x) := \sum_{j=1}^{\infty} x(j)y(j)$ for $x := (x(1), x(2), \ldots) \in c_0$. Clearly, $f_y \in (c_0)'$, and $\|f_y\| \leq \|y\|_1$. For $n \in \mathbb{N}$, define $x_n := (\operatorname{sgn} y(1), \ldots, \operatorname{sgn} y(n), 0, 0, \ldots) \in c_0$. Now $\|x_n\|_\infty \leq 1$, and $\|f_y\| \geq |f_y(x_n)| = \sum_{j=1}^{n} |y(j)|$ for all $n \in \mathbb{N}$. Hence $\|f_y\| \geq \|y\|_1$. Define $\Phi(y) : \ell^1 \to (c_0)'$ by $\Phi(y) := f_y$, $y \in \ell^1$. Then Φ is a linear isometry. To show that Φ is onto, let $f \in (c_0)'$. Then $f(x) = f\big(\sum_{j=1}^{\infty} x(j)e_j\big) = \sum_{j=1}^{\infty} x(j) f(e_j)$ for $x \in c_0$. Define $y := (f(e_1), f(e_2), \ldots)$. Then $\sum_{j=1}^{n} |y(j)| = \sum_{j=1}^{n} x_n(j)y(j) = f(x_n) \leq \|f\|$ for all $n \in \mathbb{N}$. Thus $y \in \ell^1$, and $f = f_y$. (Compare the case $p := \infty$ of Example 3.12, and also Example 4.18(iii).)
(iii) If $p \in \{1, 2\}$, then c_{00} is a dense subspace of ℓ^p, and if $p := \infty$, then c_{00} is a dense subspace of c_0 (Exercise 2.3). By Proposition 4.13(i), the dual space of $(c_{00}, \|\cdot\|_p)$ is linearly isometric to $(\ell^p)'$, that is, to ℓ^q if $p \in \{1, 2\}$, and to $(c_0)'$, that is, to ℓ^1 if $p := \infty$.

4.12. (i) It is clear that $\Psi : c' \to (c_0)'$ is linear, and $\|\Psi(x')\| \leq \|x'\|$ for all $x' \in c'$. Since c_0 is a closed subspace of c, and $e_0 := (1, 1, \ldots) \in c \backslash c_0$, there is $x' \in c'$ such that $x'(y) = 0$ for all $y \in c_0$ and $x'(e_0) = \|e_0\|_\infty = 1$. Then $x' \neq 0$, but $\Psi(x') = 0$. Hence Ψ is not an isometry.
(ii) Let $y \in \ell^1$. Then $|f_y(x)| \leq \big(|y(1)| + \sum_{j=1}^{\infty} |y(j+1)|\big)\|x\|_\infty = \|y\|_1 \|x\|_\infty$ for all $x \in c$, and so $\|f_y\| \leq \|y\|_1$. On the other hand, define $x_n := (\operatorname{sgn} y(2), \ldots, \operatorname{sgn} y(n), \operatorname{sgn} y(1), \operatorname{sgn} y(1), \ldots)$ for $n \in \mathbb{N}$. Then $x_n \in c$, $\|x_n\|_\infty \leq 1$, and

$$\|f_y\| \geq |f_y(x_n)| = \left| |y(1)| + \sum_{j=1}^{n-1} |y(j+1)| + \sum_{j=n}^{\infty} \operatorname{sgn} y(1) y(j+1) \right|$$

$$\geq \sum_{j=1}^{n} |y(j)| - \sum_{j=n+1}^{\infty} |\operatorname{sgn} y(1)|\, |y(j)| \quad \text{for all } n \in \mathbb{N}.$$

Since $\sum_{j=n+1}^{\infty} |y(j)| \to 0$, we see that $\|f_y\| \geq \|y\|_1$. To show that Φ is onto, let $f \in c'$. For $n \in \mathbb{N}$, let $u_n := \text{sgn}\, f(e_1)e_1 + \cdots + \text{sgn}\, f(e_n)e_n$. Then $u_n \in c$, $\|u_n\|_\infty \leq 1$, and $|f(e_1)| + \cdots + |f(e_n)| = f(u_n) \leq \|f\|$ for all $n \in \mathbb{N}$. Hence the series $\sum_{j=1}^{\infty} f(e_j)$ converges in $\mathbb{K}$. Let $s_f \in \mathbb{K}$ denote its sum. Consider $x \in c$, and let $x(j) \to \ell_x$. Then $x = \ell_x e_0 + \sum_{j=1}^{\infty}(x(j) - \ell_x)e_j$ (Exercise 2.27), and so $f(x) = \ell_x f(e_0) + \sum_{j=1}^{\infty} \big(x(j) - \ell_x\big) f(e_j) = \ell_x\big(f(e_0) - s_f\big) + \sum_{j=1}^{\infty} x(j) f(e_j)$. Define $y := (f(e_0) - s_f, f(e_1), f(e_2), \ldots)$. Then $\sum_{j=1}^{n} |y(j)| = |f(e_0) - s_f| + \sum_{j=1}^{\infty} |f(e_j)| \leq |f(e_0) - s_f| + \|f\| < \infty$. Thus $y \in \ell^1$, and $f = f_y$. Hence Φ is a linear isometry from ℓ^1 onto c'.

4.13. Let $H := W^{1,2}$. Fix $y \in H$. Then $f_y(x) = \langle x, \overline{y} \rangle_{1,2}$ for $x \in H$. (See Example 2.28(iv).) As in the proof of Theorem 4.14, $f_y \in H'$ and $\|f_y\| = \|\overline{y}\|_2 = \|y\|_2$. Thus the map $\Phi(y) := f_y$, $y \in H$, gives a linear isometry from H to H'. Next, if $f \in H'$, then by Theorem 4.14, $f = f_y$, where $y := \overline{y_f}$. Hence Φ is a linear isometry from H onto H'.

4.14. Let $p \in \{1, 2\}$. Then $(C([a, b]), \|\cdot\|_p)$ is a dense subspace of $L^p([a, b])$. Hence the dual space of $(C([a, b]), \|\cdot\|_p)$ is linearly isometric to the dual space of $L^p([a, b])$, that is, to $L^q([a, b])$, where $(1/p) + (1/q) = 1$.

4.15. Let X be a reflexive normed space, and let $Y := X'$. Since J is a linear isometry from X onto X'', and since $X'' = BL(Y, \mathbb{K})$ is a Banach space, X is a Banach space. Suppose X is separable as well. Then $X'' = Y'$ is separable, and so $Y = X'$ is separable. Let $X := \ell^1$. Then X is a separable Banach space. But X is not reflexive. Otherwise X' would be separable, but X' is isometric to ℓ^∞ which is not separable.

4.16. Suppose $x_n \xrightarrow{w} x$ and $x_n \xrightarrow{w} \tilde{x}$ in X. Then $x'(\tilde{x}) = \lim_{n\to\infty} x'(x_n) = x'(x)$ for all $x' \in X'$. Hence $\tilde{x} = x$ by Proposition 4.6(i).
(i) Let $E := \{x_n : n \in \mathbb{N}\}$. Suppose $x_n \xrightarrow{w} x$ in X. Then $x'(E)$ is a bounded subset of $\mathbb{K}$ for every $x' \in X'$. Hence E is a bounded subset of X.
Conversely, suppose there is $\alpha > 0$ such that $\|x_n\| \leq \alpha$ for all $n \in \mathbb{N}$, and there is a subset D of X' whose span is dense in X' and $x'(x_n) \to x'(x)$ for every $x' \in D$, that is, $J(x_n)(x') \to J(x)(x')$ for every $x' \in D$. Since $J(x_n)$, $J \in BL(X', \mathbb{K})$, and $\|J(x_n)\| = \|x_n\| \leq \alpha$ for $n \in \mathbb{N}$, Exercise 3.17 shows that $J(x_n)(x') \to J(x)(x')$ for every $x' \in X'$, that is, $x_n \xrightarrow{w} x$ in X.
(ii) If $x_n \to x$, then $|x'(x_n) - x'(x)| \leq \|x'\| \|x_n - x\| \to 0$ for every $x' \in X'$, and so $x_n \xrightarrow{w} x$ in X. Suppose X is an inner product space, $x_n \xrightarrow{w} x$ in X, and $\|x_n\| \to \|x\|$. Then $\|x_n - x\|^2 = \|x_n\|^2 - 2\text{Re}\,\langle x_n, x\rangle + \|x\|^2 \to 0$.
(iii) Let (x_n) be a bounded sequence in X. Then the bounded sequence $(\langle x_1, x_n\rangle)$ in $\mathbb{K}$ has a convergent subsequence $(\langle x_1, x_{1,n}\rangle)$ by the Bolzano–Weierstrass theorem for $\mathbb{K}$. Next, the bounded sequence $(\langle x_2, x_{1,n}\rangle)$ has a convergent subsequence $(\langle x_2, x_{2,n}\rangle)$, and so on. Define $u_n := x_{n,n}$ for $n \in \mathbb{N}$. The diagonal sequence (u_n) is a subsequence of (x_n). For each fixed $m \in \mathbb{N}$, the sequence $(\langle x_m, u_n\rangle)$ is convergent, and so the sequence $(\langle y, u_n\rangle)$ is convergent for every $y \in \text{span}\,\{x_1, x_2, \ldots\}$. Let Y denote the closure of $\text{span}\,\{x_1, x_2, \ldots\}$.

It can be seen that for every $y \in Y$, $(\langle y, u_n \rangle)$ is a Cauchy sequence, and hence it is convergent in $\mathbb{K}$. Further, if $z \in Y^\perp$, then $\langle z, u_n \rangle = 0$ for all $n \in \mathbb{N}$. Since $X = Y \oplus Y^\perp$, it follows that the sequence $(\langle x, u_n \rangle)$ is convergent for every $x \in X$. Define $f(x) := \lim_{n\to\infty} \langle x, u_n \rangle$ for $x \in X$. Then $f \in X'$, and so, there is $u \in X$ such that $f(x) = \langle x, u \rangle$ for all $x \in X$. Thus $\langle x, u_n \rangle \to \langle x, u \rangle$ for every $x \in X$, that is, $x'(u_n) \to x'(u)$ for every $x' \in X'$. Thus $u_n \xrightarrow{w} u$ in X.
Suppose $(\langle x_n, \tilde{x} \rangle)$ is convergent in $\mathbb{K}$ for every $\tilde{x} \in X$. For $n \in \mathbb{N}$, let $f_n(\tilde{x}) := \langle \tilde{x}, x_n \rangle$, $\tilde{x} \in X$. Then $f_n \in X'$ and $\|f_n\| = \|x_n\|$. Define $f(\tilde{x}) := \lim_{n\to\infty} f_n(\tilde{x})$, $\tilde{x} \in X$. By the Banach–Steinhaus theorem, $(\|f_n\|)$ is bounded, that is, $(\|x_n\|)$ is bounded, and so $f \in X'$. As above, there is $x \in X$ such that $x_n \xrightarrow{w} x$ in X.
(iv) Let $x_n \xrightarrow{w} 0$ in X. If $F \in BL(X, Y)$, then $y' \circ F(x_n) \to 0$ for every $y' \in Y'$, and so $F(x_n) \xrightarrow{w} 0$ in Y. Now let $F \in CL(X, Y)$, and assume for a moment that $F(x_n) \not\to 0$ in Y. By passing to a subsequence, if necessary, we may assume that there is $\delta > 0$ such that $\|F(x_n)\| \ge \delta$ for all $n \in \mathbb{N}$. Since (x_n) is a bounded sequence and F is a compact linear map, there is a subsequence (x_{n_k}) such that $F(x_{n_k}) \to y$ in Y. But since $F(x_{n_k}) \xrightarrow{w} 0$ in Y, we see that $y = 0$. This is impossible since $\|F(x_{n_k})\| \ge \delta$ for all $k \in \mathbb{N}$.
(v) Let (u_n) be an orthonormal sequence in the Hilbert space X. As a consequence of the Bessel inequality, $\langle x, u_n \rangle \to 0$ for every $x \in X$, that is, $u_n \xrightarrow{w} 0$ in X. Let $F \in CL(X, Y)$. Then $F(u_n) \to 0$ in Y by (iv) above.
In particular, let $X = Y := \ell^2$, and let an infinite matrix M define a map $F \in CL(\ell^2)$. Then $F(e_j) \to 0$, that is, $\alpha_2(j) \to 0$. Also, the transpose M^t of M defines a map $F^t \in CL(\ell^2)$, and so $F^t(e_i) \to 0$, that is, $\beta_2(i) \to 0$.

4.17. (i) $|x'(x)| \le \|x\|_1$ for all $x \in \ell^1$, and so $x' \in (\ell^1)'$. Clearly, $x'(e_n) = 1 \not\to 0$.
(ii) The span of the set $E := \{e_j : j \in \mathbb{N}\}$ is dense in ℓ^2, which is linearly isometric to $(\ell^2)'$. Hence the result follows from Exercise 4.16(i).
(iii) Without loss of generality, we let $x := 0$. Assume for a moment that $x_n \overset{w}{\not\to} 0$ in X. Then there is $x' \in X'$ such that $x'(x_n) \not\to 0$, and so there is $\delta > 0$ and there are $n_1 < n_2 < \cdots$ in $\mathbb{N}$ such that $|x'(x_{n_k})| \ge \delta$ for all $k \in \mathbb{N}$. For $m \in \mathbb{N}$, define $u_m := \operatorname{sgn} x'(x_{n_1})\, x_{n_1} + \cdots + \operatorname{sgn} x'(x_{n_m})\, x_{n_m}$. Since $|u_m(j)| \le |x_{n_1}(j)| + \cdots + |x_{n_m}(j)| \le \sum_{n=1}^\infty |x_n(j)| \le \alpha$ for all $j \in \mathbb{N}$, we see that $\|u_m\|_\infty \le \alpha$ for all $m \in \mathbb{N}$, and so $m\,\delta \le |x'(x_{n_1})| + \cdots + |x'(x_{n_m})| = |x'(u_m)| \le \alpha \|x'\|$ for all $m \in \mathbb{N}$, which is impossible. Thus $x_n \xrightarrow{w} 0$ in ℓ^∞. In particular, let $x_n := e_n$ for $n \in \mathbb{N}$, and observe that $\sum_{n=1}^\infty |e_n(j)| = 1$ for all $j \in \mathbb{N}$. Hence $e_n \xrightarrow{w} 0$ in ℓ^∞.
(iv) Let $x' \in X'$. By Exercise 4.11(ii), there is $y \in \ell^1$ such that $x'(x) = \sum_{j=1}^\infty x(j)y(j)$ for all $x \in c_0$, and so $x'(x_n) = \sum_{j=1}^n y(j) \to \sum_{j=1}^\infty y(j)$. Assume for a moment that $x_n \xrightarrow{w} x$ in c_0. Fix $j \in \mathbb{N}$. Then $x_n(j) \to x(j)$. But $x_n(j) = 1$ for all $n \ge j$, and so $x = (1, 1, \ldots)$ which is not in c_0. Thus there is no $x \in c_0$ such that $x_n \xrightarrow{w} x$.
(v) Let $x_n \xrightarrow{w} x$ in $C([a, b])$. Then (x_n) is bounded in $(C([a, b]), \|\cdot\|_\infty)$ by

Exercise 4.16(i). For $t \in [a,b]$, define $x_t'(x) := x(t)$, $x \in X$, so that $x_t' \in X'$. Hence $x_n(t) = x_t'(x_n) \to x_t'(x) = x(t)$ for each $t \in [a,b]$.
Conversely, suppose (x_n) is uniformly bounded on $[a,b]$, and $x_n(t) \to x(t)$ for each $t \in [a,b]$. Let $x' \in X'$. There is $y \in BV([a,b])$ such that $x'(x) = \int_a^b x\,dy$ for all $x \in X$. There are nondecreasing functions y_1, y_2, y_3, y_4 defined on $[a,b]$ such that $y = y_1 - y_2 + i(y_3 - y_4)$, and further, $\int_a^b x_n\,dy_i \to \int_a^b x\,dy_i$ for $i = 1, \ldots, 4$ by the bounded convergence theorem for the Riemann–Stieltjes integration. Thus $x'(x_n) = \int_a^b x_n\,dy \to \int_a^b x\,dy = x'(x)$.

4.18. (i)$\Longrightarrow$(ii)$\Longrightarrow$(iii) by the projection theorem.
(iii)$\Longrightarrow$(iv): If $Y^\perp = \{0\}$, then $Y = Y^{\perp\perp} = \{0\}^\perp = X$.
(iv)$\Longrightarrow$(v): Let $Y := Z(f)$ in the proof of Theorem 4.14.
(v)$\Longrightarrow$(i): Let (y_n) be a Cauchy sequence in X. For $n \in \mathbb{N}$, define $f_n(x) := \langle x, y_n \rangle$, $x \in X$. Then $f_n \in X'$, and $\|f_n - f_m\| = \|y_n - y_m\|$ for all $n, m \in \mathbb{N}$. Since X' is a Banach space, there is $f \in X'$ such that $\|f_n - f\| \to 0$. Let $y \in X$ be such that $f(x) := \langle x, y \rangle$, $x \in X$. Then $\|y_n - y\| = \|f_n - f\| \to 0$.

4.19. Let H denote the completion of X.
(i) Let span $\{u_\alpha\}$ be dense in X. Since X is dense in H, we see that span $\{u_\alpha\}$ is dense in H. By Theorem 2.31, $x = \sum_n \langle x, u_n \rangle u_n$
for every $x \in H$, and in particular for every $x \in X$. The converse follows as in the proof of (iv)$\Longrightarrow$(i) of Theorem 2.31.
(ii) Let G denote the closure of Y in H. Then $G \cap X = Y$ since Y is closed in X. Let $\langle\langle \cdot\,, \cdot \rangle\rangle$ be the inner product on H/G which induces the quotient norm on H/G (Exercise 2.40). Let $\langle\langle x_1 + Y, x_2 + Y \rangle\rangle := \langle\langle x_1 + G, x_2 + G \rangle\rangle$ for $x_1, x_2 \in X$. If $x \in X$ and $\langle\langle x + Y, x + Y \rangle\rangle = \langle\langle x + G, x + G \rangle\rangle = 0$, then $x \in G \cap X = Y$, that is $x + Y = 0 + Y$. It follows that $\langle\langle \cdot\,, \cdot \rangle\rangle$ is an inner product on X/Y, and $\langle\langle x + Y, x + Y \rangle\rangle = \langle\langle x + G, x + G \rangle\rangle = |||x + G|||^2 = d(x, G)^2 = d(x, Y)^2 = |||x + Y|||^2$ for all $x \in X$.
(iii) Let $\langle \cdot\,, \cdot \rangle'$ denote the inner product on H which induces the norm on H', as in Corollary 4.16(i). Let $f_0, g_0 \in X'$. By Proposition 3.17(ii), there are unique $f, g \in H'$ such that $f(x) = f_0(x)$ and $g(x) = g_0(x)$ for all $x \in X$, $\|f\| = \|f_0\|$ and $\|g\| = \|g_0\|$. Define $\langle f_0, g_0 \rangle' := \langle f, g \rangle'$. Then $\langle \cdot\,, \cdot \rangle'$ is an inner product on X', and $\langle f_0, f_0 \rangle' = \langle f, f \rangle' = \|f\|^2 = \|f_0\|^2$.

4.20. If there are $x_1', \ldots, x_m'$ in X' and $y_1, \ldots, y_m$ in Y such that $F(x) = \sum_{i=1}^m x_i'(x) y_i$ for $x \in X$, then $R(F) \subset \text{span}\,\{y_1, \ldots, y_m\}$, and so F is of finite rank. Conversely, suppose F is of finite rank. Let $y_1, \ldots, y_m$ be a basis for $R(F)$. Then there are unique functions $f_1, \ldots, f_m$ from X to $\mathbb{K}$ such that $F(x) = f_1(x)y_1 + \cdots + f_m(x)y_m$ for $x \in X$. Fix $i \in \{1, \ldots, m\}$. Clearly, f_i is linear. Let $Y_i := \text{span}\,\{y_j : j = 1, \ldots, m \text{ and } j \neq i\}$ and $d_i := d(y_i, Y_i)$. Since $d_i > 0$ and $|f_i(x)| d_i \le \|F(x)\| \le \|F\| \|x\|$ for all $x \in X$, f_i is continuous. Let $x_i' := f_i \in X'$ for $i = 1, \ldots, m$.
Suppose $F(x) := \sum_{i=1}^m x_i'(x) y_i$ for $x \in X$ as above, and let $y' \in Y'$. Then $F'(y')(x) = y'(F(x)) = \sum_{i=1}^m x_i'(x) y'(y_i) = \left(\sum_{i=1}^m y'(y_i) x_i'\right)(x)$ for all $x \in X$, that is, $F'(y') = \sum_{i=1}^m y'(y_i) x_i'$.

4.21. Let $p, r \in \{1, 2\}$, and $(1/p) + (1/q) = 1$ and $(1/r) + (1/s) = 1$. The transpose M^t of M defines a map $F^t \in BL(\ell^s, \ell^q)$, which can be identified with F'. Since F is compact, F' is compact, and so is F^t. Since $s, q \in \{2, \infty\}$, the sequence of the columns of M^t, which is the sequence of the rows of M, tends to 0 in ℓ^q by Exercise 3.34.
Let $p \in \{1, 2, \infty\}$ and $r := \infty$. f $M := [k_{i,j}]$, where $k_{i,1} := 1$ for $i \in \mathbb{N}$, and $k_{i,j} := 0$ otherwise, then M defines $F \in CL(\ell^p, \ell^\infty)$, but $\beta_r(i) = 1$, $i \in \mathbb{N}$.

4.22. If $p \in \{1, 2, \infty\}$, then $\|A(x)\|_p \le \|x\|_p$ for all $x \in L^p$, and so $A \in BL(L^p)$. Now let $p \in \{1, 2\}$, and $(1/p) + (1/q) = 1$. For $y \in L^q$, let

$$\Phi(y)(x) := \int_0^\infty x(t)y(t)dm(t), \quad x \in L^p.$$

Now $\Phi : L^q \to (L^p)'$ is a linear isometry, and it is onto. (Compare Examples 4.19(ii) and 4.24(ii).) Define $A^t : L^q \to L^q$ by $A^t := (\Phi)^{-1}A'\Phi$. Thus $A' \in BL((L^p)')$ can be identified with $A^t \in BL(L^q)$. Fix $y \in L^q$. Let $z(s) := 0$ if $s \in [0, 1)$, and $z(s) := y(s-1)$ if $s \in [1, \infty)$. Then $z \in L^q$, and

$$\begin{aligned}\Phi(A^t(y))(x) &= A'(\Phi(y))(x) = \Phi(y)(A(x)) \\ &= \int_0^\infty x(t+1)y(t)dm(t) = \int_1^\infty x(s)y(s-1)dm(s) \\ &= \int_0^\infty x(s)z(s)dm(s) = \Phi(z)(x).\end{aligned}$$

for all $x \in L^p$. Hence $\Phi(A^t(y)) = \Phi(z)$, and in turn, $A^t(y) = z$, as desired.

4.23. Let $n \in \mathbb{N}$. For $y \in \ell^q$, let $\Phi(y)(x) := \sum_{j=1}^\infty x(j)y(j)$, $x \in \ell^p$. Then Φ is an isometry from ℓ^q onto $(\ell^p)'$. Let $P_n^t := (\Phi)^{-1}P_n'\Phi$. Thus $P_n' \in BL((\ell^p)')$ can be identified with $P_n^t \in BL(\ell^q)$. Clearly, $(P_n^t)^2 = P_n^t$. Fix $y \in \ell^q$, and let $y_n := (y(1), \ldots, y(n), 0, 0, \ldots)$. Then $y_n \in \ell^q$, and

$$\Phi(P_n^t(y))(x) = P_n'(\Phi(y))(x) = \Phi(y)(P_n(x)) = \sum_{j=1}^n x(j)y(j) = \Phi(y_n)(x)$$

for $x \in \ell^p$. Hence $\Phi(P_n^t(y)) = \Phi(y_n)$, and in turn, $P_n^t(y) = y_n$, as desired.

4.24. For $x' \in X'$, let $y' := P'(x')$. Then $(P')^2(x')(x) = P'(y')(x) = y'(P(x)) = x'(P(P(x))) = x'(P(x)) = P'(x')(x)$ for all $x \in X$. Hence $(P')^2 = P'$. Also, $R(P') = \{x' \in X' : P'(x') = x'\} = \{x' \in X' : x'(P(x) - x) = 0$ for $x \in X\} = Z^0$ and $Z(P') = \{x' \in X' : P'(x')(x) = 0$ for $x \in X\} = \{x' \in X' : x'(P(x)) = 0$ for $x \in X\} = Y^0$.

4.25. Let $x \in X$ and $y' \in Y'$. Then

$$F''\big(J_X(x)\big)(y') = J_X(x)\big(F'(y')\big) = F'(y')(x) = y'(F(x)) = J_Y\big(F(x)\big)(y').$$

Hence $F''\big(J_X(x)\big) = J_Y\big(F(x)\big)$ for all $x \in X$, that is, $F''J_X = J_Y F$. Also, $\|F''\| = \|(F')'\| = \|F'\| = \|F\|$.
Let X_c denote the closure of $J_X(X)$ in X'', and let Y_c denote the closure of $J_Y(Y)$ in Y''. Let $x'' \in X_c$. We show that $F''(x'') \in Y_c$. Let (x_n) be a sequence in X such that $J_X(x_n) \to x''$ in X''. Then $J_Y F(x_n) = F''J_X(x_n) \to F''(x'')$ in Y'', where $J_Y F(x_n) \in J_Y(Y)$. Thus $F''(x'') \in Y_c$. Define $F_c : X_c \to Y_c$ by $F_c(x'') = F''(x'')$, $x'' \in X_c$. Then F_c is linear, and $F_c(J_X(x)) = F''(J_X(x)) = J_Y F(x)$ for all $x \in X$, that is, $F_c J_X = J_Y F$. Also, $\|F\| \le \|F_c\| \le \|F''\| = \|F\|$. The uniqueness of $F_c \in BL(X_c, Y_c)$ follows by noting that $J_X(X)$ is dense in X_c.

4.26. Suppose $F' \in CL(Y', X')$. Then $F'' \in CL(X'', Y'')$. By Exercise 4.25, $F''J_X = J_Y F$, and so $J_Y(F(\overline{U})) \subset \{F''(x'') : x'' \in X'' \text{ and } \|x''\| \le 1\}$. The latter set is a totally bounded subset of Y'' since $F'' = (F')'$ is compact (Exercise 3.38), and so $J_Y(F(\overline{U}))$ is also a totally bounded subset of Y''. Since J_Y is an isometry, $F(\overline{U})$ is a totally bounded subset of Y. Since Y is a Banach space, $F \in CL(X, Y)$ (Exercise 3.38).

4.27. Let $x_n \to x_0$ in H and $A(x_n) \to y_0$ in G. Then for every $y \in Y$, $\langle A(x_n), y\rangle \to \langle y_0, y\rangle$ on one hand, and on the other hand, $\langle A(x_n), y\rangle = \langle x_n, B(y)\rangle \to \langle x_0, B(y)\rangle = \langle A(x_0), y\rangle$, and so $\langle A(x_0), y\rangle = \langle y_0, y\rangle$. It follows that $y_0 = A(x_0)$. Thus A is a closed map. By the closed graph theorem, $A \in BL(H, G)$.
Aliter: Let $E := \{A(x) : x \in H \text{ and } \|x\| \le 1\} \subset G$. Consider $y' \in G'$.
There is $y_0 \in G$ such that $y'(y) = \langle y, y_0\rangle$ for all $y \in G$. Hence $|y'(A(x)| = |\langle A(x), y_0\rangle| = |\langle x, B(y_0)\rangle| \le \|B(y_0)\|$ for all $x \in H$ satisfying $\|x\| \le 1$. By the resonance theorem, E is a bounded subset of G, that is, $A \in BL(H, G)$. Similarly, $B \in BL(G, H)$. By the uniqueness of the adjoint, $B = A^*$.

4.28. $R(A^*) \perp R(B)$ if and only if $\langle A^*(y_1), B(y_2)\rangle = 0$ for all $y_1, y_2 \in G$. But $\langle A^*(y_1), B(y_2)\rangle = \langle y_1, AB(y_2)\rangle$ for $y_1, y_2 \in G$, and $\langle y_1, AB(y_2)\rangle = 0$ for all $y_1, y_2 \in G$ if and only if $AB(y_2) = 0$ for all $y_2 \in G$, that is, $AB = 0$.

4.29. $A^*(G^\perp) \subset G^\perp$ if and only if $\langle A^*(x), y\rangle = 0$ for all $x \in G^\perp$ and $y \in G$. But $\langle A^*(x), y\rangle = \langle x, A(y)\rangle$ for $x \in G^\perp$ and $y \in G = (G^\perp)^\perp$, and $\langle x, A(y)\rangle = 0$ for all $x \in G^\perp$ and $y \in G^{\perp\perp}$ if and only if $A(G) = A(G^{\perp\perp}) \subset (G^\perp)^\perp = G$.

4.30. Suppose $A \in BL(H, G)$ is one-one and onto. Replacing A by A^* in Theorem 4.27(i), we see that $R(A^*)$ is dense in H. Also, by Theorem 4.27(ii), A^* is bounded below. Let $\beta > 0$ be such that $\beta\|y\| \le \|A^*(y)\|$ for all $y \in G$. To show A^* is onto, consider $x \in H$. Since $R(A^*)$ is dense in H, there is a sequence (y_n) in G such that $A^*(y_n) \to x$ in H. Then $(A^*(y_n))$ is a Cauchy sequence in H, and $\beta\|y_n - y_m\| \le \|A^*(y_n) - A^*(y_m)\|$ for all $n, m \in \mathbb{N}$. Hence (y_n) is a Cauchy sequence in G. Since G is complete, there is $y \in Y$ such that $y_n \to y$ in G. Then $A^*(y_n) \to A^*(y)$. Hence $x = A^*(y) \in R(A^*)$. Thus A^* is onto. Since A^* is also one-one, consider $B := (A^*)^{-1} : H \to G$. Then B is linear. Also, if $x \in H$ and $x = A^*(y)$, then $\|B(x)\| = \|y\| \le \|A^*(y)\|/\beta = \|x\|/\beta$, and so $B \in BL(H, G)$. Hence $A^{-1} = B^* \in BL(G, H)$.
Open mapping theorem: Suppose $A \in BL(H, G)$ is onto. Let $Z := Z(A)$, and

define $\widetilde{A} : H/Z \to G$ by $\widetilde{A}(x+Z) := A(x)$ for $x \in H$. Then $\widetilde{A}$ is one-one and onto. As we saw above, $\widetilde{A}^{-1} \in BL(G, H/Z)$. Hence $\widetilde{A}$ is an open map, and so is $A = \widetilde{A} \circ Q$ since $Q : H \to H/Z$ is an open map.
The closed graph theorem can be deduced from the open mapping theorem as in Exercise 3.27(iii).

4.31. (i) Let $A \in CL(H, G)$. Then $A^* \in BL(G, H)$, and so $A^*A \in CL(H)$. Conversely, let $A^*A \in CL(H)$. Consider a bounded sequence (x_n) in H, and let $\alpha > 0$ be such that $\|x_n\| \leq \alpha$ for all $n \in \mathbb{N}$. Since A^*A is compact, there is a subsequence (x_{n_k}) such that $(A^*A(x_{n_k}))$ converges in G. But

$$\begin{aligned}\|A(x_{n_k}) - A(x_{n_j})\|^2 &= |\langle A^*A(x_{n_k} - x_{n_j}), x_{n_k} - x_{n_j}\rangle| \\ &\leq 2\alpha\|A^*A(x_{n_k}) - A^*A(x_{n_j})\| \quad \text{for all } k, j \in \mathbb{N}.\end{aligned}$$

Now the Cauchy sequence $(A(x_{n_k}))$ converges in G. Thus $A \in CL(H, G)$.
(ii) Let $A \in CL(H, G)$. Then $(A^*)^*A^* = AA^* \in CL(G)$, and by (i) above, $A^* \in CL(G, H)$. (Note: Theorem 4.21 of Schauder is not used.)
(iii) Let $A \in BL(H, G)$ be a Hilbert–Schmidt map. Let $\{u_1, u_2, \ldots\}$ be a countable orthonormal basis for H such that $\sum_n \|A(u_n)\|^2 < \infty$. Let $\{v_1, v_2, \ldots\}$ be an countable orthonormal basis for G. Then

$$\sum_m \|A^*(v_m)\|^2 = \sum_m \sum_n |\langle A^*(v_m), u_n\rangle|^2 = \sum_n \sum_m |\langle v_m, A(u_n)\rangle|^2$$

which is equal to $\sum_n \|A(u_n)\|^2$. Hence A^* is a Hilbert–Schmidt map.
(This proof shows that if $\{\tilde{u}_1, \tilde{u}_2, \ldots\}$ is another countable orthonormal basis for H, then $\sum_n \|A(\tilde{u}_n)\|^2 = \sum_m \|A^*(v_m)\|^2 = \sum_n \|A(u_n)\|^2$.)

4.32. Let $A \in BL(H)$. Define $B := (A + A^*)/2$ and $C := (A - A^*)/2$. Then B is hermitian, C is skew–hermitian, and $A = B + C$. Suppose $A = B_1 + C_1$, where B_1 is hermitian and C_1 is skew–hermitian. Then $A^* = B_1 - C_1$, and so $B_1 = (A + A^*)/2 = B$ and $C_1 = (A - A^*)/2 = C$.
Note that $BC = (A^2 - (A^*)^2 - AA^* + A^*A)/4$ and $CB = (A^2 - (A^*)^2 - A^*A + AA^*)/4$. Hence $BC = CB$ if and only if $A^*A = AA^*$, that is, A is normal. Also, $C = 0$ if and only if $A^* = A$, that is, A is hermitian, and $B = 0$ if and only if $A^* = -A$, that is, A is skew–hermitian. Finally, note that $B^2 - C^2 = (AA^* + A^*A)/2$. Hence $BC = CB$ and $B^2 - C^2 = I$ if and only if $A^*A = AA^*$ and $A^*A + AA^* = 2I$, that is, $A^*A = I = AA^*$.

4.33. Let $A \in BL(H)$. Clearly, $A^*A - AA^*$ is self-adjoint. Hence A is hyponormal if and only if $\langle A^*A(x), x\rangle \geq \langle AA^*(x), x\rangle$, that is, $\|A(x)\|^2 \geq \|A^*(x)\|^2$ for all $x \in H$. Since A is normal if and only if $\|A^*(x)\| = \|A(x)\|$ for all $x \in H$, and A^* is hyponormal if and only if $\|A^*(x)\| \geq \|A(x)\|$ for all $x \in H$, A is normal if and only if A and A^* are both hyponormal.
Let A denote the right shift operator on ℓ^2. Then A^* is the left shift operator

on ℓ^2. Hence $A^*A = I$ and $AA^*(x) = (0, x(2), x(3), \ldots)$ for all $x \in \ell^2$. Thus $A^*A \geq AA^*$, but $A^*A \neq AA^*$.

4.34. For x in H, let $B(x)(j) := x(j+1)$ for all $j \in \mathbb{Z}$. Then $\langle x, B(y)\rangle = \sum_{j=-\infty}^{\infty} x(j)\overline{y(j+1)} = \sum_{j=-\infty}^{\infty} x(j-1)\overline{y(j)} = \langle A(x), y\rangle$ for all $x, y \in H$. Hence $A^* = B$, the left shift operator on H. Also, it is easy to see that $A^*A(x) = x = AA^*(x)$ for $x \in H$. Hence A is a unitary operator on H.

4.35. Note: $\omega_n^n = 1$, but $\omega_n^p \neq 1$ for $p = 1, \ldots, n-1$, $|\omega_n| = 1$ and $\overline{\omega_n} = \omega_n^{-1}$. We show that the n columns of M_n form an orthonormal subset of $\mathbb{C}^n$. Let $j, \ell \in \{1, \ldots, n\}$. Then the inner product of the jth and ℓth columns is equal to $\left(\sum_{p=1}^n \omega_n^{(p-1)(j-1)}\omega_n^{-(p-1)(\ell-1)}\right)/n = \left(\sum_{p=1}^n \omega_n^{(p-1)(j-\ell)}\right)/n$, which is equal to $(1+\cdots+1)/n = 1$ if $j = \ell$, and which is equal to 0 if $j \neq \ell$ since $\omega_n^{j-\ell} \neq 1$ and $(1-\omega_n^{j-\ell})(1+\omega_n^{j-\ell}+\cdots+\omega_n^{(n-1)(j-\ell)}) = 1-\omega_n^{(j-\ell)n} = 0$.
Hence $\overline{M_n^t}M_n = I = M_n\overline{M_n^t}$, and so the operator A is unitary. Also, $M_n^t = M_n$ since $k_{j,p} := \omega_n^{(j-1)(p-1)}/\sqrt{n} = k_{p,j}$ for $p, j = 1, \ldots, n$.

4.36. (i) Let $x \in H$. For $n \in \mathbb{N}$, let $a_n(x) := \langle A_n(x), x\rangle$. Then $(a_n(x))$ is a monotonically increasing sequence in $\mathbb{R}$, and it is bounded above by $\alpha\langle x, x\rangle$. Hence it is a Cauchy sequence in $\mathbb{R}$. For $m \geq n$, define $B_{m,n} := A_m - A_n$. Then $0 \leq B_{m,n} \leq \alpha I - A_1$ for all $m \geq n$, and

$$\|B_{m,n}\| = \sup\{\langle B_{m,n}(x), x\rangle : x \in H \text{ and } \|x\| \leq 1\} \leq |\alpha| + \|A_1\|.$$

By the generalized Schwarz inequality, for all $m \geq n$,

$$\begin{aligned}\|B_{m,n}(x)\| &= \langle B_{m,n}(x), x\rangle^{1/4}\langle B_{m,n}^2(x), B_{m,n}(x)\rangle^{1/4} \\ &\leq \langle B_{m,n}(x), x\rangle^{1/4}\|B_{m,n}\|^{3/4}\|x\|^{1/2} \\ &\leq \langle B_{m,n}(x), x\rangle^{1/4}(|\alpha| + \|A_1\|)^{3/4}\|x\|^{1/2}.\end{aligned}$$

It follows that $(A_n(x))$ is a Cauchy sequence in H. Let $A_n(x) \to y$ in H, and define $A(x) := y$. Clearly, $A : H \to H$ is linear. Also, since $A_1 \leq A_n \leq \alpha I$, we see that $\|A_n\| \leq |\alpha| + \|A_1\|$ for all $n \in \mathbb{N}$, and so $\|A(x)\| \leq (|\alpha| + \|A_1\|)\|x\|$ for all $x \in H$. Thus $A \in BL(H)$. Since $A_n(x) \to A(x)$, we see that the monotonically increasing sequence $(\langle A_n(x), x\rangle)$ converges to $\langle A(x), x\rangle$ for each $x \in H$. Thus A is self-adjoint, and $A_n \leq A$ for all $n \in \mathbb{N}$. Finally, suppose $\widetilde{A}$ is self-adjoint, and $A_n \leq \widetilde{A}$ for all $n \in \mathbb{N}$. Then $\langle A(x), x\rangle = \lim_{n\to\infty}\langle A_n(x), x\rangle \leq \langle \widetilde{A}(x), x\rangle$ for all $x \in H$. Hence $A \leq \widetilde{A}$. The uniqueness of A is obvious.
(ii) Let $A_n^- := -A_n$, so that $A_n^- \leq A_{n+1}^- \leq -\beta I$ for all $n \in \mathbb{N}$. Then the desired result follows from (i) above.

4.37. For $x, y \in H$, define $\langle x, y\rangle_A := \langle A(x), y\rangle$. Then $\langle\cdot\,,\cdot\rangle_A : H \times H \to \mathbb{K}$ is linear in the first variable, is conjugate symmetric, and satisfies $\langle x, x\rangle_A \geq 0$ for all $x \in H$ since A is a positive operator. In Exercise 2.13, let $X := H$ and replace $\langle\cdot\,,\cdot\rangle$ by $\langle\cdot\,,\cdot\rangle_A$. Then $G = \{x \in H : \langle x, x\rangle_A = 0\}$, and G is a subspace of

H. In fact, G is closed since A is continuous. For $x + G$, $y + G \in H/G$, let $\langle\langle x + G, y + G\rangle\rangle := \langle x, y\rangle_A$. It follows that $\langle\langle\cdot, \cdot\rangle\rangle$ is an inner product on H/G. In particular, $|\langle x, y\rangle_A|^2 \le \langle x, x\rangle_A\langle y, y\rangle_A$, that is, $|\langle A(x), y\rangle|^2 \le \langle A(x), x\rangle\langle A(y), y\rangle$ for all $x, y \in H$.

4.38. Let $n, m \in \mathbb{N}$, $m \ne n$. Since $P_n P_m = 0$, we see that $R(P_m) \subset Z(P_n)$, and since P_n is an orthogonal projection operator, $Z(P_n) \perp R(P_n)$. Hence for all $m \ne n$, $R(P_n) \perp R(P_m)$. For $m \in \mathbb{N}$, let $Q_m := P_1 + \cdots + P_m$. Then ${Q_m}^* = {P_1}^* + \cdots + {P_m}^* = P_1 + \cdots + P_m = Q_m$, and so Q_m is an orthogonal projection operator. By Exercise 3.6(i), $\|Q_m\| = 0$ or $\|Q_m\| = 1$. Let $x \in H$. Then the Pythagoras theorem shows that for each $m \in \mathbb{N}$,

$$\|P_1(x)\|^2 + \cdots + \|P_m(x)\|^2 = \|P_1(x) + \cdots + P_m(x)\|^2 = \|Q_m(x)\|^2 \le \|x\|^2.$$

By Exercise 2.30, $\sum_n P_n(x)$ is summable in H. Let $P(x) := \sum_{n=1}^{\infty} P_n(x)$. Clearly, P is linear, and $\|P(x)\| \le \|x\|$ for all $x \in H$. Hence $\|P\| \le 1$. Let $x \in H$. Since $P_n(P(x)) = P_n(x)$ for all $n \in \mathbb{N}$, we obtain $P(P(x)) = \sum_n P_n(P(x)) = \sum_n P_n(x) = P(x)$. Thus P is an orthogonal projection operator on H (Exercise 3.6(i)).

Let G denote the closure of the linear span of $\bigcup_{n=1}^{\infty} R(P_n)$. Let $x \in H$. Then $P_n(x) \in R(P_n) \subset G$ for all $n \in \mathbb{N}$. Since G is a closed subspace of H, it follows that $P(x) \in G$. Hence $R(P) \subset G$. Conversely, $P_n(x) = P(P_n(x)) \in R(P)$ for all $n \in \mathbb{N}$. Hence $R(P_n) \subset R(P)$. Since $R(P)$ is a closed subspace of H, it follows that $G \subset R(P)$. Thus $R(P) = G$. If $x \in H$ and $P_n(x) = 0$ for all $n \in \mathbb{N}$, then clearly $P(x) = 0$. Conversely, if $x \in H$, and $P(x) = 0$, then $P_n(x) = P_n(P(x)) = P_n(0) = 0$. Thus $Z(P) = \bigcap_{n=1}^{\infty} Z(P_n)$.

4.39. Let $n \in \mathbb{N}$, and let $x_n := (1, e^{-i\theta}, e^{-2i\theta}, \ldots, e^{-(n-1)i\theta}, 0, 0, \ldots)/\sqrt{n}$, where $\theta \in (-\pi, \pi]$. Then $A(x_n) = (0, 1, e^{-i\theta}, e^{-2i\theta}, \ldots, e^{-(n-1)i\theta}, 0, 0, \ldots)/\sqrt{n}$. Clearly, $x_n \in X$ and $\|x_n\|_2 = 1$. Hence $\langle A(x_n), x_n\rangle = (n-1)e^{i\theta}/n \in \omega(A)$. Letting $n = 1$, we see that $0 \in \omega(A)$. Next, let $k \in \mathbb{K}$ satisfy $0 < |k| < 1$. Then $k = re^{i\theta}$, where $0 < r < 1$ and $\theta \in (-\pi, \pi]$. There is $n \in \mathbb{N}$ such that $r < (n-1)/n$. Since 0 and $(n-1)e^{i\theta}/n$ belong to $\omega(A)$, and since $\omega(A)$ is a convex subset of $\mathbb{K}$, we see that $k \in \omega(A)$. Thus $\{k \in \mathbb{K} : |k| < 1\} \subset \omega(A)$. Since $\|A\| = 1$, it follows that $\omega(A) \subset \{k \in \mathbb{K} : |k| \le 1\}$. We show that if $k \in \mathbb{K}$ and $|k| = 1$, then $k \notin \omega(A)$. Let $x \in X$. Then $|\langle A(x), x\rangle| \le \sum_{j=1}^{\infty} |x(j)||x(j+1)| \le \frac{1}{2}\sum_{j=1}^{\infty}(|x(j)|^2 + |x(j+1)|^2)$. Thus $|\langle A(x), x\rangle| \le \|x\|^2$. Assume for a moment that $\|x\|_2 = 1 = |\langle A(x), x\rangle|$. Since $|x(j)||x(j+1)| \le (|x(j)|^2 + |x(j+1)|)^2/2$ for every $j \in \mathbb{N}$, $|x(j)| = |x(j+1)|$ for each $j \in \mathbb{N}$. This is impossible since $0 < \sum_{j=1}^{\infty} |x(j)|^2 < \infty$. Thus $\omega(A) = \{k \in \mathbb{K} : |k| < 1\}$.

4.40. For $x, y \in X$, $\langle A(x+y), x+y\rangle - \langle A(x-y), x-y\rangle = 2\langle A(x), y\rangle + 2\langle A(y), x\rangle$ and $\langle A(x+iy), x+iy\rangle - \langle A(x-iy), x-iy\rangle = -2i\langle A(x), y\rangle + 2i\langle A(y), x\rangle$. Multiplying the second equality by i and adding it to the first, we obtain the generalized polarization identity.

Let $H \neq \{0\}$ be a Hilbert space over $\mathbb{C}$, $A \in BL(H)$, and $\omega(A) \subset \mathbb{R}$. Then $4\langle A(y), x\rangle = \langle A(y+x), y+x\rangle - \langle A(y-x), y-x\rangle + i\langle A(y+ix), y+ix\rangle - i\langle A(y-ix), y-ix\rangle$ for $x, y \in H$. Note that $\langle A(z), z\rangle \in \mathbb{R}$ for every $z \in H$. Hence $4\langle x, A(y)\rangle = \overline{4\langle A(y), x\rangle} = 4\langle A(x), y\rangle$. Thus A is self-adjoint.

Chapter 5

5.1. Suppose $k \in \mathbb{K}$, $k \neq 0$. If $k \notin \sigma(A)$, then $(A^{-1} - k^{-1}I)\big((A-kI)^{-1}A\big) = -k^{-1}(I - kA^{-1})(I - kA^{-1})^{-1} = -k^{-1}I$ and $\big((A-kI)^{-1}A\big)(A^{-1} - k^{-1}I) = (I - kA^{-1})^{-1}(-k^{-1})(I - kA^{-1}) = -k^{-1}I$, and so $k^{-1} \notin \sigma(A^{-1})$. Replacing A by A^{-1}, and k by k^{-1}, we see that if $k^{-1} \notin \sigma(A^{-1})$, then $k \notin \sigma(A)$. Hence $\sigma(A^{-1}) = \{\lambda^{-1} : \lambda \in \sigma(A)\}$.

5.2. Let $p(t) := a_n t^n + a_{n-1}t^{n-1} + \cdots + a_1 t + a_0$, where $n \in \mathbb{N}, a_n, \ldots, a_0 \in \mathbb{K}$, and $a_n \neq 0$. Suppose $\lambda \in \sigma(A)$. Assume for a moment that $p(\lambda) \notin \sigma(p(A))$, that is, $p(A) - p(\lambda)I$ is invertible. For $m \in \{1, \ldots, n\}$, let $q_m(A) := A^{m-1} + \lambda A^{m-2} + \cdots + \lambda^{m-2}A + \lambda^{m-1}I$, so that $A^m - \lambda^m I = (A - \lambda I)q_m(A) = q_m(A)(A - \lambda I)$. Define $q(A) := a_n q_n(A) + \cdots + a_1 q_1(A)$. It follows that $p(A) - p(\lambda)I = (A - \lambda I)q(A) = q(A)(A - \lambda I)$. Hence $A - \lambda I$ is invertible. This contradiction shows that $p(\lambda) \in \sigma(p(A))$.
Next, let $\mathbb{K} := \mathbb{C}$, and $\mu \in \mathbb{C}$. Then there are $\lambda_1, \ldots, \lambda_n \in \mathbb{C}$ such that $p(t) - \mu = a_n(t - \lambda_1)\cdots(t - \lambda_n)$. Suppose $\mu \in \sigma(p(A))$. Then $p(A) - \mu I = a_n (A - \lambda_1 I)\cdots(A - \lambda_n I)$. If $A - \lambda_j I$ is invertible for each $j = 1, \ldots, n$, then so would be $p(A) - \mu I$. Hence there is $\lambda_j \in \sigma(A)$ such that $\mu = p(\lambda_j)$.

5.3. Let $I - AB$ be invertible. Then $(I - BA)\big(I + B(I - AB)^{-1}A\big) = I - BA + B\big((I - AB)^{-1} - AB(I - AB)^{-1}\big)A = I - BA + B(I - AB)(I - AB)^{-1}A = I - BA + BA = I$. Similarly, $\big(I + B(I - AB)^{-1}A\big)(I - BA) = I$. (This formula is conceived as follows: $(I - BA)^{-1} = I + BA + BABA + BABABA + \cdots = I + B\big(I + AB + ABAB + \cdots\big)A = I + B(I - AB)^{-1}A$.)
Let $k \in \mathbb{K}$ be nonzero, and let $\widetilde{A} := A/k$. Then $AB - kI = -k(I - \widetilde{A}B)$ is invertible if and only if $-k(I - B\widetilde{A}) = BA - kI$ is invertible.

5.4. Let $\lambda \in \sigma(A) = \sigma_e(A)$. Then there is nonzero $x := (x(1), \ldots, x(n)) \in \mathbb{K}^n$ such that $A(x) - \lambda x = 0$. Suppose $\|x\|_\infty = |x(i)|$. We show that $\lambda \in D_i$. Now $k_{i,1}x(1) + \cdots + (k_{i,i} - \lambda)x(i) + \cdots + k_{i,n}x(n) = 0$, that is,

$$k_{i,i} - \lambda = -k_{i,1}\frac{x(1)}{x(i)} - \cdots - k_{i,i-1}\frac{x(i-1)}{x(i)} - k_{i,i+1}\frac{x(i+1)}{x(i)} - \cdots - k_{i,n}\frac{x(n)}{x(i)}.$$

Since $|x(j)|/|x(i)| \leq 1$ for all $j \in \{1, \ldots, n\}$, we obtain $|k_{i,i} - \lambda| \leq r_i$.

5.5. Suppose $A(x) = \lambda x$ and $u = (g_1(x), \ldots, g_n(x))$. Then $\sum_{j=1}^n g_j(x)x_j = \lambda x$, and hence $\sum_{j=1}^n g_j(x)g_i(x_j) = \lambda g_i(x)$, that is, $\sum_{j=1}^n u(j)g_i(x_j) = \lambda u(i)$ for

each $i = 1, \dots, n$. Thus $Mu^t = \lambda u^t$. Also, $x = \big(\sum_{j=1}^n u(j)x_j\big)/\lambda$.
Conversely, suppose $Mu^t = \lambda u^t$ and $x = \big(\sum_{j=1}^n u(j)x_j\big)/\lambda$. Then for each $i = 1, \dots, n$, $\sum_{j=1}^n g_i(x_j)u(j) = \lambda u(i)$, and so

$$A(x) = \sum_{i=1}^n g_i(x)x_i = \frac{1}{\lambda}\sum_{i=1}^n \Bigg(\sum_{j=1}^n u(j)g_i(x_j)\Bigg)x_i$$
$$= \frac{1}{\lambda}\sum_{i=1}^n \lambda u(i)x_i = \sum_{i=1}^n u(i)x_i = \lambda x.$$

Also, $\lambda u(i) = \sum_{j=1}^n g_i(x_j)u(j) = g_i\big(\sum_{j=1}^n u(j)x_j\big) = g_i(\lambda x) = \lambda g_i(x)$, and so $u(i) = g_i(x)$ for all $i = 1, \dots, n$, that is, $u = (g_1(x), \dots, g_n(x))$.
Also, $x = 0$ if and only if $u = 0$. Thus x is an eigenvector of A corresponding to λ if and only if u^t is an eigenvector of M corresponding to λ.

5.6. Suppose $\lambda \in \sigma_e(A)$. Then there is nonzero $x \in X$ such that $A(x) = \lambda x$, that is, $\big(x_0(t) - \lambda\big)x(t) = 0$ for all $t \in [a, b]$. Let $t_0 \in [a, b]$ be such that $x(t_0) \neq 0$. Since x is continuous at t_0, there is $\delta > 0$ such that $x(t) \neq 0$ for all $t \in I := [a, b] \cap (t_0 - \delta, t_0 + \delta)$. Hence $x_0(t) = \lambda$ for all $t \in I$.
Conversely, suppose $\lambda \in \mathbb{K}$, and $x_0(t) = \lambda$ for all t in a nontrivial subinterval I of $[a, b]$. Then there is $t_0 \in [a, b]$, and there is $\delta > 0$ such that $(t_0 - \delta, t_0 + \delta) \subset I$. For $t \in [a, b]$, define $x(t) := 1$ if $|t - t_0| \le \delta/2$, $x(t) := 2(t - t_0 + \delta)/\delta$ if $t_0 - \delta < t < t_0 - \delta/2$, $x(t) := 2(t_0 - t + \delta)/\delta$ if $t_0 + \delta/2 < t < t_0 + \delta$, and $x(t) := 0$, if $|t - t_0| \ge \delta$. Then $x \in X$ and $x \neq 0$. Since $x_0(t) - \lambda = 0$ for all $t \in [a, b]$ satisfying $|t - t_0| < \delta$, and $x(t) = 0$ for all $t \in [a, b]$ satisfying $|t - t_0| \ge \delta$, we see that $\big(x_0(t) - \lambda\big)x(t) = 0$ for all $t \in [a, b]$, that is, $A(x) = \lambda x$.

5.7. Let E denote the essential range of x_0. Suppose $\lambda \in E$. For $n \in \mathbb{N}$, let $S_n := \{t \in [a, b] : |x_0(t) - \lambda| < 1/n\}$, and let x_n denote the characteristic function of S_n. Then $\|x_n\|_2^2 = m(S_n) > 0$, and

$$\|A(x_n) - \lambda x_n\|_2^2 = \int_{S_n} |x_0 - \lambda|^2 |x_n|^2 dm \le \frac{m(S_n)}{n^2} = \frac{\|x_n\|_2^2}{n^2}.$$

Hence $A - \lambda I$ is not bounded below, that is, $\lambda \in \sigma_a(A)$. Thus $E \subset \sigma_a(A)$.
Conversely, suppose $k \notin E$, that is, there is $\epsilon > 0$ such that $m(\{t \in [a, b] : |x_0(t) - k| < \epsilon\}) = 0$. Then $|x_0(t) - k| \ge \epsilon$ for almost all $t \in [a, b]$, and the function $1/(x_0 - k)$ belongs to X. For $y \in X$, define $B(y) := y/(x_0 - k)$. Then $B \in BL(X)$ and $(A - kI)B = I = B(A - kI)$. This shows that $\sigma(A) \subset E$. Since $\sigma_a(A) \subset \sigma(A)$, we obtain $\sigma_a(A) = E = \sigma(A)$.
Next, suppose $\lambda \in \sigma_e(A)$. Then there is nonzero $x \in X$ such that $A(x) = \lambda x$, that is, $(x_0(t) - \lambda)x(t) = 0$ for almost all $t \in [a, b]$. Since x is nonzero, $m(\{t \in [a, b] : x(t) \neq 0\}) > 0$, and so $m(\{t \in [a, b] : x_0(t) = \lambda\}) > 0$.

Conversely, let $\lambda \in \mathbb{K}$, $S := \{t \in [a,b] : x_0(t) = \lambda\}$, and suppose $m(S) > 0$. Let x denote the characteristic function of the set S. Then $x \in X$ and $x \neq 0$. Since $(x_0(t) - \lambda)x(t) = 0$ for all $t \in [a,b]$, $\lambda \in \sigma_e(A)$.

5.8. For $n \in \mathbb{N}$, $A^n - \lambda^n I = B(A - \lambda I)$, where $B := A^{n-1} + \lambda A^{n-2} + \cdots + \lambda^{n-2} A + \lambda^{n-1} I$. Let $\lambda \in \sigma_a(A)$. Then there is a sequence (x_n) in X such that $\|x_n\| = 1$ for every $n \in \mathbb{N}$, and $A(x_n) - \lambda x_n \to 0$. Consequently, $A^n(x_n) - \lambda^n x_n = B(A(x_n) - \lambda x_n) \to 0$, and so $\lambda^n \in \sigma_a(A^n)$. Hence $|\lambda^n| \leq \|A^n\|$, and so $|\lambda| \leq \inf\left\{\|A^n\|^{1/n} : n \in \mathbb{N}\right\}$.

5.9. (i) Let $\alpha > 0$ be such that $\|A(x)\| \leq \alpha\|x\|$ for all $x \in X$. If $k \in \mathbb{K}$ and $|k| > \alpha$, then $\|A(x) - kx\| \geq |k|\|x\| - \|A(x)\| \geq (|k| - \alpha)\|x\|$ for all $x \in X$, and so $k \notin \sigma_a(A)$. Thus $\sigma_a(A) \subset \{k \in \mathbb{K} : |k| \leq \alpha\}$.
(ii) Let $\beta > 0$ be such that $\|A(x)\| \geq \beta\|x\|$ for all $x \in X$. If $k \in \mathbb{K}$ and $|k| < \beta$, then $\|A(x) - kx\| \geq \|A(x)\| - |k|\|x\| \geq (\beta - |k|)\|x\|$ for all $x \in X$, and so $k \notin \sigma_a(A)$. Thus $\sigma_a(A) \subset \{k \in \mathbb{K} : |k| \geq \beta\}$.
(iii) If A is an isometry, let $\alpha = \beta := 1$ in (i) and (ii) above.

5.10. For $x \in \ell^1$, $\|A(x)\| \leq 2\|x\|_1$, and $\|A(e_2)\|_1 = \|2e_3\|_1 = 2$. Hence $\|A\| = 2$. Further, $A^2(x) = (0, 0, 2x(1), 2x(2), \ldots)$, and $\|A^2(e_1)\|_1 = \|2e_3\| = 2$. Hence $\|A^2\| = 2$. Since ℓ^1 is a Banach space, $|\lambda| \leq \|A^2\|^{1/2} = \sqrt{2}$ for every $\lambda \in \sigma(A)$.

5.11. Let $\alpha := \sup\{|\lambda_n| : n \in \mathbb{N}\}$. Then $\|A(x)\|_p \leq \alpha\|x\|_p$ for all $x \in X$. Hence $\|A\| \leq \alpha$. In fact, $\|A\| = \alpha$ since $A(e_j) = \lambda_j e_j$ for each $j \in \mathbb{N}$. This also shows that $\lambda_j \in \sigma_e(A)$ for each $j \in \mathbb{N}$. Conversely, let $\lambda \in \sigma_e(A)$, and $A(x) = \lambda x$ for a nonzero $x \in X$. Then there is $j \in \mathbb{N}$ such that $x(j) \neq 0$. Since $\lambda_j x(j) = \lambda x(j)$, we obtain $\lambda = \lambda_j$. Thus $\sigma_e(A) = \{\lambda_j : j \in \mathbb{N}\}$.
Let E denote the closure of $\{\lambda_j : j \in \mathbb{N}\}$. Since $\sigma_a(A)$ is closed in $\mathbb{K}$, $E \subset \sigma_a(A)$. Conversely, suppose $k \in \mathbb{K} \setminus E$, and let $\delta := d(k, E) > 0$. Define $B(y) := (y(1)/(k - \lambda_1), y(2)/(k - \lambda_2), \ldots)$ for $y \in X$. For $y \in X$, $|B(y)(j)| = |y(j)|/|k - \lambda_j| \leq |y(j)|/\delta$ for all $j \in \mathbb{N}$, and so $B(y) \in X$, and $\|B(y)\|_p \leq \|y\|_p/\delta$. Thus $B \in BL(X)$. It is easy to check that $(A - kI)B = I = B(A - kI)$. Hence $\sigma(A) \subset E$. Thus $\sigma_a(A) = \sigma(A) = E$.

5.12. Since $\mathbb{K}$ is a separable metric space, so is its subset E. Let $\{\lambda_j : j \in \mathbb{N}\}$ be a countable dense subset of E. For $x := (x(1), x(2), \ldots)$ in ℓ^2, define $A(x) := (\lambda_1 x(1), \lambda_2 x(2), \ldots)$. Then $\sigma(A) = E$ by Exercise 5.11.
Suppose (λ_j) is a sequence in $\mathbb{K}$ such that $\lambda_j \to 0$, and define A as above. Also, for $n \in \mathbb{N}$, define $A_n(x) := (\lambda_1 x(1), \ldots, \lambda_n x(n), 0, 0, \ldots)$, $x \in \ell^2$. Then $A_n \in BL(\ell^2)$ is of finite rank for each $n \in \mathbb{N}$. Since $\|A - A_n\| = \sup\{|\lambda_j| : j = n+1, n+2, \ldots\} \to 0$, we see that $A \in CL(\ell^2)$. Further, $\sigma_e(A) = \{\lambda_j : j \in \mathbb{N}\}$, and $\sigma(A) = E = \{\lambda_n : n \in \mathbb{N}\} \cup \{0\}$.

5.13. Let $k \in \mathbb{K}$ with $|k| > \|A\|$, and define $\widetilde{A} := A/k$. Then $\|\widetilde{A}\| < 1$. Hence $I - \widetilde{A}$ is invertible, and $(A - kI)^{-1} = -(I - \widetilde{A})^{-1}/k = -\left(\sum_{n=0}^{\infty} \widetilde{A}^n\right)/k = -\sum_{n=0}^{\infty} A^n/k^{n+1}$. Also, $\|(A - kI)^{-1}\| \leq 1/|k|(1 - \|\widetilde{A}\|) = 1/(|k| - \|A\|)$. Further, for $n \in \mathbb{N}$, and $x \in \ell^p$, $A^n(x) = (0, \ldots, 0, x(1), x(2), \ldots)$, where the first n entries are equal to 0. Hence $(A - kI)^{-1}(y)(j) = -\sum_{n=0}^{\infty} A^n(y)(j)/k^{n+1} = -y(j)/k - \cdots - y(1)/k^j$ for $y \in \ell^p$, and $j \in \mathbb{N}$.

5.14. Clearly, $\|B\| = 1$. Hence $\sigma_e(B) \subset \sigma_a(B) \subset \sigma(B) \subset \{\lambda \in \mathbb{K} : |\lambda| \le 1\}$. Let $x \in X$. For $\lambda \in \mathbb{K}$, $B(x) = \lambda x$ if and only if $x(j+1) = \lambda x(j)$ for all $j \in \mathbb{N}$, that is, $x := x(1)(1, \lambda, \lambda^2, \ldots)$. If $X := \ell^1$, ℓ^2, or c_0, then $(1, \lambda, \lambda^2, \ldots) \in X$ if and only if $|\lambda| < 1$; if $X := \ell^\infty$, then $(1, \lambda, \lambda^2, \ldots) \in X$ if and only if $|\lambda| \le 1$, and if $X := c$, then $(1, \lambda, \lambda^2, \ldots) \in X$ if and only if $|\lambda| < 1$ or $\lambda = 1$. In all cases, $\sigma_a(B) = \{\lambda \in \mathbb{K} : |\lambda| \le 1\}$ since $\sigma_a(B)$ is a closed subset of $\mathbb{K}$. It follows that $\sigma(B) = \{\lambda \in \mathbb{K} : |\lambda| \le 1\}$.

5.15. Clearly, $\|A\| = 1$. Hence $\sigma_e(A) \subset \sigma_a(A) \subset \sigma(A) \subset \{\lambda \in \mathbb{K} : |\lambda| \le 1\}$. Let $x \in L^p$. For $\lambda \in \mathbb{K}$, $A(x) = \lambda x$ if and only if $x(t+1) = \lambda x(t)$ for almost all $t \in [0, \infty)$, that is, $x(t+j) = \lambda^j x(t)$ for almost all $t \in [0, 1)$ and all $j \in \mathbb{N}$. Let $p \in \{1, 2\}$. Then $\int_j^{j+1} |x(t)|^p dm(t) = |\lambda^j|^p \int_0^1 |x(s)|^p dm(s)$ by the translation invariance of the Lebesgue measure. It follows that $\|x\|_p^p = \big(\int_0^1 |x(s)|^p dm(s)\big) \sum_{j=0}^{\infty} |\lambda^p|^j$. The series $\sum_{j=0}^{\infty} |\lambda^p|^j$ is convergent if and only if $|\lambda| < 1$. Hence $\sigma_e(A) = \{\lambda \in \mathbb{K} : |\lambda| < 1\}$. Next, let $p = \infty$. Since $\{|\lambda|^j : j \in \mathbb{N}\}$ is a bounded subset of $\mathbb{K}$ if and only if $|\lambda| \le 1$, we see that $\sigma_e(A) = \{\lambda \in \mathbb{K} : |\lambda| \le 1\}$. In all cases, $\sigma_a(A) = \{\lambda \in \mathbb{K} : |\lambda| \le 1\}$ since $\sigma_a(A)$ is a closed subset of $\mathbb{K}$. It follows that $\sigma(A) = \{\lambda \in \mathbb{K} : |\lambda| \le 1\}$.

5.16. Let $x \in X$ and $x \ne 0$. Then $\langle r_A(x), x\rangle = \langle A(x), x\rangle - q_A(x)\langle x, x\rangle = 0$, that is, $r_A(x) \perp x$. It follows that for $k \in \mathbb{K}$, $\|A(x) - k\,x\|^2 = \|r_A(x)\|^2 + |q_A(x) - k|^2\|x\|^2 \ge \|r_A(x)\|^2$.

5.17. Since $A \in CL(X)$, $\sigma(A) = \sigma_a(A)$, which is a closed and bounded subset of $\mathbb{K}$ by Proposition 5.5. Also, if $\lambda \in \sigma_a(A)$, then $|\lambda| \le \inf\{\|A^n\|^{1/n} : n \in \mathbb{N}\}$. (See Exercise 5.8.)

5.18. For $x := (x(1), x(2), \ldots) \in X$, define $D(x) := (w_1 x(1), w_2 x(2), \ldots)$, $R(x) := (0, x(1), x(2), \ldots)$ and $L(x) := (x(2), x(3), \ldots)$. Then $A = RD$ and $B = LD$. Since $w_n \to 0$, D is a compact operator (Exercise 3.35(ii)). Hence A and B are compact operators. Let $x \in X$ and $\lambda \in \mathbb{K}$.
Now $A(x) = \lambda x$ if and only if $0 = \lambda x(1)$ and $w_j x(j) = \lambda x(j+1)$ for all $j \in \mathbb{N}$. If $\lambda \ne 0$ and $A(x) = \lambda x$, then $x = 0$, and so $\lambda \notin \sigma_e(A)$. Since A is compact, $\sigma(A) = \sigma_e(A) \cup \{0\} = \{0\}$. Let $E_0 := \{x \in X : A(x) = 0\} = \{x \in X : w_j x(j) = 0 \text{ for all } j \in \mathbb{N}\}$. Hence $E_0 \ne \{0\}$ if and only if there is $j \in \mathbb{N}$ such that $w_j = 0$. Also, $e_j \in E_0$ if and only if $w_j = 0$. Thus if $\{j \in \mathbb{N} : w_j = 0\}$ is a finite set, then $E_0 = \text{span}\,\{e_j : j \in \mathbb{N} \text{ and } w_j = 0\}$, and otherwise $\dim E_0 = \infty$.
Next, $B(x) = \lambda x$ if and only if $w_{j+1} x(j+1) = \lambda x(j)$ for all $j \in \mathbb{N}$. If $\lambda \ne 0$ and $B(x) = \lambda x$, then $x = 0$. For if there is $j_0 \in \mathbb{N}$ such that $x(j_0) \ne 0$, then $w_j \ne 0$ and $x(j) \ne 0$ for all $j \ge j_0$, and in fact $|x(j)| \to \infty$ as $j \to \infty$. Hence $\sigma_e(B) \subset \{0\}$. Since B is compact, $\sigma(B) = \sigma_e(B) \cup \{0\} = \{0\}$. Let $G_0 := \{x \in X : B(x) = 0\}$. Then $G_0 = \{x \in X : w_j x(j) = 0 \text{ for all } j \ge 2\}$. Clearly, $e_1 \in G_0$, and so $0 \in \sigma_e(B)$. Also, for $j \ge 2$, $e_j \in G_0$ if and only if $w_j = 0$.

Thus if $\{j \in \mathbb{N} : j \geq 2 \text{ and } w_j = 0\}$ is a finite set, then $G_0 = \{e_1\} \cup \text{span}\,\{e_j : j \geq 2 \text{ and } w_j = 0\}$, and otherwise $\dim G_0 = \infty$.

5.19. Let $X := L^2([a, b])$. Since $k(\cdot\,, \cdot) \in L^2([a, b]\times[a, b])$, we see that A is a compact operator on X. Let $x \in X$, and define

$$y(s) := A(x)(s) = \int_a^s x(t)dm(t), \quad s \in [a, b].$$

Since $x \in L^1([a, b])$, the fundamental theorem of calculus for Lebesgue integration shows that y is absolutely continuous on $[a, b]$, and $y' = x$ almost everywhere on $[a, b]$. Also, $y(a) = 0$.
Let $A(x) = 0$. Then $y(s) = 0$ for almost all $s \in [a, b]$. In fact, $y(s) = 0$ for all $s \in [a, b]$ since y is continuous on $[a, b]$. Hence $x(s) = y'(s) = 0$ for almost all $s \in [a, b]$. This shows that $0 \notin \sigma_e(A)$. Next, let $\lambda \in \mathbb{K}$, $\lambda \neq 0$ and $A(x) = \lambda x$. Then $x = A(x)/\lambda = y/\lambda$, and so x is absolutely continuous on $[a, b]$. By the fundamental theorem of calculus for Riemann integration, y is in $C^1([a, b])$, and $y'(s) = x(s)$ for all $s \in [a, b]$. Hence $x = y/\lambda$ is in $C^1([a, b])$, and $\lambda x'(s) = y'(s) = x(s)$ for all $s \in [a, b]$. Also, $x(a) = y(a)/\lambda = 0$. Thus x satisfies Bernoulli's differential equation $\lambda x' - x = 0$, and also the initial condition $x(a) = 0$. It follows that $x = 0$, and so $\lambda \notin \sigma_e(A)$. Thus $\sigma_e(A) = \emptyset$. Also, $\sigma_a(A) = \sigma(A) = \{0\}$ since A is compact.

5.20. First, consider $X := C([0, 1])$. Then $A \in CL(X)$ since $k(\cdot\,, \cdot)$ is continuous. Let $x \in X$, and define

$$y(s) := A(x)(s) = \int_0^s t\,x(t)\,dt + s\int_s^1 x(t)\,dt, \quad s \in [0, 1].$$

Then $y(0) = 0$. By the fundamental theorem of calculus for Riemann integration, $y \in C^1([0, 1])$, and

$$y'(s) = s\,x(s) - s\,x(s) + \int_s^1 x(t)\,dt = \int_s^1 x(t)\,dt \quad \text{for all } s \in [0, 1].$$

Then $y'(1) = 0$. Further, $y' \in C^1([0, 1])$, and $y''(s) = -x(s)$ for $s \in [0, 1]$. Thus we see that if $x \in X$ and $y := A(x)$, then $y \in C^2([0, 1])$, $y'' = -x$ and $y(0) = 0 = y'(1)$. Conversely, suppose $x \in X$, and let $y \in C^2([0, 1])$ satisfy $y'' = -x$ and $y(0) = 0 = y'(1)$. Integrating by parts,

$$\begin{aligned} A(y'')(s) &= \int_0^s t\,y''(t)dt + s\int_s^1 y''(t)dt \\ &= s\,y'(s) - 0\,y'(0) - \int_0^s y'(t)dt + s\big(y'(1) - y'(s)\big) \\ &= -\int_0^s y'(t)dt = -y(s) + y(0) = -y(s) \end{aligned}$$

for all $s \in [0, 1]$. Hence $A(y'') = -y$, that is, $A(x) = y$.
Let $x \in X$ be such that $A(x) = 0$. Then $0'' = -x$, that is, $x = 0$. Hence $0 \notin \sigma_e(A)$. Next, let $\lambda \in \mathbb{K}$ and $\lambda \neq 0$. Let $x \in X$ be such that $A(x) = \lambda x$. Then it follows that $\lambda x'' = -x$ and $\lambda x(0) = 0 = \lambda x'(1)$, that is, $\lambda x'' + x = 0$ and $x(0) = 0 = x'(1)$. Now the differential equation $\lambda x'' + x = 0$ has a nonzero solution satisfying $x(0) = 0 = x'(1)$ if and only if $\lambda = 4/(2n-1)^2\pi^2$, $n \in \mathbb{N}$. In this case, the general solution is given by $x(s) := c_n \sin(2n-1)\pi s/2$, s is in $[0, 1]$, where $c_n \in \mathbb{K}$. (If $\mathbb{K} := \mathbb{C}$, we must first show that $\lambda \in \mathbb{R}$, as in the footnote in Example 5.23(ii).) Fix $n \in \mathbb{N}$, let $\lambda_n := 4/(2n-1)^2\pi^2$, $x_n(s) := \sin(2n-1)\pi s/2$, $s \in [0, 1]$, and let $y_n := \lambda_n x_n$. Then $y_n'' = \lambda_n x_n'' = -x_n$ and $y_n(0) = 0 = y_n'(1)$. Hence $A(x_n) = y_n = \lambda_n x_n$. It follows that λ_n is an eigenvalue of A, and the corresponding eigenspace of A is spanned by the function x_n. There are no other eigenvalues of A. It follows that $\sigma_e(A) = \{1/(2n-1)^2\pi^2 : n \in \mathbb{N}\}$. Since A is a compact operator, $\sigma_a(A) = \sigma(A) = \sigma_e(A) \cup \{0\}$.
Next, consider $Y := L^p([0, 1])$, where $p \in \{1, 2, \infty\}$. The arguments in this case are exactly the same as the ones given in Example 5.23(ii).

5.21. Let $k \in \mathbb{K}$. Suppose $A + B - kI$ is invertible, and $A - kI$ is one-one. Since $A - kI = A + B - kI - B = (A + B - kI)(I - (A + B - kI)^{-1}B)$, we see that $(I - (A + B - kI)^{-1}B)$ is one-one, that is, 1 is not an eigenvalue of $(A + B - kI)^{-1}B$. Since B is compact, so is $(A + B - kI)^{-1}B$. Hence 1 is not a spectral value of $(A + B - kI)^{-1}B$, that is, $I - (A + B - kI)^{-1}B$ is invertible. It follows that $A - kI$ is invertible.

5.22. Suppose $A - kI$ is one-one, that is, $k \notin \sigma_e(A)$. Since A is compact, $A - kI$ is invertible. In particular, $A - kI$ is onto.
Conversely, suppose $A - kI$ is onto. For $n \in \mathbb{N}$, let $Z_n := Z((A - kI)^n)$. Then Z_n is a closed subspace of Z_{n+1} for all $n \in \mathbb{N}$. Assume for a moment that $Z_n \subsetneq Z_{n+1}$ for all $n \in \mathbb{N}$. Fix $n \in \mathbb{N}$. By the Riesz lemma, there is $z_{n+1} \in Z_{n+1}$ such that $\|z_{n+1}\| = 1$ and $d(z_{n+1}, Z_n) \geq 1/2$. It is easy to see that $(A - kI)(Z_{n+1}) \subset Z_n$ and $A(Z_n) \subset Z_n$. Hence for all $z \in Z_n$,

$$\|A(z_{n+1}) - A(z)\| = \|kz_{n+1} + (A - kI)(z_{n+1}) - A(z)\| \geq |k|/2 > 0.$$

In particular, $\|A(z_{n+1}) - A(z_{m+1})\| \geq |k|/2$ for all $n, m \in \mathbb{N}$ with $n \neq m$. Now (z_{n+1}) is a bounded sequence in X, but the sequence $(A(z_{n+1}))$ has no convergent subsequence. This contradicts the compactness of A. Hence there is $m \in \mathbb{N}$ such that $Z_{m+1} = Z_m$. Let $Z_0 := \{0\}$. We show that $Z_m = Z_{m-1}$. Let $y \in Z_m$. Since $A - kI$ is onto, there is $x \in X$ such that $y = (A - kI)(x)$. Now $(A - kI)^{m+1}(x) = (A - kI)^m(y) = 0$, that is, $x \in Z_{m+1} \subset Z_m$. Thus $(A - kI)^{m-1}(y) = (A - kI)^m(x) = 0$, that is, $y \in Z_{m-1}$. Similarly, $Z_{m-1} = Z_{m-2}, \ldots, Z_2 = Z_1$, and $Z_1 = Z_0$, that is, $A - kI$ is one-one. (Compare the proof of Proposition 5.20.)

5.23. If $k \notin \sigma_a(A)$, then the result follows from Lemma 5.19. Now suppose k is in $\sigma_a(A)$. Since A is compact and $k \neq 0$, we see that $k \in \sigma_e(A)$, and the corresponding eigenspace E_k is finite dimensional. Let $\{x_1, \ldots, x_m\}$ be a basis for E_k, and find $x'_1, \ldots, x'_m$ in X' such that $x'_j(x_i) = \delta_{i,j}$ for $i, j = 1, \ldots, m$. Define $Y := \bigcap_{j=1}^{m} Z(x'_j)$. Then Y is a closed subspace of X, and $X = Y \oplus Z(A - kI)$. Define $B : Y \to X$ by $B(y) := (A - kI)(y)$, $y \in Y$. Then B is one-one. In fact, arguing as in the proof of Proposition 5.18, we see that B is bounded below. Let $\beta > 0$ be such that $\beta\|y\| \le \|(A - kI)(y)\|$ for all $y \in Y$. We show that $R(B)$ is a closed subspace of X. For $n \in \mathbb{N}$, let $y_n \in Y$ be such that $(A(y_n) - ky_n)$ converges in X to, say, z. Let $\alpha > 0$ be such that $\|A(y_n) - ky_n\| \le \alpha$ for all $n \in \mathbb{N}$. Since $\beta\|y_n\| \le \|A(y_n) - ky_n\| \le \alpha$ for all $n \in \mathbb{N}$, (y_n) is a bounded sequence in Y. By Lemma 5.17, (y_n) has a convergent subsequence, and if it converges to y in Y, then $A(y) - ky = z$. Thus $z \in R(B)$, and so $R(A - kI) = R(B)$ is a closed subspace of X. (Compare the proof of Lemma 5.19.)

5.24. Suppose (i) holds. Then $1 \notin \sigma_e(A)$. In fact, $1 \notin \sigma(A)$ since A is compact. In this case, the inverse $(I - A)^{-1}$ is continuous, that is, $x := (I - A)^{-1}(y)$ depends continuously on $y \in X$.
Clearly, (ii) holds if and only if $1 \in \sigma_e(A)$. In this case, the eigenspace $E_1 := \{x \in X : A(x) = x\}$ of A corresponding to its nonzero eigenvalue 1 is finite dimensional, since A is compact.

5.25. (i) Since A is compact, $1 \in \sigma_e(A')$ if and only if $1 \in \sigma_e(A)$.
(ii) Let $y \in X$. Suppose there is $x \in X$ such that $x - A(x) = y$. If x' is in $Z(A' - I)$, that is, if $A'(x') = x'$, then $x'(y) = x'(x) - x'(A(x)) = x'(x) - A'(x')(x) = 0$. Conversely, let $\{x'_1, \ldots, x'_m\}$ be a basis for $Z(I - A')$, which is finite dimensional since A' is a compact operator, and suppose that $x'_j(y) = 0$ for $j = 1, \ldots, m$. Then $x'(y) = 0$ for all $x' \in Z(I - A')$. Assume for a moment that there is no $x \in X$ such that $x - A(x) = y$, that is, $y \notin R(I - A)$. Clearly, $y \neq 0$. Since $R(I - A)$ is a closed subspace of X (Exercise 5.23), there is $x' \in X'$ such that $x'(z) = 0$ for all $z \in R(I - A)$ and $x'(y) = \|y\|$. Then $x'(x) - A'(x')(x) = x'(x - A(x)) = 0$ for every $x \in X$, that is, $x' \in Z(I - A')$, but $x'(y) = \|y\| \neq 0$. This is a contradiction.
Next, suppose $x_0 \in X$ and $x_0 - A(x_0) = y$. Then $x \in X$ satisfies $x - A(x) = y$ if and only if $x - x_0 - A(x - x_0) = 0$, that is, $x - x_0 \in Z(A - I)$, which is finite dimensional since A is a compact operator. This is the same thing as saying $x := x_0 + k_1x_1 + \cdots + k_mx_m$, where $k_1, \ldots, k_m$ are in $\mathbb{K}$, and $\{x_1, \ldots, x_m\}$ is a basis for $Z(A - I)$.

5.26. Let J denote the canonical embedding of X into X''. By Exercise 4.25, $A''J = JA$, and so $(A'' - kI)J = J(A - kI)$ for $k \in \mathbb{K}$. It follows that if $A'' - kI$ is one-one, then $A - kI$ is one-one, and if $A'' - kI$ is bounded below, then $A - kI$ is bounded below. Hence $\sigma_e(A) \subset \sigma_e(A'')$ and $\sigma_a(A) \subset \sigma_a(A'')$.
Next, since X' is a Banach space, $\sigma(A'') = \sigma(A')$. Also, $\sigma(A') \subset \sigma(A)$.

5.27. Let A be a normal operator on a Hilbert space H. By mathematical induction on $n \in \mathbb{N}$, we show that if $\lambda \in \mathbb{K}$, $x \in X$ and $(A - \lambda I)^n(x) = 0$, then $(A - \lambda I)(x) = 0$. If $n = 1$, then this is obvious. Assume this holds for $m \in \mathbb{N}$.

Suppose $(A-\lambda I)^{m+1}(x)=0$. Let $y:=(A-\lambda I)(x)$. Then $(A-\lambda I)^m(y)=0$. By the inductive assumption, $(A-\lambda I)(y)=0$. Now

$$\begin{aligned}\|(A-\lambda I)(x)\|^2 &= \langle (A-\lambda I)(x),(A-\lambda I)(x)\rangle = \langle (A^*-\overline{\lambda}I)(A-\lambda I)(x),x\rangle\\ &\leq \|(A^*-\overline{\lambda}I)(y)\|\|x\| = \|(A-\lambda I)(y)\|\|x\|=0\end{aligned}$$

since A is normal. Thus $(A-\lambda I)(x)=0$.

5.28. Assume for a moment that $\sigma_e(A)$ is uncountable. For each $\lambda_\alpha\in\sigma_e(A)$, let u_α be a corresponding eigenvector of A with $\|u_\alpha\|=1$. Since A is normal, $\{u_\alpha\}$ is an uncountable orthonormal subset of H. Since $\|u_\alpha-u_\beta\|=\sqrt{2}$ for $\alpha\neq\beta$, no countable subset of H can be dense in H. This is a contradiction to the separability of H.

5.29. Let $A\in BL(\ell^2)$ be defined by an infinite matrix $M:=[k_{i,j}]$. If M is diagonal, then it is clear that $\overline{M^t}M=\text{diag}\,(|k_{1,1}|^2,|k_{2,2}|^2,\ldots)=M\overline{M^t}$. Hence A is a normal operator (Example 4.28(i)).

Conversely, suppose A is a normal operator, and M is upper triangular. Note that $k_{i,j}=0$ for all $i>j$. First, $A(e_1)=\sum_{i=1}^\infty k_{i,1}e_i=k_{1,1}e_1$. Since A is normal, e_1 is an eigenvector of A^* corresponding its eigenvalue $\overline{k_{1,1}}$, that is, $A^*(e_1)=\overline{k_{1,1}}e_1$. But since the matrix $\overline{M^t}$ defines the operator A^*, we see that $A^*(e_1)=\sum_{i=1}^\infty\overline{k_{1,i}}e_i$. Hence $\overline{k_{1,2}}=\overline{k_{1,3}}=\cdots=0$. Next, $A(e_2)=\sum_{i=1}^\infty k_{i,2}e_i=k_{1,2}e_1+k_{2,2}e_2=k_{2,2}e_2$, since $k_{1,2}=0$ as we have just shown. Again, since A is normal, $A^*(e_2)=\overline{k_{2,2}}e_2$. But $A^*(e_2)=\sum_{i=1}^\infty\overline{k_{2,i}}e_i=\sum_{i=2}^\infty\overline{k_{2,i}}e_i$. Hence $\overline{k_{2,3}}=\overline{k_{2,4}}=\cdots=0$. In this manner, by mathematical induction, we obtain $k_{i,i+1}=k_{i,i+2}=\cdots=0$ for every $i\in\mathbb{N}$. Thus $M=\text{diag}\,(k_{1,1},k_{2,2},\ldots)$.

If A is a normal operator, and M is lower triangular, then the normal operator A^* is defined by the upper triangular matrix $\overline{M^t}$. Hence $\overline{M^t}$ is a diagonal matrix, that is, M is a diagonal matrix.

5.30. Let $A\in BL(H)$ be unitary. Since A is normal, and A is an isometry, $\sigma(A)=\sigma_a(A)\subset\{k\in\mathbb{K}:|k|=1\}$ by Exercise 5.9(iii). Let $k\in\mathbb{K}$ with $|k|\neq 1$. If $y\in H$ and $y=(A-kI)(x)$, $x\in H$, then

$$\|y\|=\|A(x)-kx\|\geq|\,|k|-1|\,\|x\|=|\,|k|-1|\,\|(A-kI)^{-1}(y)\|$$

since $\|A(x)\|=\|x\|$. Hence $\|(A-kI)^{-1}\|\leq 1/|\,|k|-1|$.

5.31. Let $k\in\mathbb{K}\backslash\overline{\omega(A)}$, and $\beta:=d\big(k,\omega(A)\big)$. Since $\sigma(A)\subset\overline{\omega(A)}$, $A-kI$ is invertible. If $x\in H$ and $\|x\|=1$, then $\|A(x)-kx\|\geq|\langle A(x)-kx,x\rangle|=|\langle A(x),x\rangle-k|\geq\beta$. Let $y\in H$. If $y=(A-kI)(x)$, where $x\in H$, then

$$\|y\|=\|A(x)-kx\|\geq\beta\|x\|=\beta\|(A-kI)^{-1}(y)\|.$$

Hence $\|(A-kI)^{-1}\|\leq 1/\beta$.

Let $A\in BL(H)$ be self-adjoint. Since $(m_A,M_A)\subset\omega(A)\subset[m_A,M_A]$, we

see that $\overline{\omega(A)} = [m_A, M_A]$. Let $k \in \mathbb{K}\backslash[m_A, M_A]$. Clearly, $\beta = |\text{Im}\, k|$ if $\text{Re}\, k \in [m_A, M_A]$, $\beta = |k - m_A|$ if $\text{Re}\, k < m_A$, and $\beta = |k - M_A|$ if $\text{Re}\, k > M_A$. Further, if $\mathbb{K} := \mathbb{R}$, and $k \in \mathbb{R}\backslash\sigma(A)$, then $(A - kI)^{-1}$ is self-adjoint, and so $\|(A - kI)^{-1}\| = \sup\{|\mu| : \mu \in \sigma((A - kI)^{-1})\} = \sup\{|(\lambda - k)^{-1}| : \lambda \in \sigma(A)\} = 1/d$, where $d := d(k, \sigma(A))$ (Exercise 5.1).
(Note: If $\mathbb{K} := \mathbb{C}$, A is normal, and $k \in \mathbb{C}\backslash\sigma(A)$, then $(A - kI)^{-1}$ is normal, and so the above proof works since $\|B\| = \sup\{|\mu| : \mu \in \sigma(B)\}$ for a normal operator B on a Hilbert space H over $\mathbb{C}$.)

5.32. Let $A \in BL(H)$ be self-adjoint. Then $\pm i \notin \sigma(A)$. Since $A^* = A$, we obtain $T(A)^* = (A - iI)^{-1}(A + iI) = (A + iI)(A - iI)^{-1}$. It follows that $(T(A))^*T(A) = I = T(A)(T(A))^*$, that is, $T(A)$ is unitary. Also, $1 \notin \sigma(T(A))$ since $T(A) - I = (A - iI - A - iI)(A + iI)^{-1} = -2i(A + iI)^{-1}$ is invertible.
Next, let $B \in BL(H)$ be unitary, and suppose $1 \notin \sigma(B)$. Since $B^* = B^{-1}$, we obtain $S(B)^* = -i(I - B^*)^{-1}(I + B^*) = -i(I - B^{-1})^{-1}(I + B^{-1}) = i(I + B)(I - B)^{-1} = S(B)$, that is, $S(B)$ is self-adjoint.
Further, it can be easily checked that $S(T(A)) = i(I + T(A))(I - T(A))^{-1} = A$ and $T(S(B)) = (S(B) - iI)(S(B) + iI)^{-1} = B$.

5.33. If $A \geq 0$, then $\sigma(A) \subset \overline{\omega(A)} \subset [0, \infty)$. Conversely, suppose $\sigma(A) \subset [0, \infty)$. Then $\inf \omega(A) = m_A \in \sigma_a(A) = \sigma(A) \subset [0, \infty)$, and so $\omega(A) \subset [0, \infty)$.
If $0 \in \sigma_e(A)$, then $0 \in \omega(A)$ since $\sigma_e(A) \subset \omega(A)$. Conversely, suppose $A \geq 0$, and $0 \in \omega(A)$, that is, there is $x \in H$ such that $\|x\| = 1$ and $\langle A(x), x\rangle = 0$. By the generalized Schwarz inequality, $\|A(x)\| \leq \langle A(x), x\rangle^{1/4}\langle A^2(x), A(x)\rangle^{1/4}$, and so $A(x) = 0$. Hence $0 \in \sigma_e(A)$.

5.34. Define $B(x) := (x(1)\cos\theta + x(2)\sin\theta, -x(1)\sin\theta + x(2)\cos\theta)$ for $x := (x(1), x(2)) \in \mathbb{R}^2$. It is easy to check that $\langle A(x), y\rangle = \langle x, B(y)\rangle$ for all $x, y \in \mathbb{R}^2$. Hence $A^* = B$. Also, $\|A(x)\|_2^2 = (x(1)\cos\theta - x(2)\sin\theta)^2 + (x(1)\sin\theta + x(2)\cos\theta)^2 = x(1)^2 + x(2)^2 = \|x\|_2^2$ for all $x \in \mathbb{R}^2$. Hence $A : \mathbb{R}^2 \to \mathbb{R}^2$ is a linear isometry. Since $\mathbb{R}^2$ is finite dimensional, A is onto. Thus A is a unitary operator. Clearly, A is defined by the 2×2 matrix $M := [k_{i,j}]$, where $k_{1,1} := \cos\theta$, $k_{1,2} := -\sin\theta$, $k_{2,1} := \sin\theta$ and $k_{2,2} := \cos\theta$. Now $\det(M - tI) = t^2 - 2t\cos\theta + 1$, and it is equal to 0 if and only if $t = \cos\theta \pm (\cos^2\theta - 1)^{1/2} \in \mathbb{R}$. Hence $\sigma(A) = \sigma_e(A) = \{1\}$ if $\theta := 0$, $\sigma(A) = \sigma_e(A) = \{-1\}$ if $\theta := \pi$, and $\sigma(A) = \sigma_e(A) = \emptyset$ otherwise.

5.35. Since A is a Hilbert–Schmidt operator on H, it is a compact operator (Exercise 3.40(i)). Hence $\sigma_e(A)$ is countable. Also, the eigenspace corresponding to each nonzero eigenvalue of A has a finite orthonormal basis. Further, since A is a normal operator, any two eigenspaces of A are mutually orthogonal. Let (λ_n) be the sequence of nonzero eigenvalues of A, each eigenvalue being repeated as many times as the dimension of the corresponding eigenspace. Then there is a countable orthonormal subset $\{u_1, u_2, \ldots\}$ of H such that $A(u_n) = \lambda_n u_n$ for each $n \in \mathbb{N}$. By Exercise 4.31(iii), A^* is a Hilbert–Schmidt operator on H. Let $\{\tilde{u}_1, \tilde{u}_2, \ldots\}$ be an orthonormal basis for H such that $\sum_j \|A^*(\tilde{u}_j)\|^2 < \infty$.

Then

$$\sum_n |\lambda_n|^2 = \sum_n \|A(u_n)\|^2 = \sum_n \sum_j |\langle A(u_n), \tilde{u}_j \rangle|^2$$
$$= \sum_j \sum_n |\langle u_n, A^*(\tilde{u}_j) \rangle|^2 \le \sum_j \|A^*(\tilde{u}_j)\|^2$$

by the Parseval formula and the Bessel inequality.

5.36. Suppose A is normal, and let $\mu_1, \ldots, \mu_k$ be the distinct eigenvalues of A. For $j \in \{1, \ldots, k\}$, let $E_j := Z(A - \mu_j I)$, and let P_j denote the orthogonal projection operator on H with $R(P_j) = E_j$. If $x \in X$, then $P_j(x) \in E_j$, and so $AP_j(x) = \mu_j P_j(x)$ for $j = 1, \ldots, k$. Also, if $i \neq j$, then $E_i \perp E_j$, and so $R(P_j) = E_j \subset E_i^\perp = Z(P_i)$, that is, $P_i P_j = 0$.
Let $G := E_1 + \cdots + E_k$. Then $G^\perp = \{0\}$ as in the proof of Theorem 5.39, and so $G = H$. Consider $x \in H$. Then $x = x_1 + \cdots + x_k$, where $x_j \in E_j = R(P_j)$ for $j = 1, \ldots, k$. Thus $x = P_1(x) + \cdots + P_k(x)$ and $A(x) = AP_1(x) + \cdots + AP_k(x) = \mu_1 P_1(x) + \cdots + \mu_k P_k(x)$, that is, $I = P_1 + \cdots + P_k$ and $A = \mu_1 P_1 + \cdots + \mu_k P_k$, as desired.
The converse follows easily since every orthogonal projection operator is normal, and a linear combination of normal operators is normal.

5.37. Suppose A is a nonzero compact self-adjoint operator on H. Let $\mu_1, \mu_2, \ldots$ be the distinct nonzero eigenvalues of A. Since A is self-adjoint, each μ_j is real, and since A is compact, either the set $\{\mu_1, \mu_2, \ldots\}$ is finite or $\mu_n \to 0$. For each j, let $E_j := Z(A - \mu_j I)$, and let P_j denote the orthogonal projection operator on H with $R(P_j) = E_j$. Since A is compact, each P_j is of finite rank. If $x \in X$, then $P_j(x) \in E_j$, and so $AP_j(x) = \mu_j P_j(x)$ for each j. Also, if $i \neq j$, then $E_i \perp E_j$, and so $R(P_j) = E_j \subset E_i^\perp = Z(P_i)$, that is, $P_i P_j = 0$.
Let G denote the closure of span $(\cup_j E_j)$, and let P denote the orthogonal projection operator on H with $R(P) = G$. By Exercise 4.38, $P(x) = \sum_j P_j(x)$ for all $x \in H$. Also, $H = Z(A) \oplus G$, as in the proof of Theorem 5.40. Let P_0 denote the orthogonal projection operator on H with $R(P_0) = Z(A)$. Then for $x \in H$, $x = P_0(x) + P(x) = P_0(x) + \sum_j P_j(x)$ and $A(x) = AP_0(x) + \sum_j AP_j(x) = \sum_j \mu_j P_j(x)$.
In fact, $A = \sum_j \mu_j P_j$. This is obvious if the set $\{\mu_1, \mu_2, \ldots\}$ is finite. Suppose now that this set is infinite. First note that $\|x\|^2 = \|P_0(x)\|^2 + \|P(x)\|^2 = \|P_0(x)\|^2 + \sum_{j=1}^{\infty} \|P_j(x)\|^2$, and so $\sum_{j=1}^{n} \|P_j(x)\|^2 \le \|x\|^2$ for all $x \in H$ and all $n \in \mathbb{N}$. For $n \in \mathbb{N}$, define $A_n := \sum_{j=1}^{n} \mu_j P_j$. Given $\epsilon > 0$, find $n_0 \in \mathbb{N}$ such that $|\mu_j| < \epsilon$ for all $n > n_0$. Then for all $n \ge n_0$,

$$\|A(x) - A_n(x)\|^2 = \left\| \sum_{j=n+1}^{\infty} \mu_j P_j(x) \right\|^2 \leq \sum_{j=n+1}^{\infty} |\mu_j|^2 \|P_j(x)\|^2$$
$$< \epsilon^2 \sum_{j=n+1}^{\infty} \|P_j(x)\|^2 \leq \epsilon^2 \|x\|^2, \quad x \in H.$$

Thus $\|A - A_n\| < \epsilon$ for all $n \geq n_0$. Hence $A = \sum_{j=1}^{\infty} \mu_j P_j$ in $BL(H)$.
For the converse, note that each P_j is self-adjoint since it is an orthogonal projection operator, each $\mu_j \in \mathbb{R}$, and so $A^* = \sum_j \overline{\mu_j} P_j{}^* = \sum_j \mu_j P_j = A$. Also, A is compact since $A_n := \sum_{j=1}^{n} \mu_j P_j$ is a bounded operator of finite rank for each n and $\|A_n - A\| \to 0$.

5.38. The kernel $k(s, t) := \min\{1 - s, 1 - t\}$, $s, t \in [0, 1]$, is a real-valued continuous function on $[0, 1]\times[0, 1]$, and $k(t, s) = k(s, t)$ for all $s, t \in [0, 1]$. Hence A is a compact self-adjoint operator on H. Also, $A \neq 0$. Let λ be a nonzero eigenvalue of A, and let $x \in H$ be a corresponding eigenvector of A. Since $A(x) \in C([0, 1])$, $x = A(x)/\lambda$ is continuous on $[0, 1]$. Now

$$y(s) := A(x)(s) = (1 - s)\int_0^s x(t)\,dt + \int_s^1 (1 - t)x(t)\,dt, \quad s \in [0, 1].$$

Clearly, $y(1) = 0$. By the fundamental theorem of calculus for Riemann integration, $y \in C^1([0, 1])$, and

$$y'(s) = (1 - s)x(s) - \int_0^s x(t)\,dt - (1 - s)x(s) = -\int_0^s x(t)\,dt, \quad s \in [0, 1].$$

Hence $y'(0) = 0$. Further, $y' \in C^1([0, 1])$, and $y''(s) = -x(s)$ for $s \in [0, 1]$. Thus $y \in C^2([0, 1])$, $y'' = -x$ and $y'(0) = 0 = y(1)$. Since $y = A(x) = \lambda x$, we see that $\lambda x'' = -x$ and $\lambda x'(0) = 0 = \lambda x(1)$, that is, $\lambda x'' + x = 0$ and $x'(0) = 0 = x(1)$. Now the differential equation $\lambda x'' + x = 0$ has a nonzero solution satisfying $x'(0) = 0 = x(1)$ if and only if $\lambda = 4/(2n - 1)^2\pi^2$, $n \in \mathbb{N}$. In this case, the general solution is given by $x(s) := c_n \cos(2n - 1)\pi s/2$ for $s \in [0, 1]$, where $c_n \in \mathbb{K}$. (If $\mathbb{K} := \mathbb{C}$, we must first show that $\lambda \in \mathbb{R}$, as in the footnote in Example 5.23(ii).) Conversely, fix $n \in \mathbb{N}$, let $\lambda_n := 4/(2n - 1)^2\pi^2$, $x_n(s) := \cos(2n - 1)\pi s/2$, $s \in [0, 1]$, and let $y_n := \lambda_n x_n$. Clearly, $y_n'' = \lambda_n x_n'' = -x_n$ and $y_n'(0) = 0 = y_n(1)$. Integrating by parts,

$$A(y_n'')(s) = (1 - s)\int_0^s y_n''(t)dt + \int_s^1 (1 - t)y_n''(t)dt$$
$$= (1 - s)(y_n'(s) - y_n'(0)) - (1 - s)y_n'(s) + y_n(1) - y_n(s)$$
$$= -y_n(s) \quad \text{for } s \in [0, 1].$$

Thus $A(y_n'') = -y_n$, that is, $A(x_n) = y_n = \lambda_n x_n$. It follows that λ_n is in fact an eigenvalue of A, and the corresponding eigenspace of A is spanned by the function x_n. Hence $A(x) = \sum_{n=1}^{\infty} \lambda_n \langle x, u_n \rangle u_n$, where $\lambda_n := 4/(2n-1)^2\pi^2$ and $u_n(s) := \sqrt{2}\cos(2n-1)\pi s/2,\ s \in [0, 1]$, for $n \in \mathbb{N}$.

5.39. The kernel $k(s, t) := \min\{s, t\},\ s, t \in [0, 1]$, is a real-valued continuous function on $[0, 1] \times [0, 1]$, and $k(t, s) = k(s, t)$ for all $s, t \in [0, 1]$. Hence A is a compact self-adjoint operator on $L^2([0, 1])$. Also, $A \neq 0$. By Exercise 5.20, the nonzero eigenvalues of A are given by $\lambda_n := 4/(2n-1)^2\pi^2$, $n \in \mathbb{N}$, and the eigenspace of A corresponding to λ_n is span $\{x_n\}$, where $x_n(s) := \sin(n - 1/2)\pi s,\ s \in [0, 1]$. Let $u_n(s) := \sqrt{2}\sin(n-1/2)\pi s,\ s \in [0, 1]$. Then $\{u_n : n \in \mathbb{N}\}$ is an orthonormal basis for $L^2([0, 1])$ consisting of eigenvectors of A. For $x \in L^2([0, 1])$, $A(x) = \sum_{n=1}^{\infty} \lambda_n \langle x, u_n \rangle u_n$, that is,

$$\int_0^s t\,x(t)\,dm(t) + s\int_s^1 x(t)\,dm(t) = \frac{8}{\pi^2}\sum_{n=1}^{\infty} \frac{s_n(x)}{(2n-1)^2}\sin(2n-1)\frac{\pi s}{2},$$

where $s_n(x) := \int_0^1 x(t)\sin(n-1/2)\pi t\,dm(t)$, and the series on the right side converges in $L^2([0, 1])$. We shall use Theorem 5.43.

Let $y \in L^2([0, 1])$ and $\mu \in \mathbb{K},\ \mu \neq 0$. Consider the integral equation

$$x(s) - \mu\left(s\int_0^s t\,x(t)\,dm(t) + s\int_s^1 x(t)\,dm(t)\right) = y(s),\quad s \in [0, 1].$$

Let $\mu_n := \lambda_n^{-1} = (2n-1)^2\pi^2/4$ for $n \in \mathbb{N}$. If $\mu \neq \mu_n$ for any $n \in \mathbb{N}$, then there is a unique $x \in L^2([0, 1])$ satisfying $x - \mu A(x) = y$. In fact, since $\mu/(\mu_n - \mu) = 4\mu/\big((2n-1)^2\pi^2 - 4\mu\big)$ and $\langle y, u_n \rangle = \sqrt{2}\,s_n(y)$ for $n \in \mathbb{N}$,

$$x(s) = y(s) + 8\mu\sum_{n=1}^{\infty} \frac{s_n(y)}{(2n-1)^2\pi^2 - 4\mu}\sin(2n-1)\frac{\pi s}{2},\quad s \in [0, 1].$$

Further, $\|x\| \leq \alpha\|y\|$, where $\alpha := 1 + 4|\mu|/\min_{n\in\mathbb{N}}\{|(2n-1)^2\pi^2 - 4\mu|\}$.
Next, suppose $\mu := (2n_1 - 1)^2\pi^2/4$, where $n_1 \in \mathbb{N}$. There is x in $L^2([0, 1])$ satisfying $x - \mu A(x) = y$ if and only if $x \perp u_{n_1}$, that is, $s_{n_1}(y) = 0$. In this case, since $\mu/(\mu_n - \mu) = (2n_1 - 1)^2/4(n - n_1)(n + n_1 - 1)$ for $n \neq n_1$,

$$x(s) = y(s) + \frac{(2n_1-1)^2}{2}\sum_{n \neq n_1} \frac{s_n(y)}{(n-n_1)(n+n_1-1)}\sin(2n-1)\frac{\pi s}{2}$$
$$+ k_1\sin(2n_1-1)\frac{\pi s}{2},\quad s \in [0, 1],\ \text{where } k_1 \in \mathbb{K}.$$

5.40. Suppose A is a nonzero compact operator on H. Then A^*A is a compact self-adjoint operator on H. Also, since $\|A(x)\|^2 = \langle A^*A(x), x \rangle$ for all x in H, we

see that $Z(A^*A) = Z(A)$. In particular, $A^*A \neq 0$. Then H has an orthonormal basis $\{u_\alpha\}$ consisting of eigenvectors of A^*A. Let $A^*A(u_\alpha) = \lambda_\alpha u_\alpha$ for each α. By Theorem 5.36(iii), the set $S := \{u_\alpha : \lambda_\alpha \neq 0\}$ is countable. Let $S := \{u_1, u_2, \ldots\}$ and $A^*A(u_n) = \lambda_n u_n$ for each n. If S is in fact denumerable, then $\lambda_n \to 0$. Also, $\lambda_n \geq 0$ for each n since A^*A is positive. Let $s_n := \sqrt{\lambda_n} > 0$ and $v_n := A(u_n)/s_n$ for each n. Then $s_n s_m \langle v_n, v_m \rangle = \langle A(u_n), A(u_m) \rangle = \langle A^*A(u_n), u_m \rangle = \lambda_n \langle u_n, u_m \rangle$ for all n, m. It follows that $\{v_1, v_2, \ldots\}$ is an orthonormal subset of H. If $\{v_1, v_2, \ldots\}$ is denumerable, then so is S, and $s_n \to 0$. Let $x \in H$. Let $s_n \leq s$ for all n. Then $\sum_n |s_n|^2 |\langle x, u_n \rangle|^2 \leq s^2 \|x\|^2$, and so $\sum_n s_n \langle x, u_n \rangle v_n$ converges in H. Define $B(x) := \sum_n s_n \langle x, u_n \rangle v_n$ for $x \in H$. Then $B(u_n) = s_n v_n = A(u_n)$ for all n. Also, if $u_\alpha \in Z(A)$, then $u_\alpha \perp u_n$ for all n, and so $B(u_\alpha) = 0 = A(u_\alpha)$. Hence $A(x) = B(x) = \sum_n s_n \langle x, u_n \rangle v_n$ for all $x \in H$.
Conversely, let $A(x) = \sum_n s_n \langle x, u_n \rangle v_n$ for all $x \in H$, where $\{u_1, u_2, \ldots\}$ and $\{v_1, v_2, \ldots\}$ are countable orthonormal subsets of H, and $s_1, s_2, \ldots$ are positive numbers such that $s_n \to 0$ if the set $\{v_1, v_2, \ldots\}$ is denumerable. First, suppose $\{v_1, v_2, \ldots\}$ is finite. Then there is $m \in \mathbb{N}$ such that $A(x) = \sum_{n=1}^{m} s_n \langle x, u_n \rangle v_n$ for all $x \in H$, and so $R(A) = \operatorname{span}\{v_1, \ldots, v_m\}$. Since A is a bounded operator of finite rank, it is compact. Next, suppose $\{v_1, v_2, \ldots\}$ is denumerable. Let $\epsilon > 0$. There is $m_0 \in \mathbb{N}$ such that $|s_n| < \epsilon$ for all $n > m_0$. For $m \in \mathbb{N}$, let $A_m(x) := \sum_{n=1}^{m} \lambda_n \langle x, u_n \rangle v_n$, $x \in H$. Then

$$\begin{aligned}\|A(x) - A_m(x)\|^2 &= \Big\| \sum_{n=m+1}^{\infty} s_n \langle x, u_n \rangle v_n \Big\|^2 = \sum_{n=m+1}^{\infty} |s_n|^2 |\langle x, u_n \rangle|^2 \\ &\leq \epsilon^2 \sum_{n=m+1}^{\infty} |\langle x, u_n \rangle|^2 \leq \epsilon^2 \|x\|^2 \quad \text{for } x \in H \text{ and } m \geq m_0.\end{aligned}$$

Thus $\|A - A_m\| \leq \epsilon$ for all $m \geq m_0$. Since $A_m \to A$ in $BL(H)$, and each A_m is a bounded operator of finite rank, A is compact.
In this case, $\langle A^*(x), y \rangle = \langle x, A(y) \rangle = \sum_n s_n \langle x, v_n \rangle \langle u_n, y \rangle = \langle C(x), y \rangle$ for all $x, y \in H$, where $C(x) := \sum_n s_n \langle x, v_n \rangle u_n$ for $x \in H$. Hence $A^* = C$. Let $x \in H$ and $y := A(x)$. Then $\langle y, v_n \rangle = s_n \langle x, u_n \rangle$ for all n, and so $A^*A(x) = A^*(y) = C(y) = \sum_n s_n \langle y, v_n \rangle u_n = \sum_n s_n^2 \langle x, u_n \rangle u_n$.
Also, $\|A(u_n)\|^2 = \langle A^*A(u_n), u_n \rangle = \langle \lambda_n u_n, u_n \rangle = s_n^2$ for all n, and so $\sum_n s_n^2 = \sum_n \|A(u_n)\|^2$.
Now suppose H is a separable Hilbert space. Then the orthonormal basis $\{u_\alpha\}$ for H consisting of eigenvectors of A^*A is countable. Also, $A^*A(u_\alpha) = 0$ if and only if $A(u_\alpha) = 0$. Hence $\sum_\alpha \|A(u_\alpha)\|^2 = \sum_n \|A(u_n)\|^2$. By Exercise 4.31(iii), A is a Hilbert–Schmidt operator if and only if $\sum_n \|A(u_n)\|^2 < \infty$, that is, $\sum_n s_n^2 < \infty$.

References

1. M. Ahues, A. Largillier and B.V. Limaye, *Spectral Approximation for Bounded Operators*, Chapman & Hall/CRC, Boca Raton, Florida, 2001.
2. F. Altomare and M. Campiti, *Korovkin-type Approximation Theory and its Applications*, de Gruyter, Berlin, New York, 1994.
3. S. Banach,*Théorie des Opérations Linéaires*, Monografje Matematyczne, Warsaw, 1932.
4. E.A. Coddington, *An Introduction to Ordinary Differential Equations*, Prentice-Hall, Englewood Cliffs, N. J., 1961.
5. R. Courant and D. Hilbert, *Methods of Mathematical Physics*, Vol. I, Interscience, New York, 1953.
6. R.A. Devore and G.G. Lorentz, *Constructive Approximation*, Springer-Verlag, Berlin, 1991.
7. J. Diestel, *Sequences and Series in Banach Spaces*, Springer-Verlag, New York, 1984.
8. J. Dieudonné, *Foundations of Modern Analysis*, Academic Press, New York, 1969.
9. P. Enflo, A counter-example to the approximation problem in Banach spaces, *Acta Math.* **103** (1973), pp. 309–317.
10. G. Fichtenholtz and L. Kantorovich, Sur les opérations linéaires dans l'espace des fonctions bornées, *Studia Math.* **5** (1934), pp. 69-98.
11. S.R. Foguel, On a theorem of A.E. Taylor, *Proc. Amer. Math. Soc.* **9** (1958), p. 325.
12. S.R. Ghorpade and B.V. Limaye, *A Course in Calculus and Real Analysis*, Undergraduate Texts in Mathematics, Springer, New York, 2006.
13. S.R. Ghorpade and B.V. Limaye, *A Course in Multivariable Calculus and Analysis*, Undergraduate Texts in Mathematics, Springer, New York, 2010.
14. K. Gustafson, The Toeplitz-Hausdorff theorem for linear operators, *Proc. Amer. Math. Soc.* **25** (1970), pp. 203–204.
15. P.R. Halmos and V. S. Sunder, *Bounded Integral Operators on L^2 spaces*, Springer-Verlag, New York, 1978.
16. E. Hewitt and K. Stromberg, *Real and Abstract Analysis*, Graduate Texts in Mathematics, Springer-Verlag, New York, 1965.
17. K. Hoffman, *Banach Spaces of Analytic Functions*, Prentice-Hall, Englewood Cliffs, N. J., 1962.
18. R.A. Horn and C. R. Johnson, *Matrix Analysis*, Cambridge Univ. Press, Cambridge, 1985.
19. P.P. Korovkin, *Linear Operators and Theory of Approximation*, Hindustan Publ. Corp., Delhi, 1960.
20. R.G. Kuller, *Topics in Modern Analysis*, Prentice-Hall, Englewood Cliffs, N.J., 1969.
21. B.V. Limaye, Operator approximation, *Semigroups, Algebras and Operator Theory*, Springer Proceedings in Mathematics and Statistics, Vol. 142, Springer India, 2015, pp. 135–147.

© Springer Science+Business Media Singapore 2016, corrected publication 2023

B.V. Limaye, *Linear Functional Analysis for Scientists and Engineers*,
https://doi.org/10.1007/978-981-10-0972-3

22. J. Lindenstrauss and L. Tzafriri, On the complemented subspaces problem, *Israel J. Math.* **9** (1971), pp. 263–269.
23. F. Riesz and B. Sz.-Nagy, *Functional Analysis*, Frederick Ungar, New York, 1955.
24. H.L. Royden, *Real Analysis*, third ed., Macmillan, New York, 1988.
25. W. Rudin, *Principles of Mathematical Analysis*, third ed., McGraw-Hill, New York, 1976.
26. W. Rudin, *Real and Complex Analysis*, third ed., McGraw-Hill, New York, 1986.
27. F.C. Sánchez and J.M.F. Castillo, Isometries of finite-dimensional normed spaces, *Extracta Mathematicae* **10** (1995), pp. 146–151.
28. A.E. Taylor and D.C. Lay, *Introduction to Functional Analysis*, second ed., Wiley, New York, 1980.
29. D.S. Watkins, *Fundamentals of Matrix Computations*, second ed., Wiley, New York, 2002.
30. R. Whitley, Projecting m onto c_0, *Amer. Math. Monthly* **73** (1966), pp. 285–286.
31. P.P. Zabreiko, A theorem for semiadditive functionals, *Functional Analysis and its Applications*, **3** (1969), pp. 86–88. (Translated from the original Russian version in Funksional'nyi Analiz Ego Prilozheniya)

Index

© Springer Science+Business Media Singapore 2016, corrected publication 2023

B.V. Limaye, *Linear Functional Analysis for Scientists and Engineers*,
https://doi.org/10.1007/978-981-10-0972-3

Printed in the United States
by Baker & Taylor Publisher Services